Common U.S. Customary Units, Their SI Equivalents, and Common SI Prefixes Used in Static Structural Analysis

U.S. Customary Units and SI Equivalents

Measure	U.S. Customary Unit	SI Equivalent
Length	in	25.4 mm
	ft	0.3048 m
Area	in^2	645.2 mm^2
	ft^2	0.0929 m^2
Volume	in^3	16.39 cm^3
	ft^3	0.02832 m^3
Force	lb	4.448 N
	kip	4.448 kN
Stress	lb/in^2	6.895 N/m^2 = 6.895 Pa
	lb/ft^2	47.88 N/m^2 = 47.88 Pa
Moment of force	in · lb	0.1130 N · m
	kip · ft	1.356 kN · m
Moment of inertia (of an area)	in^4	41.62×10^4 mm^4

Common SI Prefixes

Factor	Prefix	Symbol
10^9	giga	G
10^6	mega	M
10^3	kilo	k
10^{-3}	milli	m
10^{-6}	micro	μ
10^{-9}	nano	n

STRUCTURAL ANALYSIS

Also Available from McGraw-Hill

Schaum's Outline Series in Civil Engineering

Most outlines include basic theory, definitions, and hundreds of solved problems and supplementary problems with answers.

Titles on the Current List Include:

Advanced Structural Analysis
Basic Equations of Engineering
Descriptive Geometry
Dynamic Structural Analysis
Engineering Mechanics, 4th edition
Fluid Dynamics
Fluid Mechanics & Hydraulics
Introduction to Engineering Calculations
Introductory Surveying
Mathematical Handbook of Formulas & Tables
Mechanical Vibrations
Reinforced Concrete Design, 2d edition
Space Structural Analysis
State Space & Linear Systems
Statics and Strength of Materials
Strength of Materials, 2d edition
Structural Analysis
Structural Steel Design, LFRD Method
Theoretical Mechanics

Schaum's Solved Problems Books

Each title in this series is a complete and expert source of solved problems containing thousands of problems with worked out solutions.

Related Titles on the Current List Include:

3000 Solved Problems in Calculus
2500 Solved Problems in Differential Equations
2500 Solved Problems in Fluid Mechanics & Hydraulics
3000 Solved Problems in Linear Algebra
2000 Solved Problems in Numerical Analysis
700 Solved Problems in Vector Mechanics for Engineers: Statics
800 Solved Problems in Vector Mechanics for Engineers: Dynamics

Available at your College Bookstore. A complete list of Schaum titles may be obtained by writing to:
        Schaum Division
        McGraw-Hill, Inc.
        Princeton Road, S-1
        Hightstown, NJ 08520

STRUCTURAL ANALYSIS

LOUIS C. TARTAGLIONE
UNIVERSITY OF LOWELL

McGraw-Hill, Inc.
New York   St. Louis   San Francisco   Auckland   Bogotá   Caracas
Hamburg   Lisbon   London   Madrid   Mexico   Milan   Montreal
New Delhi   Paris   San Juan   São Paulo   Singapore
Sydney   Tokyo   Toronto

STRUCTURAL ANALYSIS

Copyright © 1991 by McGraw-Hill, Inc. All rights reserved. Printed in the United States of America. Except as permitted under the United States Copyright Act of 1976, no part of this publication may be reproduced or distributed in any form or by any means, or stored in a data base or retrieval system, without the prior written permission of the publisher.

2 3 4 5 6 7 8 9 0   DOC DOC   9 5 4 3 2 1

P/N 062936-6
PART OF
ISBN 0-07-909703-0

This book was set in Times Roman.
The editors were Jack Maisel and B. J. Clark;
the designer was Mel Haber;
the production supervisor was Janelle S. Travers.
R. R. Donnelley & Sons Company was printer and binder.

Cover photo by Neil Kveberg, Minnesota Department of Transportation. A night shot of the Smith Avenue High Bridge, St. Paul, Minnesota. Taken on July 25, 1987.

Library of Congress Cataloging-in-Publication Data

Tartaglione, Louis, C.
    Structural analysis/Louis C. Tartaglione.
       p.   cm.
    Includes bibliographical references.
    ISBN 0-07-909703-0 (set)
    1. Structural analysis (Engineering)   I. Title.
TA645.T36   1991
624.1'71—dc20   90-34316

About the Author

Louis C. Tartaglione is a professor of civil engineering at the University of Lowell. He received an undergraduate degree in civil engineering from Manhattan College and a masters degree in civil engineering from the University of Connecticut and has studied in the doctorate program at the University of New Hampshire. During the past 20 years, Professor Tartaglione has taught mainly graduate and undergraduate courses in structural analysis, structural design, and related subjects. He is a registered professional engineer and a Fellow in ASCE, serves as New England district councillor for Chi Epsilon, and is a member of many other professional organizations. He has served on numerous professional committees for the American Society of Civil Engineers and the Commonwealth of Massachusetts and has authored over 90 research and technical papers and reports in the field of structural analysis and design. In addition, he maintains a consulting engineering practice.

Prior to joining academia, Tartaglione served in the professional mainstream for 14 years in a variety of structural engineering activities, which included analysis and design of nuclear submarines, numerous aerospace structures such as Mercury, Gemini, and the Apollo Command Module. He has also served as a consultant on the space shuttle program and numerous other aerospace programs.

Contents

Preface — xv

CHAPTER 1
INTRODUCTION

1.1	Structural Analysis—A Link to Efficient Design	1
1.2	The Role of the Structural Engineer	2
1.3	Types of Structures	5
1.4	The Analysis Procedure	14
1.5	Design Loads	15
	1.5.1 Design Codes	15
	1.5.2 Dead Loads	16
	1.5.3 Live Loads	17
	1.5.4 Wind Loads	18
	1.5.5 Snow Loads	22
	1.5.6 Rain Loads	22
	1.5.7 Earthquake Loads	24
	1.5.8 Vehicular Highway Loads	24
	1.5.9 Railway Loads	25
	1.5.10 Live-Load Dynamic Impact	26
	1.5.11 Load Factors and Factors of Safety	27
1.6	U.S. Customary Units and Their SI Equivalents	28
1.7	Computation Accuracy and Checking	29
1.8	Use of Computers	30
	Problems	31
	References	31

CHAPTER 2
REACTIONS

2.1	Static Equilibrium Equations	32
2.2	Types of Support Restraint	34
2.3	Geometric Stability and Determinacy	36
2.4	Using the Free-Body Diagram	37
2.5	Beam Reactions by Equilibrium Equations	38
2.6	Frame Reactions by Equilibrium Equations	41
2.7	Truss Reactions by Equilibrium Equations	43

2.8	Reactions by Use of IMAGES and the Personal Computer	45
2.9	Interior Hinges in Construction	50
	2.9.1 Introduction	50
	2.9.2 Three-Hinged Arch Construction	50
	2.9.3 Cantilever Construction	55
2.10	Cable Construction	59
2.11	Principle of Superposition	61
	Problems	62

CHAPTER 3
PLANE TRUSSES

3.1	Introduction	68
3.2	Types of Trusses	70
	3.2.1 Simple Truss	70
	3.2.2 Compound Truss	71
	3.2.3 Complex Truss	72
	3.2.4 Roof Trusses	72
	3.2.5 Bridge Trusses	72
3.3	Ideal Assumptions and Design Concepts	74
3.4	Sign Convention	76
3.5	Truss Notations	76
3.6	Method of Joints	77
3.7	Methods of Section	83
	3.7.1 Method of Shears	83
	3.7.2 Method of Moments	84
3.8	Stability and Determinacy	90
3.9	Computer Solutions	94
3.10	Closure	97
	Problems	98

CHAPTER 4
SPACE TRUSSES

4.1	Introduction	103
4.2	Basic Concepts	103
4.3	Static Equilibrium Equations	106
4.4	Types of Design Supports	106
4.5	Stability and Determinacy	106
4.6	Special Joint Observations	108
4.7	Illustrative Example Problems	109
4.8	Computer Solutions	114
	Problems	115

CHAPTER 5
SHEAR AND BENDING-MOMENT DIAGRAMS

5.1	Introduction	118
5.2	Shears and Bending-Moment in Beams	119
5.3	Sign Convention	121
5.4	Shear and Bending-Moment Functions	122
5.5	Relationships among Load, Shear, and Moment	125
5.6	Shear and Bending-Moment Diagrams for Beams	128
5.7	Shear and Bending-Moment Diagrams for Frames	132
5.8	V and M Diagrams by Method of Superposition	137
5.9	Computer Applications	139
	Problems	143

CHAPTER 6
INFLUENCE LINES—BEAMS

6.1	The Influence Line—A Design Aid	147
	6.1.1 Introduction	147
	6.1.2 The Influence Line	148
6.2	Beam Support Reactions	148
6.3	Beam Shears	151
6.4	Beam Moments	155
6.5	Qualitative Influence Lines for Beams	157
6.6	Bridge Deck and Building Floor Systems	160
	6.6.1 Influence Lines for Bridge Deck Girders	161
	6.6.2 Influence Lines for Building Girders	164
6.7	Use of Influence Lines for Concentrated Loads	164
6.8	Use of Influence Lines for Uniform Loads	166
6.9	Applications for Design-Load Combinations	168
	6.9.1 Railroad-Bridge Live Loads	168
	6.9.2 Highway-Bridge Live Loads	170
6.10	Moving-Load Placement for Maximum Effect	172
	6.10.1 Singular Concentrated Loads	172
	6.10.2 Uniform Distributed Load	172
	6.10.3 Series of Concentrated Loads—Shear	172
	6.10.4 Series of Concentrated Loads—Bending Moment	177
	6.10.5 Absolute Maximum Moment	181
	Problems	184
	References	188

CHAPTER 7
INFLUENCE LINES—PLANE TRUSSES

7.1	Load-Path Idealization for Axially Loaded Truss Members	189
7.2	Parallel Chord Truss Members	190

7.3	Nonparallel Chord Truss Members	193
7.4	Applications for Design-Load Combinations	195
7.5	Closing Remarks on Influence Lines	198
	Problems	199

CHAPTER 8
APPROXIMATE METHODS FOR STATICALLY INDETERMINATE STRUCTURES

8.1	Introduction to Statically Indeterminate Structures	204
8.2	Use of Approximate Analyses	206
8.3	Industrial (Mill) Building Analysis	206
8.4	Trusses with Double Diagonals	212
	8.4.1 Truss with Slender Double Diagonals in Each Panel	212
	8.4.2 Truss with Stiff Double Diagonals in Each Panel	213
8.5	Continuous Beams under Gravity Loads	213
8.6	Building Frames Subjected to Lateral Loads	216
	8.6.1 Portal Method	217
	8.6.2 Cantilever Method	222
8.7	Comparison with Computer Results	225
8.8	Vierendeel Truss	225
	Problems	229
	References	233

CHAPTER 9
DEFORMATIONS: BEAMS

9.1	Introduction: Why Compute Deflections?	234
9.2	Double Integration Method	236
9.3	Method of Moment Area	243
9.4	Method of Conjugate Beam	253
9.5	Special Application for Fixed-End Beams	261
	Problems	264
	Reference	267

CHAPTER 10
DEFORMATIONS: VIRTUAL WORK

10.1	Introduction to Virtual Work and Energy Methods	268
10.2	Bernoulli's Principle of Virtual Work	272
10.3	Truss Deflections by Virtual Work	273
10.4	Beam Displacements by Virtual Work	278
10.5	Beam Rotations by Virtual Work	284
10.6	Frame Deflections	286

10.7	Computer Applications and Comparisons	288
10.8	Maxwell-Betti Theorem of Reciprocal Deflections	290
10.9	Closure on Virtual Work	292
10.10	Historical Reference—Castigliano's Theorems	293
	Problems	300

CHAPTER 11
STATICALLY INDETERMINATE STRUCTURES

11.1	Introduction	306
11.2	Method of Consistent Deformations—Beams	309
	11.2.1 Beam with One Redundant	309
	11.2.2 Beam with Two or More Redundants	312
11.3	Support Settlement	314
11.4	Method of Consistent Deformations—Frames	318
11.5	Method of Consistent Deformations—Trusses	319
	11.5.1 Externally Indeterminate Truss with One Redundant Reaction Component	319
	11.5.2 Internally Indeterminate Truss with One Redundant Member	321
	11.5.3 Indeterminate Truss with Two or More Redundants	323
11.6	Castigliano's Theorem of Least Work	328
11.7	Influence Lines for Indeterminate Structures	333
	11.7.1 Quantitative Influence Lines—Continuous Beams	334
	11.7.2 Quantitative Influence Lines—Indeterminate Trusses	338
	11.7.3 Qualitative Influence Lines—Müeller-Breslau	339
	11.7.4 Critical Live-Load Placement	340
	11.7.5 Moment Envelope Curve	341
11.8	Qualitative Influence Lines by Computer	343
	Problems	344
	References	350

CHAPTER 12
SLOPE DEFLECTION

12.1	Introduction	351
12.2	Slope-Deflection Equations	353
12.3	Beam Applications	357
12.4	Simple End-Support Modification	363
12.5	Frames without Sidesway	365
12.6	Awareness of Symmetry and Antisymmetry	369
12.7	Frames with Sidesway	369
12.8	Frames with Support Settlement	373
12.9	Frames with Sloping Leg Members	374
	Problems	376
	References	380

CHAPTER 13
MOMENT DISTRIBUTION

13.1	Introduction	381
13.2	Basic Assumptions and Fundamental Concepts	383
13.3	Applications: Beams without Support Settlement	388
13.4	Stiffness Modifications	392
	13.4.1 Simple End Modification	392
	13.4.2 Symmetry and Antisymmetry Modifications	395
13.5	Beams with Support Settlement	397
13.6	Frames without Sidesway	399
13.7	Frames with Sidesway	402
13.8	Nonprismatic Beams	412
	Problems	428
	References	434

CHAPTER 14
MATRIX STRUCTURAL ANALYSIS

14.1	Introduction	435
14.2	Matrix Algebra	437
	14.2.1 Definitions	437
	14.2.2 Types of Matrices	438
	14.2.3 Matrix Operations	439
	A. Equality of Matrices	439
	B. Addition and Subtraction of Matrices	439
	C. Scalar Multiplication of a Matrix	439
	D. Multiplication of Matrices	439
	E. Transposed Matrix	441
	F. Determinant of a Square Matrix	441
	G. Adjoint Matrix	444
	H. Inverse of a Matrix	446
	I. Partitioning of Matrices	446
	14.2.4 Solution of Simultaneous Linear Equations	447
14.3	Matrix Structural Analysis	449
	14.3.1 Basic Concepts of Force (Flexibility) Method	449
	14.3.2 Global and Local Coordinate Systems	452
	14.3.3 Degrees of Freedom	453
	14.3.4 Basic Concepts of Displacement (Stiffness) Method	453
	A. Spring Analogy	453
	B. Stiffness Matrix for a Horizontal Truss Member	454
	C. Stiffness Matrices for Truss Members in Local Coordinates	457
	D. Transformation of Truss Stiffness Matrix	458
	E. Formulation of the Joint-Equilibrium Equations	460
	F. Assembly of the Structure Stiffness Matrix	462
	G. Applications for Truss Analysis	466
	H. Settlement, Temperature Change, and Fabrication Errors	472

	14.3.5	Beam Analysis by the Stiffness Method	474
		A. Formation of Beam-Element Stiffness Matrix	476
		B. Element Matrix for a Beam without Axial Force	478
		C. Structure Idealization	478
		D. Equivalent Nodal Forces	480
		E. Three-Dimensional Beam Stiffness Equations	483
	14.3.6	Plane-Frame Analysis by the Stiffness Method	484
14.4	Symmetry and Antisymmetry		488
14.5	Closing Remarks		489
	Problems		490
	References		492

APPENDICES

Appendix 1—A Tutorial for Use of the IMAGES Program by Celestial Software, Inc.	493
Appendix 2—Geometric Properties of Area	519
Appendix 3—Answers To Selected Problems	520

INDEX 525

Preface

This book attempts to provide a clear and simple introduction to classical methods of structural analysis and personal computer applications for the civil engineering undergraduate student. The computer revolution of the past 40 years has significantly altered and improved the procedures the modern engineer uses in the execution of analysis and design of civil engineering structures. The reasonably inexpensive personal computer (PC) and the availability of relatively inexpensive software has inspired both large, and especially smaller, engineering firms to use PCs in the performance of their daily engineering tasks. Advances made during the past decade in microcomputer development have encouraged the development of PC software capable of solving complex engineering problems that were dependent on an expensive mainframe or minicomputer for solution in the past.

The major goal of this text is to introduce many of the traditional methods used for the analysis of typical civil engineering structures. Another goal is to clearly present analysis methods as primary links directed toward efficient structural design. Two minor goals are (1) to introduce students to the use of the PC and related software which can serve as solution checks to the numerous problems encountered during their first courses in structural analysis, and (2) to provide an awareness of and appreciation for the PC and its potential use in their future careers. This text is sufficient for at least one academic year of undergraduate study.

Educators and contemporary writers of structural analysis differ in their opinion of what the first course should present: classical methods alone, a vehicle to develop the principles using matrix methods immediately, or a combination of both classical and matrix approaches. I have found that many students have difficulty in grasping the basic principles and analytical skills of structural analysis which rely upon familiar prerequisite mathematical skills. Therefore, the added introduction of the operations of matrix analysis to students who lack a prior background in matrix algebra may impede their efforts to learn and understand the basic principles of structural analysis. Consequently, it was decided that this book would focus its attention on the introduction of classical methods using traditional engineering mathematics. The methods presented here focus exclusively on linear elastic structural behavior. Also, the addition of PC applications to the solution process should excite the students' curiosity and realization of their need for future study of modern methods of analysis that utilize high-speed computers. Moreover, I believe that undergraduate students seeking further enrichment of this subject will gravitate toward additional course study in matrix and finite element analysis.

A simplified two-dimensional version of a computer program called *IMAGES* has been customized and copyrighted by Celestial Software Inc., Berkeley, California. It has been included as App 1 at the end of the text to assist the students in their understanding of structural analysis. It is particularly helpful since the program can perform analyses of typical beam, frame, and truss structures and provide graphic displays of structural displacements for various load conditions. Numerous text example results are accompanied by a verification check solution output from the IMAGES computer program. Floppy discs containing the program are found in a jacket attached to the inner rear cover. This user-friendly menu-driven program can run on an IBM PC, IBM XT, IBM AT, or IBM compatible system. An overview of the program and information on how to use the interactive tutorial software are provided in the appendices. Other PC software is also cited in the text to demonstrate the versatility of analysis procedures by use of PCs. Instructors should have little or no difficulty in the instructive use of IMAGES. Moreover, a minimum amount of effort is required by the student to become adept in the use of IMAGES. In most cases, the user-friendly menus and on-screen help aids are sufficient to easily guide a new user to a successful solution.

Some traditional methods that have provided good service to the engineering community for decades, such as column analogy, graphic statics, subdivided trusses, moment distribution analysis of multistory frames with sidesway, and others, are not included in this book. Some traditional methods contained in this text, which may be judged to be obsolete or less important than those abandoned, are offered with the opinion that they serve to develop and enhance the students' understanding of structural behavior.

Each instructor approaches this subject with different goals and objectives. Hopefully, this text provides reasonable coverage of most areas of elementary structural analysis to serve as a springboard from which the instructor can direct subject treatment. The examples illustrated in text are intended to provide direct and simple application of the principles. However, the problem sets at the end of each chapter offer a range of easy-to-difficult exercises. The presentation of more challenging problems is left to the election of the instructor.

This author firmly believes that students can best achieve a good grasp of structural behavior through an earnest study of the fundamental principles of elementary structural analysis accompanied by extensive practice through hand calculations of problem sets.

Although the author elects to exclude matrix methods from the introductory course in structural analysis, an elementary chapter on matrix operations and matrix structural analysis concludes the text. This chapter should allow students to appreciate more fully the activities that exist during the running of the IMAGES program, and serve to excite their curiosity for additional study. This chapter may also assist those instructors who wish to use matrix methods as a concluding subject to a first course and the main subject of a second (or third) course in structural analysis. Clearly, the chapter can also serve those instructors who elect to introduce matrix algebra at the onset of the course to establish a clearer understanding of the interactive program routines.

Acknowledgments

The author wishes to express his sincere appreciation and thanks to many individuals, friends, and organizations for their assistance, advice, insights, and/or encouragement throughout this endeavor.

Initially, my thanks go out to Professor Anthony DeLuzio of Merrimack College who introduced me to the IMAGES-2D computer program five years ago and encouraged me to use it in my undergraduate course in structural analysis. I also thank Tony for his early review of and helpful commentary on Chapt. 14 of this text. Much appreciation for the interest and cooperation of Mr. Keith Leung and his staff at Celestial Software, Inc., in meeting the author's needs and modifications of the IMAGES-2D program for use with this text. I am grateful to the University of Lowell for the approval of a sabbatical leave which enabled me to devote full attention to the completion of this project. Added thanks to the civil engineering department at Merrimack College, in particular Dr. Francis Griggs, for use of their structural slide collection from which many of the text photographs were derived and to International Structural Slides, Berkeley, California, for permission to use these slides as text photos. I have enjoyed the professional association with the staff at McGraw-Hill, particularly my principal editor Mr. B. J. Clark and his very able and patient assistant Ms. Judy Pietrobono, both of who have made this exciting yet overwhelming task easier to manage and complete. I thank my students and engineering associates who, over the years, have improved my perception and understanding of structural behavior. My deepest gratitude goes out to my professional colleagues who reviewed this text and who offered sincere and valuable insights and recommendations which improved the quality of the text for the benefit of its readers. They are Fred A. Akl, Ohio University; Tomasz Arciszewski, Wayne State University; Alfred G. Bishara, Ohio State University; Arthur P. Boresi, University of Wyoming; Terry L. Kohutek, Texas A&M University; Narendra Taly, California State University—Los Angeles; and Vernon B. Watwood, Michigan Technological University. Finally, I must confess that this arduous task was feasible because of the constant support, sacrifice, and enormous patience of my wife, Dorothy—the equilibrating force in my life.

Louis C. Tartaglione

STRUCTURAL ANALYSIS

Colosseum, Rome. (*Courtesy of Francis E. Griggs.*)

Pont du Gard, Paris. (*Courtesy of Francis E. Griggs.*)

CHAPTER ONE

Introduction

1.1. STRUCTURAL ANALYSIS—A LINK TO EFFICIENT DESIGN

The main purpose of this text is to introduce civil engineering students to classical methods of structural analysis. The undergraduate who begins the first course in structural analysis recognizes bridges, buildings, roads, highways, tunnels, towers, and dams as typical forms of civil engineering structures. In earlier studies of statics and strength of materials, a student develops logical problem-solving skills and an understanding of engineering materials and their behavior for various loads and environments. These courses offer problem routines that introduce the student to basic structural elements, namely, beams, columns, bars, cables, trusses, and frames. The study of structural analysis will enhance the importance of the principles of statics and enrich the student with many analytical approaches for the evaluation of basic structural systems. Moreover, a firm grasp of the basic methods of structural analysis will help to develop an understanding of structural behavior and serve to assist future engineers to produce safe, functional, and cost-effective designs.

The fundamental purpose of a structural analysis is to determine the magnitudes of force and displacement for each element of a design system for a given set of design loads. Although the subject of structural analysis is often taught separately from design courses, it should be realized that analysis and design are interrelated and that structural analysis is an important link to efficient design.

Technological advancements in microcomputer hardware and software, and the growing use of personal computer (PC) systems in engineering practice have inspired the author to present structural analysis methods with PC applications. A version of a microcomputer program called *IMAGES-2D* has been customized by Celestial Software,[1] as an appendage to the text, to assist students in their studies of structural analysis. The program can be used to check the solutions of two-dimensional structural analysis problems. Floppy disks containing the program are found in a jacket attached to the inner rear cover; tutorial instructions on the use of IMAGES are found in App. 1. This user-friendly, menu-driven program can be run on an IBM PC, IBM XT, IBM AT, or IBM compatible system. IMAGES can perform analyses of typical beam, frame, and truss structures and display structural displacements for various load conditions.

Throughout the text, numerous example problems are done by classical methods and verified by computer solutions; also, many example problems are accompanied by explanations of the input and output data formats to IMAGES.

1.2. THE ROLE OF THE STRUCTURAL ENGINEER

An engineering enterprise is a multifaceted adventure in which the structural engineer is often expected to perform several tasks and make many technical decisions. From the onset to the completion of the project, the structural engineer will interact with a body of people which may include owners, architects, other engineers, contractors, inspectors, building suppliers, tradespeople, and so forth. Before the formal studies of structural analysis begins, a review of some of the tasks, concerns, and responsibilities that the structural engineer may encounter are presented.

Initial studies and planning Assume that a municipality engages an engineering company to design a highway bridge that is needed to improve the flow of commuter traffic into and out of the business center. The bridge must span a body of water. The engineering firm must clearly recognize the general design objectives and study all possible sites where the structure can be erected. To this end, traffic studies are performed and several possible locations are selected where a bridge can provide the needed traffic service. At each potential site location, the soil and foundation conditions are investigated to determine if the site is suitable to support the bridge stucture and can readily accommodate its piers and abutments. Probable design loads and load combinations on the bridge must be determined. These loads will include the bridge weight as well as snow, ice, wind, automobiles, and trucks on the bridge. Information relative to the depth of water, tides, water currents, and wind force history at each potential site location are also collected and studied. The choice of possible site locations will also be

---

[1] Celestial Software, Inc., Berkeley, Calif., copyright publisher of numerous versions of IMAGES.

affected by social, political, environmental, and economic issues, as well as by the conclusions and recommendations from the engineering studies.

After potential sites are selected and approved for further study, a team of engineers will proceed to establish one or more conceptual bridge designs at each site. In addition to satisfying the basic design objectives, each conceptual design must be aesthetically pleasing, compatible with the surrounding environment, acceptable to the community at large, and economically feasible.

Planning studies will be made to estimate time schedules for preliminary analysis, design and review, reanalysis and redesign. Since the reality of the final design choice is achieved through construction, construction planning and management studies are also required. These studies will consider various methods of fabrication and construction, the availability of skilled labor, the types of excavation and erection equipment required, sources of materials, the ease of construction, safety, time schedules, and the related costs associated with each activity. Construction management planning can provide an organized approach to facilitate the construction process and result in a cost-effective project.

Preliminary design and analysis The engineering objectives of a preliminary design are to select and study various geometric forms and structural systems, material options, member sizes, and member shapes which translate a conceptual design into a serviceable, safe, and economical design. These objectives are discussed more fully in the following sections.

Safety and serviceability A properly designed structure is proportioned and assembled from materials which can adequately support real or specified design loads (service loads) and will not fail or deform excessively

Construction of segmental bridge over Houston Ship Canal, Houston. (*Courtesy of American Concrete Institute*.)

Cantilever structure under construction. (*Courtesy of Nelson Steel Erectors*.)

under all anticipated load combinations. To reduce the probability of material failure, structures are proportioned from design loads which include *factors of safety* so that the allowable material stress of its members is well below the material strength. In the case of a slender compression member, the maximum compression stress derived from all load combinations must be less than the stress level that will cause the member to buckle. Typical design loads for civil engineering structures and load factors are presented in Sec. 1.5. It is also vital that the structure is properly supported and able to sustain erection loads during the various stages of its construction.

In addition, every structure must be able to support service loads without causing excessive deflection or vibration. Several cases of high wind forces exerted against tall buildings have been known to produce excessive lateral sway that results in noticeable vibration, motion sickness to occupants, or cracks in windows, plaster ceilings, and/or masonry walls. Many design techniques have been developed to restrict the motion of tall buildings under lateral wind loads (e.g., shear walls, panel frame bracing systems, tubular design with perimeter columns acting as windbracing, bundled tube construction, and mechanical damping isolators).

Other examples which demonstrate the importance of deflection and vibration control are listed below:

1 Floor systems are frequently designed to support sensitive equipment or machinery which will function properly only if deflections and vibrations are sustained within established acceptable limits.

2 Steel tension members are designed to limited slenderness ratio values to prevent excessive deflection or vibration.

3 Floor and roof beams that support plastered ceilings are proportioned to limit the maximum live-load deflection to an empirical value of 1/360 of the span in order to prevent plaster cracking.

4 Long-span bridge deck designs must have adequate stiffness to prevent excessive deck displacements and vibrations, and to ensure vehicle ride quality, passenger comfort, and safety. Bridges must also be able to support aerodynamic wind forces.

At various stages of structural analysis and design, the structural engineer will consult numerous available guides, recommendations, and standards of good design practice—many of which are found in design codes referenced in Sec. 1.5.

Design economics The structural engineer is responsible for providing a safe, serviceable, and economical design. It may take many years of experience before an engineer can be regarded as an efficient structural analyst or a good design engineer. It is certain, however, that a good structural engineer is adept in engineering mechanics and strength of materials, and has a keen insight and knowledge of how structures behave and transmit forces for various load environments. For the student, this development process is continued with the study of structural analysis, an adventure which I hope is rewarding, enjoyable, and recognized as a link to safe and efficient design.

Design-analysis-review cycles The hypothetical example of a highway bridge discussed above will undoubtedly proceed through many cycles of design, analysis, and critical review before a final design is accepted and eventually constructed. The important studies of structural design, construction methods and practices, and construction management are not contained in this text.

1.3. TYPES OF STRUCTURES

Various types of civil engineering structures abound—many are visible and others hidden from view. Coffer dams, pile systems, tunnels, offshore drilling platforms, culverts, and most foundation work are normally not in view, whereas bridges, buildings, roads, dams, domes, and towers are clear marks of civil engineering structures. The basic forms of structure are usually typified as trusses, frames, arches, membranes, plates, shells, or cable-supported, such as guyed towers, suspension, and cable-stayed bridges. However, many structures exist which combine the benefits of several basic forms into a complex but unified system. A brief description of each basic structural form follows.

Figure 1.1 Construction of Cobo Conference and Exhibit Center, Detroit. (*Courtesy of Modern Steel Corporation.*)

Figure 1.2 Speedy construction of light industrial building by use of open web steel joist and long span girder members. (*Courtesy of Vulcraft.*)

Truss A truss is a geometrically stable arrangement of slender members which primarily support axial load. The truss members are usually arranged in triangular patterns to form a light, stiff, and stable structure, and are joined together at gusset plates by bolt or weld connections. Two-dimensional trusses are often used as roof or floor beams to span long distances since they are able to experience small displacements under load. Three-dimensional trusses are commonly used for transmission towers, cranes, derricks, flat roofs, curved stadium roofs, and so forth. Examples of two- and three-dimensional truss configurations are shown in Figs. 1.1 to 1.3.

The analysis of a truss is simplified by use of the following idealizations: (1) All members are straight, (2) all members are joined together at their ends by frictionless pins, (3) at each joint, the lines of action of all joining truss members converge at a common point, (4) loads are applied or are transmitted directly to truss pin joints, and (5) trusses are supported at pin-joint locations. These idealizations will result in axial tension or compression as the primary stress force in each truss member.

Frame A frame is a stable structural form consisting of two or more flexural members which can resist bending moment, shear, and axial forces. The steel skeleton of a high-rise building under construction is a typical structural frame. A frame is classified as a *rigid frame* when its members are joined together by *moment resisting connections*; that is, where joint trans-

7 Types of Structures

Figure 1.3 Public auditorium, Pittsburg. This retractable domed steel structure has a 417-ft diameter and a rise of 109 ft. It was designed to serve as a convention hall and an open-air amphitheater that seats 13,600 people. (*Courtesy of International Structural Slides, Berkeley.*)

Eiffel Tower, Paris (1899). This classic tower is 984 ft high and made of open-lattice steelwork. Designed by Alexandre Gustave Eiffel. (*Courtesy of International Structural Slides, Berkeley.*)

lation and rotation at the joints can occur without relative rotation between joint members. Structural frames often use semirigid, and pin end connections at some member joints. *Semirigid frame connections* have notable moment resistance capability, but to a lesser degree than found at rigid joint connections. Also, semirigid joints can experience some degree of relative rotation between joint members. *Pin connection* joints provide little or no moment resistance and allow the pin connected members freedom of

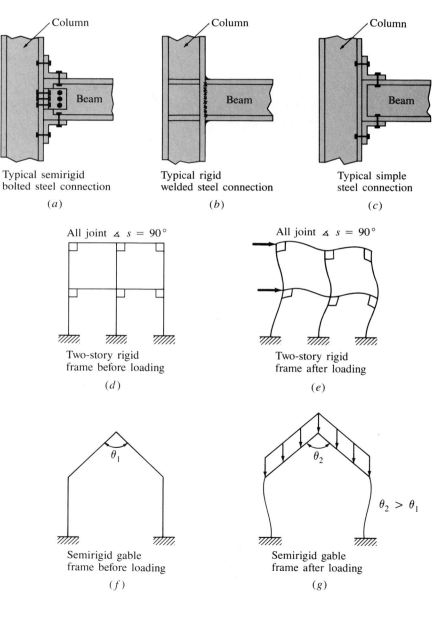

Figure 1.4

Typical semirigid bolted steel connection
(a)

Typical rigid welded steel connection
(b)

Typical simple steel connection
(c)

Two-story rigid frame before loading
(d)

Two-story rigid frame after loading
(e)

Semirigid gable frame before loading
(f)

Semirigid gable frame after loading
(g)

rotation under load. Figure 1.4 illustrates typical forms of rigid, semirigid, and pin (simple) steel frame connections. Various forms of building frame construction are shown in Figs. 1.5 to 1.7.

Figure 1.5 Expansion and remodeling project of the Albert Sabin Convention Center, Cincinnati. Three types of framing support roofs over the lobbies. Barrel vaults are used over the second-floor lobbies. (*Courtesy of THP Limited Consulting Engineers, Cincinnati.*)

Figure 1.6 Cincinnati Convention Center Expansion Project. (*Courtesy of THP Limited Consulting Engineers, Cincinnati.*)

Figure 1.7 Cincinnati Convention Center Expansion Project. (*Courtesy of THP Limited Consulting Engineers, Cincinnati.*)

Arch The arch can be defined as a curved structural shape that is usually configured to support gravity loads in a manner that results in uniform compressive resistance. The gravity loading will tend to flatten the arch and push its supports outward. Therefore, both support ends must provide horizontal as well as vertical force resistance. Arches are frequently used for masonry portals in buildings and are visible in highway construction, bridge designs, airplane hangers, churches, and cathedrals as shown in Figs. 1.8 to 1.11.

Figure 1.8 Timber arch construction, Back Bay Station, Boston. (*Joeffry Thomas/Castien Production.*)

Figure 1.9 St. Louis Priory Church, St. Louis. The church groups 20 peripheral chapels under 21-ft-high thin concrete parabolic arches. (*Courtesy of Portland Cement Association & The American Concrete Institute.*)

11 Types of Structures

Figure 1.10 Riverside Viaduct, New York City. An example of a complex structural form. (*Courtesy of Goodkind & O'Dea Consulting Engineers and Planners.*)

Figure 1.11 Central Avenue Bridge, Ithaca, New York, after reconstruction (1987). This stone and brick arch bridge serves as a main entrance to the Cornell University campus. (*Courtesy of Clarke & Rapuano, Inc.*)

The ancient Egyptians, who were restricted to building with stone and masonry materials, used the arch form. However, it was the Romans who mastered the construction of stone arch structures and presented its form in aqueducts, bridges, and buildings throughout the Roman Empire.

Membranes, plates, and shells Membranes are thin-walled structures such as air supported stadium roofs and weather balloons which provide tensile resistance in two directions.

Flat plate structures, unlike membranes, can provide bending, tensile, and compressive force resistance. Many floor slabs are designed as plate structures. A popular form of civil engineering plate construction is that of folded plate roofs.

Shells are often defined as curved plates. Roof domes, water and fuel storage tanks, and grain silos typify some forms of shell structures. Figures 1.12 to 1.14 present design applications of plate-, shell-, and membrane-type structures. However, these structural forms are not studied in this text.

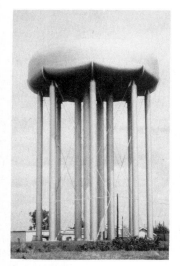

Figure 1.12 Water tower near Urbana, Illinois. (*Courtesy of International Structural Slides, Berkeley.*)

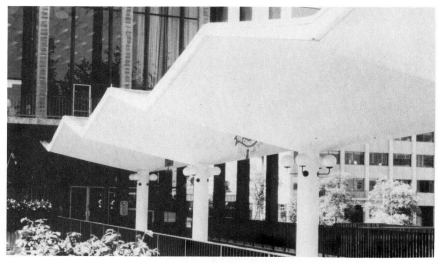

Figure 1.13 Hypar shell roof structure, University of California, Berkeley. (*Courtesy of International Structural Slides, Berkeley.*)

Figure 1.14 Domed shell roof structure of a passenger-waiting building at McCarran Airport Terminal, Las Vegas. (*Courtesy of ACI & Expanded Shale, Clay and Slate Institute, Washington, D.C.*)

Cable Cable supported structures are very common and often expressed in the form of suspension and cable-stayed bridge construction. The Brooklyn Bridge of New York City (1883) was the first suspension bridge in the world to use steel-wire cable [1.1].[1] The Golden Gate Bridge of San Francisco (1937) is another notable cable suspension bridge of the United States. Cable support systems are also used for large roof construction such as found in enclosed sports stadiums. Figures 1.15 and 1.16 demonstrate two popular cable supported bridge structures.

[1] A double number within brackets [ ] denotes a reference document cited at the end of each chapter.

Figure 1.15 Brooklyn Bridge, New York City. The first bridge supported by steel suspension cables. A double cable system supports two roadways in the vertical direction. A central walkway (located at the top level of the trusses) divides the two roadways. Completed in 1883. Designed by John A. Roebling.

Figure 1.16 Sunshine Skyway (cable-stayed) Bridge, Tampa Bay, Florida. Total bridge length is approximately 22,000 ft. (*Courtesy of Figg & Muller, Engineers, Inc.*)

Complex Structures Structures which combine many of the basic forms identified above may be classified as complex structures. Numerous bridges consist of arch-shaped or rigid frame-shaped piers, trussed girders and bracing systems, cable supports, and a composite deck with many plate elements. Buildings and covered stadiums are other types of complex structures where designers have elected to combine the benefits of several structural forms into a unified system.

1.4. THE ANALYSIS PROCEDURE

The student of structural analysis should observe that most structures are of three-dimensional form and can be subjected to applied forces in the x, y, and z directions. Figure 1.17 shows a schematic of a light storage structure

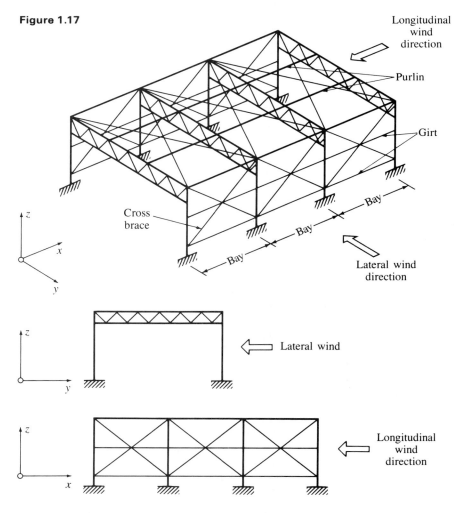

Figure 1.17

which must be able to support gravity loading (including its weight plus design loads such as snow, ice, and ponding water). Cross-bracing members, along with longitudinal members (called girts and purlins), unite the building frames and provide lateral and longitudinal wind force resistance.

An attempt to perform a three-dimensional analysis of the structure by hand computations will prove to be awkward and time consuming. A set of easier two-dimensional analyses can be performed on the structure which considers a rational distribution of load components acting on separate planar elements. For example, it is reasonable to assume that a typical interior frame (shown in the y–z plane of Fig. 1.17) will resist the loads acting on wall and roof areas that extend a one-half bay width on both sides of the frame.

In general, the methods developed in this text are applied to two-dimensional (planar) structural analysis; one exception is found in Chap. 4, where three-dimensional space trusses are treated.

1.5. DESIGN LOADS

1.5.1. Design Codes

The structural engineer will usually find design loads specified in a design code or a building code. Codes, in general, are collections of basic guides aimed at regulating design and construction for the safety, health, and general welfare of its users. When adopted by a governmental body, these codes become law. A typical building code will regulate construction, maintenance, repair, and demolition of buildings and also specify minimum structural design loads. Some of the prominent *building codes* are as follows:

Basic Building Code, Building Officials and Code Administrators International, Inc. (BOCA)

American National Standards Building Code, American National Standards Institute, Inc. (ANSI)

Uniform Building Code, International Conference of Building Officials (ICBO)

Standard Building Code, Southern Building Code Congress (SBCC)

Design codes tend to be specification-type codes that provide minimum design loads, factors of safety, allowable stresses, allowable loads, design formulas, as well as construction details for specific types of structures. Some of the prominent *structural design codes* follow:

Manual of Steel Construction, American Institute of Steel Construction (AISC)

Building Code Requirements for Reinforced Concrete, American Concrete Institute (ACI)

Timber Construction Manual, American Institute of Timber Construction (AITC)

Standard Specifications for Highway Bridges, American Association of State Highway and Transportation Officials (AASHTO)

Design codes are guides for general design practice based on sound engineering principles and engineering experience. They are subject to periodic review and revision to reflect upon new construction materials, innovative systems, research and test findings, improved load factors and empirical design equations, updated safety provisions, and so forth. Design codes also specify that the engineer is not restricted to the standard practices prescribed if he or she can verify by tests or analysis that a safe design can be realized.

1.5.2. Dead Loads   Dead loads include stationary loads of constant magnitude such as the weight of the structural members and any items that are permanently attached to the structure. In a multistory building, the dead loads include the weights of all beams, columns, floor slabs, exterior walls and glass, stairwells, elevator shafts, plus plumbing and electrical fixtures. At the start of an analysis, the dead weight of the support members is initially estimated since the member dimensions, including floor slab thicknesses, are unknown. The actual dead loads are established after a structural analysis is performed and the members are properly sized to support the design loads. The ability to estimate structural weights will generally improve with repeated design experience. Table 1.1 provides some unit dead-load data used to estimate minimum design dead loads for various members and materials used in buildings and other structures.

TABLE 1.1
Minimum Unit Dead Loads for Design [1.3]

	psf
Concrete slabs	
Concrete, reinforced-stone, per inch of thickness	$12\frac{1}{2}$
Concrete, reinforced, lightweight, per inch of thickness	9
Concrete, plain stone, per inch of thickness	12
Concrete, plain, lightweight, per inch of thickness	$8\frac{1}{2}$
Floor finish	
Linoleum	2
Asphalt tile	2
Underflooring, per inch of depth	3
Hardwood flooring, per inch of depth	4
Linoleum on stone-concrete fill	33
Floor finish tile, per inch of depth	12

Roof and wall coverings

Asphalt shingles	2
Wood shingles	3
Roman clay tile	12
Spanish tile	20
5-ply felt and gravel	6
Copper or tin	1
Rigid insulation, $\frac{1}{2}$ in	$\frac{3}{4}$
Fiberboard, $\frac{1}{2}$ in	$\frac{3}{4}$
Slate, $\frac{1}{4}$ in	10
Wood sheathing, per inch of thickness	3

Walls

8-in clay brick, medium absorption	79
8-in concrete brick, heavy aggregate	89
8-in concrete brick, light aggregate	68
8-in concrete block, heavy aggregate	55
12-in concrete block, heavy aggregate	85
8-in concrete block, light aggregate	35
12-in concrete block, light aggregate	55
Partition walls:	
Wood studs 2 × 4, unplastered	4
Wood studs 2 × 4, plastered, one side	12
Wood studs 2 × 4, plastered, two sides	20

Suspended ceilings

Cement on wood lath	12
Cement on metal lath	15
Plaster on tile or concrete	5
Plaster on wood lath	8
Suspended metal lath and gypsum plaster	10
	pcf

Materials (density)

Normal weight plain concrete	144
Normal weight reinforced concrete	150
Structural steel	490
Timber (nominal)	25–45

1.5.3. Live Loads

Live loads are loads which can vary with reference to magnitude, position, and time. Typical live loads on a classroom floor are students, desks, and chairs, all of which can vary in number and position at different times. In the day-to-day operations in a manufacturing building, live loads may exist in the form of crane girders transporting heavy objects within the facility, or masses moving along assembly production lines. Automobiles, trucks, trains, and pedestrians, are representative forms of live loads on highway and bridge structures. Mother Nature imposes live-load forces on exposed structures in the form of wind, snow, ice, rain, and earthquakes. Other examples of structures under live loading include (1) dams which must contain a variable depth of water supply, (2) highway retaining walls to resist earth pressure forces that can include a surcharge pressure due to traffic loads, and (3) off-shore oil drilling structures and seawalls which undergo dynamic pounding from waves and wind forces. The student will find that this text will generally deal with live loads associated with buildings and bridge structures.

Building-floor live loads Recommended minimum values of live loads can be found in existing design building codes. The live loads used to design building floors are based on the occupancy or use of the building. Table 1.2 contains typical minimum live loads for use in building design; these loads are expressed in pound-force per square foot (psf) and are generally considered as uniformly distributed when performing structural analysis. It is worth noting that some interior partition walls are movable and therefore may be classified as live loads; partition walls are also expressed as uniformly distributed loads. Other types of live loads are presented in greater detail in the paragraphs that follow.

1.5.4. Wind Loads

All above ground structures must be able to sustain the effects of wind loading. However, the evaluation of wind forces on structures is a complex aerodynamic problem. Some of the factors that have been found to influence the magnitude and distribution of wind forces on a building structure

TABLE 1.2

Minimum Design Live Loads [1.3]

Occupancy or use	Live load (psf)
Assembly halls, fixed seats	60
Assembly halls, movable seats	100
Garages (passenger cars only)	50
Manufacturing (light)	125
Manufacturing (heavy)	250
Office buildings:	
Offices	50
Lobbies	100
Corridors above first floor	80
Residential:	
Multifamily private apartments	40
Dwellings: First floor	40
Second floor and habitable attics	30
Uninhabitable attics (see BOCA [1.10] footnotes)	20
Schools: Classrooms	40
Storage warehouse (light)	125
Storage warehouse (heavy)	250

include wind speed and direction, shape and surface texture of the exterior walls, geometric bearing of each exterior wall, building height above ground level, the slopes of the roof, and its geographic location. The American National Standards Institute (ANSI) provides a standard which can be used by engineers to determine wind loads on buildings and other structures [1.3]. The ANSI A58.1-1982 standard includes a basic wind-speed contour map for the United States which is established from periodic collections of wind-speed data at weather stations, and by local, state, and federal agencies throughout the United States (see Fig. 1.18). This contour map is statistically determined from data that corresponds to a height of 10 m (33 ft) above ground for a mean recurrence interval of 50 years (that is, an annual probability of 0.02).

If the wind velocity pressure q is ideally expressed by its kinetic energy as $\frac{1}{2}pV^2$, where p is the density of air at sea level (0.765 lb/ft^3), and V is the basic wind velocity expressed in miles per hour, then

$$q = 0.00256V^2 \qquad (q \text{ in units of lb/ft}^2)$$

ANSI A58.1-1982 presents a modified version of this equation such that the velocity pressure q_z at height z above ground level is computed from the formula

$$q_z = 0.00256K_z(IV)^2 \qquad (q \text{ in units of lb/ft}^2)$$

where I is an importance factor and K_z is an exposure coefficient that depends upon the exposure category and varies with z, the height of the building above ground level. The basic wind velocity used in this equation may be determined from Fig. 1.18. The K_z factor is used to adjust the basic wind velocity for building heights greater and less than 33 ft above ground level. Factor K_z also depends on the building-site exposure category (see Fig. 1.18). ANSI also provides a factor of safety I (called an importance factor) that is used to reflect the seriousness of a consequent structural failure. Variation in the dynamic wind pressure due to wind gusts may also have to be factored into a design wind-load analysis (refer to ANSI A58.1-1982 above).

Figure 1.19 graphically presents the fractional portion of the dynamic wind pressure that is recommended for design wind loading on symmetrical gabled-roof buildings. This data was presented by an American Society of Civil Engineers (ASCE) Task Committee on Wind Forces of the Committee on Loads and Stresses of the Structural Division in two ASCE journal publications entitled "Wind Forces on Structures" [1.4] and [1.5]. A study of Fig. 1.19 reveals that the wind exerts pressure on the windward side and suction on both the leeward side and leeward slope of a gable building; the force on the windward slope is either pressure or suction depending on the slope angle. The following example will demonstrate the use of Figs. 1.18 and 1.19 to determine wind forces on a gabled-roof building.

20 Introduction

Category*	Importance Factor, I (Wind Loads)	
	100 Miles from Hurricane Oceanline, and in Other Areas	At Hurricane Oceanline
I	1.00	1.05
II	1.07	1.11
III	1.07	1.11
IV	0.95	1.00

The building and structure classification categories are listed in Table 1 of A58.1-1982.

Velocity Pressure Exposure Coefficient, K_z

Height above Ground Level, z (feet)	Exposure A	Exposure B	Exposure C	Exposure D
0–15	0.12	0.37	0.80	1.20
20	0.15	0.42	0.87	1.27
25	0.17	0.46	0.93	1.32
30	0.19	0.50	0.98	1.37
40	0.23	0.57	1.06	1.46
50	0.27	0.63	1.13	1.52
60	0.30	0.68	1.19	1.58
70	0.33	0.73	1.24	1.63
80	0.37	0.77	1.29	1.67
90	0.40	0.82	1.34	1.71
100	0.42	0.86	1.38	1.75
120	0.48	0.93	1.45	1.81
140	0.53	0.99	1.52	1.87
160	0.58	1.05	1.58	1.92
180	0.63	1.11	1.63	1.97
200	0.67	1.16	1.68	2.01
250	0.78	1.28	1.79	2.10
300	0.88	1.39	1.88	2.18
350	0.98	1.49	1.97	2.25
400	1.07	1.58	2.05	2.31
450	1.16	1.67	2.12	2.36
500	1.24	1.75	2.18	2.41

For values of height z greater than 500 ft, K_z may be calculated from Eq. A in the Appendix of A58.1-1982.

Basic Wind Speed, V

Location	V (mi/h)
Hawaii	80
Puerto Rico	95

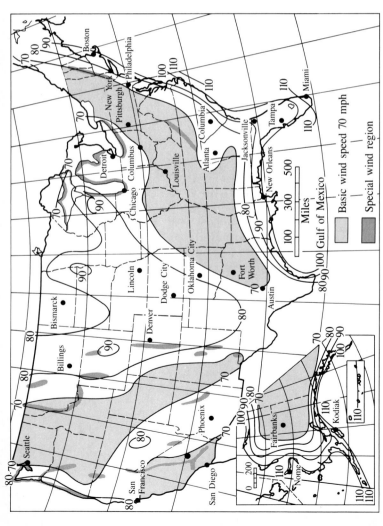

Figure 1.18 Basic wind speed, mi/h

Notes: 1. Values are fastest-mile speeds at 33 ft (10 m) above ground for exposure category C and are associated with an annual probability of 0.02.
2. Linear interpolation between wind speed contours is acceptable.
3. Caution in the use of wind speed contours in mountainous regions of Alaska is advised.

Exposure A — Large city centers with at least 50 percent of the building having a height in excess of 70 feet.
Exposure B — Urban and suburban areas, wooded areas, or other terrain with numerous closely spaced obstructions having the size of single-family dwellings or larger.
Exposure C — Open terrain with scattered obstructions having heights generally less than 30 feet. This category includes flat, open country and grasslands.
Exposure D — Flat, unobstructed coastal areas directly exposed to wind flowing over large bodies of water.

EXAMPLE 1.1

A gable frame structure is assumed to experience a basic wind speed of 84 mph, as directed. If $I = 0.95$ (category IV on Fig. 1.18 which corresponds to buildings and structures that represent low hazard to human life in the event of failure; e.g., storage shed) and the structure has a C exposure, determine the equivalent static wind forces on a typical interior frame using the wind-load data of Figs. 1.18 and 1.19.

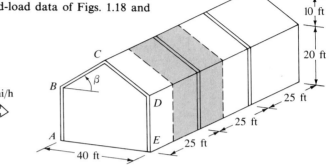

Solution

$$q_z = 0.00256 \, K_z \, (IV)^2 \quad \text{for exposure C, } K_z = 0.98 \text{ (see Fig. 1.18)}$$
$$q_z = 0.00256 \cdot 0.98 \, (0.95 \times 84)^2 = 16 \text{ psf}$$

A typical interior frame is assumed to support wind loads of one full bay spacing (see shaded area above).
Using ASCE wind-curve data of Fig. 1.19 where $\beta = 26.6°$, static wind loads are as follows:

$$P_{AB} = +0.80 \, q_z \cdot 25 \text{ ft} = +0.80 \cdot 16 \cdot 25 = +320 \text{ plf}$$
$$S_{BC} = -0.25 \, q_z \cdot 25 \text{ ft} = -0.25 \cdot 16 \cdot 25 = -100 \text{ plf}$$
$$S_{CD} = -0.60 \, q_z \cdot 25 \text{ ft} = -0.60 \cdot 16 \cdot 25 = -240 \text{ plf}$$
$$S_{DE} = -0.50 \, q_z \cdot 25 \text{ ft} = -0.50 \cdot 16 \cdot 25 = -200 \text{ plf}$$

Note: $P \sim$ pressure force
$S \sim$ suction force

Resultant wind loading

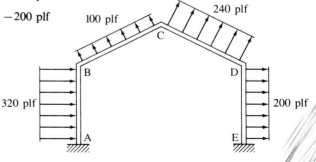

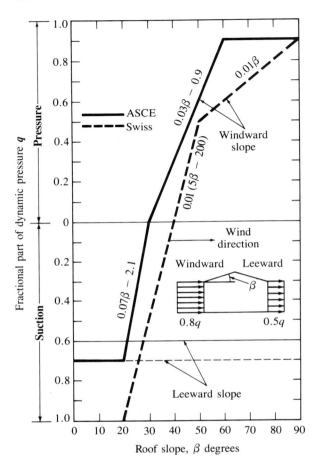

Figure 1.19

1.5.5. Snow Loads

The BOCA Basic Building Code provides recommended snow loads for consideration in the design of buildings and other structures in the United States. The design snow load in pounds-force per square foot (psf) is determined from maps of the United States which show isoclines of ground snow for 25-, 50-, or 100-year mean recurrence intervals. The 50-year mean recurrence interval snow-load map of Fig. 1.20 is prescribed for buildings used for business, factory, mercantile, and single- or multifamily dwelling unit structures. The design snow load is obtained by multiplying the ground snow load isocline value for a geographical area by factors which consider wind exposure, slope of roof, snow-load distribution, and so forth. The American National Standards Institute (see ANSI A58.1-1982), as well as state and local building codes are additional sources from which recommended design snow-load data can be obtained.

1.5.6. Rain Loads

A pond of rainwater may accumulate on a roof structure if the roof drainage system becomes blocked. Therefore, a roof structure should be designed

23 Design Loads

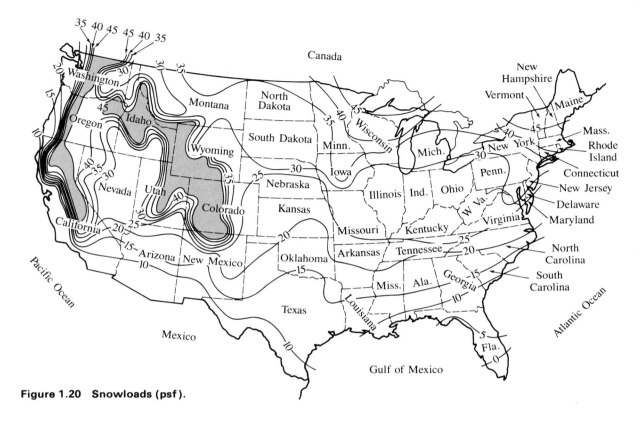

Figure 1.20 Snowloads (psf).

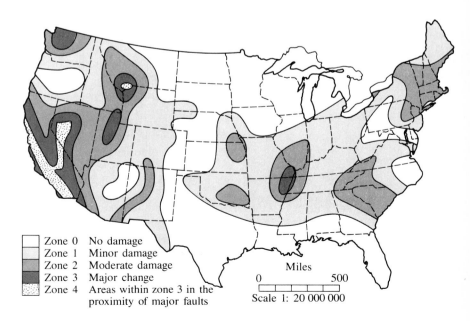

Figure 1.21 Seismic zones.

Zone 0 No damage
Zone 1 Minor damage
Zone 2 Moderate damage
Zone 3 Major change
Zone 4 Areas within zone 3 in the proximity of major faults

Miles
0   500
Scale 1: 20 000 000

to support as much rainwater as it can hold (ponding load). ANSI A58.1-1982 states that roofs are to be designed to preclude instability from ponding loads.

1.5.7. Earthquake Loads

Earthquakes are defined as a series of random dynamic ground excitations that result in vertical and horizontal components of acceleration, velocity, and displacement of the attached structures. The vertical earthquake motion is usually of negligible effect on building design. Yet, the horizontal earthquake response may result in significant shear forces in the building columns. Figure 1.21 shows a map of the United States that divides the country into five risk zones predicated on past earthquake history and the probability of future risk. The zone factor (Z) equal to a value of 4 is used for the California coastal region where high frequencies of intense earthquakes have occurred and are predicted to continue to occur. A zone factor of 0 demonstrates a low-risk region where no evidence of earthquake history exists, such as found in regions from Texas to Florida.

The *Uniform Building Code* [1.8] provides a simple design formula based on the seismic code developed by the Structural Engineers Association of California (SEAOC) that can be used to estimate equivalent static horizontal column shears of building structures resulting from potential horizontal earthquake motion [1.9].

The formula is

$$V = ZICW/R_w$$

where V = the total base shear force

Z = a zone factor

I = an importance factor which relates to the intended occupancy use

C = a coefficient which considers both the natural vibration of the structure and its foundation soil characteristics

W = the total seismic dead load

R_w = a lateral force resistance factor based on the type of structural system used

1.5.8. Vehicular Highway Loads

The AASHTO [1.2] has established a system of live loads to be used for highway bridge design that includes a standard two-axle truck, a three-axle truck (cab plus semitrailer), and uniform lane loading to simulate a line of traffic. One specified truck loading is an H20–44, where 20 designates a 20-ton two-axle truck (8 kips on the front axle and 32 kips on the rear axle), and 44 signifies the year of the published specification. An HS20 truck designation represents a three-axle truck (tractor truck with semitrailer)

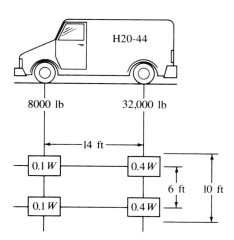 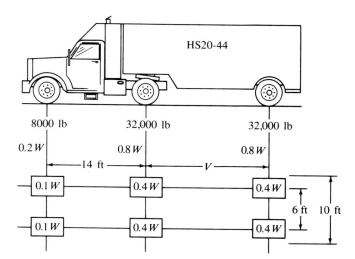

W = Combined weight on the first two axles which is the same as for the corresponding H truck.

Figure 1.22

Variable spacing—14 to 30 ft inclusive. Spacing to be used is that which produces maximum stresses.

with 8 kips on the front axle and 32 kips on each of the two rear axles. An HS15 designation indicates that the load is 75 percent of the HS20 loading. Figure 1.22 shows a schematic sketch of the HS20 type of loading where it is noted that the wheel spacing on an axle is 6 ft and the vehicle is assumed to occupy a lane width of 10 ft.

The *uniform lane loading* consists of a uniform load that represents a line of trucks plus a concentrated load to account for an extra heavy truck somewhere in the line of traffic. The concentrated load is positioned for maximum effect. The lane loads are expressed in terms of the truck weight in the following manner:

> The uniform load per foot is equal to 0.016 times the truck weight whereas the concentrated load is equal to 0.65 times the truck weight for shear, or 0.45 times the truck weight for bending moment. For the H20–44 truck described above, the lane loading would consist of a uniform load of 0.64 kips/ft plus a 26-kip concentrated load for shear, or an 18-kip concentrated load for moment.

An alternative live-load system to be considered for the design of structures on the interstate highway system consists of two 24-kip axle loads spaced 4 ft apart. The interstate highway system loading is usually found to be critical for short spans.

1.5.9. Railway Loads The American Railway Engineering Association (AREA) has established live-load systems for use in the design of railway bridges. One system called *Cooper loads* derives its name from Theodore Cooper who devised a series of concentrated axle loads at fixed distances apart to simulate the forward two engines of a train representative of the 1890s. Cooper simulated the line of freight and passenger cars that trail behind the engine cars as a uniformly distributed loading. The classical E–40 Cooper loading is shown in Fig. 1.23a, where the 40 corresponds to the load of the main wheels of the locomotive expressed in units of kips (1 kip equals 1000 pounds). However, the trains of Cooper's time are no longer in public use today. Therefore, the Cooper loadings have been updated by loads representative of heavier modern trains which bear designations such as E–72, E–80, and E–90. The modern Cooper loadings consider the identical locomotive wheel spacings of the E–40 loading. However, the wheel-load intensities and the trailing distributed loading are proportionally greater than the E–40 loading by a factor of 72/40 for the E–72 load, 80/40 for the E–80 load, and 90/40 for the E–90 load. Figure 1.23b shows the commonly used E–72 Cooper loading.

1.5.10. Live-Load Dynamic Impact

The live loads associated with moving vehicles on a highway bridge or trains on railroad structures are dynamic forces that are applied suddenly and can result in significant vibration to a structure. All of us have experienced impact force while traveling in a car over a pothole or over a rough road and know the effect of sudden starting or stopping while traveling in trains, autos, or trucks. Dead loads and some live loads (e.g., snow) are applied gradually to a structure and do not produce an impact effect. However, the dynamic effect associated with suddenly applied loading

Figure 1.23

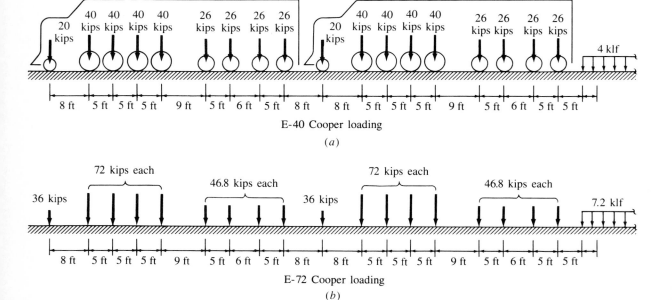

results in a load increase on the structure. Consequently, impact factors are applied to dynamic design loads to anticipate the effects of these live-load events. To account for potential moving-vehicle impact on bridge decks of highway bridge structures, the AASHTO specifies a statically equivalent increase of the design live load by use of a simple empirical impact factor [1.2]. This conservative factor is expressed as

$$I = 50/(L + 125) \leqslant 0.30$$

where I is the impact factor and L is that portion of the span length, in feet, to be loaded to provide a maximum stress condition. The use of this factor will be realized more fully after a study of influence lines is completed in Chap. 6. Impact factors of greater magnitude are found in the AREA specifications for use in the design of railroad bridges where heavy, fast moving trains could result in greater impact force as compared to bridge impact from highway vehicles. In addition, the AISC contains impact provisions for steel building design which include factors for the design of elevator supports and for heavy and light machinery supports (e.g., motor-driven crane girder supports).

1.5.11. Load Factors and Factors of Safety

Factors of safety Factors of safety are usually introduced to limit the computed stresses or deflections to acceptable levels at which the structure or member can support load and be serviceable. The factors are established by a standard or code related to a particular material and method of design and are statistically based on testing, research, and past experience. A usual objective in a *limit state design*, as found in the steel, reinforced concrete, and timber design codes, is that a structure or member behaves elastically when subjected to service loads. Codes often strive to meet this objective by establishing the limiting stress as a percentage of the material yield strength. For example, in steel design the maximum allowable bending stress is equal to 66 percent of the material yield strength (factor of safety = $1/0.66 = 1.52$).

Load factors Load-factor designs use separate factors for each type of load, and for different possible load combinations. The American Concrete Institute (ACI) assigns a different factor for each load type. Each factor is influenced by the degree of accuracy in estimating the load and the potential load variation that may occur during the life of the structure. For the combination of dead (D) and live (L) loads, the ACI required strength (U) must at least be equal to the following:

$$U = 1.4D + 1.7L$$

Wind load represents one form of live load. However, the simultaneous occurrence of dead (D) and wind (W) in combination with other live loads (L) is less probable. Therefore, the ACI suggests a reduced load combination for study as follows:

$$U = 0.75(1.4D + 1.7L + 1.7W)$$

Load-factor designs also include strength reduction factors to account for unforeseen overloads, probable understrength of structural members due to variations in material strength and dimensions, importance of each member in the structure, uncertainties in the analysis or design equations, and the like. The ACI specifies strength reduction factors (ϕ) for the design of reinforced concrete members (e.g., 0.90 for bending, 0.75 for shear, and 0.70 for tied columns). The AISC defines resistance factors (ϕ) for steel member design which are similar to the ACI strength reduction factors [1.6].

1.6. U.S. CUSTOMARY UNITS AND THEIR SI EQUIVALENTS

American students of engineering realize that prior to 1960, most countries of the world used a metric system for weights and measures, while common

TABLE 1.3

Common U.S. Customary Units, Their SI Equivalents, and SI Prefixes Used in Static Structural Analysis

U.S. Customary Units and SI Equivalents		
Measure	U.S. Customary Unit	SI Equivalent
Length	in	25.4 mm
	ft	0.3048 m
Area	in^2	645.2 mm^2
	ft^2	0.0929 m^2
Volume	in^3	16.39 cm^3
	ft^3	0.02832 m^3
Force	lb	4.448 N
	kip	4.448 kN
Stress	lb/in^2	6.895 N/m^2 = 6.895 Pa
	lb/ft^2	47.88 N/m^2 = 47.88 Pa
Moment of force	in · lb	0.1130 N · m
	kip · ft	1.356 kN · m
Moment of inertia (of an area)	in^4	41.62 × 10^4 mm^4
Common SI Prefixes		
Factor	Prefix	Symbol
10^9	giga	G
10^6	mega	M
10^3	kilo	k
10^{-3}	milli	m
10^{-6}	micro	μ
10^{-9}	nano	n

practice in the United States was and is based on a system of units of pound (lb), foot (ft), and second (s) that correspond to force, length, and time, respectively. The International System of Units (SI), derived from the French "Systéme International d'Unités," was developed by the General Conference of Weights and Measures (CGPM) to represent a universal, coherent system of measurement for worldwide scientific and technological use. The SI system is considered to be an absolute system with base units of kilogram (kg), meter (m), and second (s) that correspond to mass, length, and time, respectively. The principal features of SI, as noted by the U.S. Department of Commerce, National Bureau of Standards [1.7] are as follows:

1 It recognizes only one unit for each physical quantity (e.g., kilogram, meter, liter, second, Newton).

2 All units in the system relate to each other on a one to one basis (e.g., one Newton of force (N) = $kg \cdot m/s^2$; stress or elastic modulus = N/m^2; linear velocity = m/s).

3 A set of prefixes can be attached to the units to form preferred multiples and submultiples in powers of 1000.

Problems in this text are presented in both U.S. Customary and SI units. Table 1.3 provides U.S. Customary units, their SI equivalents, and SI prefixes that are commonly used in structural engineering practice. The contents of Table 1.3 are also presented on the front inside cover of the text.

1.7. COMPUTATION ACCURACY AND CHECKING

It is important for students to realize that analyses of most civil engineering structures do not require eight significant digits of accuracy—the number of digits displayed on a hand-held calculator and on computer printouts. The solutions are, at best, only as accurate as the input data used. Methods of analysis generally incorporate theorems and principles based on ideal assumptions. However, a degree of uncertainty always exists with reference to the magnitude and direction of load, material behavior, and the true support restraints. These factors plus the acceptable tolerances for material properties, shape dimensions, manufacturing, construction, and assembly practices serve to justify that computations to three significant digits is satisfactory for most civil engineering structural analyses.

The reader should be aware that the methods of analysis presented in this book are referred to as *first-order* analysis; that is, it is assumed that loads are supported by the original geometric shape of the structure under investigation. A more complex *second-order* analysis, which considers the deformed shape of the structure in the formulation of the equilibrium equations, will generally provide more accurate results; however, changes in structural geometry are often found to be minor and do not justify second-order analysis.

Making mistakes is an unappreciated, yet inevitable part of our humanity that we strive to minimize. Therefore, it is recommended that students develop the good habit of carrying out *independent checks* of their analyses to insure reasonable accuracy. This text presents numerous methods of analysis that can be used to solve a particular problem. One method can be used to independently check the results obtained from another method. In addition, a computer solution by use of the IMAGES program can also serve as a check solution for comparison with hand calculated results. However, there are many occasions where a reliable computer program yields erroneous results due to human input errors or poor judgment in modeling a structural system. Therefore, the classical methods of this text, which involve hand calculations, will often serve to verify computer solutions.

1.8. USE OF COMPUTERS

To the structural analyst, the salient features that modern computers provide are (1) the opportunity to solve complex structures that were practically unsolvable to past generations of engineers, (2) the opportunity to study the behavior of a structure for a vast number of environmental and load conditions, and (3) the ability to perform analyses of simple or complex structures accurately, efficiently, and economically.

Many computer programs are available from nonproprietary sources for mainframe systems, minicomputers, and most recently for microcomputers. Major companies and corporations are likely to utilize all forms of computers whereas economics will usually guide the smaller firms to use PCs and PC network systems. It is a fair estimate that 90 percent or more of the practicing engineering community restrict their computer activities to that of program user as compared with the remaining smaller group of engineers who are actively engaged in writing computer programs or developing new modeling techniques. *Each user has the responsibility to carefully review and understand the full nature of each computer program to be used including the principles, limitations, and restrictions on which it was based.* It is always advisable to check the accuracy of a computer program before using it for new unsolved problems. This can be done by initially using the computer program to solve many classical problems which have established solutions. A mature analyst will also recognize that there are simple problems that can be solved as easily and possibly quicker by traditional hand calculation methods when compared to use of computer programs. Both approaches merit our respect and careful use.

Problems

1.1 Determine the wind-load forces for the gable frame structure shown below using the ASCE curve of Fig. 1.19. Let $V = 80$ mph. Assume $K_z = I = 1.0$.

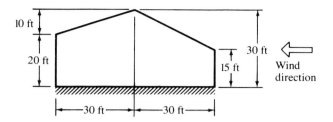

1.2 Using Figs. 1.18 and 1.19, determine the wind-load forces on the gable structure shown in Prob. 1.1 above if the wind direction is reversed and the structure is located in New Orleans, Louisiana. Exposure D. Let $I = 1.0$.

1.3 If the gable structure of Prob. 1.1 consists of frames spaced at 8-ft 0-in c–c, determine the wind force in plf on (*a*) a typical interior frame and (*b*) an exterior frame. Use wind direction shown in Prob. 1.1.

References

1.1 David G. McCullough, *The Great Bridge*, Simon and Schuster, New York, 1972.

1.2 *Standard Specifications for Highway Bridges*, 12th ed., American Association of State Highway and Transportation Officials, Washington, D.C., 1977.

1.3 "American National Standards Building Code Requirement for Minimum Design Loads in Buildings and Other Structures," American National Standards Institute, Inc., New York, A58.1-1982.

1.4 "Wind Forces on Structures," American Society of Civil Engineers, Structural Division, Technical Report 84,ST4, 1958.

1.5 "Wind Forces on Structures," *Trans. American Society of Civil Engineers*, no. 126, pt. II, 1961.

1.6 "Guide to Load and Resistance Factor Design," American Institute of Steel Construction, Inc., Chicago, 1986.

1.7 "Recommended Practice for the Use of Metric (SI) Units in Building Design and Construction," U.S. Department of Commerce, National Bureau of Standards, NBS Technical Note 938, April 1977.

1.8 *Uniform Building Code*, 8th ed., May 1988.

1.9 *Recommended Lateral Force Requirements and Tentative Commentary*, 5th ed., Structural Engineers Association of California (SEAOC), San Francisco, 1988.

1.10 *The BOCA Basic Building Code*, 7th ed., Building Officials and Code Administrators International, Inc., Chicago, 1978, p. 203.

CHAPTER TWO

Reactions

2.1. STATIC EQUILIBRIUM EQUATIONS

The three fundamental laws of Sir Isaac Newton (1642–1727) form the basis of engineering mechanics and the principles on which this subject rests. The first law of Newton states that a body initially at rest remains at rest if the resultant force acting on the body is zero; this condition is known as static equilibrium. Newton's third law indicates that the forces of action and reaction between interacting bodies are equal in magnitude, opposite in direction, and have the same line of action. The second law of Newton concludes that if the sum of all active and reactive forces is not zero, the body will move at an acceleration (a) in the direction of the resultant force (F) as defined by the expression $F = ma$, where m is the mass of the body. Since a body (or structure) in equilibrium is usually subjected to a system of forces, it may be clearer to express Newton's second law as $\Sigma F = ma$, where ΣF represents the summation of all active and reactive forces on the structure.

In the study of dynamics, moving structures, such as cars, trains, boats, and airplanes, are often studied by using the equation above as an expression of dynamic equilibrium, namely, $\Sigma F - ma = 0$. However, most civil engineering structures are considered to be stationary bodies which experience small member deformations and minor support displacements under normal design-load conditions. Thus, most civil engineering structures are expected to remain at rest and in a state of static equilibrium (zero acceleration) so that the expression above reduces to $\Sigma F = 0$. In this equation, F represents both linear and rotational (moment) force components.

33 Static Equilibrium Equations

Olympic Oval, Calgary, Alberta. (*Courtesy of Prestressed Concrete Institute.*)

For the sake of clarity, F will be used to represent a linear force component, and M will be used to signify a moment force component, simply called *moment*.

If the active and reactive forces on a structure are expressed in terms of their components in a rectangular x, y, z Cartesian coordinate system (see Fig. 2.1), static equilibrium can be defined by

$$\Sigma F_x = 0 \qquad \Sigma F_y = 0 \qquad \Sigma F_z = 0$$
$$\Sigma M_x = 0 \qquad \Sigma M_y = 0 \qquad \Sigma M_z = 0 \qquad (2.1)$$

where ΣF_x, ΣF_y, and ΣF_z denote force summation in the x, y, and z direction, respectively; ΣM_x, ΣM_y, and ΣM_z represent moment summation about the respective x, y, and z axes.

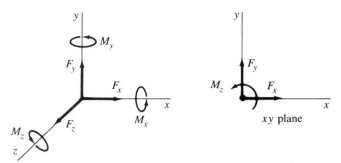

Figure 2.1

For a two-dimensional structure defined in an xy plane, the effective equations of equilibrium are as follows:

$$\Sigma F_x = 0 \qquad \Sigma F_y = 0 \qquad \Sigma M_z = 0 \tag{2.2}$$

2.2. TYPES OF SUPPORT RESTRAINT

Common types of coplanar structural restraints include cable, link, roller, hinge (pin), and fixed supports. The term *restraint* implies that translation and/or rotation is not permitted to occur at points of support of a structure. Restraints are also called supports or constraints and are usually expressed in terms of their reaction components. Table 2.1 contains various

TABLE 2.1
Types of Coplanar Supports

Support Type	Support Symbol	Reaction	Number of Unknowns
Cable	Cable	F, θ	1
Link	Link	F	1
Roller	or	F	1
Rocker		F	1
Frictionless pin or hinge		F_y, F_x	2
Fixed end		F_x, F_y, M	3

35 Types of Support Restraint

types of coplanar supports shown with their respective reaction components and free-body diagram symbols.

Cable, roller, and link supports provide restraint in a known direction. Therefore, each of these support types have only one unknown reaction.

Cables, particularly thin cables, can only provide tensile force resistance in the direction of the taut cable.

Links are two-force, pin-ended members that can only provide tensile or compressive force resistance in the member direction.

A *roller* support permits limited freedom of motion (no restraint) in the direction parallel to its support surface but is restrained in the direction perpendicular to its support surface. The symbolic circle used to represent the roller support in schematic free-body diagrams should not be misconstrued as an unattached round log under a wood plank. Figure 2.2 demonstrates that *the roller restraint can occur in either direction perpendicular to the support path*, depending on the load position on the beam.

Rocker supports and members in contact with a smooth surface also display one unknown reaction component which is directed perpendicular to the support surface and toward the structure.

Since rollers, cables, links, and rockers provide a known direction of restraint, only the force magnitude is unknown. In statics problems, it is customary to assume the direction of an unknown force; if the force magnitude is found to be of the positive sign, the assumed direction is correct; if negative, the magnitude remains correct but of the opposite direction. Thus

Roller bearing support of Potomac River Highway Bridge, Washington, D.C. (*Courtesy of International Structural Slides, Berkeley*.)

Detail of rocker bearing support of bridge structure near Rio Vista, California. (*Courtesy of International Structural Slides, Berkeley*.)

Hinged bearing support on bridge structure near Rio Vista, California. (*Courtesy of International Structural Slides, Berkeley*.)

Figure 2.2

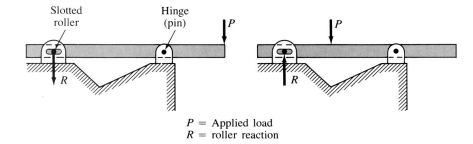

P = Applied load
R = roller reaction

rollers, cables, links, and rockers have only one unknown reaction to be determined.

A *hinge* (or *pin*) support has two unknown reaction-force components to be determined since it prevents both horizontal and vertical translation. Ideally, we presume that members are joined to a hinge support by a frictionless pin which permits rotational freedom about it (zero moment restraint). Visualize the motion of a typical hinge-supported door. The resultant force direction of a hinge support will vary for different load conditions on its attached structure; in a typical xy plane, the hinge resultant force can easily be described in terms of its horizontal (x) and vertical (y) components.

The *fixed support*, like the hinge, ideally prevents translation but also prevents rotation (complete moment restraint), such as, a telephone pole or flag pole anchored into the ground. Thus, the fixed support has three unknown reaction components, namely, two reaction components that prevent translation, and an unknown moment reaction that resists rotation.

2.3 GEOMETRIC STABILITY AND DETERMINACY

Geometrically stable structure A structure at rest may be considered *geometrically stable* if it remains at rest after static loading and the original orientation of all its members remain virtually unchanged.

Statically determinate structure A structure is said to be *statically determinate* if the structure is geometrically stable and the number of unknown reaction components and available equations of statics are equal. The unknown reaction components for statically determinate coplanar structures are determined by use of Eqs. (2.2). Clearly, the unknown reactions are external to the structure. Therefore, when the number of unknown reaction components and the number of equilibrium equations are equal, the structure is defined as *statically determinate externally*.

Statically indeterminate structure Most large building and bridge structures contain many support reactions. In general, when the number of

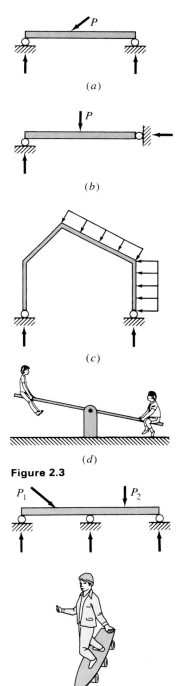

Figure 2.3

Figure 2.4

reaction components exceed the number of equilibrium equations, the structure is said to be *statically indeterminate externally*. Chapters 11 to 14 present numerous methods of analysis for the study of statically indeterminate planar structures.

Geometrically unstable structure If the number of reaction components is less than the number of equilibrium equations, the structure is *geometrically unstable externally*, that is, the supports are not properly constrained and excessive motion of the structure may occur when loads are applied. The roller-supported beams, the frame, and the hinged teeter-totter of Fig. 2.3 each show the ready state of geometric instability. Since the roller supports of Figs. 2.3a and 2.3c provide only vertical force resistance, both structures are unable to support horizontal load components and maintain horizontal equilibrium.

In Fig. 2.3b, it is apparent that the roller support reactions of the beam do not provide proper constraint since they are concurrent at the left support point (note that the beam will immediately start to rotate when a load P is placed on the beam). In general, improper support constraint exists when all support reactions are concurrent at a point or all reactions are parallel to each other. Replacement of the roller at the left end by a pin support will not improve the situation. Proper support is provided when the lines of action of the reactions and the resultant load vector pass through the same point. This can be achieved by replacing the roller at the right end of the beam with a pin support.

It is possible for a structure to be geometrically unstable externally even though the number of reaction components are equal to the number of static equations. For example, the roller-supported beam of Fig. 2.4 has three parallel and vertical reactions. Therefore, the beam is unable to resist any horizontal load component (visualize the resemblance to a skateboard or roller skates). Clearly, the beam of Fig. 2.4 requires an added horizontal constraint to satisfy horizontal equilibrium.

In a structural analysis, we strive to evaluate the internal member forces as well as the external reactions due to a load condition. However, structures often have more members than available equations for their direct solution. Therefore, structures can also be classified as *internally* determinate, indeterminate, or geometrically unstable. A study of these classifications of structures is presented in Sec. 3.8.

2.4 USING THE FREE-BODY DIAGRAM

One of the most important engineering aids developed during the study of statics is the construction of the free-body diagram. The solution of unknown external reactions and internal member force(s) of a statically determinate structure are easily determined from Eqs. (2.2) with the visual aid of a free-body diagram (FBD). Figure 2.5a shows an FBD model of a

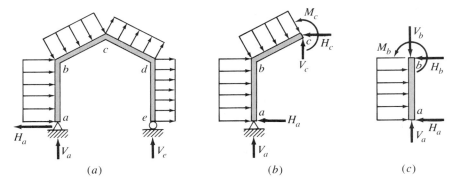

Figure 2.5

wind-loaded gable frame structure that is used to determine support reactions. Figure 2.5b shows an FBD model for a section of the frame that is used to compute the internal forces at the roof apex (point c). Figure 2.5c provides an FBD for the windward vertical member ab that can be used to compute the internal forces at point b. The importance and proper use of the FBD in structural analysis cannot be overstated.

2.5. BEAM REACTIONS BY EQUILIBRIUM EQUATIONS

Beam reactions can be determined from static equilibrium equations in the following manner:

1 Draw an FBD of the beam showing all external loads. Add to the FBD all unknown reaction-force components, clearly showing each force in an assumed direction.

2 Apply moment equilibrium ($\Sigma M = 0$) about a point where all except one of the unknown reaction components converge; then, directly determine the remaining unknown reaction force.

3 Use the remaining equilibrium equations ($\Sigma H = 0$ and $\Sigma V = 0$) and the result of step 2 above to find the other unknown reactions.

4 Choose a point (on or off the beam) other than the point used in step 2 to verify moment equilibrium. If $\Sigma M = 0$ about this point, you have established an *independent check* of the reaction computations.

Step 4 may precede step 3, followed by $\Sigma V = 0$ to find the last unknown reaction. Then, $\Sigma H = 0$ can be used as the check equation to verify the results. This approach is presented in the example problems that follow. Beyond this chapter, the independent check may not be shown to save space; yet, it is both an important and necessary step of the complete analysis procedure.

39 Beam Reactions by Equilibrium Equations

InterFirst Tower, Fort Worth. (*Courtesy of American Concrete Institute, photo by Frank S. Kelly.*)

In Example 2.1, the roller *support* is horizontal. Thus, the roller *reaction* is vertical since a roller reaction is known to be directed perpendicular to its support. The calculations reveal that all reaction component directions were assumed correctly; the magnitude and direction of V_L are verified by the independent check equation. Observe that the positive sign conventions of F_x, F_y, and M are arbitrary selections which should be clearly defined in all analyses. The check equation of $\Sigma M_R = 0$ used a positive sign convention which is opposite to that used for $\Sigma M_L = 0$ simply to demonstrate that the choice of direction is arbitrary. However, the author recommends using one consistent set of sign conventions in a given problem.

EXAMPLE 2.1

Find all reaction components for the simple beam.

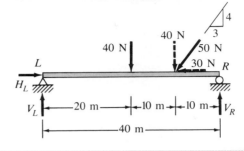

Solution

$\circlearrowleft\ +\Sigma M_L = 0 = -40(20) - 40(30) + 40\ V_R \qquad V_R = 50\text{N}\uparrow$ ◀

$+ \rightarrow \Sigma F_x = 0 = H_L - 30 \qquad H_L = 30\text{N} \rightarrow$ ◀

$+ \uparrow \Sigma F_y = 0 = V_L - 40 - 40 + V_R \qquad V_L = 30\text{N}\uparrow$ ◀

Check

$\circlearrowright\ +\Sigma M_R = 0 = V_L(40) - 40(20) - 40(10)$

$\qquad V_L = 30\text{N}\uparrow$ ✓

EXAMPLE 2.2

Calculate all reaction components for the overhanging beam.

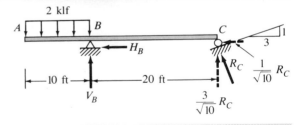

Solution

$\circlearrowleft\ +\Sigma M_B = 0 = (2 \times 10)\left(\dfrac{10}{2}\right) + \dfrac{3}{\sqrt{10}} R_C (20\text{ ft})$

Therefore $\quad R_C = -5.27 \text{ kips } {}^3\!\!\diagdown_{\!1}$ ◀

$+ \rightarrow \Sigma F_x = 0 = -H_B - \dfrac{1}{\sqrt{10}} R_C \quad \text{or} \quad H_B = -\dfrac{1}{\sqrt{10}} R_C$

$H_B = \dfrac{-1}{\sqrt{10}}(-5.27) = +1.67 \text{ kips} \quad \therefore \quad H_B = 1.67 \text{ kips} \leftarrow$ ◀

$+ \uparrow \Sigma F_y = 0 = -2(10) + V_B + \dfrac{3}{\sqrt{10}} R_C = -20 + V_B + \dfrac{3}{\sqrt{10}}(-5.27)$

$\qquad V_B = 25 \text{ kips}\uparrow$ ◀

Check

$\zeta + \Sigma M_A = 0 = -(2 \times 10)\left(\dfrac{10}{2}\right) + 10V_B + \dfrac{3}{\sqrt{10}} R_C(30) = -100 + 250 - 150 = 0$ ✓

EXAMPLE 2.3

Calculate all reaction components for the cantilever beam with rigid extension bar.

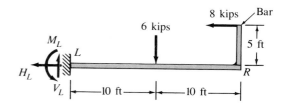

Solution

$\zeta + \Sigma M_L = 0 = M_L - 6(10) + 8(5)$ $M_L = 20$ kip·ft ◂

$+ \rightarrow \Sigma F_x = 0 = -H_L - 8$ $H_L = -8$ kips, $H_L = 8$ kips → ◂

$+ \uparrow \Sigma F_y = 0 = V_L - 6$ $V_L = 6$ kips↑ ◂

Check

$\zeta + \Sigma M_R = 0 = M_L - 20V_L + 6(10) + 8(5)$

$= 20 - 20(6) + 60 + 40 = -120 + 120 = 0$ ✓

In Example 2.2, the roller support at C is on an incline. Thus, the roller reaction is directed normal to the incline. The first equation reveals that initially the roller reaction was assumed in the wrong direction. Therefore, the roller reaction is actually directed downward for the load condition given. It is worth noting that the check equation, $\Sigma M_A = 0$, is referenced to a point on the beam other than a reaction support point.

Example 2.3 shows a cantilever beam with a 5-ft bar that is rigidly attached and normal to the free end of the beam. Reaction components at the fixed-end support, point L, are easily found by the equations of statics and verified by use of the equilibrium equation $\Sigma M_R = 0$.

2.6 FRAME REACTIONS BY EQUILIBRIUM EQUATIONS

Examples 2.4 and 2.5 also demonstrate the use of the free-body diagram and the principles of statics to determine support reactions of structural frames. In Example 2.6, force equilibrium is defined by $\Sigma H = 0$ and $\Sigma V = 0$

in place of $\Sigma F_x = 0$ and $\Sigma F_y = 0$, respectively, simply to demonstrate that practicing engineers often use different letter notations for force, moment, displacements, and stresses; however, the structural analyst is expected to clearly define the symbolic notations used.

EXAMPLE 2.4

Calculate all reaction components of the frame.

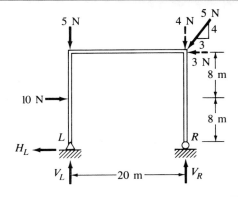

Solution

$\circlearrowleft + \Sigma M_L = 0 = 20V_R + 3(16) - 4(20) - 10(8)$ $\quad V_R = 5.6$ N↑ ◀

$+ \uparrow \Sigma F_y = 0 = V_L + V_R - 5 - 4$ $\quad V_L = 3.4$ N↑ ◀

$+ \rightarrow \Sigma F_x = 0 = 10 - H_L - 3$ $\quad H_L = 7$ N ← ◀

Check

$\circlearrowleft + \Sigma M_R = 0 = 3(16) + 5(20) - 10(8) - 20V_L$

$\qquad 148 - 148 = 0$ ✓

EXAMPLE 2.5

Find all reactions for the frame structure.

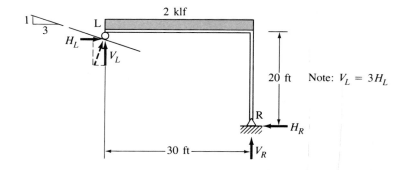

Note: $V_L = 3H_L$

Solution

$\circlearrowleft + \Sigma M_R = 0 = V_L(30) + H_L(20) - (2 \cdot 30)15$

$\qquad 0 = 3H_L(30) + H_L(20) - 60(15) \qquad H_L = 8.18 \text{ kips} \rightarrow$ ◀

$+ \uparrow \Sigma F_y = 0 = V_L - (2 \cdot 30) + V_R \qquad \therefore V_L = 3H_L = 24.5 \text{ kips} \uparrow$ ◀

$\qquad V_R = 35.5 \text{ kips} \uparrow$ ◀

$+ \rightarrow \Sigma F_x = 0 = H_L - H_R \qquad \therefore H_R = 8.18 \text{ kips} \leftarrow$ ◀

Check

$\circlearrowleft + M_L = 0 = (2 \cdot 30)15 + H_R(20) - 30V_R = 0$ ✓

EXAMPLE 2.6

Find all frame reaction forces.

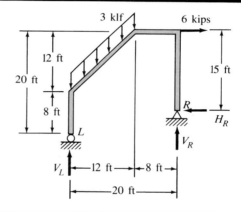

Solution

$\circlearrowleft + \Sigma M_R = 0 = V_L(20) - (3 \cdot 12)(6 + 8) + 6(15) \qquad V_L = 20.7 \text{ kips} \uparrow$ ◀

$+ \uparrow \Sigma V = 0 = V_L - (3 \cdot 12) + V_R \qquad V_R = 15.3 \text{ kips} \uparrow$ ◀

$+ \rightarrow \Sigma H = 0 = 6 - H_R \qquad H_R = 6 \text{ kips} \leftarrow$ ◀

Check

$\circlearrowleft + \Sigma M_L = 0 = (3 \cdot 12)(6) + 6(20) - 5H_R - 20V_R$

$\qquad = 336 - 336 = 0$ ✓

2.7 TRUSS REACTIONS BY EQUILIBRIUM EQUATIONS

Examples 2.7 and 2.8 continue to demonstrate the use of the basic principles of statics to find support-reaction components. An independent check is made by $\Sigma M = 0$ about point C in Examples 2.7 and 2.8.

EXAMPLE 2.7

Determine the truss reaction forces.

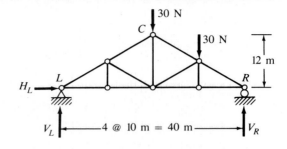

Solution

$\zeta + \Sigma M_L = 0 = 30(20) + 30(30) - 40V_R \qquad V_R = 37.5 \text{ N}\uparrow$ ◀

$\rightarrow + \Sigma H = 0 = H_L \qquad H_L = 0$ ◀

$\uparrow + \Sigma V = 0 = V_L + V_R - 30 - 30 \qquad V_L = 22.5 \text{ N}\uparrow$ ◀

Check

$\zeta + \Sigma M_C = 0 = V_L(20) + 30(10) - V_R(20)$

$\qquad = 750 - 750 = 0$ ✓

EXAMPLE 2.8

Find all reaction components for the cantilever truss.

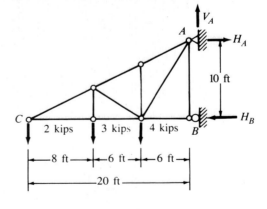

Solution

$+ \uparrow \Sigma V = 0 = V_A - 2 - 3 - 4 \qquad V_A = 9 \text{ kips}\uparrow$ ◀

$\zeta + \Sigma M_A = 0 = H_B(10) - 2(20) - 3(12) - 4(6) \qquad H_B = 10 \text{ kips} \leftarrow$ ◀

$+ \rightarrow \Sigma H = 0 = H_A - H_B \qquad H_A = 10 \text{ kips} \rightarrow$ ◀

Check

$$\circlearrowleft + \Sigma M_C = 0 = H_A(10) - 20V_A + 3(8) + 4(14)$$
$$= 180 - 180 = 0 \quad \checkmark$$

2.8 REACTIONS BY USE OF IMAGES[1] AND THE PERSONAL COMPUTER

An alternative way of checking the magnitudes and directions of structural support reactions is achieved by use of a computer program. To this end, the analyst may choose to write a computer routine or elect to use an available program. In the paragraphs below, a commercially available program is introduced which performs analyses of various structures—including the determination of support reaction forces.

Introduction to IMAGES

A customized version of IMAGES, a personal computer program that performs static structural analysis, accompanies this text in the form of two floppy diskettes found in a jacket on the inner rear text cover. Tutorial instructions on the running of IMAGES is found in App. 1. This program can run on an IBM PC, XT, AT, or IBM compatible system and will be useful in solving beam, truss, and frame structure problems. It is assumed that the student has been introduced to the personal computer, is able to format diskettes and perform the normal interactive operations of a PC. This program will offer the student an additional approach to check solutions of most problems encountered in this text and should prove helpful in independent study. However, it must be realized that an engineering computer program such as IMAGES, best serves the individual who fully comprehends both the principles of the engineering subject and the associated mathematical routines it uses for efficient computer solutions. *Consequently, students should be aware that their primary mission is to learn the basic principles and methods of structural analysis presented in this text and should view IMAGES and other computer programs as auxiliary aids for the success of their mission.* Students, instructors, and practicing engineers alike should find their experiences with IMAGES and the PC both interesting and rewarding.

Applications

IMAGES provides four main menus and numerous submenus that are used to perform a structural analysis. The initial menu seen on the monitor is the *Program Menu* which prompts for a selection as follows:

---

[1] Celestial Software, Inc., Berkeley, Calif., copyright publisher of various versions of IMAGES.

46 Reactions

1 Geometry Definition
2 Static Analysis
3 Dynamic or Seismic Analysis (not included)
4 Restart IMAGES
5 End Session
?

Selection 1 provides a *Geometry Menu* which prompts for a choice from the following:

1 Create/Edit Geometry
2 Check Geometry
3 Renumber Nodes
4 Plot Geometry
5 Return to Program Menu

Since a typical problem begins with the definition of the geometry, the choice of selection 1 is made which reveals a submenu called *Create/Edit Geometry* where the following list of required input data appears:

1 Enter Problem Title
2 Define Material Properties
3 Define Node Points
4 Define Elements
5 Define Cross-Section Properties
6 Define Restraints
7 Save Geometry on Disk
8 Get Old Geometry File
9 Return to Geometry Menu

As the submenu title suggests, one may enter a new geometry title or alter a previous geometry file that was saved on disk. After selection 1 is prompted, a new or existing title is specified. If the title refers to an existing file, the monitor will immediately prompt you to *press "8" to get old geometry file* before proceeding. *If you don't get it, you lose it!*

Example Problem 1

Consider the simply supported beam shown in Fig. 2.6. The object of this problem is to check the reactions of Example 2.1. A computer model of the beam is drawn so that points where loads or reactions exist are signified by node numbers 1 to 4. Circled numbers between nodes represent the three elements of the beam. Two sets of orthogonal *xy* axes are used to define a structure geometry, namely, (1) a local coordinate system where *x* is oriented along each structural member length (from node *i* to node *j*) and *y* is perpendicular to the member, and (2) a global coordinate *xy* system to which the node coordinates of all members of the structure are referenced. In Fig. 2.6 the local beam *xy* axis coincides with the global *xy* axis. Although reactions alone are of immediate interest, IMAGES also computes member forces, stresses, and node displacements, and requires material and cross-section property input data.

For material properties, one set of material property values is input for *each* beam material (only one in this example). The program will accept any consistent set of units; yet, the user should note that most default values are specified in inches, pounds, and seconds.

Following all menu selections of IMAGES, the word "HELP" or "H" can be entered after which the monitor displays the data-entry format required. Node point coordinates (*x*, *y*) are easily defined as (0, 0), (20, 0), (30, 0), and (40, 0) for nodes 1 to 4, respectively.

When selection 4 (Define Elements) of the Create/Edit Geometry Menu is chosen, another set of menu choices appears on the monitor, namely,

1 Define Beams
2 Define Springs
3 Return to Geometry Menu

After selecting Define Beams, the following program prompt appears:

    Next Beam 1
    ?

Entering "HELP" or "H" instructs the user to enter the *I* node, *J* node, Prop, Mat, Pin *I*, Pin *J* of beam element no. 1, each separated by a comma; *I* and *J* may be any valid node numbers, Prop is a number that identifies a set of cross-section property data, Mat is a number that identifies a set of material property data, and Pin *I* and Pin *J* are member end-release codes which are discussed in IMAGES Example Problem 4 and require no input for this problem. All beam elements must be defined afterwhich the word "EXIT" or "E" is entered on the keyboard to return to the Geometry Menu.

Choosing *Define Cross-Section Properties* from the Geometry Menu, and employing the HELP prompt, reveals the format of area property input data. One set of properties is chosen to represent all three beam elements of the example problem.

Moving on to *Define Restraints*, with a little "help," will prompt for a node to be restrained, and the degree of freedom (DOF) at the restrained node. The DOF is defined in the global coordinate directions for ground-support node points (see App. 1).

Figure 2.6

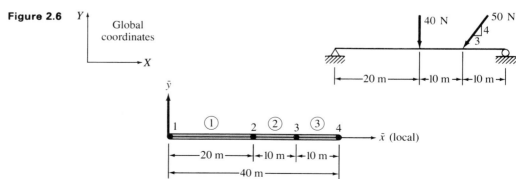

The *DOF global restraints* are defined as follows:

$$1 = X \text{ translation}$$
$$2 = Y \text{ translation}$$
$$3 = Z \text{ rotation}$$

In this example, node 1 is a hinge support with the DOF restraints = 12, and node 4 is a roller support with a DOF restraint = 2.

After the geometry data is completely entered, the next operation is to return to the Create/Edit Geometry Menu and *Save Geometry on Disk*. A warning will appear if the data has not been saved. After "saving," return to Geometry Menu. Now, the geometry must be checked for modeling and coding errors, the nodes must be renumbered (internally) for the efficient computer solution of the problem, and the geometry can be plotted for visual verification.

After the geometry is completely defined, saved, and checked, return to the Program Menu and select *Static Analysis*. Immediately, a *Static Menu* appears on the PC monitor from which *Define Concentrated Loads* is chosen. Then, each of the three X and Y load components on the beam of Fig. 2.6 are entered individually. The concentrated loads are entered at the nodes and referenced to the global coordinate system chosen. A concentrated load is positive or negative as defined by the sign of its magnitude: for example, at node 2, the 40 N load is in the negative Y direction and is input as 2, 2, −40. After all loads have been defined, enter EXIT and return to the Static Menu where the selection command *Save Loads on Disk* must be made. Then, prompt *Return to Static Menu*.

Assemble Stiffness Matrix and *Solve Displacements* are two *compulsory* routines to be entered for solving problems by IMAGES. Since the IMAGES computer program is based upon matrix structural analysis and finite elements methods, these two routines will be better understood after the principles of structural analysis are grasped and Chap. 14 is studied. The key words of these routines correctly suggest that the equilibrium equations of the structural system, in matrix form, are efficiently *assembled* and *solved* with the aid of a microcomputer.

Finally, by prompting the PC to print reactions, the results are as follows:

At node 1: $F_x = 0.00000E + 00$
$$F_y = 2.92500E + 01$$
At node 4: $F_y = 5.07500E + 01$

These results provide a reasonable check of the accurate reactions obtained in Example 2.1 by statics.

Example Problem 2

Example 2.2 presents an interesting dilemma worth some attention. A careful study of the two support reactions reveals that it is possible for them (1) to act parallel to each other or (2) to coincide at some point in space; either possibility may result in a geometrically unstable beam (refer to Sec. 2.3). Computer solutions of geometrically unstable structures have been known to provide inconsistent and erroneous results; often, a computer program will detect an unstable geometry or improper restraints, come to an abrupt halt, and provide an error message to the operator.

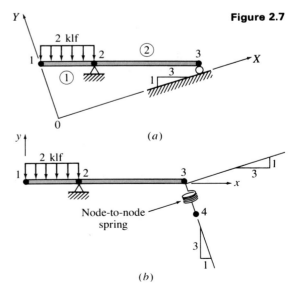

Figure 2.7

49 Reactions by Use of IMAGES and the Personal Computer

```
================ I M A G E S  2 D ================
= Copyright (c) 1983   Celestial Software Inc. =
================================================

SOLVE RESTRAINT REACTIONS      Version 3.0
```

LOAD CASE 1

RESTRAINT REACTIONS

Node	Direction	Reaction
2	F_x	6.28260E+03
2	F_y	2.42322E+04
3	F_y	-5.27047E+03

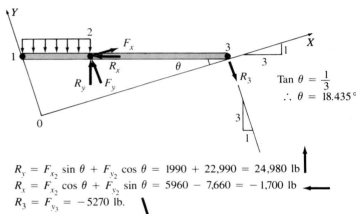

$$R_y = F_{x_2} \sin\theta + F_{y_2} \cos\theta = 1990 + 22{,}990 = 24{,}980 \text{ lb}$$
$$R_x = F_{x_2} \cos\theta + F_{y_2} \sin\theta = 5960 - 7{,}660 = -1{,}700 \text{ lb}$$
$$R_3 = F_{y_3} = -5270 \text{ lb}.$$

Figure 2.8

IMAGES offers two approaches to represent the inclined roller support (see Fig. 2.7). Figure 2.7a shows the global X axis directed along the inclined support at node 3 where restraint is easily input in the global Y direction. Students are recommended to always align rollers along a global axis. Fortunately, rollers used in civil engineering structures are usually on level supports; thus, the X axis usually proves to be the favorable choice for a global axis. This problem was solved by IMAGES in the use of this approach.

Figure 2.7b shows the use of an advanced approach that is not recommended for introductory level students. Here the global X axis is directed along the beam axis (x) and a node point 4 is added to form a line with node 3 that is perpendicular to the inclined support. From App. 1, ground springs and node-to-node spring elements are discussed. A node-to-node spring can be placed between nodes 3 and 4 to simulate the roller reaction. However, the proper selection of the spring stiffness that will provide the correct reaction force is a complex matter beyond the scope of this text. Consequently, the model of Fig. 2.7a was used to determine beam reactions.

The computer printout of global restraint reactions is shown in Fig. 2.8. The roller reaction is equal to that found by statics and with the aid of trigonometry, the hinge reaction components are easily verified. Also note that the uniform distributed load on element 1–2 is defined in the local y direction such that a positive input value results in the load in the negative y direction (see App. 1).

50 Reactions

Example Problem 3

A model of the structural frame of Example 2.6 is shown in Fig. 2.9. The *Additional Options* selection from the Create/Edit Loads Menu permits uniform distributed loads to be entered in the selected global direction. The 3 klf uniform gravity load acting on the frame is given in terms of horizontal projected length. In IMAGES, the gravity load is expressed in terms of load per linear foot along the member length. Therefore, the 3 klf load is the equivalent input value of 2.12 klf in IMAGES. The reaction components computed from this model provide an excellent check when compared to the solution values from statics.

Figure 2.9

Member length ② = $12\sqrt{2} \approx 17$ ft

Distributed load per slope length of ② = 3 klf $\frac{12 \text{ ft}}{17 \text{ ft}} \approx 2.12$ klf

Note: Nodes are numbered and shown as unfilled circles on the rigid frame above.

2.9 INTERIOR HINGES IN CONSTRUCTION

2.9.1. Introduction

Interior hinges (pins) are often used to join flexural members at points other than support points. In Fig. 2.10, interior pins are used to connect two halves of an arch frame structure. For arches of sizable dimensions, it is practical to design the arch as two halves since it will simplify their manufacture, transportation, and erection. In addition, interior hinge connections can be used to join segments of a continuous beam or a multitruss cantilever bridge structure (see Fig. 2.11).

In Secs. 2.9.2 and 2.9.3 that follow, two significant benefits of interior-hinge construction are realized, namely, (1) proper placement of interior-hinge joints in flexural systems can result in reduced bending moments, displacements, and stresses, and (2) the presence of interior-hinge member connections can provide a statically determinate structure that can simplify the structural analysis procedure.

2.9.2. Three-Hinged Arch Construction

Arch structures are usually formed to support gravity loads which tend to flatten the arch shape and thrust the supported ends outward. Hinge or fixed-end supports are generally used to provide the horizontal restraint to keep the arch geometrically stable. The horizontal thrust forces at the supports acting with the vertical loading on the arch tend to develop counteracting moments that result in low bending stresses. Consequently, masonry and stone arch construction was successfully used in ancient times. A two-hinged arch is shown in Fig. 2.12*a* where it is evident that the arch is statically indeterminate. Figure 2.12*b* shows a hingeless arch that is highly

Figure 2.10 High Bridge (in foreground) spans over Harlem River, New York City. In background are Alexander Hamilton Bridge and Washington Traffic Bridge, respectively. Hinge connection of arch frame members of High Bridge, New York City. (*Courtesy of International Structural Slides, Berkeley.*)

indeterminate since both support ends are fully fixed. Figure 2.13 illustrates a free-body diagram of a three-hinged arch, similar to the arch construction shape of Fig. 2.12a. However, the three-hinged arch structure shown in Fig. 2.13 is statically determinate.

Example 2.9 shows that the vertical reactions at points A and B are easily found from equations of statics (i.e., $\Sigma M_A = 0$ and $\Sigma V = 0$). Construction of a free-body diagram of one arch member, such as member AC, permits

Figure 2.11

(*a*) Cantilever construction

(*b*) Cantilever truss construction

52 Reactions

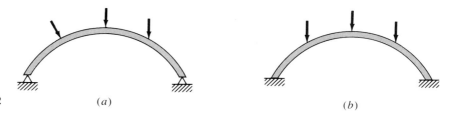

Figure 2.12 (a) (b)

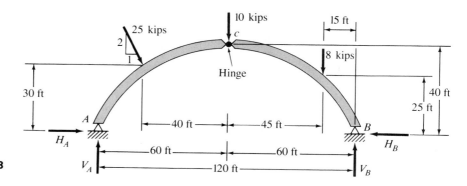

Figure 2.13

solution of the horizontal reaction at A by use of the equation $\Sigma M_C = 0$. The remaining unknown horizontal reaction at C is found from $\Sigma H = 0$. The results are verified by summation of moments about point C ($\Sigma M_C = 0$).

EXAMPLE 2.9

For the three-hinged arch shown in Fig. 2.13, determine all support reactions.

Solution

Using the Entire Arch (FBD ACB):

$$\curvearrowleft + \Sigma M_A = 0 = \left(\frac{2}{\sqrt{5}} \cdot 25 \text{ kips}\right)(20 \text{ ft}) + \left(\frac{1}{\sqrt{5}} \cdot 25 \text{ kips}\right)(30 \text{ ft}) + 10 \text{ kips}(60 \text{ ft})$$

$$+ 8 \text{ kips}(105 \text{ ft}) - 120 \, V_B \qquad V_B = 18.52 \text{ kips}\uparrow \;\blacktriangleleft$$

$$+\uparrow \Sigma V = 0 = V_A + V_B - \left(\frac{2}{\sqrt{5}} \cdot 25 \text{ kips}\right) - 10 \text{ kips} - 8 \text{ kips} \qquad V_A = 21.84 \text{ kips}\uparrow \;\blacktriangleleft$$

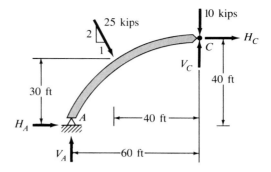

Using the FBD of member AC:

$\circlearrowleft + \Sigma M_C = 0 = 60\ V_A - 40\ H_A - \left(\dfrac{2}{\sqrt{5}} \cdot 25\text{ kips}\right)(40\text{ ft}) - \left(\dfrac{1}{\sqrt{5}} \cdot 25\text{ kips}\right)(10\text{ ft}) = 0$

$0 = 60(21.84) - 40H_A - 1006 = 0 \qquad H_A = +7.61\text{ kips} \rightarrow$ ◀

Using the FBD of the entire arch structure shown in Fig. 2.13,

$+ \rightarrow \Sigma H = 0 = H_A + \dfrac{1}{\sqrt{5}} \cdot 25\text{ kips} - H_B$

$\qquad = 7.61 + \dfrac{25}{\sqrt{5}} - H_B \qquad H_B = +18.79\text{ kips} \leftarrow$ ◀

Check (Using entire arch ACB):

$\circlearrowleft + \Sigma M_C = 0 = 60V_A - 40H_A - \left(\dfrac{2}{\sqrt{5}} \cdot 25\text{ kips}\right)(40\text{ ft}) - \left(\dfrac{1}{\sqrt{5}} \cdot 25\text{ kips}\right)(10\text{ ft})$

$\qquad + 8\text{ kips}(45\text{ ft}) + 40H_B - 60V_B$

$0 = 60(21.84) - 40(+7.61) - \left(\dfrac{2}{\sqrt{5}} \cdot 25\text{ kips}\right)(40\text{ ft}) - \left(\dfrac{25}{\sqrt{5}}\right)(10\text{ ft})$

$\qquad + 8(45) + 40(18.79) - 60(18.52)$

$0 = 2117 - 2117 \quad \checkmark$

EXAMPLE 2.10

The gable frame below consists of two rigid segments joined by a pin connection at point C. Determine all support reactions.

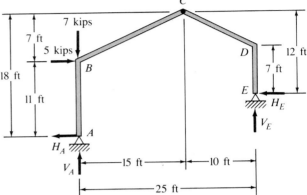

From the FBD above,

$\circlearrowleft + \Sigma M_A = 0 = 5 \text{ kips}(11 \text{ ft}) - 25V_E - (18 - 12)H_E$ or

$$25V_E + 6H_E = 55 \tag{1}$$

Using an FBD or frame element CDE alone,

$\circlearrowleft + \Sigma M_C = 0 = 12H_E - 10V_E$ or $10V_E - 12H_E = 0$ (2)

Multiplying Eq. (1) by two and adding the resulting equation to Eq. (2) yields,

$60V_E = 110 \qquad V_E = 1.83 \text{ kips}\uparrow$ ◀

From Eq. (2),

$10V_E - 12H_E = 0 = 10(1.83) - 12H_E \qquad H_E = 1.53 \text{ kips} \leftarrow$ ◀

Recalling the entire frame FBD above,

$+ \uparrow \Sigma V = 0 = V_A + V_E - 7 = 0 = V_A + 1.83 - 7 \qquad V_A = 5.17 \text{ kips}\uparrow$ ◀
$+ \rightarrow \Sigma H = 0 = 5 - H_A - H_E = 5 - H_A - 1.53 \qquad H_A = 3.47 \text{ kips} \leftarrow$ ◀

Check equation using FBD of entire three-hinged frame structure.

$\circlearrowleft + \Sigma M_E = 0 = 25V_A + (18 - 12)H_A - 7 \text{ kips}(25 \text{ ft}) + 5 \text{ kips}(12 \text{ ft} - 7 \text{ ft})$

$0 = 25(5.17) + 6(3.47) - 7(25) + 5(5)$

$0 = 175 - 175 \quad \checkmark$

Example 2.10 shows a three-hinged gable frame structure with end supports at different elevations. Two equations are written in terms of the two unknown reaction components at E as follows: $\Sigma M_A = 0$ using the FBD of the entire structure, and $\Sigma M_C = 0$, using the FBD of the right frame sector CDE. These two equations are solved simultaneously in terms of H_E and V_E.

2.9.3. Cantilever Construction

The maximum bending moment in a simply supported beam under uniform distributed load (w) occurs at midspan and is equal to $wL^2/8$ where L is the span length. Since the moment increases rapidly with increasing span length, longer spans result in deeper, more complex, and more expensive members. Cantilever construction represents a design concept that can be used for long span structures. If spans are properly proportioned, cantilever construction can develop smaller values of bending moment, deflection, and stress when compared with simple support construction. In Fig. 2.14, three alternate ways of supporting a three-span highway overpass structure are shown. Figure 2.14a considers three individual simple beams and shows representative bending-moment diagrams due to uniform distributed

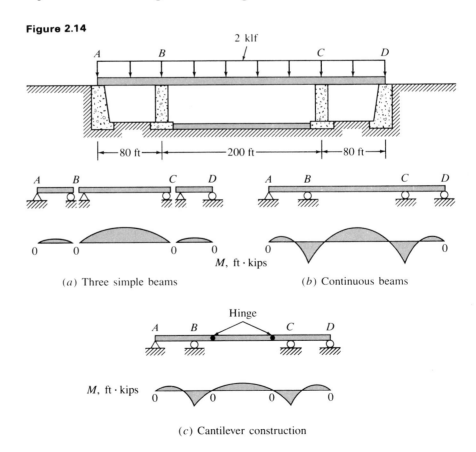

Figure 2.14

(a) Three simple beams

(b) Continuous beams

(c) Cantilever construction

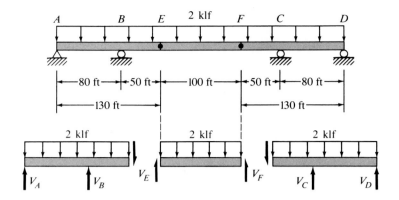

Note: 1 $M_E = M_F = 0$ at interior hinges and
$M_A = M_D = 0$ at simple end supports
2 $M_B \neq 0$ and $M_C \neq 0$ at interior simple supports

Figure 2.15

loading. Figure 2.14b shows a *continuous* beam over several supports and an associated bending-moment diagram for uniform loading; this beam is statically indeterminate. Figure 2.14c shows a form of cantilever construction consisting of three beam segments; the two outer segments are simple beams with overhanging cantilevers, and the center segment is hinge connected to the cantilevered ends of the outer segments. This form of cantilever construction is statically determinate.

In Fig. 2.15, free-body diagrams of three beam segments of a cantilevered construction are shown. The center beam is treated as a simply supported beam since the hinges at both of its ends provide zero-moment restraint. Therefore, the end reactions of the center beam are initially found by statics equations. Then, using the FBD of the end segment beams with known center-beam reactions applied, equations of statics are written to find the remaining support reactions.

Example 2.11 presents the solution for the 360-ft beam span under a uniform loading of 2 klf. The bending-moment diagram for the beam is shown in Fig. 2.16; comparable bending-moment diagrams are also shown using the same span and loading for a three-simple-beam segment design and for a continuous beam design. Figure 2.16a shows the maximum bending moment equal to 10,000 kip·ft for the three-simple-beam configuration. In Fig. 2.16b, the maximum bending moment is found to be 5600 kip·ft for the continuous beam configuration. Figure 2.16c shows a maximum bending moment of 7500 kip·ft for the cantilevered configuration which is 25 percent less than the maximum moment of 10,000 kip·ft for the simple beam configuration of Fig. 2.16a.

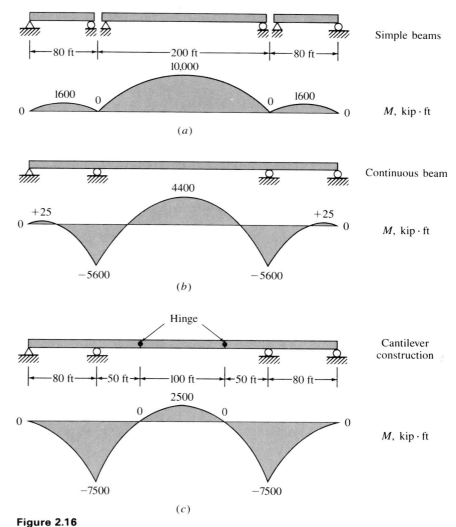

Figure 2.16

CONCLUSIONS

1 The bending-moment magnitudes can vary for the cantilever construction of Fig. 2.16c depending on the placement of the interior hinge points whereas Figs. 2.16a and 2.16b show bending-moment values that a⁓ unvarying for the 2 klf uniform loading.

2 Ideal placement of the interior hinges in Fig. 2.16c can be shown t
in a maximum bending moment of 5000 kip·ft for the 2 k˙
which is about 10 percent less than the maximum moment ⸺
for the continuous beam of Fig. 2.16b. Ideal placement ⸺
rior hinges are approximately placed at 109 ft from b⸺

EXAMPLE 2.11

The highway girder of Figs. 2.14 and 2.15 are to be used to determine all vertical support reactions. Neglect horizontal reactions.

For center beam EF:

$\zeta + \Sigma M_E = 0 = 100V_F - (2 \times 100)50 \qquad V_F = 100 \text{ kips} \uparrow$ ◀

$+ \uparrow \Sigma V = 0 = V_E + V_F - (2 \times 100) \qquad V_E = 100 \text{ kips} \uparrow$ ◀

It is evident from *symmetry* that $V_B = V_C$ and $V_A = V_D$.
For FBD ABE:

$\zeta + \Sigma M_A = 0 = -2 \text{ klf}(130 \text{ ft})\left(\dfrac{130 \text{ ft}}{2}\right) - 130V_E + 80V_B$

$0 = -2(130)\left(\dfrac{130}{2}\right) - 130(100) + 80V_B \qquad V_B = 373.8 \text{ kips} \uparrow = V_C$ ◀

$+ \uparrow \Sigma V = 0 = V_A + V_B - V_E - 2(130)$

$0 = V_A + 373.8 - 100 - 260 \qquad V_A = -13.8 \text{ kips} \downarrow = V_D$ ◀

With all reactions determined use entire FBD $ABEFCD$ below to check results.

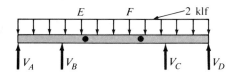

Check

$\zeta + \Sigma M_E = 0 = V_D(230 \text{ ft}) + V_C(150 \text{ ft}) - 2 \text{ klf}(360 \text{ ft})(50 \text{ ft})$

$\qquad - V_B(50 \text{ ft}) - V_A(130 \text{ ft})$

$0 = (-13.8 \text{ kips})(230 \text{ ft}) + 373.8 \text{ kips}(150 \text{ ft}) - 2(360)(50)$

$\qquad - 373.8(50) - (-13.8)(130)$

$0 = 57864 - 57864 \quad \checkmark$

Example Problem 4 (Use of Pin Codes)

The IMAGES computer program can be used to check the results of Example 2.11. A schematic of the IMAGES beam-element model is presented in Fig. 2.17. The program can define interior hinge connections at nodes 3 and 4 by means of member end releases (pin codes). The details of these pin-code computer inputs are described as follows:

Enter the Define Elements selection of the Create/Edit Geometry and select 1. "Define Beams." Use of the help prompt provides the input format which includes Pin I and Pin J. Pin I and Pin J are codes that define how beam elements are joined together at common nodes. IMAGES assumes that axial, shear, and moment continuity exist at member joints unless specified otherwise. The pin codes define member end release of restraint in terms of local coordinates (x, y, and θ_z) and are represented by a three-digit code as follows:

100 = axial (x) release
010 = shear (y) release
001 = moment (θ_z) release

A zero digit represents restraint of one or more degrees of freedom (DOF). A "one" represents the released DOF. A default pin-code value is equal to 000 and signifies that the joining members are fully connected to each other. Combined codes are valid:

101 can represent a roller joint

Note Entered pin codes are displayed on the PC monitor as a 0 or 1 unless two or three releases are entered at I or J.

Returning to the example problem, beam-element 3, which is defined from node $I = 3$ to node $J = 4$, is hinge supported at both nodes. Therefore, the pin-code entries for beam element number 3 are Pin $I = 001$ and Pin $J = 001$. Pin-code entries are entered for only one of the members that join at nodes 3 and 4. Pin codes could be entered for beam member 2 at node 3, and for beam 4 at node 4 *in place of* the entries to beam 3 cited above. A *general rule* is: At a node with "n" beams connecting to it, apply only pin-code releases to ($n - 1$) beams for a particular DOF release.

Figure 2.17 IMAGES beam element model of Example 2.11

2.10 CABLE CONSTRUCTION

Cables generally used in construction are made of high-strength steel wire. Since cables are very flexible members, it is assumed they can support direct tension but cannot resist compression or bending. When a supported cable

60 Reactions

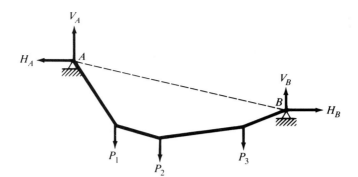

Figure 2.18 Cable under vertical concentrated loads.

is subjected to transverse loads, it undergoes noticeable sag or deformation unlike other structural elements which assume negligible elastic deformations. The *sag* curve of the cable is a function of the loading. Figure 2.18 shows the typical polygonal shape a cable will assume under the action of several concentrated loads and negligible cable weight. An applied uniform distributed load on a horizontal cable will assume a parabolic shape. If the sag deformations are known, the cable support reaction can be found and the tension determined along the cable. Example 2.12 provides a numerical solution of a cable of negligible weight under the action of two concentrated loads.

EXAMPLE 2.12

A taut cable placed between hinge supports at A and D sags under the application of two vertical concentrated loads at B and C as shown. The measured sag at points B and C are both equal to 0.60 ft. Determine the cable reactions and cable tension at the support points.

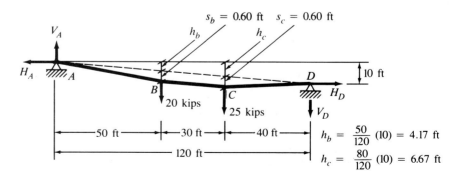

$$\curvearrowleft + \Sigma M_A = 0 = 10H_D - 120V_D - 25 \text{ kips}(80 \text{ ft}) - 20 \text{ kips}(50 \text{ ft})$$

$$0 = 10H_D - 120V_D - 3000 \quad (1)$$

Since cable assumes no bending resistance, an FBD of section CD yields,

$$\circlearrowleft + \Sigma M_C = 0 = -40V_D - H_D(h_C + S_C - 10 \text{ ft})$$

$$0 = -40V_D - H_D(6.67 + 0.60 - 10)$$

or

$$0 = -40V_D + 2.73H_D \qquad (2)$$

Solving Eqs. (1) and (2) simultaneously,

$H_D = 1657$ kips $\rightarrow$ and $V_D = 113$ kips$\downarrow$ ◀

Tension at support $D = \sqrt{H_D^2 + V_D^2} = 1661$ kips

$\rightarrow \Sigma H = 0 = H_D - H_A$ $\quad \therefore \quad H_A = H_D = 1657$ kips $\leftarrow$ ◀

$\uparrow \Sigma V = 0 = V_A - 20 - 25 - 113$ $\quad \therefore \quad V_A = 158$ kips$\uparrow$ ◀

Tension at support $A = \sqrt{H_A^2 + V_A^2} = 1665$ kips

Note

The horizontal force component in the cable remains constant while the resultant tensile force varies along the cable length.

2.11 PRINCIPLE OF SUPERPOSITION

Linear elastic structural analyses are often simplified by application of the principle of superposition. The principle states that the total effect of a system of forces acting on a structure is equivalent to the sum of the effects

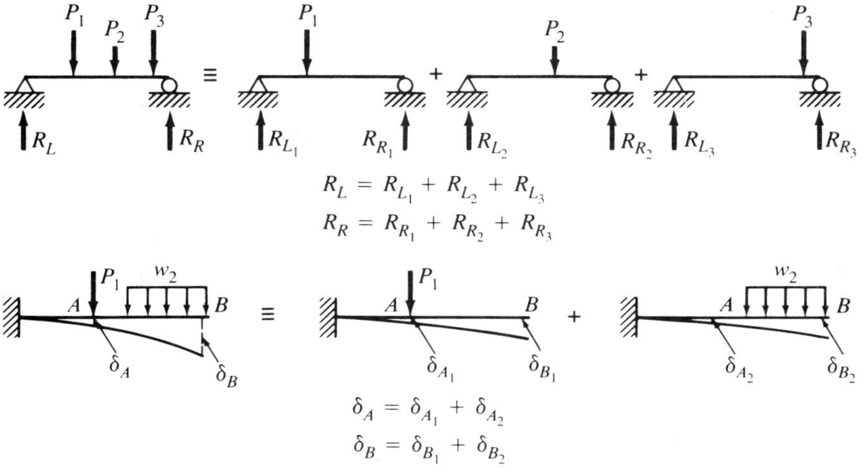

P_1 P_2 P_3 ≡ P_1 + P_2 + P_3
R_L R_R R_{L_1} R_{R_1} R_{L_2} R_{R_2} R_{L_3} R_{R_3}
$R_L = R_{L_1} + R_{L_2} + R_{L_3}$
$R_R = R_{R_1} + R_{R_2} + R_{R_3}$
P_1 w_2 A B ≡ P_1 A B + w_2 A B
δ_A δ_B δ_{A_1} δ_{B_1} δ_{A_2} δ_{B_2}
$\delta_A = \delta_{A_1} + \delta_{A_2}$
$\delta_B = \delta_{B_1} + \delta_{B_2}$

Figure 2.19 Principle of superposition.

caused by each individual force if (1) the geometry of the structure undergoes negligible change during the application of loads, and (2) the system of loads results in linear elastic behavior of the structure. In Example 2.8 of Sec. 2.7, support reactions are found due to three loads placed on the truss shown. However, individual reaction components at the supports could be determined for *each* of the three loads; then the sum of the three individual reactions are found equal to the total reaction at each support as determined in Example 2.8. Similarly, total stresses or displacements can be determined in the same manner as long as the structure material exhibits linear elastic behavior under load, and the geometry does not change appreciably (see Fig. 2.19). The student will find the principle of superposition to be very valuable in many structural analysis computations.

Problems

Determine reaction components for the structures shown below.

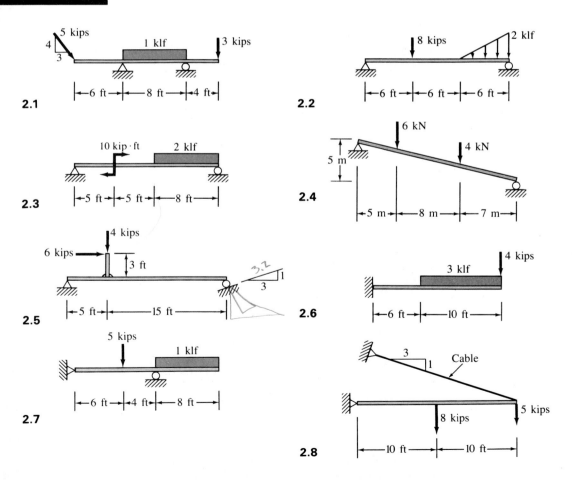

63 Problems

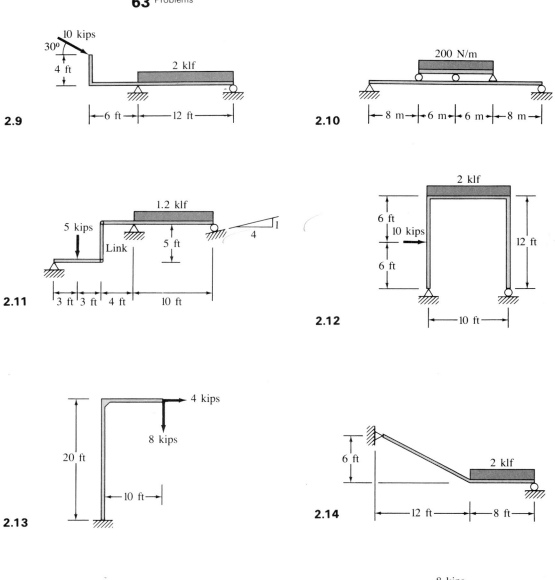

2.9

2.10

2.11

2.12

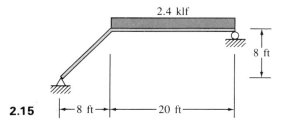

2.13

2.14

2.15

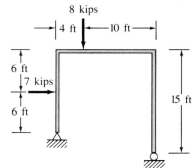

2.16

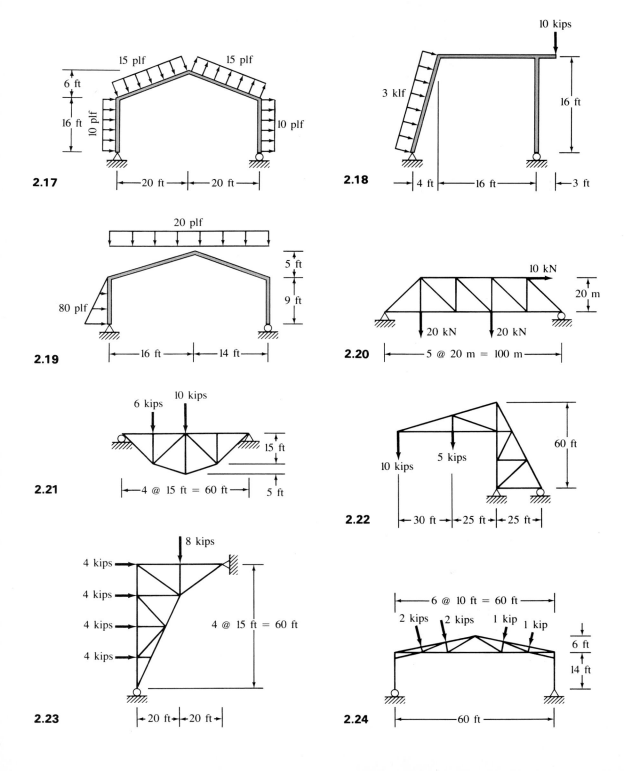

65 Problems

2.25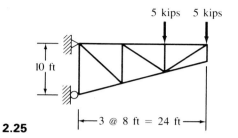

2.26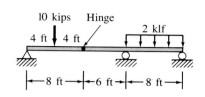

2.27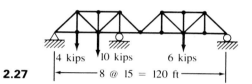

2.28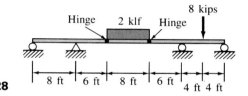

2.29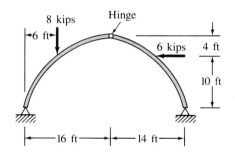

2.30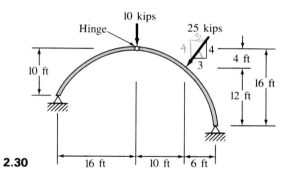

2.31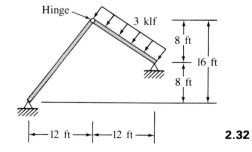

2.32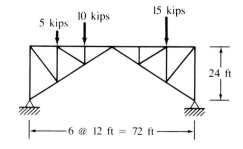

66 Reactions

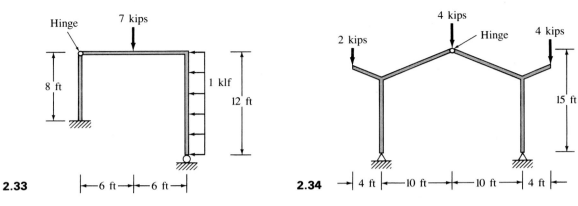

2.33

2.34

2.35 Compute the reactions, sag in cable at the 12-kip load, and the maximum cable force.

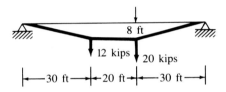

2.36 Compute the reactions, sag in cable at the 40 kN load, and the maximum cable force.

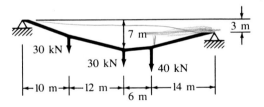

2.37 Determine which structures below are statically determinate, indeterminate (state degree of indeterminacy), and unstable relative to external loads.

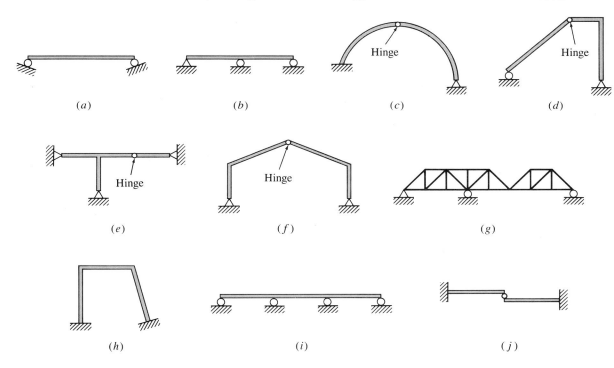

CHAPTER THREE

Plane Trusses

Figure 3.1 Crumlin Railway Viaduct, Wales. Constructed entirely from pin-connected iron members in the 1840s. (*Courtesy of International Structural Slides, Berkeley.*)

3.1. INTRODUCTION

Trusses are usually made by joining slender members into triangular patterns to form a light, stiff, and geometrically stable structure (see Figs. 3.1 and 3.2). Truss members are primarily designed to resist only axial force. Ideally, the members are assumed to be straight and joined together (at member ends) by frictionless pins. It is intuitive from Fig. 3.3 that the triangle is the only pin-connected structural form that can maintain a stable geometric shape when subjected to elastic loads. Consequently, trusses provide a stable arrangement that can support heavy loads without significant displacement relative to its span length. This feature makes the truss an attractive long-span support member that is suitable for many structures, such as bridges, towers, domes, flat roofs, derricks, and the like. The favorable characteristics of light weight and significant stiffness make the truss form a popular choice for open-web floor joists and for braced wall components in tall building construction (see Figs. 3.4 to 3.6).

In truss analysis, all external loads and reactions are assumed to occur at frictionless pin connections. Therefore, all truss members are two-force members; that is, able to transmit only axial tension or compression force. These member forces are called *primary forces* and their resultant stresses are called *primary stresses*. In reality, truss members are usually fastened together at gusset plates by bolts, rivets, or weld connections and can develop moment resistance at their joints. However, structural analysis is simplified by the assumption of pin-connected truss joints.

The dead weight of nonvertical truss members can cause *secondary* shear and bending forces to develop at the truss joints. Additional secondary forces can develop when the centroidal axes of all connecting members at a

Figure 3.2 Lift Bridge, Sacramento River Delta, near Rio Vista, California. (*Courtesy of International Structural Slides. Berkeley*.)

Figure 3.3

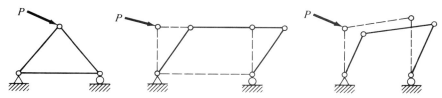

Figure 3.4 Open-web floor joists in a three-story office building in Franklin, Massachusetts. (*Courtesy of Daigle Engineers, Inc., Methuen, Mass*.)

Figure 3.5 Alcoa building, San Francisco. Outer building frame displays an exposed triangular truss arrangement. (*Courtesy of International Structural Slides, Berkeley.*)

Figure 3.6 Franklin Mills retail center, Philadelphia. Integrated use of gable and arch frames with open web steel joist roof members. (*Courtesy of Western Development Corporation of Washington, D.C., developer of Franklin Mills and Morse/Diesel, Inc.*)

truss joint do not coincide at the joint; obviously, convergence of member forces at a joint results in zero moment about the joint, whereas force eccentricity at the joint results in bending. Good detailing and design practices can minimize the development of secondary stresses. Since secondary forces are usually small in comparison to the primary forces, truss analysis methods are developed for primary forces alone based on the ideal assumptions defined in Sec. 3.3.

3.2. TYPES OF TRUSSES

Trusses can be classified by their geometric construction or by their use group. Various truss classifications are presented and described in the paragraphs below.

3.2.1. Simple Truss

A simple truss is made by joining three straight members into a triangular shape with frictionless pins. A truss continues to be defined as simple if two members and one new joint are added to form another triangle. Figure 3.7 demonstrates the formation of one simple truss from another. Figure 3.8 shows numerous simple truss forms where some display the name of the engineer or architect credited with their development. This chapter will present analyses of plane simple trusses.

71 Types of Trusses

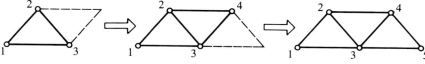

Figure 3.7

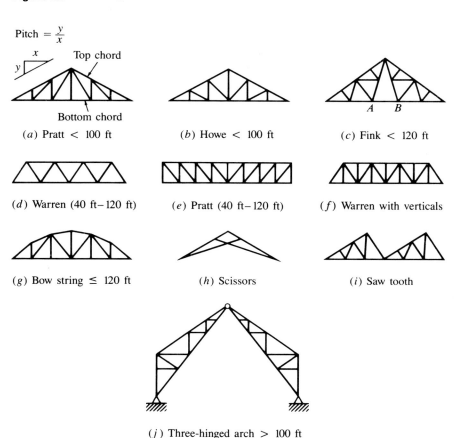

Figure 3.8

3.2.2. Compound Truss

A compound truss consists of connecting two or more simple trusses together while maintaining a rigid and stable form. In Fig. 3.8c, the Fink truss is developed from two other simple trusses that have one common joint and one link member *AB*. Figure 3.9a provides another example where link member *BD* joins two existing simple trusses (shown shaded) that have a common joint at *C*. The compound truss of Fig. 3.9b is rigidly formed by joining simple trusses with three added members which are nonparallel and nonconcurrent to avoid instability. Figure 3.9c demonstrates a condition where two simple trusses are connected together by a third simple truss. Figure 3.9d shows that two or more joints and members can connect simple trusses to form a compound truss.

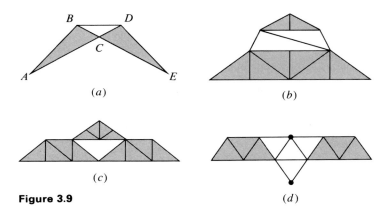

Figure 3.9

3.2.3. Complex Truss

When a truss cannot be classified as a simple or compound truss, it is referred to as a complex truss. Examples of complex trusses are shown in Fig. 3.10. Calculation of member forces in a complex truss is a difficult task if one resorts to hand calculation methods without the assistance of a digital computer.

3.2.4. Roof Trusses

Roof trusses must be able to support wind, rain, and snow loads, as well as its own dead weight plus attachments, such as lighting fixtures, duct work, and interior ceilings. Some forms of roof trusses have been presented in Fig. 3.8. The continuity of the upper members shown in Fig. 3.8a is referred to as the *top chord*, whereas the continuity of the lower members is called the *bottom chord*; inclined members are called *diagonals* and vertical members are simply called *verticals*. If the top chord is inclined, the y/x slope ratio is called the *pitch* of the roof. The Pratt or Howe trusses of Figs. 3.8a and b, are frequently used for roof trusses; the Fink truss of Fig. 3.8c is often selected if a greater roof pitch is desired. Flat roof (or slightly pitched) trusses, such as the Warren and Pratt trusses of Figs. 3.8d to f, are often used, with favor given to the slightly pitched variety when rain runoff is a concern. When top and bottom truss chords are horizontal, the truss is referred to as a *parallel chord truss*. The bowstring, scissors, and sawtooth trusses of Figs. 3.8g to i have names reflecting their shape.

The bowstring truss has been used for light industrial and commercial buildings and for small airplane hangers whereas the sawtooth truss is commonly used where uniform natural light is desired. The scissors truss is generally used for short spans requiring a steep roof. The three-hinged arch roof truss (see Fig. 3.8j) provides a design alternative that has been used for large spans in excess of 100 ft.

3.2.5. Bridge Trusses

In recent decades, a diminishing interest in truss bridge design has been evident. Modern bridge designs include cable stayed decks and welded plate girder deck systems for large spans, with growing attention toward prestressed and reinforced concrete for short spans. Yet, it is important for engineers to study truss-bridge structures since (1) many future truss bridges will be designed and constructed, and (2) the engineering profession is endowed with the responsibility of maintenance and upgrade of thousands of truss bridges that remain in public service.

It is generally stated that economical bridge-truss design can be achieved if diagonals are placed at 45° angles, panel spacing maintained ⩽ 30–40 ft, and the ratio of the depth (distance between top and bottom chord) to span length is within the range of 1/5 to 1/10. The Pratt, Howe, and Warren trusses of Figs. 3.11a to c are parallel chord trusses suitable for moderate length spans. The Parker truss of Fig. 3.11d provides an increasing depth from end to center span that is economical for longer length spans. However, increasing the depth between top and bottom chords often results in long diagonal lengths that may tend to buckle under compressive loads. A study of the Baltimore, Pettit (or Pennsylvania), and K trusses of Figs. 3.11e to g are called *subdivided trusses* since the diagonal member lengths are divided by introducing subdiagonal and subvertical members. The division of long members into shorter length members, by the addition of subverticals or subdiagonals, increases their capacity to support compressive force. As bridge spans increase, the arch, cantilever, and suspension trusses

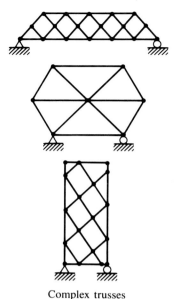

Complex trusses

Figure 3.10

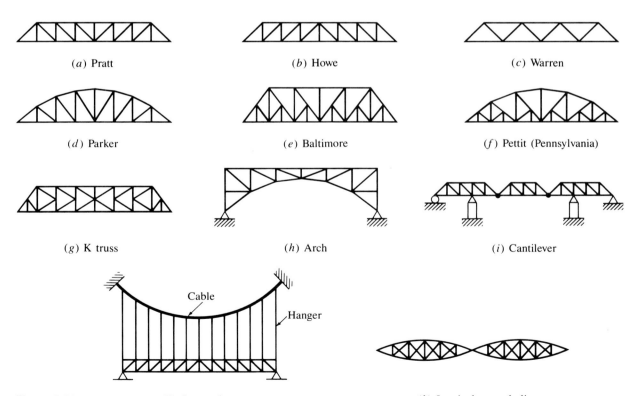

Figure 3.11

Figure 3.12 Lenticular truss bridge spanning the Merrimack River, Lowell, Massachusetts. (Photo taken by the author.)

of Figs. 3.11h to j may provide a more favorable design selection. For pure architectural flavor, Fig. 3.11k shows a lenticular-parabolic truss that is very striking and functional, but not a likely candidate for future design due to high fabrication costs. Figure 3.12 shows a lenticular truss bridge spanning the Merrimack River in Lowell, Massachusetts.

A special way of classifying truss-bridge structures is with reference to the position of the floor system that supports the traffic loads. When the trusses are totally visible as one proceeds across a bridge, and the top chords are tied overhead by cross-bracing members, the bridge is called a *through bridge*. When the deck is supported by the top chords of the truss bridge and the trusses are not visible as a traveler passes over the bridge, the bridge is called a *deck bridge*. Another type of bridge, referred to as a *half-through* or *pony bridge*, is usually identified by visible truss top chords which are not cross-braced (usually due to inadequate overhead clearance), and by a deck that is located between the top and bottom chords. Bracing is provided beneath the roadway deck for half-through truss bridges.

3.3. IDEAL ASSUMPTIONS AND DESIGN CONCEPTS

Assumptions Used in Truss Analysis

1 All members are connected at both ends by smooth frictionless pins.

2 All members are straight.

3 All loads are applied at joints.

4 All support reactions occur at joints.

5 All load conditions satisfy Hooke's law.

6 Centroids of all joint members coincide at the joint.

7 Member weight is negligible when compared with loads.

The first assumption implies that truss members are only able to resist load in axial tension or compression (two-force members). The second assumption eliminates the occurrence of secondary bending due to axial compression force in members. Assumptions 3, 4, and 6 consider that member bending at joints is negligible under loading. Assumption 5 proposes that the truss geometry will undergo minor change due to elastic load deformations; therefore, an analysis which is based on the original undeformed truss geometry is reasonably correct.

Design concepts The roof-truss design concept of Fig. 3.13 shows that the roof loads are transmitted directly to the truss joints through a system of members called *purlins*. The knee braces shown provide added rigidity to the column connections.

The bridge-truss design concept of Fig. 3.14 shows the main elements of a classical truss-bridge structure. Floor beams are shown attached between trusses at lower chord joints. Intermittent members called *stringers* are shown joined to the floor beams. The stringers are selected to support design loads and the roadway deck. Added rigidity is provided by use of portal and sway braces and by lateral bracing between panels of truss top chords. Floor bracing is also shown attached between truss bottom chords. Conceptually, the design load path is as follows: (1) Traffic loads are placed on the roadway deck, (2) transferred to the stringer supports, and (3) transmitted to their respective floor beams. (4) Then, the stringer end reactions on the floor beams are transmitted to the beam-truss joints and (5) transferred to the truss support joints.

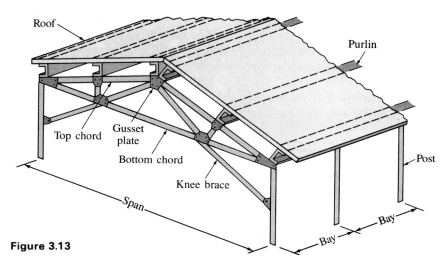

Figure 3.13

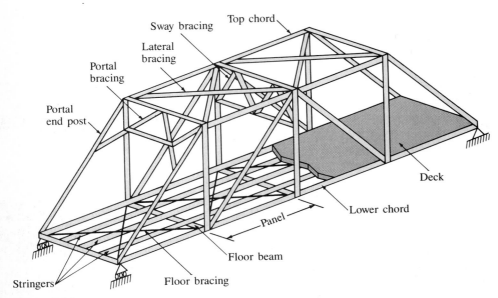

Figure 3.14

3.4. SIGN CONVENTION

Assume *all* truss members act in tension (T). Therefore, when a member force is calculated to be of positive sign, it means that the original tension direction is correct; if negative, the member acts in compression (C).

3.5. TRUSS NOTATIONS

Two commonly used methods of notation are mentioned. First, in modeling trusses for computer solutions, it is customary to define truss joints (nodes) by integer values starting with 1. For classical hand calculation methods, upper chord joints are noted by U_n and lower chord joints by L_n, where n represents a joint number starting with 0. Figure 3.15a demonstrates the later notation where it is observed that upper chord members are defined as $U_n U_{n+1}$ and lower chord members as $L_n L_{n+1}$; vertical members are noted as $U_n L_n$ (e.g., $U_3 L_3$), and diagonal members as either $U_n L_{n+1}$ or $L_n U_{n+1}$ (e.g., $U_1 L_2$ or $L_2 U_3$). Figure 3.15b contains no vertical members and shows that joint numbers at both upper and lower chords progress by $n + 2$. It is also common to specify truss joints by letter notation (see Fig. 3.17).

Figure 3.15

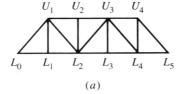

(a)

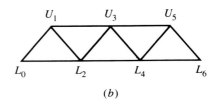

(b)

3.6. METHOD OF JOINTS

The method of joints is based on the principles of statics. When a truss is in static equilibrium, each member and each joint is in equilibrium. Figure 3.16a shows a free-body diagram (FBD) of joint no. 1 with both members acting in the assumed tension direction; the members appear to be pulling away from the joint. The circled FBD of the joint pin shows that both member forces are concurrent at the joint. Therefore, only equations $\Sigma H = 0$ and $\Sigma V = 0$ remain to be satisfied to ensure equilibrium at the joint. Simple truss member forces are readily solved by proceeding to draw the FBDs of each joint and employing the two equations above. *The analysis begins and continues by selecting a joint where* only two **unknown** *member forces exist or remain to be found.* Figures 3.16b and 3.16c serve to demonstrate that the FBD of a joint must include known reactions (R_6) and/or applied load forces (P) at the joint.

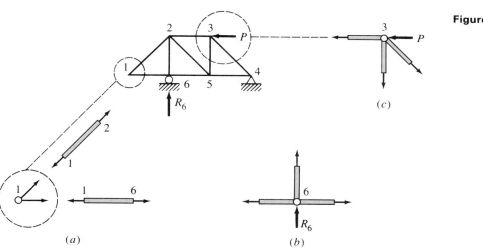

Figure 3.16

EXAMPLE 3.1

Determine the force in each truss member by the method of joints.

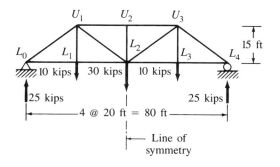

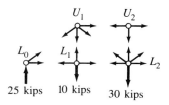

FBD of joints $L_0 \rightarrow L_2$, U_1 and U_2

78 Plane Trusses

After support reactions are found from equations of statics, five joint FBDs left of the line of symmetry are drawn. Joint L_0 alone shows two unknown member forces.

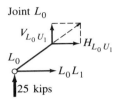

In general, the resultant force (R), horizontal component (H), and vertical component (V) are related by

$$R^2 = H^2 + V^2 \quad \text{or} \quad R = \sqrt{H^2 + V^2}$$

Also, these forces are proportional to their true and projected lengths, that is,

$$\frac{R}{l} = \frac{H}{l_h} = \frac{V}{l_v}$$

where l = the true member length
l_h = the horizontal projected length
l_v = the vertical projected length

From the FBD,

$$\Sigma V = 0 = 25 \text{ kips} + V_{L_0 U_1} \quad \text{or} \quad V_{L_0 U_1} = -25 \text{ kips (compression)}$$

since $\dfrac{H}{l_h} = \dfrac{V}{l_v} \quad H_{L_0 U_1} = \left(\dfrac{l_h}{l_v}\right) V_{L_0 U_1}$

$$= \left(\frac{20 \text{ ft}}{15 \text{ ft}}\right)(-25 \text{ kips}) = -33.3 \text{ kips}$$

$$\Sigma H = 0 = L_0 L_1 + H_{L_0 U_1} \quad L_0 L_1 = -H_{L_0 U_1}$$

$$= -(-33.3 \text{ kips}) = +33.3 \text{ kips (tension)}$$

or the resultant force of $L_0 U_1$ can easily be determined by

$$R = \left(\frac{l}{l_h}\right) H = \left(\frac{25 \text{ ft}}{20 \text{ ft}}\right)(-33.3 \text{ kips}) \cong -41.7 \text{ kips}$$

or $$R = \left(\frac{l}{l_v}\right) V = \left(\frac{25 \text{ ft}}{15 \text{ ft}}\right)(-25 \text{ kips}) \cong -41.7 \text{ kips (C)}$$

79 Method of Joints

Proceeding to the FBD of joint L_1:

For joint U_1, observe that *known* member forces are shown in their determined manner as tension or compression. Also note that all diagonal forces are replaced by components for ease of solution. Unknown forces are shown in an assumed tensile direction.

$\Sigma V = 0 = 25 \text{ kips} - 10 \text{ kips} - V_{U_1L_2}$ $\therefore V_{U_1L_2} = +15 \text{ kips} (T)$

$H_{U_1L_2} = \left(\dfrac{20}{15}\right)(+15 \text{ kips}) = +20 \text{ kips} (T)$

$R_{U_1L_2} = U_1L_2 = \left(\dfrac{25}{15}\right)(+15 \text{ kips}) = +25 \text{ kips} (T)$

$\Sigma H = 0 = U_1U_2 + H_{U_1L_2} + 33.3$

$U_1U_2 = -33.3 - H_{U_1U_2} = -33.3 - 20 = -53.3 \text{ kips} (C)$

$\Sigma V = 0 = U_2L_2$ $\therefore U_2L_2 = 0$

$\Sigma H = 0 = U_2U_3 + 53.3 \text{ kips}$ $\therefore U_2U_3 = -53.3 \text{ kips} (C)$

Also, from symmetry it is apparent that $U_1U_2 = U_2U_3$.

$\Sigma V = 0 = V_{L_2U_3} + 15 \text{ kips} - 30 \text{ kips}$ $\therefore V_{L_2U_3} = +15 \text{ kips} (T)$

$\Sigma H = 0 = L_2L_3 + H_{L_2U_3} - 20 \text{ kips} - 33.3 \text{ kips}$

$L_2L_3 = 53.3 - H_{L_2U_3} = 53.3 - \left(\dfrac{20}{15}\right)(+15 \text{ kips})$

$L_2L_3 = 53.3 - 20 = +33.3 \text{ kips} (T)$

These values are also verified by symmetry where $L_1L_2 = L_2L_3$ and $U_1L_2 = L_2U_3$. The student should easily identify other truss-member forces by use of symmetry.

An examination of the parallel-chord truss reveals symmetry of geometry, loads, and reactions. Therefore, it is only necessary to evaluate the member forces on one side of the axis of symmetry. A study of the free-body diagrams of the five joints at or left of the line of symmetry shows that joint L_0 alone has two unknown member forces.

Since the method of joints uses the $\Sigma H = 0$ and $\Sigma V = 0$ equilibrium equations, it is convenient to replace each diagonal member force by its horizontal and vertical components as follows: The resultant force (R), horizontal component (H), and vertical component (V) are related by the equation

$$R^2 = H^2 + V^2 \quad \text{or} \quad R = (H^2 + V^2)^{1/2}$$

Also, a resultant and its force components are related by the proportions of the total member length and the projected member lengths as follows:

$$R/l = H/l_h = V/l_v \quad \text{since} \quad l^2 = l_h^2 + l_v^2$$

where l is the total member length, and l_h and l_v are the respective horizontal and vertical projected lengths.

Observe that *after* both member forces are determined at joint L_0, one proceeds to the FBD of joint L_1 where only two unknown forces are to be found. Also note that the computed tensile force of member $L_0 L_1$ is properly applied at pin L_1. The FBD of joint U_1 shows the force of member $L_0 U_1$ properly applied in terms of its known compressive components. Once a force component of a diagonal member is found, the resultant and the remaining component forces are easily found from the relationships given above.

EXAMPLE 3.2

By the method of joints, determine all member forces of the roof truss below.

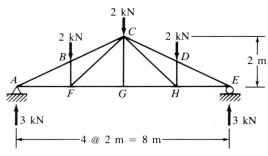

From symmetry,

$AB = DE$

$AF = HE$

$BC = CD$

$BF = DH$

$FC = CH$

$FG = GH$

Method of Joints

Joint A: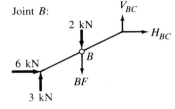

$\Sigma V = 0 = V_{AB} + 3 \quad \therefore \quad V_{AB} = -3 \text{ kN } (C)$

$H_{AB} = 2V_{AB} = -6 \text{ kN } (C)$

$R_{AB} = \sqrt{5}(V_{AB}) = -6.71 \text{ kN } (C)$

$\Sigma H = 0 = AF + H_{AB} \quad AF = -H_{AB} = -(-6) = +6 \text{ kN } (T)$

Joint B: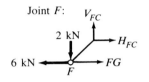

$\Sigma H = 0 = 6 + H_{BC} \quad \therefore \quad H_{BC} = -6 \text{ kN } (C)$

$V_{BC} = \frac{1}{2}(H_{BC}) = -3 \text{ kN } (C)$

$\Sigma V = 0 = 3 \text{ kN} - 2 \text{ kN} - BF + V_{BC}$

$BF = V_{BC} + 1 \text{ kN} = -3 \text{ kN} + 1 \text{ kN} = -2 \text{ kN } (C)$

$\Sigma V = 0 = V_{FC} - 2 \text{ kN} \quad \therefore \quad V_{FC} = +2 \text{ kN } (T)$

Joint F: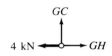

$H_{FC} = V_{FC} = +2 \text{ kN } (T)$

$\Sigma H = 0 = FG + H_{FC} - 6 \text{ kN}$

$FG = 6 - H_{FC} = 6 - 2 = +4 \text{ kN } (T)$

$\Sigma V = 0 = GC \quad\quad GC = 0$

Joint G:

$\Sigma H = 0 = GH - 4 \text{ kN} \quad GH = +4 \text{ kN } (T)$

A common practice is to present the member forces on a schematic diagram of the truss as shown below.

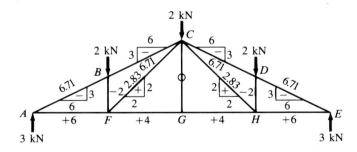

The joints for the roof truss are defined by letter notation. The analysis should be self-explanatory. Although the force of member GC is zero for this load case, member GC provides added rigidity and geometric stability to the truss.

Zero-force members The method of joints becomes easier if one can first visually identify members of zero force. For example, member GC of Example 3.2 is identified as a zero-force member since there is no vertical

82 Plane Trusses

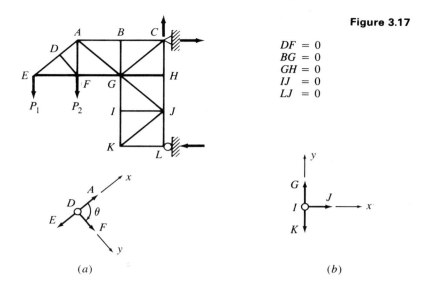

Figure 3.17

load applied at joint G, and GC is the only member at G with a vertical force component. Therefore, $\Sigma V = 0$ is satisfied by member $GC = 0$. Figure 3.17 presents a truss with several zero-force members. Since applied external loads are not present at joints B, D, H, and I, orthogonal axes can be established to identify zero force in members DF, BG, GH, and IJ (see Figs. 3.17a and 3.17b). It is important to note that member DF shown in Fig. 3.17a is a zero-force member for the range of θ values between 0 and 180 degrees. A visual inspection of joint L reveals a horizontal roller reaction, a horizontal member KL, and a vertical member JL; by applying the equation $\Sigma V = 0$ at joint L, one can easily identify that JL is a zero-force member. If the applied load at joint E is removed, both ED and EF must be zero force members to satisfy the $\Sigma H = 0$ and $\Sigma V = 0$ joint equilibrium equations. Therefore, certain truss members may have zero force for some load cases and develop force resistance for other load conditions. Also be aware that zero-force members often serve to subdivide long compression members to increase their load capacity.

When using the method of joints, some analysts prefer to express the joint equilibrium equations in terms of the trigonometric components of the unknown member forces. To illustrate this point, recall the FBD of joint A of Example 3.2 where member AF is horizontal and member AB is a diagonal; the angle between AB and AF is hereby defined as α. The equations of equilibrium at joint A can be expressed as

$$\Sigma H = 0 = AF + AB \cdot \cos \alpha \quad \text{and} \quad \Sigma V = 0 = 3 \text{ kN} + AB \cdot \sin \alpha$$

where $\tan \alpha = 1 \text{ m}/2 \text{ m} = 0.5$, $\alpha \approx 26.6°$, and member forces AB and AF are easily determined.

3.7. METHODS OF SECTION

There are many truss configurations where truss member forces cannot be determined by the method of joints alone. The methods of section which follow, will provide the analyst with additional techniques to evaluate the member forces.

3.7.1. Method of Shears If a truss is in equilibrium, then any free-body diagram (FBD) section of the truss must be in equilibrium. In essence, this is done when a circular section is made around a joint and the FBD of the joint section is isolated to perform the method of joints. In Fig. 3.18, an imaginary line 1–1 is shown to cut through the truss, and an FBD of the left section is isolated and presented in Fig. 3.18a. In addition to the external loads and reaction components, an unknown tensile force is assumed for each member cut by section line 1–1. Applying $\Sigma V = 0$ to the left section,

$$\Sigma V = 0 = V_{dh} + 8 \text{ kips} - 10 \text{ kips} - 10 \text{ kips} \qquad V_{dh} = +12 \text{ kips } (T)$$

Another section line 2–2 is passed through the truss where an FBD of the right section is given as Fig. 3.18b. Similarly, if the $\Sigma V = 0$ equation is employed, V_{dj} is easily found equal to 3 kips (T); V_{dj} could also be found from an FBD section taken to the left of line 2–2. This approach of dividing a truss into two sections followed by summation of vertical-force equilibrium on either FBD section is called the *method of shears*. For the

Figure 3.18

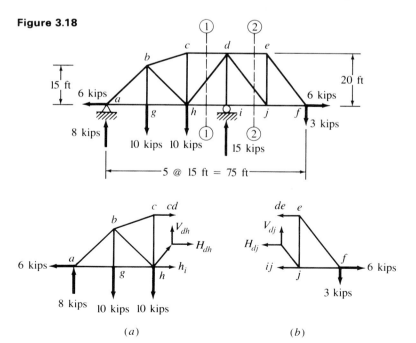

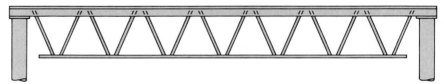

Figure 3.19

analysis of parallel chord trusses, such as the steel-bar joist girder shown in Fig. 3.19, the method of shears provides an easy solution of each diagonal vertical component and its subsequent resultant force.

3.7.2. Method of Moments

The analysis of a bowstring truss is presented in Example 3.3 where the methods of section are mainly used to determine member forces from three interesting sections. The model is defined for later solution by the IMAGES computer program. The first FBD is obtained by passing a section line through members 16, 10, 5, and 2. The left FBD section shows that four unknown member forces are to be determined of which three force vectors pass through joint 2. Therefore, summation of moment equilibrium about joint 2 permits easy solution of the force in member 16 (F_{16}). This approach is called the method of moments. The remaining FBD sections of Example 3.3 assist the analyst to evaluate the other member forces by means of the methods of joints and shears. The second FBD section is established by passing a section line through members 16, 10, and 1 whereas the third section line is a vertical cut through the middle left panel.

EXAMPLE 3.3

By the method of sections, determine the forces in members ②, ⑥, ⑪, ⑩, and ⑯.

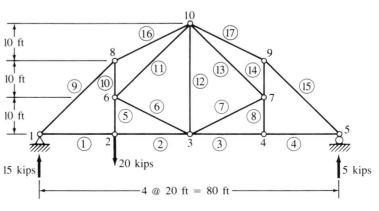

(a)

Consider the FBD section shown in Fig. b by using $\Sigma M_2 = 0$. Members ②, ⑤, and ⑩ have forces which pass through joint 2. The horizontal component of force in member ⑯ and the 15 kip reaction produce moments about joint 2.

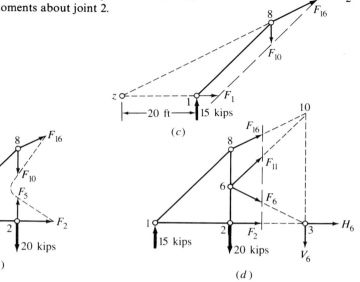

$$\therefore \quad \circlearrowleft + \Sigma M_2 = 0 = 15 \text{ kips}(20 \text{ ft}) + H_{16}(20 \text{ ft})$$

$$H_{16} = -15 \text{ kips } (C)$$

$$V_{16} = \frac{10}{20}(H_{16}) = -7.5 \text{ kips } (C) \quad \text{and} \quad F_{16} = -16.8 \text{ kips } (C) \blacktriangleleft$$

Then, $\Sigma H = 0 = F_2 + H_{16}$ or $F_2 = -H_{16} = +15 \text{ kips } (T)$ ◀

For the FBD shown in Fig. c,

$$\circlearrowleft + \Sigma M_z = 0 = F_{10}(40 \text{ ft}) - 15 \text{ kips}(20 \text{ ft})$$

$$F_{10} = +7.5 \text{ kips } (T) \blacktriangleleft$$

From the method of joints, it is visually obvious that $F_5 = +20$ kips (T).

$$\circlearrowleft + \Sigma M_{10} = 0 = 15 \text{ kips}(40 \text{ ft}) - 20 \text{ kips}(20 \text{ ft}) - F_2(30 \text{ ft}) - (30 \text{ ft})H_6$$

$$0 = 600 - 400 - 15 \text{ kips}(30 \text{ ft}) - 30H_6$$

$$H_6 = -8.33 \text{ kips } (C) \qquad \therefore \quad V_6 = \frac{10}{20}(H_6) = -4.17 \text{ kips } (C) \blacktriangleleft$$

Using the method of shears and the FBD shown in Fig. d.

$$\uparrow + \Sigma Y = 0 = 15 \text{ kips} - 20 \text{ kips} + V_{16} + V_{11} - V_6$$

$$0 = 15 - 20 - 7.5 + V_{11} - (-4.17); \ V_{11} = +8.33 \text{ kips } (T)$$

$$H_{11} = V_{11} = +8.33 \text{ kips } (T) \quad \text{and} \quad F_{11} = \sqrt{2}V_{11} = +11.8 \text{ kips } (T) \blacktriangleleft$$

EXAMPLE 3.4

The method of moments is known to utilize the important transmissibility principle of statics which states that a force can be applied at any point on its line of action without a change in the external effects. Consider the FBD of Fig. 3.18a: Members dh and hi both pass through point h. Since the section must be in equilibrium, $\Sigma M = 0$ about joint h involves only one unknown member, namely, member cd. The equation reads:

$$\circlearrowleft + \Sigma M_h = 0 = cd \cdot 20 \text{ ft} + 8 \text{ kips} \cdot 30 \text{ ft} - 10 \text{ kips}(15 \text{ ft} + 0 \text{ ft})$$

from which $cd = -4.5$ kips (C) ◀

Figure 3.20a shows the same free-body section of Fig. 3.18a where the lines of action of force hd and force cd pass through a point in space coincident with point d. Moment equilibrium about point d yields:

$$\circlearrowleft + \Sigma M_d = 0 = 6 \text{ kips} \cdot 20 \text{ ft} + 8 \text{ kips} \cdot 45 \text{ ft}$$
$$- 10 \text{ kips}(30 \text{ ft} + 15 \text{ ft}) - hi \cdot 20 \text{ ft}$$

from which member force $hi = +1.5$ kips (T).

By method of shears, $V_{hd} = +12$ kips (T) and $H_{hd} = 3/4(V_{hd}) = +9$ kips.

Figure 3.20

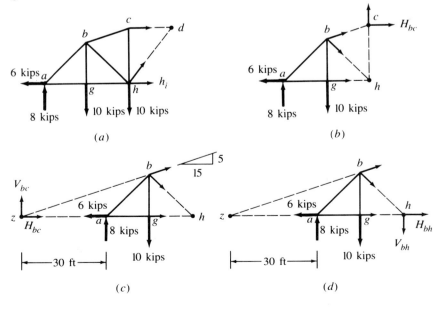

(a) (b)
(c) (d)

A verification of these last two force calculations can be made by establishing if horizontal-force equilibrium of the FBD is satisfied. That is,

$$+ \rightarrow \Sigma H = 0 = cd + H_{hd} + hi - 6 \text{ kips} = 0$$
$$-4.5 \text{ kips} + 9 \text{ kips} + 1.5 \text{ kips} - 6 \text{ kips} = 0 \quad \text{OK}$$

If another section line is passed through members bc, bh, and gh of Fig. 3.18, the left-hand section would appear as seen in Figure 3.20b. Since there are two unknown diagonal forces to be found, the method of shears is not of immediate help. However, there are many ways to use the method of moments to evaluate the three unknown forces of Fig. 3.20b. Some of these follow:

1 Using $\Sigma M_b = 0$, determine member force gh.

2 From Fig. 3.20b, use $\Sigma M_h = 0$ to determine member force bc. To accomplish this act, the force in bc is transmitted to a point in space coincident with point c where it is seen that the H_{bc} component alone produces a moment about point h.

3 Another choice is to transmit the forces in bc and gh to an imaginary point of intersection in space, shown as point z on Fig. 3.20c. Then, application of $\Sigma M_h = 0$ provides the solution of V_{bc} directly.

4 In Fig. 3.20d, it is shown that the force in member bh can be found by transmitting force bh to point h in space. Then, by use of $\Sigma M_z = 0$, V_{bh} is found directly.

Other points of force transmittal are available about which the method of moments can be used to compute member forces.

Applications

From Fig. 3.20b, determine the force in member bc.

$$\circlearrowleft + \Sigma M_h = 0 = H_{bc} \cdot 20 + 8 \text{ kips} \cdot 30 - 10 \text{ kips} \cdot 15$$
$$H_{bc} = -4.5 \text{ kips } (C)$$

Then

$$V_{bc} = 1/3(H_{bc}) = -1.5 \text{ kips} \quad \text{and} \quad R_{bc} = -4.74 \text{ kips}$$

From Fig. 3.20c, determine the force in member bc.

$$\circlearrowleft + \Sigma M_h = 0 = V_{bc} \cdot 60 + 8 \text{ kips} \cdot 30 - 10 \text{ kips} \cdot 15$$
$$V_{bc} = -1.5 \text{ kips } (C)$$

From Fig. 3.20d, determine the force in member bh.

$$\circlearrowleft + \Sigma M_z = 0 = V_{bh} \cdot 60 + 10 \text{ kips} \cdot 45 - 8 \text{ kips} \cdot 30$$
$$V_{bh} = -3.5 \text{ kips } (C)$$

88 Plane Trusses

Then

$$H_{bh} = V_{bh} = -3.5 \text{ kips} \quad \text{and} \quad R_{bh} = \sqrt{2}(V_{bh}) = -4.95 \text{ kips (C)}$$

Examples 3.5 and 3.6 present studies of Fink and K trusses, respectively, where special "section cuts" are presented which permit the ready computation of some member forces.

EXAMPLE 3.5

Determine the numbered bar forces for the Fink truss loaded as shown below. The remaining bar forces are found by use of the principle of symmetry.

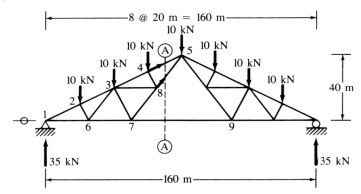

Solution

The member forces at joints 1, 2, and 6 are easily found by the methods of joints and moments. However, three or more unknowns exist at joint nos. 3, 7, 4, and 8. Consider passing a section A–A as shown above through members 4–5, 5–8, and 7–9.

By $\Sigma M_5 = 0$: (use left FBD)

$$35 \text{ kN}(80 \text{ m}) - 10 \text{ kN}(20 + 40 + 60)\text{m} - F_{7-9}(40 \text{ m}) = 0$$

$$\therefore \quad F_{7-9} = +40 \text{ kN}$$

The remaining members at joint 7 are determined by the method of joints. Subsequently, all other unknown members are also found by method of joints. The results are:

$F_{1-2} = -78.3$ kN $\quad F_{2-6} = -8.9$ kN $\quad F_{3-7} = -17.9$ kN $\quad F_{4-8} = -8.9$ kN $\quad F_{7-9} = +40$ kN

$F_{1-6} = +70$ kN $\quad F_{3-4} = -69.3$ kN $\quad F_{3-8} = +10$ kN $\quad F_{6-7} = +60$ kN $\quad F_{8-5} = +30$ kN

$F_{2-3} = -73.8$ kN $\quad F_{3-6} = +10$ kN $\quad F_{4-5} = -64.8$ kN $\quad F_{7-8} = +20$ kN

EXAMPLE 3.6

Determine the force components in members a and b of the K truss shown in Fig. a.

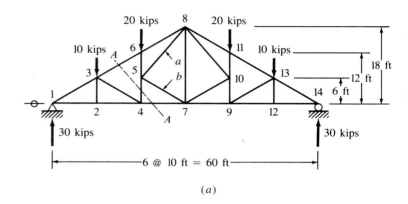

(a)

Solution

Initially compute member forces in 4–5 and 5–6. By the method of joints at node 6:

$\Sigma F_y = 0: F_{5-6} = -20$ kips

Next, determine F_{4-5} by passing a section A–A as shown in Fig. a and use method of sections (moments). Using the FBD of the left truss section:

$\circlearrowleft + \Sigma M_1 = 0: F_{4-5}(20 \text{ ft}) - 10 \text{ kips}(10 \text{ ft}) = 0 \quad \therefore \quad F_{4-5} = +5$ kips

Using the FBD of joint 5 shown in Fig. b.

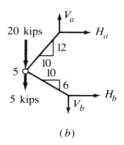

(b)

$+ \rightarrow \Sigma F_x = 0: H_a + H_b = 0 \quad \therefore \quad H_a = -H_b$ (1)

$\uparrow + \Sigma F_y = 0: V_a - V_b - 20 \text{ kips} - 5 \text{ kips} = 0 \quad \text{or} \quad V_a - V_b = 25$ (2)

The force components in members a and b are related as follows:

$$\frac{H_a}{V_a} = \frac{10}{12} \quad \text{or} \quad H_a = 0.833 V_a$$

and $\quad \dfrac{H_b}{V_b} = \dfrac{10}{6} \quad$ or $\quad H_b = 1.66 V_b$

which substituted into Eq. (1) yields

$$0.833 V_a + 1.66 V_b \quad \therefore \quad V_a = -2 V_b \tag{3}$$

Substitution of Eq. (3) into Eq. (2) results in

$$-2V_b - V_b = 25 \quad \text{or} \quad V_b = -8.3 \text{ kips}$$

Then, from Eq. (2), $V_a = 25 + V_b = 25 - 8.3 = +16.7$ kips

Finally, from the proportional relationships shown above

$$H_a = 0.833 V_a = 0.833(+16.7) = +13.9 \text{ kips}$$

and $\quad H_b = 1.66 V_b = 1.66(-8.3) = -13.8$ kips

$$a = [(13.9)^2 + (16.7)^2]^{1/2} = +21.7 \text{ kips} \quad \text{and} \quad b = -[(13.8)^2 + (8.3)^2]^{1/2} = -16.1 \text{ kips}$$

3.8 STABILITY AND DETERMINACY

The initial activity of a truss analysis should begin with a study to establish if the truss is statically determinate and geometrically stable. A plane truss with three reaction components is said to be *statically determinate externally* since there are three available equations and three unknown reactions; if four or more reaction components exist, three equations of statics are not sufficient to determine the reactions, and the truss structure is said to be *statically indeterminate externally*; for four unknown reactions and three equations, the truss is said to be statically indeterminate to the first degree; for five unknown reactions and three equations, the truss is statically indeterminate to the second degree, and so forth.

Special conditions arise in a structural design where additional equilibrium equations can be written to solve what may apper to be a statically indeterminate problem. These equations are referred to as *equations of condition*. In Chap. 2, it is found that a planar three-hinged arch has four unknown support-reaction components which cannot be determined from the three available equations of statics alone. However, the three-hinged arch design joins the two arch halves by a frictionless pin. Since moment resistance cannot be provided at the interior pin, a static equation of condition arises which demands that the summation of moments of force, from either side of the pin, about the interior pin, must equal zero. This equation of condition provides the added equation necessary to evaluate the four reactions and declare the arch as statically determinate.

A structure with less than three reaction components is not properly restrained against motion and is declared *unstable*. Also, if all reactive forces of a structure are parallel to each other, or concur at a point, the structure is unstable since it cannot satisfy all static equilibrium equations and may deform excessively and/or collapse.

After it has been established that a truss structure is geometrically stable externally, it is necessary to evaluate if the truss arrangement is stable and statically determinate internally. In Sec. 3.1, it was shown that a simple truss is constructed by three members assembled into a triangle and connected by pins at their joints. The basic triangular form is a rigid form in the sense that it is able to support elastic loads without significant movement and change of its geometric orientation. In Fig. 3.21b, we observe that a simple truss remains rigid after two added members are connected to one new joint and attached to the existing framework to form a new triangle. The final member arrangement of the truss is obtained by the continued addition of two members and one new joint. However, a truss may be geometrically unstable internally if all the members do not form into triangular patterns. Many complex trusses display a geometric form which does not consist of triangular member patterns. Moreover, complex trusses often contain too many members which make it difficult to establish if the truss is geometrically stable or unstable by visual inspection. A fundamental approach to resolve this matter considers the equilibrium of the truss as follows:

At each truss joint, all forces are concurrent. Therefore, two equations ($\Sigma V = 0$ and $\Sigma H = 0$) are available to satisfy equilibrium and thereby solve for unknown joint forces. If j represents the number of joints in a truss, then $2j$ independent equilibrium equations can be written. If the basic three-member truss of Fig. 3.21a is expanded to include the addition of two members and one joint to form another truss (see Fig. 3.21b), the new truss framework provides two more equations as well as two more unknown member forces. Similarly, the progressive truss expansion shown in Figs. 3.21c and d will add two equilibrium equations and two more unknown

Figure 3.21

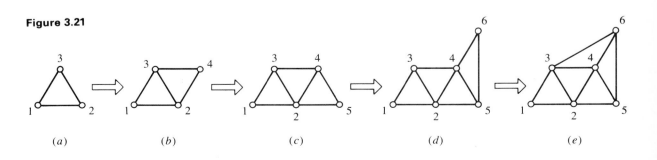

member forces. Now, if one member is added without introducing a new joint as shown in Fig. 3.21e, the truss has one more member than available equations of statics for member solution and the truss will prove to be *statically indeterminate internally* to the first degree.

A necessary but not sufficient relationship between the members, reactions, and joints can be expressed as follows: Let r represent the least number of reaction components required to maintain external stability (but not greater than the actual number provided). The value of r for a truss improperly supported by two rollers is 2; for a simple truss with two hinge supports, the value of r is 3; for a three-hinged arch truss, the value of r is 4 since four reaction components are required to maintain external stability. Let m represent the total number of truss member forces to be determined, and j represent the total number of truss joints. At each joint, two equations of statics can be written. Therefore, if

$$m + r < 2j \quad \text{truss is unstable}$$

$$m + r = 2j \quad \text{truss is statically determinate}$$
$$\text{(perform visual check for stability)}$$

$$m + r > 2j \quad \text{truss is statically indeterminate}$$
$$\text{(perform visual check for stability)}$$

This expression can be used for simple, compound, and complex trusses; several truss arrangements are studied as shown in Fig. 3.22. When $(m + r) \geqslant 2j$, a visual inspection is still required to determine if the truss

Figure 3.22

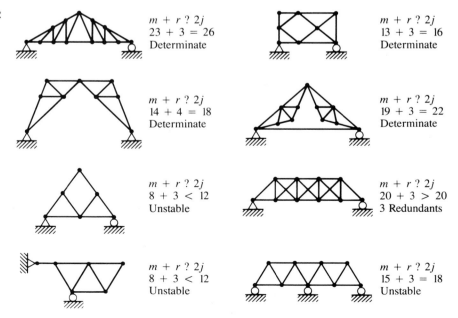

$m + r ? 2j$
$23 + 3 = 26$
Determinate

$m + r ? 2j$
$13 + 3 = 16$
Determinate

$m + r ? 2j$
$14 + 4 = 18$
Determinate

$m + r ? 2j$
$19 + 3 = 22$
Determinate

$m + r ? 2j$
$8 + 3 < 12$
Unstable

$m + r ? 2j$
$20 + 3 > 20$
3 Redundants

$m + r ? 2j$
$8 + 3 < 12$
Unstable

$m + r ? 2j$
$15 + 3 = 18$
Unstable

arrangement is stable. For $(m + r) > 2j$, there are obviously more unknowns than available equations and these extra unknowns are referred to as *redundants*.

The stability of complex trusses is not easily verified from the $(m + r) = 2j$ expression alone since a complex truss often contains non-triangular member patterns which may cause the truss to be unstable. A truss with unstable geometric patterns is said to have *critical form*. Moreover, analysis of unstable trusses always yields force results that are inconsistent or noncompatible. Therefore, a complex truss is stable if a force analysis of the truss provides a unique set of results.

It is possible for an unstable truss to have $(m + r) \geq 2j$. A study of the trusses of Fig. 3.23 reveals the following:

1 Figure 3.23a shows a statically determinate and stable truss (both externally and internally).

2 Figure 3.23b shows the same truss with the diagonal in panel no. 2 removed and placed in panel no. 3 so that $(m + r) = 2j$. It appears that the truss is statically determinate both internally and externally, but the truss is unstable since the horizontal members of panel no. 2 can easily deform. The addition of another diagonal in panel no. 4 will make $(m + r) > 2j$, but the truss will remain unstable since panel no. 2 is unstable.

3 The addition of two roller supports to the rectangular panel no. 2 will prevent the panel from "racking" and will make the truss stable (see Fig. 3.23c). Since $(m + r)$ exceeds $2j$ by two, the truss is said to be statically indeterminate internally to the second degree; that is, the truss has two redundant components.

Figure 3.23

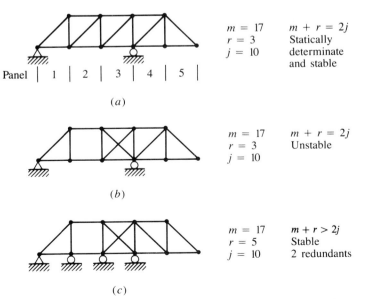

It can be summarized that a plane truss is geometrically stable if

1 It has a member arrangement that is internally stable.

2 It has at least three nonparallel and nonconcurrent reaction components.

3.9. COMPUTER SOLUTIONS

Example Problem 5

Solutions of statically determinate plane trusses can easily be obtained by use of an established computer program such as IMAGES-2D. The bowstring truss of Example 3.3 was modeled for solution by IMAGES-2D as given in Fig. 3.24 with node points defined. By use of the right-hand rule, the positive local x axis of each member is directed from i to j to which the y axis is normal. Truss members are derived in IMAGES-2D from beam elements by use of pin codes to remove the moment restraint at the nodes to simulate frictionless pins.

However, it should be remembered that for n members connected at a truss joint, apply the pin-code releases to $(n-1)$ members at the joint. Figure 3.25 provides computer output of the truss member forces (tabulated as local loads) for load case 1, and concludes with a maximum load summary table; the shear and bending forces are evidently seen to be negligible. Four truss members are drawn in Fig. 3.25, showing their respective local coordinates and containing their axial force values from the local loads table. IMAGES-2D can perform up to five static load cases during one session for the same truss geometry. Other program limits are defined in App. 1.

If an attempt is made to solve an unstable truss by IMAGES, either (1) a singularity message will be flashed on the monitor and the program will halt until modifications are made to make the truss geometrically stable, or (2) an output solution will be obtained with inconsistent results (force equilibrium should be verified at all joints).

Solution of simultaneous linear equations (SOSLE) Using the method of joints, two equations can be written at each joint of a statically determinate stable truss. These joint equations are expressed in terms of the unknown truss-member forces, unknown reaction components, and known applied loads. Since $(m + r) = 2j$, there will exist n equations in n unknowns that can be solved simultaneously. Example 3.7 illustrates the formulation and assembly of the joint equations of a simple truss for the solution of simultaneous linear equations (SOSLE). Numerous mathematical procedures are available to solve $n \times n$ linear equations; also, many computer programs exist to execute an SOSLE routine both for mainframe as well as for microcomputers. Chapter 14

Figure 3.24

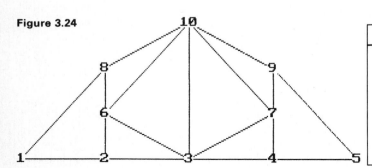

		Elements			
Member	i	j	Member	i	j
1	1	2	9	1	8
2	2	3	10	6	8
3	3	4	11	6	10
4	4	5	12	3	10
5	6	2	13	10	7
6	6	3	14	7	9
7	3	7	15	9	5
8	7	4	16	8	10
			17	10	9

SOLVE BEAM LOADS AND STRESSES   Version 3.0   08/15/85

LOAD CASE 1

LOCAL LOADS

Element	Node	Axial	Shear	Bending
1	i= 1	-1.50000E+01	-8.88281E-19	0.00000E+00
1	j= 2	1.50000E+01	8.88281E-19	0.00000E+00
2	2	-1.50000E+01	3.57904E-19	0.00000E+00
2	3	1.50000E+01	-3.57904E-19	0.00000E+00
3	3	-5.00000E+00	1.71664E-19	0.00000E+00
3	4	5.00000E+00	-1.71664E-19	0.00000E+00
4	4	-5.00000E+00	3.58713E-19	0.00000E+00
4	5	5.00000E+00	-3.58713E-19	0.00000E+00
5	i= 6	-2.00000E+01	-2.34361E-10	0.00000E+00
5	j= 2	2.00000E+01	2.34361E-10	-6.06890E-09
6	6	9.31695E+00	0.00000E+00	0.00000E+00
6	3	-9.31695E+00	0.00000E+00	0.00000E+00
7	3	-1.86339E+00	0.00000E+00	0.00000E+00
7	7	1.86339E+00	0.00000E+00	0.00000E+00
8	7	-1.77636E-15	2.75762E-10	0.00000E+00
8	4	1.77636E-15	-2.75762E-10	2.75762E-09
9	1	2.12132E+01	9.53674E-07	1.44775E-09
9	8	-2.12132E+01	-9.53674E-07	0.00000E+00
10	6	-7.50000E+00	-1.61070E-18	0.00000E+00
10	8	7.50000E+00	1.61070E-18	0.00000E+00
11	6	-1.17851E+01	-4.76837E-07	-1.21839E-09
11	10	1.17851E+01	4.76837E-07	0.00000E+00
12	3	-3.33333E+00	1.95096E-20	0.00000E+00
12	10	3.33333E+00	-1.95096E-20	0.00000E+00
13	10	2.35702E+00	0.00000E+00	0.00000E+00
13	7	-2.35702E+00	0.00000E+00	0.00000E+00
14	7	-2.50000E+00	1.07214E-10	1.07214E-09
14	9	2.50000E+00	-1.07214E-10	0.00000E+00
15	9	7.07107E+00	0.00000E+00	0.00000E+00
15	5	-7.07107E+00	0.00000E+00	1.33674E-09
16	i= 8	1.67705E+01	4.76837E-07	-4.09380E-09
16	j= 10	-1.67705E+01	-4.76837E-07	0.00000E+00
17	10	5.59017E+00	0.00000E+00	-4.12838E-10
17	9	-5.59017E+00	0.00000E+00	-4.12838E-10

provides a matrix algebra approach commonly used to solve an SOSLE. Although numerous truss analysis software packages and computer programs are available, the author repeats his belief that a solid grasp of this subject is best achieved with practice by classical hand solutions, followed by verification checks by computer solutions.

Maximum Load Summary

Beam	End	Comp.	Load
9	I	Axial	2.121E+01
9	I	Shear	9.537E-07
16	I	Bend.	-4.094E-09
9	J	Axial	-2.121E+01
9	J	Shear	-9.537E-07
5	J	Bend.	-6.069E-09

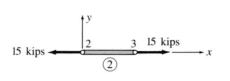

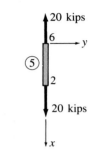

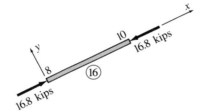

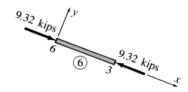

Figure 3.25

EXAMPLE 3.7

Using the method of joints, establish a system of equations to determine all unknown member forces and reactions.

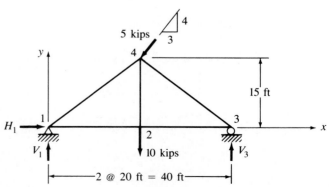

A positive coordinate x–y axes system is assumed as shown.
Unknown reactions are defined as H_1, V_1, and V_3.
Unknown member resultant forces are defined by F_{ij} and assumed to act in tension.

At joint 1:

$$\Sigma H = 0 = F_{12} + F_{14}\left(\frac{4}{5}\right) + H_1 \qquad \Sigma V = 0 = V_1 + F_{14}\left(\frac{3}{5}\right)$$

At joint 2:

$$\Sigma H = 0 = -F_{12} + F_{23} \qquad \Sigma V = 0 = F_{24} - 10 \text{ kips}$$

At joint 3:

$$\Sigma H = 0 = -F_{23} - F_{34}\left(\frac{4}{5}\right) \qquad \Sigma V = 0 = V_3 + F_{34}\left(\frac{3}{5}\right)$$

At joint 4:

$$\Sigma H = 0 = -F_{14}\left(\frac{4}{5}\right) + F_{34}\left(\frac{4}{5}\right) - \left(\frac{4}{5}\right)(5 \text{ kips})$$

$$\Sigma V = 0 = -F_{14}\left(\frac{3}{5}\right) - F_{24} - F_{34}\left(\frac{3}{5}\right) - \frac{3}{5}(5 \text{ kips})$$

Reassembly of these equations, in decimal form with known constants transferred to the right side of the equation yields,

Joint 1 $\Sigma H = 0$: $F_{12} +$ $0.80F_{14} + 0.0F_{24} + 0.0F_{23} + 0.0F_{34} + H_1$ ... ... $= 0$
$\Sigma V = 0$: $0.0F_{12} +$ $0.60F_{14} + 0.0F_{24} + 0.0F_{23} + 0.0F_{34} + \cdots + V_1$ ... $= 0$

Joint 2 $\Sigma H = 0$: $-F_{12}$ ...   ... $+$ F_{23} ...   . $= 0$
$\Sigma V = 0$:   ...   ... F_{24} ...   . $= 10$ kips

Joint 3 $\Sigma H = 0$:   ...   . $-F_{23} - 0.8F_{34}$ ...   ... $= 0$
$\Sigma V = 0$:   ...   . $+ 0.60F_{34}$ ... $+ V_3 = 0$

Joint 4 $\Sigma H = 0$:   ... $-0.80F_{14}$ ...   ... $+ 0.80F_{34}$ ...   ... $= 4$ kips
$\Sigma V = 0$:   ... $-0.60F_{14}$ $-F_{24}$ ... $-0.60F_{34}$ ...   ... $= 3$ kips

A computer program can easily be utilized to solve this 8×8 set of equations.

3.10. CLOSURE

The assumptions and idealizations introduced in this chapter have provided the analyst with methods to determine primary forces in two-dimensional truss members. If the members of a truss are joined together by moment resistant connections, then the structure is truly a rigid frame that is capable of transmitting shear and bending forces as well as axial force. Moreover, the presence of rigid joints may eliminate the potential instability usually associated with nontriangular patterns as discussed in Sec. 3.8; however, the number of unknowns will increase due to the additional bending and shear member forces and will result in many redundants. In that event, the IMAGES-2D program can easily solve the rigid joint-truss structure by elimination of the use of pin-code releases at member ends.

Problems

3.1 to 3.26 Determine forces (or force components) in all truss bars.

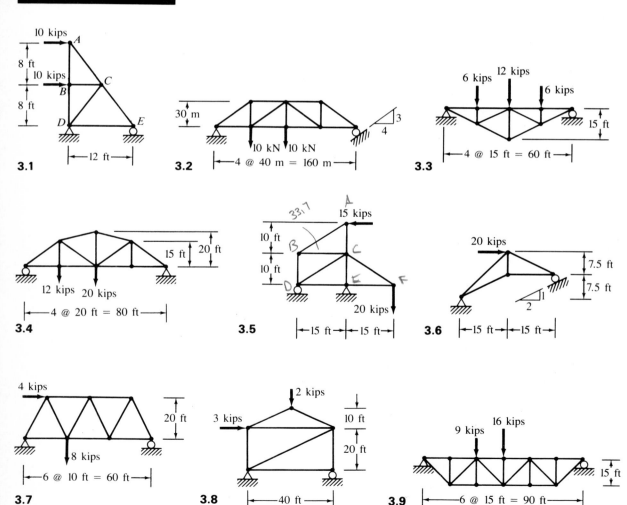

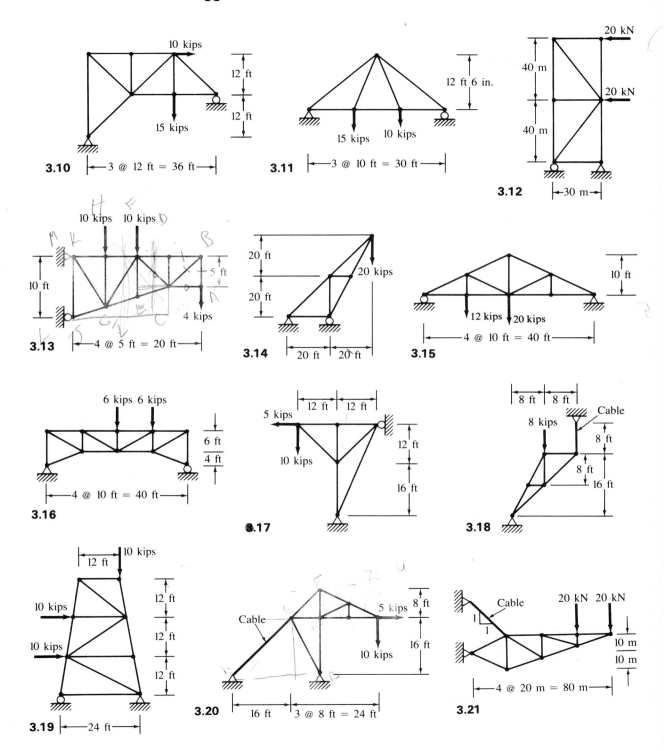

100 Plane Trusses

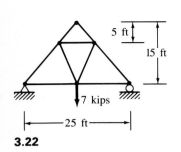

3.22

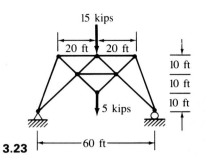

3.23

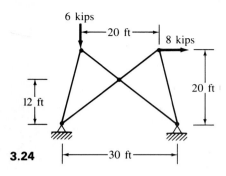

3.24

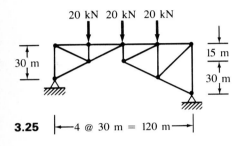

3.25

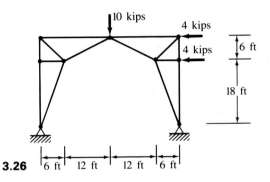

3.26

3.27 to 3.39 Compute the forces in the designated truss bars.

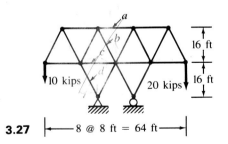

3.27

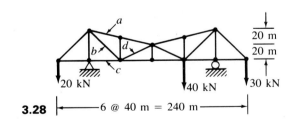

3.28

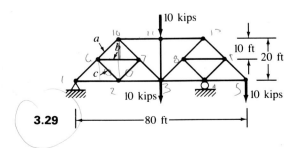

3.29

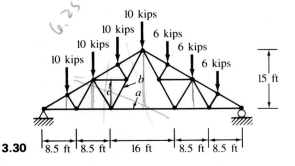

3.30

101 Problems

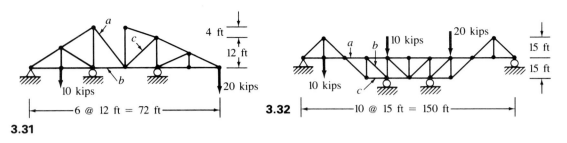

3.31

3.32

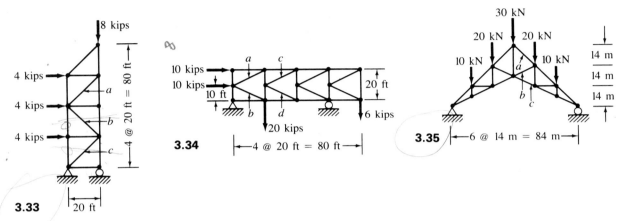

3.33

3.34

3.35

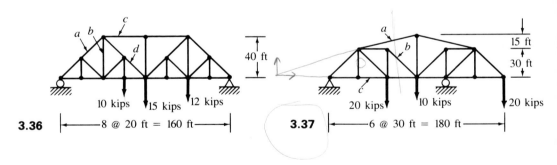

3.36

3.37

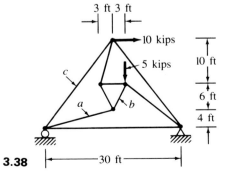

3.38

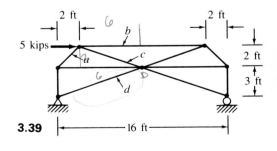

3.39

3.40 Classify the truss structures as to external and internal determinacy and geometric stability (indicate degree of indeterminacy where appropriate).

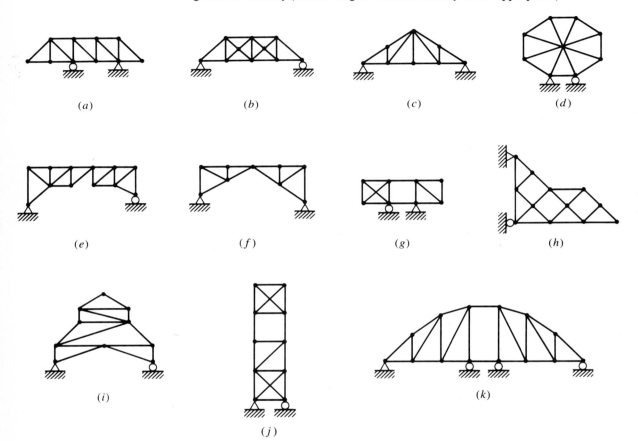

CHAPTER FOUR

Space Trusses

4.1. INTRODUCTION

Preliminary structural analyses of civil engineering structures are often performed as two-dimensional (planar) studies even though structures, such as buildings, bridges, domes, and towers are three dimensional. Consider the light industrial building frame of Fig. 1.17; one planar analysis will evaluate the action of gravity and lateral wind forces on the main frames in the yz plane, while another study will consider the structural behavior of the wall bracing system in the xz plane due to longitudinal wind forces. For many truss-bridge structures, it is also common practice to perform planar analyses of the parallel trusses under gravity loads, as well as for the top and bottom chord bracing truss systems subjected to lateral wind loads. A two-dimensional analysis is generally much simpler to perform than a three-dimensional analysis, especially if the effects of out-of-plane forces (e.g., torsion) can easily be superimposed. However, the structural behavior of many three-dimensional systems, particularly space trusses, geodesic domes, and the like, cannot be properly evaluated by planar studies. This chapter provides elementary concepts and methods for the three-dimensional analysis of space trusses.

4.2. BASIC CONCEPTS

A space truss is a three-dimensional structure that is usually constructed from the same fundamental triangular form associated with a plane truss. If

104 Space Trusses

Minnesota Center, Minneapolis. Recipient of the 1988 Grand Award from the Minnesota Consulting Engineers Council. (*Courtesy of American Institute of Steel Construction*.)

Geodesic dome space frame, Caesar's Palace Hotel, Las Vegas. (*Courtesy of International Structural Slides, Berkeley*.)

105 Basic Concepts

we add three new members to the basic three-member plane truss, one to each of the existing joints, and attach the other ends to a new out-of-plane joint, the newly formed structure is a simple space truss known as a tetrahedron (see Fig. 4.1). The process of continued addition of three members and mutual connection at a new joint will result in other simple and stable space trusses. The frictionless pin member connections in plane trusses are replaced with smooth ball and socket joints in space trusses to ensure that all members are two-force members. A summary of the basic assumptions associated with space-truss analysis are as follows:

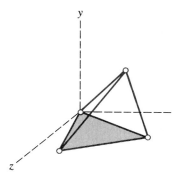

Figure 4.1

1. All members are straight.
2. All loads are applied at joints.
3. All members are axially loaded (two-force members) due to ideal ball and socket joint connections.
4. All support reactions occur at member joints and do not transmit moment.
5. All load conditions satisfy Hooke's law.
6. All joint member axes coincide at a common point.
7. Member weight is negligible with comparison to loads.

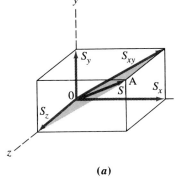

Figure 4.2

In space trusses, the primary load resistance is from the axial member forces. Since ideal joint conditions cannot be fully realized, some secondary torsion, shear, and bending-member forces may develop. However, if careful attention is given to design details and fabrication, the secondary forces can be negligible in comparison to the primary forces. Space-truss analysis in this text will only consider primary force evaluation.

A three-dimensional xyz Cartesian coordinate system can be selected to define a space-truss geometry. In Fig. 4.2a, a truss member force is shown in three-dimensional space to extend from origin O to point A and is assumed to experience a tension value S which is composed of three perpendicular components S_x, S_y, and S_z. An important force relationship between the resultant and its components is developed as follows: The resultant force in the xy plane, S_{xy}, is found from the relationship $S_{xy}^2 = S_x^2 + S_y^2$. In Fig. 4.2a, S_{xy} and S_z lie in the shaded plane from which

$$S^2 = S_{xy}^2 + S_z^2 = (S_x^2 + S_y^2) + S_z^2 = S_x^2 + S_y^2 + S_z^2 \tag{4.1}$$

Similarly, the total member length l, can be expressed in terms of its projected lengths, l_x, l_y, and l_z as shown in Fig. 4.2b, namely,

$$l^2 = l_x^2 + l_y^2 + l_z^2 \tag{4.2}$$

Consequently, it can be established that

$$S/l = S_x/l_x = S_y/l_y = S_z/l_z \qquad (4.3)$$

Therefore, the resultant force of a truss member can be readily found from the force-to-length ratio relationships of Eq. (4.3) if one of the component forces is known.

4.3. STATIC EQUILIBRIUM EQUATIONS

In three-dimensional space, six equations are required to satisfy static equilibrium of a space truss, namely

$$\begin{array}{ll} \Sigma F_x = 0 & \Sigma M_x = 0 \\ \Sigma F_y = 0 & \Sigma M_y = 0 \\ \Sigma F_z = 0 & \Sigma M_z = 0 \end{array} \qquad (4.4)$$

The methods of joints and sections developed in Chap. 3 are also applicable to space-truss analysis. When the method of joints is used, three equations are available at each joint, namely, $\Sigma F_x = 0$, $\Sigma F_y = 0$, and $\Sigma F_z = 0$; since the forces coincide at the joint, the moment equilibrium equations are trivial. However, moment equilibrium equations are used to determine support reactions and to compute some member forces using the method of sections (see Examples 4.1 and 4.2).

4.4. TYPES OF DESIGN SUPPORTS

Three support types are generally identified with space trusses, each of which is assumed to provide no moment resistance. The *ball and socket* support has reaction components which prevent motion in each of three mutually perpendicular directions as shown in Fig. 4.3a. The *roller* or *ball* support has one reaction component that is directed perpendicular to its support plane (see Fig. 4.3b). The *guided* or *slotted roller* support has two reaction components; in Fig. 4.3c, the support is shown to permit free movement in the z direction, but prevents movement in both the x and y directions.

4.5. STABILITY AND DETERMINACY

External determinacy In Chap. 3, we learned that it is good engineering practice to precede a truss analysis with a study to determine if the truss is statically determinate, indeterminate, or geometrically unstable. This task

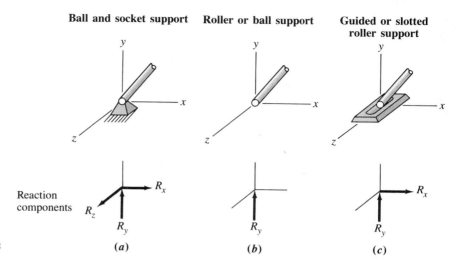

Figure 4.3

may become more difficult for space trusses. One of the necessary conditions to determine if a space truss is statically determinate is to compare the number of available equilibrium equations [six from Eqs. (4.4) above] with the number of reaction components, r. That is, if

$r = 6$, the truss is statically determinate externally

$r > 6$, the truss is statically indeterminate externally

$r < 6$, the truss is geometrically unstable

Another necessary condition to ensure that a truss is stable, is for the reaction components to be properly placed to resist both translation along each of the three orthogonal axes and rotation about each axis. Although the wedge-shaped space truss of Fig. 4.4 has six reaction supports, it is unstable since it cannot resist forces in the z direction. At times, a visual inspection of the applied and reactive forces in each of the various planes of

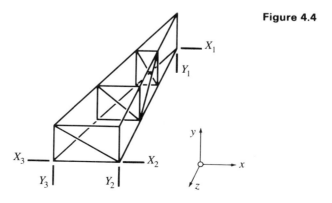

Figure 4.4

a space truss will reveal if the truss is stable. However, cases may arise where it is difficult to detect if the arrangement of the six reactions results in an unstable structure. If the truss structure is stable, a unique set of reaction values exist for a given load condition. In a matrix analysis (see Chap. 14), the unstable nature of the structure is revealed when the determinant of the coefficient matrix of the equilibrium equations is zero and no inverse exists. Consequently, when a computer program is used to perform the analysis of an unstable structure, it will usually halt and provide an error message announcing the unstable nature of the structure.

Internal determinacy In Sec. 3.8, a criterion based on equilibrium was formed to determine if a plane truss is internally determinate, indeterminate, or stable. The criterion can be logically expanded to evaluate space-truss determinacy as follows:

> By the method of joints, three equilibrium equations can be written at each joint of a space truss. Let m represent the number of truss members and j represent the number of joints. Then, $(m + r)$ is the total number of unknowns, $3j$ is the total number of available equations, and the following expressions reveal that for
>
> $(m + r) = 3j$, the truss is statically determinate internally
>
> $(m + r) > 3j$, the truss is statically indeterminate internally
>
> $(m + r) < 3j$, the truss is unstable

The expressions above represent a necessary but not sufficient condition to assess the space-truss determinacy. External stability of space trusses also requires that support reactions are not parallel to one another, and that at least three nonconcurrent reaction components must exist in any one plane. In addition to using the expressions above, a careful visual inspection of various planes of the space truss should be made to detect if a mechanism exists which will make the structure unstable. The analyst should be aware that a force analysis of a stable truss will result in a unique solution, whereas a force analysis of an unstable truss will result in inconsistent member forces. Another approach that has been used to detect if a truss is stable is called the zero-load test which is not included in this text, but is readily found in many structural analysis texts.

4.6 SPECIAL JOINT OBSERVATIONS

The equations of statics for space trusses are used to develop theorems that can be applied from special joint observations.

Theorem 1: If all but one member at a joint lie in the same plane, and no external force exists at the joint, that member must have zero force. If an external force exists at the joint, the external force component normal to the plane is equal to the member force component normal to the plane.

Theorem 2: If all but two members at a joint are in equilibrium or have zero force, and no external force exists at the joint, the two remaining members must have zero force unless they are colinear. If the two remaining members are colinear, they are equal in magnitude and opposite in direction.

4.7 ILLUSTRATIVE EXAMPLE PROBLEMS

The following examples demonstrate the use of statics to find reaction components and member forces of space trusses. In Example 4.1, the truss is initially found to be statically determinate, after which member forces are solved from statics; use of the force/length ratio relationships simplify member force solution. The results are summarized in a table and the analysis concludes with an independent check. Examples 4.2 and 4.3 emphasize the benefit of identifying zero member forces by use of the two theorems presented in Sec. 4.6, and illustrate the continuing use of the basic principles of statics.

EXAMPLE 4.1

The simple tetrahedron of Fig. *b* is supported by a ball and socket at B, a slotted roller at C, and a roller at D where B, C, and D lie in the xz plane. Two external loads are applied at A, namely, $A_x = 10$ kips and $A_y = -18$ kips. Determine all reaction components and member forces.

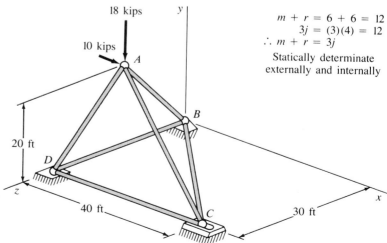

$m + r = 6 + 6 = 12$
$3j = (3)(4) = 12$
$\therefore m + r = 3j$
Statically determinate externally and internally

110 Space Trusses

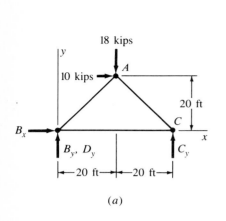
(a)

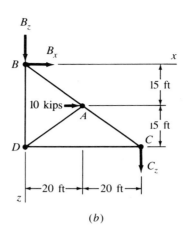
(b)

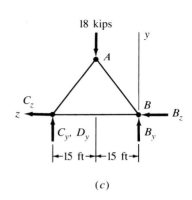
(c)

Solution for reactions

From Fig. *a*:

$\Sigma M_{z \text{ axis thru } B \text{ and } D} = 0 = 10 \text{ kips}(20 \text{ ft}) + 18 \text{ kips}(20 \text{ ft}) - 40 C_y \qquad C_y = 14 \text{ kips} \uparrow$ ◀

From Fig. *b*:

$\Sigma M_{y \text{ axis thru } B} = 0 = 40 C_z - 10 \text{ kips}(15 \text{ ft}) \qquad C_z = 3.75 \text{ kips} \downarrow$ ◀

$\Sigma M_{y \text{ axis thru } D} = 0 = 30 B_x + 10 \text{ kips}(15 \text{ ft}) + 40 C_z$

$\qquad 0 = 30 B_x + 150 + 40(3.75 \text{ kips}) \qquad B_x = -10 \text{ kips} \leftarrow$ ◀

$\Sigma F_z = 0 = B_z + C_z = B_z + 3.75 \text{ kips} \qquad B_z = -3.75 \text{ kips} \uparrow$ ◀

From Fig. *c*:

$\Sigma M_{x \text{ axis thru } C \text{ and } D} = 0 = 30 B_y - 18 \text{ kips}(15 \text{ ft}) \qquad B_y = 9 \text{ kips} \uparrow$ ◀

$\Sigma F_y = 0 = B_y + C_y + D_y - 18 \text{ kips}$

$\qquad 0 = 9 \text{ kips} + 14 \text{ kips} + D_y - 18 \text{ kips} \qquad D_y = -5 \text{ kips} \downarrow$ ◀

Member force solution

Using method of joints,

@ joint B: $\quad \Sigma F_y = 0 = S_{ABy} + B_y \qquad S_{ABy} = -B_y = -9 \text{ kips (C)}$ ◀

@ joint C: $\quad \Sigma F_y = 0 = S_{ACy} + C_y \qquad S_{ACy} = -C_y = -14 \text{ kips (C)}$ ◀

@ joint D: $\quad \Sigma F_y = 0 = S_{ADy} + D_y \qquad S_{ADy} = -D_y = +5 \text{ kips (T)}$ ◀

Recalling that $\dfrac{S}{l} = \dfrac{S_x}{l_x} = \dfrac{S_y}{l_y} = \dfrac{S_z}{l_z}$,

an effective way to determine the remaining reaction components and their resultants are determined in a tabular format as follows:

Member	Projected Lengths, ft			Total Length, ft	Member Force Components, kips			Member Resultant Force	Sense
	l_z	l_x	l_y	l	S_x	S_y	S_z	S (kips)	(T) or (C)
AB	20	20	15	32	−9	−9	−6.75	−14.4	C
AC	20	20	15	32	−14	−14	−10.5	−22.4	C
AD	20	20	15	32	5	5	3.75	8.0	T
BC	40	0	30	50	19	0	14.3	23.8	T
BD	0	0	30	30	0	0	3.85	3.85	T
CD	40	0	0	40	−5	0	0	5.0	C

@ joint B (Fig. b):

$\Sigma F_x = 0 = B_x + S_{ABx} + S_{BCx} = -10 \text{ kips} - 9 \text{ kips} + S_{BCx}$

$S_{BCx} = 19 \text{ kips } T$ ◀

Then,

$S_{BCy} = 0; \quad S_{BCz} = \dfrac{30}{40}(S_{BCx}) = \dfrac{30}{40}(19) = 14.3 \text{ kips } T;$ ◀

$S_{BC} = \dfrac{50}{30}(19) = 23.8 \text{ kips } T$ ◀

@ joint B (Fig. b):

$\Sigma F_z = 0 = S_{BD} + B_z + S_{ABz} + S_{BCz}$

$0 = S_{BD} - 3.75 - 14.4 + 14.3 \quad S_{BD} = +3.85 \text{ kips } T$ ◀

@ joint D (Fig. b):

$\Sigma F_x = 0 = S_{ADx} + S_{CD} \quad S_{CD} = -S_{ADx} = -5 \text{ kips } C$ ◀

Check
Joint C (Fig. b):

$\Sigma F_x = 0 = -S_{ACx} - S_{BCx} - S_{CDx} = 14 - 19 + 5 = 0 \quad$ OK

$\Sigma F_y = 0 = S_{ACy} + C_y = -14 + 14 = 0 \quad$ OK

$\Sigma F_z = 0 = -S_{ACz} - S_{BCz} + C_z = 10.5 - 14.3 + 3.8 = 0 \quad$ OK

112 Space Trusses

EXAMPLE 4.2

Determine all truss reactions and member forces. All supports are guided rollers: A has freedom in x direction, B and C have freedom in z direction.

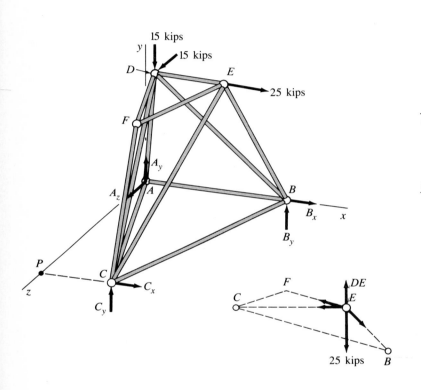
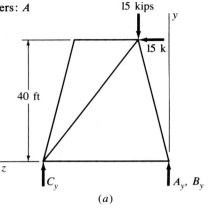
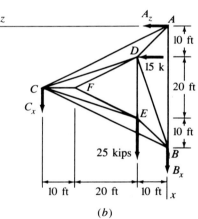

Reaction solutions

From Fig. a:

$\Sigma M_{x \text{ axis along } A \text{ and } B} = 0$:

$40C_y - 15 \text{ kips}(10 \text{ ft}) - 15 \text{ kips}(40 \text{ ft}) \qquad C_y = 18.8 \text{ kips}\uparrow$ ◀

$\Sigma M_{z \text{ axis along } A} = 0$:

$C_y(20 \text{ ft}) + B_y(40 \text{ ft}) - 15 \text{ kips}(10 \text{ ft}) - 25 \text{ kips}(40 \text{ ft})$

$18.8(20) + 40B_y - 1150 = 0 \qquad B_y = 19.4 \text{ kips}\uparrow$ ◀

$\Sigma F_y = 0 = A_y + B_y + C_y - 15 \text{ kips}$

$0 = A_y + 19.4 + 18.8 - 15 \qquad A_y = -23.2 \text{ kips}\downarrow$ ◀

From Fig. b:

$\Sigma F_z = 0 = A_z + 15$ kips $A_z = -15$ kips → ◀

$\Sigma M_{y \text{ axis thru } A} = 0$:

$C_x(40 \text{ ft}) + 25 \text{ kips}(10 \text{ ft}) - 15 \text{ kips}(10 \text{ ft}) = 0$ $C_x = -2.5$ kips ← ◀

$\Sigma F_x = 0 = B_x + C_x + 25$ kips

$0 = B_x - 2.5 + 25$ $B_x = -22.5$ kips ← ◀

Member solutions

Zero members: Consider joint F.

1. First observe the plane formed by members FC and FD. Since member FE is not in the same plane, and no external load exists at point F, by theorem 1, $FE = 0$.
2. Also, view the plane at F formed by members FC and FE. Notice that member FD is not in the plane. By theorem 1 again, member $FD = 0$. At joint F, only FC remains and must $= 0$.
3. Following step 1, only FC and FD remain and are not colinear, thus by theorem 2, $FC = FD = 0$.
4. Construct the FBD at point E. Consider $\Sigma M = 0$ about a line in space coincident with C and B as shown. Members EC and EB pass through the line CB producing zero moment about it. Member FE (although already proven $= 0$) is parallel to line CB and would not produce a moment if it were nonzero. Since member DE is colinear with the 25 kips external force at joint E, they prove to be equal and opposite in sense by virtue of moment equilibrium about imaginary line BC.
5. The force DE and the 25-kip external force at joint E are equal in magnitude, opposite in direction (sense), and colinear. Also, member $FE = 0$. Thus only member forces EC and EB remain to be found. By theorem 2, $EC = EB = 0$ since they are not colinear.
6. At joint B (note: all forces assumed in tension):

$\Sigma F_y = 0 = S_{BEy} + S_{BDy} + B_y = 0 + S_{BDy} + 19.4$ kips $S_{BDy} = -19.4$ kips (C)

Since

$l_{BD}^2 = 30^2 + 40^2 + 10^2 = 2600$ $l_{BD} = 51$ ft   ∴ $S_{BD} = 51/40(-19.4) = -24.7$ kips (C)

$\Sigma F_z = 0 = S_{BCz} + S_{BDz} + S_{BEz} = S_{BCz} + 10/51(-24.7) + 0$ ∴ $S_{BCz} = +4.84$ kips (T)

Since

$l_{BC}^2 = 20^2 + 40^2 + 0^2 = 2000$ or $l_{BC} = 44.7$ ft

$S_{BC} = 44.7/40(4.84) = 5.4$ kips (T)

$\Sigma F_x = 0 = B_x - S_{AB} - S_{BDx} - S_{BEx} - S_{BCx}$

$0 = -22.5$ kips $- S_{AB} - 30/51(-24.7$ kips$) - 0 - 20/44.7(5.4)$

$0 = -22.5$ kips $- S_{AB} + 14.5$ kips $- 2.4$ kips   ∴ $S_{AB} = +10.4$ kips (T)

By the method of joints the remaining members are:

$S_{AC} = +10.3$ kips (T), $S_{AD} = +24.5$ kips (T) and $S_{CD} = -23.9$ kips (C)

EXAMPLE 4.3

Determine the zero-force members of the truss. Ball and socket supports exist at A and B, slotted roller at C (as shown), and ball support at D.

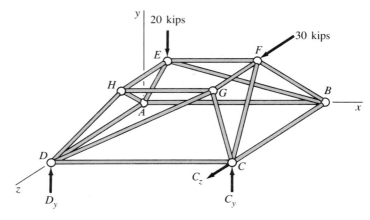

1. At joint H, members HE, HA, and HD lie in the same plane. Member force HG does not lie in the referenced plane and no load is applied at H. By theorem 1, $HG = 0$.
2. Similarly, at joint G, members GH, GD, and GC lie in the same plane with GF not in the plane. Therefore, $GF = 0$.
3. Consider an FBD of joint F:
   Members FC and FB intersect the line BC in space. Member GF and the 30 kips applied force at F are parallel to the line BC. Therefore, if moment equilibrium of the forces at joint F is established about the line AB in space, member $FE = 0$.
4. Since HG and $GF = 0$, at joint G only members GC and GD are unknown. Since they are not colinear, from theorem 2, $GC = GD = 0$.
5. The remaining members transmit axial tensile, and axial compressive forces.

4.8 COMPUTER SOLUTIONS

Statically determinate and stable space trusses are solvable through the use of many existing proprietary and commercially available computer codes. Moreover, most students realize that similar computer codes are available at their campus computer facilities. You can also perform a truss analysis in the following manner: After a space truss is found to be stable and statically determinate, the method of joints can be used to write three equations at each joint in terms of the externally applied loads and the unknown reaction and member forces; this will result in a set of n equations in n unknowns. These simultaneous equations are now ready for solution. The student who is well acquainted with matrix algebra may elect to write a BASIC or FORTRAN program using the matrix form $[X] = [A]^{-1} \cdot [B]$ to solve the equations, where $[A]^{-1}$ is the inverse matrix of the coefficients of the unknowns, $[X]$ is the column matrix of the unknowns, and $[B]$ is the column matrix of constant values. Most PCs are equipped with a form of BASIC or BASICA computer programming language. Those students unfamiliar but interested in an immediate introduction to matrix algebra should read Sec. 14.2 of Chap. 14. Others may choose to use available computer programs for the solution of simultaneous linear equations (SOSLE) that may be available at your computer center, or can be found in the appendices of modern text books (see App. B of Harold I. Laursen, *Structural Analysis*, 3d ed., McGraw-Hill, New York, 1988).

Problems

Determine all reaction components and member forces of each space truss. Solid lines indicate direction of reaction vector components.

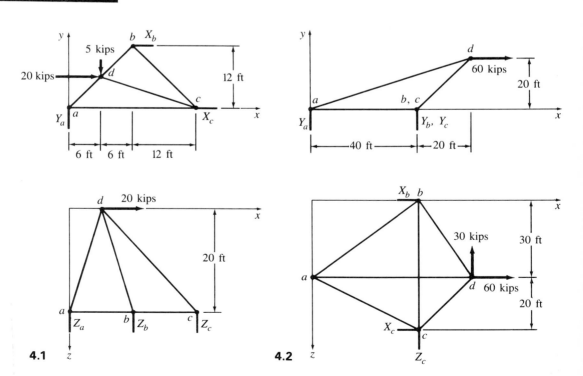

4.1

4.2

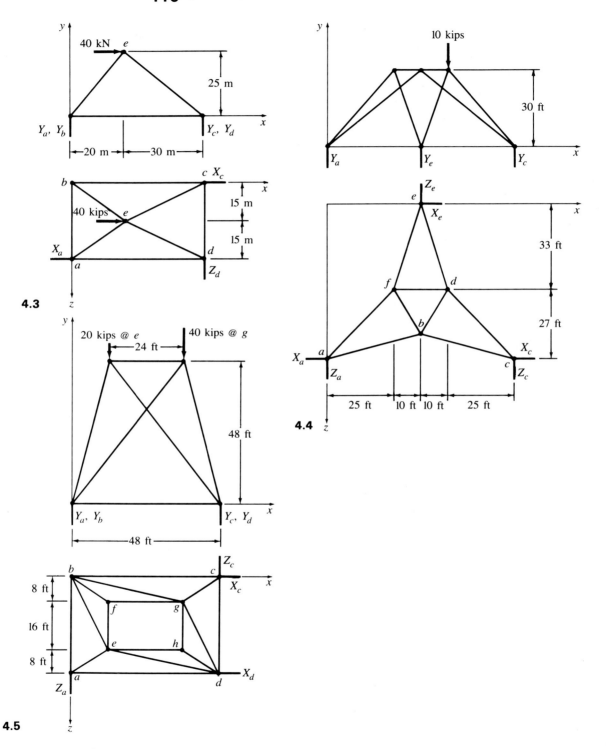

117 Problems

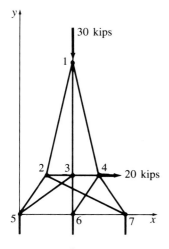

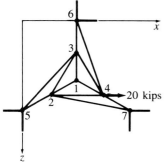

4.6 to 4.11 Solve each truss cited by writing joint equilibrium equations at each node in terms of known loads and unknown member force and reaction components. Solve these equations using an available solution of simultaneous equation computer routine.

4.6 Nodal coordinates:

Joint no.	x	y	z
1	20.0	80.0	23.1
2	10.0	30.0	28.9
3	20.0	30.0	11.6
4	30.0	30.0	28.9
5	0.0	0.0	34.6
6	20.0	0.0	0.0
7	40.0	0.0	34.6

4.7 Redo Prob. 4.1.
4.8 Redo Prob. 4.2.
4.9 Redo Prob. 4.3.
4.10 Redo Prob. 4.4.
4.11 Redo Prob. 4.5.

CHAPTER FIVE

Shear and Bending-Moment Diagrams

5.1. INTRODUCTION

The ability to construct shear and bending-moment diagrams is one of the most essential tools the structural engineer must possess. In truss analysis, the assumed frictionless pin connections develop axial force resistance in the members and negligible transverse shear and bending forces. Yet, if a truss has either rigid connections or loads applied on members at points other than joint locations, the shear and bending forces become significant to the final design; in that case, the structure more appropriately is called a frame and the need to determine the shear and bending-moment resistance throughout the structure becomes important. Graphical plots of the variation of shear and bending moment in a structure are called shear and bending-moment diagrams. Aside from beam and frame structures, these diagrams are used to assist in the design of building roof and floor systems, bridge decks, and all structural members subjected to transverse loading.

From studies of mechanics of materials, the shear and bending-moment diagrams are generally developed for various types of load. Some applied transverse loads occur as concentrated forces or local moments at specific locations on a structure (e.g., beam reactions on a girder or moments at beam-column connections in a steel building). Civil engineering structures also experience distributed loads which can be uniform or variable in magnitude. Some types of distributed load include wind force on a gable frame, soil or water pressure against a retaining wall or dam, and the weight of a wood floor or concrete slab supported by its floor beams.

Pont de Bercy, River Seine, Paris. Modern masonry arch bridge. (*Courtesy of International Structural Slides, Berkeley.*)

The shear (V) and bending-moment (M) diagrams provide a visual display of the variations of V and M magnitudes and identify those locations on a structure where maximum values occur. These diagrams are important visual aids to design.

5.2. SHEAR AND BENDING MOMENT IN BEAMS

Type of supports Consider the schematic drawings of various types of beams shown in Fig. 5.1. The *simple beam* of Fig. 5.1a has a hinge support and a roller support at opposite beam ends; the hinge support prevents translation at its end; the roller support provides limited freedom of translation parallel to its support surface but prevents translation normal to the support surface. In addition, the beam can rotate freely at both the hinge and the roller end locations (i.e., no end-moment resistance). However, the presence of a hinge support or roller beam support at an interior location will usually result in a nonzero moment value.

The *cantilever beam* of Fig. 5.1b is free at one end and fixed at the other end; the fixed-end support prevents rotation and translation, that is, it develops axial, transverse, and moment resistance.

The *continuous beam* of Fig. 5.1c has simple end supports similar to Fig. 5.1a, plus an interior roller support; this beam is statically indeterminate. It was shown in Chap. 2 that placement of an interior hinge in a continuous beam can result in a statically determinate beam (see Fig. 5.1d).

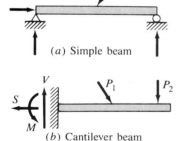

(*a*) Simple beam

(*b*) Cantilever beam

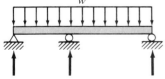

(*c*) Continuous beam

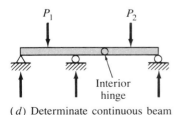

(*d*) Determinate continuous beam

Figure 5.1

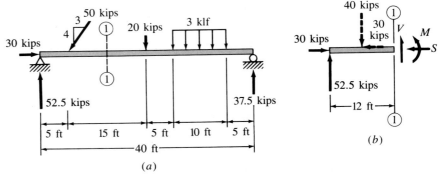

Figure 5.2

Shear and bending moment defined Figure 5.2 shows a simple beam supporting several loads with its reaction components given. If a section 1–1 is taken through the beam at any location, as shown in Fig. 5.2, and the left FBD section is drawn (see Fig. 5.2b), it is evident that internal forces must exist at the cut section to satisfy static equilibrium. The internal forces are symbolized as S (axial force), V (shear force), and M (bending moment). The *shear* at the section is defined as the *algebraic sum* of *all* force components on the FBD section perpendicular (transverse) to the beam axis. Summation of moments of *all* forces on the left FBD section about the 1–1 location must be resisted by an internal moment in the beam to maintain equilibrium; this internal resisting moment is called the *bending moment*. Obviously, the shear and bending moment can be found at any section of a beam structure using either a left or right FBD section. For the beam of Fig. 5.2, shear and bending-moment values are found as follows:

At the left end:

$$V = 52.5 \text{ kips}, M = 0, \text{ and } S = 30 \text{ kips } (C)$$

At 2 ft from the left support:

$$V = 52.5 \text{ kips}, \quad M = 52.5 \text{ kips}(2 \text{ ft}) = 105 \text{ kip} \cdot \text{ft}, \quad \text{and} \quad S = 30 \text{ kips } (C)$$

At section 1–1:

$$V = 52.5 \text{ kips} - 40 \text{ kips} = 12.5 \text{ kips} \qquad S = 0$$

and

$$M = 52.5 \text{ kips}(12 \text{ ft}) - 40 \text{ kips}(12 \text{ ft} - 5 \text{ ft}) = 350 \text{ kip} \cdot \text{ft}$$

At 22 ft from the left support:

$$V = 52.5 \text{ kips} - 40 \text{ kips} - 20 \text{ kips} = -7.5 \text{ kips} \qquad S = 0$$

and

$$M = 52.5 \text{ kips}(22 \text{ ft}) - 40 \text{ kips}(22 \text{ ft} - 5 \text{ ft})$$
$$- 20 \text{ kips}(22 \text{ ft} - 20 \text{ ft}) = 435 \text{ kip} \cdot \text{ft}$$

In Fig. 5.2, a 20-kip concentrated load exists at the beam midspan (20 ft from the left support). Note the step change in shear value that occurs to the left and right of the 20-kip load. That is,

at 19.99 ft from the left support $\quad V = 52.5 - 40 = 12.5$ kips

and

at 20.01 ft from the left support $\quad V = 12.5 - 20 = -7.5$ kips

The sudden step change is equal to the intensity of the concentrated load of 20 kips.

5.3 SIGN CONVENTION

In order to clearly establish a sense of shear and bending-moment behavior on a beam structure, the following sign convention will be used for beams in this text:

Consider a set of xy axes where y is the vertical axis and x is horizontal and directed along a typical beam-length axis. *Shear at a section is defined as positive when the algebraic sum of all force components in the y direction, as measured from the left end of the beam, is upward. Shear is also defined as positive if the algebraic normal force summation to a beam section, measured from the right end of the beam, results in a downward direction. Negative shear is opposite to positive shear.*

Consider the simple beam of Fig. 5.3a as straight and unloaded. Observe in Fig. 5.3b, the magnified deflection of the beam due to a load P applied at midspan. Parallel sets of vertical lines, which were initially drawn on the unloaded beam of Fig. 5.3a, are shown to experience shear deformation due to load as shown in Fig. 5.3c. The distorted line set to the left of midspan shows that summation of transverse beam forces from the left support is upward and is maintained in vertical equilibrium by the opposite shear force shown. This is positive shear distortion. Also, the distorted line set to the right of midspan shows that force summation from the right support is upward; this signifies negative shear. Figure 5.3c does not include bending effects.

The simple beam of Fig. 5.3a deflects as a concave upward curve along its entire length after the P force is applied (see Fig. 5.3b). The parallel line

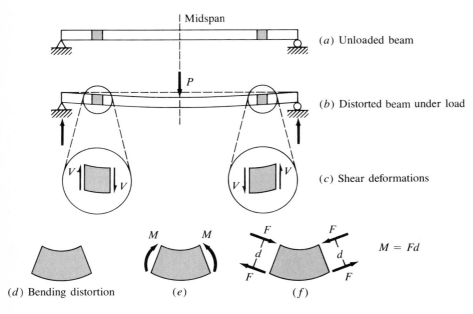

(d) Bending distortion (e) (f)

sets drawn on both sides of beam midspan before load is applied, also undergo bending deformation due to loading. The bending deformation shown in Fig. 5.3d can be visualized as the result of bending moments applied at both ends (see Fig. 5.3e or, in couple form, Fig. 5.3f). Furthermore, it is evident that a simple horizontal beam, under gravity loading, experiences compression above the beam neutral axis and tension below the beam neutral axis. From Fig. 5.3e, it is seen that both clockwise and counterclockwise moments produce the same bending effect to a beam section; therefore, the sign convention for bending moment is *not* associated with vector direction. Instead, *positive bending moment in a beam is defined as a moment that produces compression above the neutral axis and tension below the neutral axis.*

5.4. SHEAR AND BENDING-MOMENT FUNCTIONS

It is of value for an analyst to develop the facility to express the shear and bending-moment functions of a beam in algebraic form. The expressions can be used by the analyst to develop a tabular form of displaying the shear and bending-moment values at various locations along the beam. In subsequent chapters, work and energy methods are shown to utilize these functional expressions to calculate structural displacements and for solving statically indeterminate structures. Example 5.1 develops algebraic expressions for load, shear, and bending-moment for a simple beam with an overhang supporting four conventional types of loads. Concentrated loads are readily defined as constant directed values, whereas continuous loads which vary as a function of x are developed as shown in Example 5.1, Figs. b to d. Observe that a new algebraic expression for shear is required at each location where a new transverse force is introduced onto the beam. Also, note that shear values are constant between concentrated forces and variable along regions where distributed loading exists. Similarly, it is necessary to redefine a bending-moment expression in regions where either added moment-producing forces exist, or where local moments are applied. From Fig. a in Example 5.1, constant values of shear (V) exist in the following regions:

$$A \text{ to } B: \quad V = 0$$
$$B \text{ to } C: \quad V = -10 \text{ kips}$$
$$C \text{ to } D: \quad V = -10 + 25 = +15 \text{ kips}$$
$$E \text{ to } G: \quad V = -10 + 25 - 2(8) = -1 \text{ kip}$$

EXAMPLE 5.1

Determine load, shear, and bending-moment expressions.

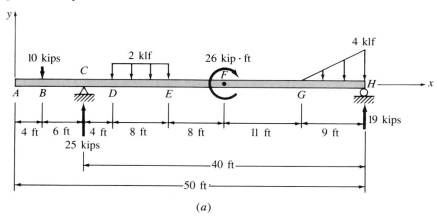

(a)

Loads

1. Concentrated loads are readily identified by their magnitudes and direction. Thus, at B, load $= -10$ kips.

2. Concentrated (local) moments are readily identified by their magnitudes. Clockwise directed moments cause positive step increases ($+M$) when x increases to the right.

3. Uniform loads: (Fig. b)

   If x is referenced from point D as shown, the uniform load is expressed as $-2x$.

4. Linear distributed load: (Fig. c)

   For x referenced from point G, from Fig. d, it is found that the linear distributed load is expressed as

   $$-\frac{1}{2}(x)\left(\frac{4x}{9}\right) = -\frac{2}{9}x^2$$

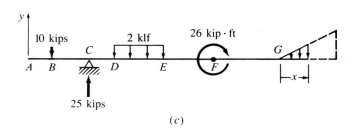

(b)

(c)

(d)

Shear

From Fig. b in Example 5.1, shear varies along the region from D toward E as a function of x by the following expression:

$$V(x) = -10 + 25 - 2x = (15 - 2x) \text{ kips}$$

From Figs. a, c, and d in Example 5.1, shear varies along the region from G to H as a function of x by the following expression:

$$V(x) = -10 + 25 - 2(8) - 2x^2/9 = (-1 - 2x^2/9) \text{ kips}$$

Bending Moments

From Fig. a in Example 5.1, a zero value of bending-moment (M) exists from A to B. The beam bending moment varies from B to H and is expressed below as a function of x, with the x origin placed at the left point of each region. Bending moment, M(x), is expressed as follows:

B to C: $M(x) = -10x$ kip·ft

C to D: $M(x) = -10(x + 6) + 25x = (15x - 60)$ kip·ft

D to E: $M(x) = -10(x + 10) + 25(x + 4) - 2x(x/2)$

$\qquad = (-x^2 + 15x)$ kip·ft

E to F: $M(x) = -10(x + 18) + 25(x + 12) - 2(8)(x + 4)$

$\qquad = (-x + 56)$ kip·ft ... to the left of point F

Right of point F:

$$M(x) = -x + 56 + 26 = (-x + 82) \text{kip} \cdot \text{ft}$$

F to G: $M(x) = -10(x + 26) + 25(x + 20) - 2(8)(x + 12) + 26$

$$= (-x + 74) \text{kip} \cdot \text{ft}$$

G to H: $M(x) = -10(x + 37) + 25(x + 31) - 2(8)(x + 23) + 26 - (2x^2/9)(x/3)$

$$= (-2x^3/27 - x + 63) \text{kip} \cdot \text{ft}$$

The shear and bending-moment diagrams presented in Fig. 5.4 are drawn by replacing x, in the expressions above, with appropriate beam dimensions from Fig. a in Example 5.1. It is worth noting that as the moment diagram is drawn from left to right, the local clockwise moment of 26 kip·ft at point F results in a sudden positive increase in bending moment.

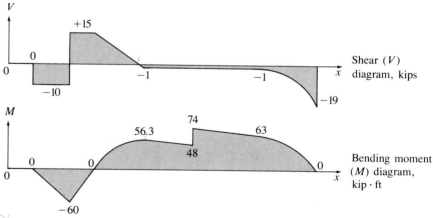

Figure 5.4

5.5 RELATIONSHIPS AMONG LOAD, SHEAR, AND MOMENT

Relationships exist among load, shear, and bending moment which simplify the construction of the shear and bending-moment diagrams. Figure 5.5a

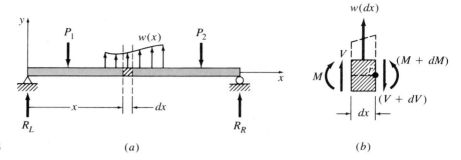

Figure 5.5   (a)   (b)

presents a simple beam plotted along an x axis where positive-force direction is assumed in the $+y$ direction. An FBD of a differential beam length, dx, is shown in Fig. 5.5b with the load, internal shear, and moment forces applied in accordance with the established positive-sign convention. The presence of load on the differential length results in forces of V and M on the left face, and $(V + dV)$ and $(M + dM)$ on the right face of the differential length. Writing the vertical force equilibrium equation yields

$$\Sigma F y = 0 = V + w(dx) - (V + dV) = w(dx) - dV \quad \text{or} \quad dV = w(dx)$$

The moment-equilibrium equation written about the neutral axis at the right face of the differential length, point r, yields

$$\Sigma M_r = 0 = V(dx) + M + w(dx)(dx/2) - (M + dM) = V(dx) - dM$$

where the term $w(dx)^2/2$ is considered negligible. This equation can be rewritten as $dM = V(dx)$.

Slope relationships for the V and M diagrams Since the beam is aligned with x, the shear and moment diagrams can be viewed as plots of V versus x, and M versus x, respectively. If the equations derived above are rewritten as follows

$$dV/dx = w \quad (5.1)$$
and
$$dM/dx = V \quad (5.2)$$

dV/dx is the slope of the shear diagram, which at any x location on the beam is equal to the load intensity at x, and dM/dx is the slope of the moment diagram, which at any x location on the beam is equal to the shear intensity at x.

Slope Eqs. (5.1) and (5.2) clearly define the shape of the shear and bending-moment diagrams, respectively. Equation (5.1) is not valid at points where concentrated forces occur due to the discontinuity in shear. Figure 5.6 illustrates that a *positive* load intensity yields a *positive* slope change of shear; likewise, a *positive* shear intensity at x results in a *positive* slope change of moment as shown. In addition, (1) negative intensities yield negative slope changes, (2) greater intensities result in greater slope inclination, (3) low intensities result in shallow slopes, and (4) zero slopes will usually occur (*a*) on moment diagrams at points of zero shear and (*b*) on shear diagrams in regions where no distributed load is present.

Shear and moment diagrams by area integration Shear and moment diagrams are often constructed by simple area integration. The original form of Eqs. (5.1) and (5.2), $dV = w(dx)$ and $dM = V(dx)$, are visually presented in Fig. 5.6, from which

$$(V_2 - V_1) = \int_1^2 w(dx) = \text{area under the distributed load diagram between points 1 and 2} \qquad (5.3)$$

and

$$(M_2 - M_1) = \int_1^2 V(dx) = \text{area under the shear diagram between points 1 and 2} \qquad (5.4)$$

In cases where area integration is complex, the V and M diagrams can always be generated directly from statics and from written algebraic expressions for shear and moment as presented earlier. Geometric properties of some areas are presented on the front cover facing page and in App. 2.

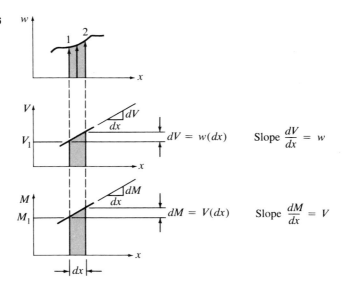

Figure 5.6

128 Shear and Bending-Moment Diagrams

Boundary conditions The success of drawing the V and M diagrams correctly will improve if careful attention is given to boundary conditions. Some of the pertinent boundary conditions are as follows:

1 At free ends of beams:
  a. Shear is zero unless acted upon by a concentrated transverse load
  b. Bending moment is zero unless acted upon by a local moment or couple.
2 At simple end supports of beams, bending moment is zero unless acted upon by a local moment or couple.
3 At interior hinges of beams (*not supports*), the bending moment is zero.

Other important observations are:

1 At fixed-end supports of beams, both shear and bending moment are usually nonzero in value.
2 At simple interior supports of beams, both shear and bending moment are usually nonzero in value.

An awareness of incorrect V and M boundary values should alert the analyst to seek out evident calculation errors and to correct the V and M diagrams.

5.6. SHEAR AND BENDING-MOMENT DIAGRAMS FOR BEAMS

Examples of shear and bending-moment diagrams for several types of beams are presented in Examples 5.2 to 5.6 with supplementary comments on their construction.

EXAMPLE 5.2

Draw V and M diagrams for the simple beam illustrated.

Comments

1. Slope of shear diagram is negative and constant over entire beam length: $dV/dx = w = -3$ klf.
2. Slope of moment diagram, $dM/dx = V$ varies with V.
3. Maximum moment occurs where shear is zero; that is, $dM/dx = V = 0$.

4. Note that moment is zero at both supports: $M_A = M_B = 0$.
5. Change in moment from left end (A) to midspan (C):

$$M_C - M_A = M_C - 0 = \int_A^C V \cdot dx$$

or

$$M_C = \frac{1}{2}\left(\frac{wL}{2}\right)\left(\frac{L}{2}\right) = \frac{wL^2}{8}$$

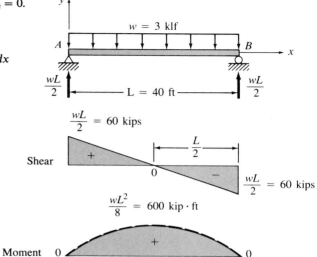

EXAMPLE 5.3

Draw shear (V) and moment (M) diagrams for the simple beam illustrated.

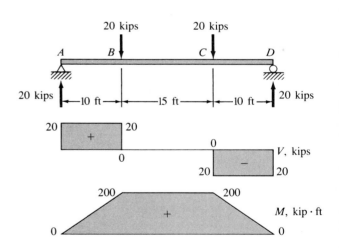

Comments

1. From A to B, shear value is positive and constant. Thus, slope to M diagram, $dM/dx = V$ from A to B is also positive and constant.

2. Since uniform load $(w) = 0$ over entire beam length, slope of shear, $dV/dx = w = 0$ over entire beam length.
3. Shear is zero from B to C. Thus, slope of M diagram $dM/dx = V = 0$ from B to C.
4. Area under V diagram from A to $B = \int_A^B V \cdot dx = M_B - M_A = M_B - 0 = M_B = 200$ kip · ft.
5. Since area under V diagram from B to $C = 0$,

$$M_C - M_B = 0; \quad \therefore \quad M_C = M_B = 200 \text{ kip} \cdot \text{ft}$$

6. Constant negative shear from C to D results in constant negative slope of M diagram from C to D.
7. From C to D:

$$M_D - M_C = \int_C^D V \cdot dx = -20 \text{ kips}(10 \text{ ft})$$

or

$$M_D = -200 + M_C \quad \therefore \quad M_D = -200 + 200 = 0$$

8. Note: Moment diagram is initiated from known boundary value at A. That is, $M_A = 0$.
9. Also, starting from A to D, we arrive at calculated value of $M_D = 0$ in step 7 above which agrees with boundary condition at point D.

EXAMPLE 5.4

Draw V and M diagrams for the cantilever beam illustrated.

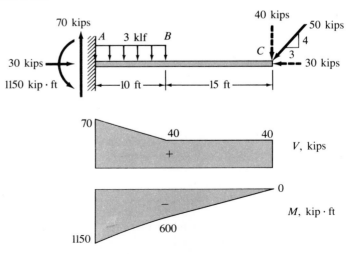

Comments

1. Fixed-end forces at point A determined by principles of statics.
2. Only perpendicular force components contribute to values of V diagram.
3. From A to B, slope of V diagram $= -3$. That is, $dV/dx = w = -3$ from A to B.
4. Slope of V diagram from B to C, $dV/dx = w = 0$.
5. $V_A = 70$ kips

   and $V_B - V_A = \int_A^B dV$ $\quad \therefore \quad V_B - V_A = \int_0^{10}(-3)dx = -30$ or $V_B = +40$ kips
6. Since $V_C - V_B = \int_B^C w \cdot dx = 0$, $V_B = V_C = +40$ kips.
7. Note: Slope of M diagram ($dM/dx = V$) is steeper at A (70 kips) than at B (40 kips).
8. Constant V from B to C yields constant slope of M diagram from B to C.
9. Boundary condition of $M_C = 0$ is verified.

EXAMPLE 5.5

Draw V and M diagrams for the simple beam with an overhang illustrated.

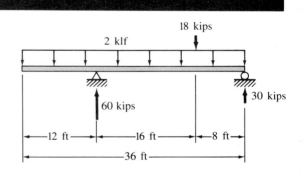

Comments

1. Sudden change of shear at left support results in sudden slope change in M diagram.
2. Moment diagram begins and ends with known boundary values of zero.
3. Maxima and minima values of moment occur at points where shear passes through zero.

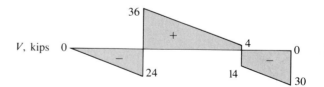

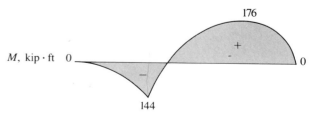

EXAMPLE 5.6

Draw the V and M diagrams for the beam illustrated below.

Comments

1. Boundary conditions: $M_A = M_C = M_D = 0$.
2. Note zero slope of shear diagram at A.
3. Note increasing negative slope of V diagram from A to B.

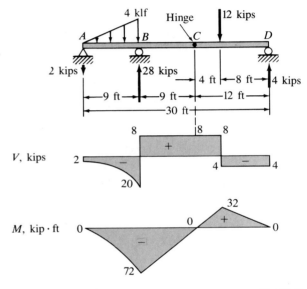

5.7 SHEAR AND BENDING-MOMENT DIAGRAMS FOR FRAMES

Shear and bending-moment diagrams are also drawn to assist in the design of frame structures. The definitions for shear and bending moment of Sec. 5.1, and the support and interior-hinge boundary conditions of Sec. 5.5 are valid for both beam and frame structures alike. However, the sign convention developed above for positive shear and bending moment in a typical horizontal beam, cannot be applied to frames consisting of many vertical members, stories, and bays as shown in Fig. 5.7; shear integration for vertical members is up or down and not left to right or right to left; also, bending-moment sense for tension of compression above or below the neutral axis is meaningless when applied to vertical frame members. The following approach is offered and applied in subsequent example frame problems:

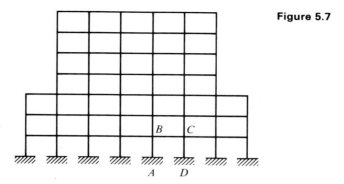

Figure 5.7

1 Consider typical members of a frame, such as *ABCD* of Fig. 5.7, redrawn as Fig. 5.8. Assume that vertical-column members above *B* and *C* only transmit compression forces to columns *AB* and *CD* and that adjacent beam members provide negligible bending at *B* and *C*. Projected normal lines are drawn from the ends of each member away from the frame as shown.

2 Then, a free-body diagram (FBD) is drawn for each frame member between the normal lines with end forces and applied loads in place.

3 Finally, two additional lines between the projected normals are drawn to serve as base lines to draw *V* and *M* diagrams. *Note:* **V** *and* **M** *labels are placed to the right of these lines to establish a similar orientation used for horizontal beams.*

Figure 5.8

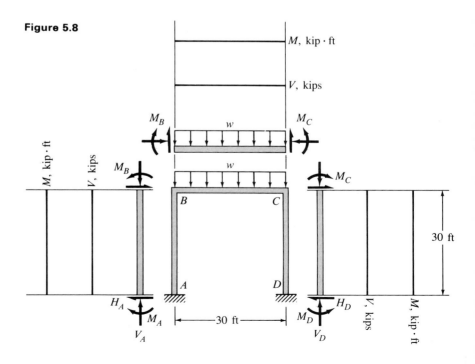

134 Shear and Bending-Moment Diagrams

The shear and moment diagrams of each member can now be constructed in the projected plane and treated as if they are horizontal. Example 5.7 demonstrates this approach for shear and bending-moment diagram construction of a simple portal frame. The V and M diagrams illustrated in Example 5.7 are often expressed in the composite forms of Fig. 5.9.

EXAMPLE 5.7

Draw the shear (V) and moment (M) diagrams for the frame of Fig. 5.8 if $w = 0.20$ klf and the reactions at A and D are given as shown.

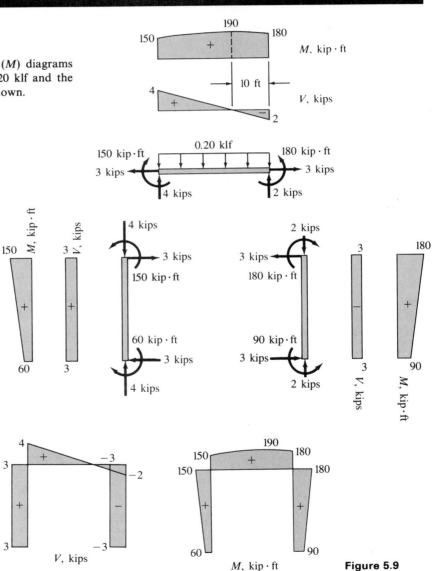

Figure 5.9

135 Shear and Bending-Moment Diagrams for Frames

Wind-load forces exerted on a gable frame are determined in Example 1.1 of Chap. 1. Example 5.8 provides a detailed development of the V and M diagrams of the wind-loaded gable frame; the frame includes hinge supports at points A and E, and an interior hinge joins the roof members at point C. Support reactions are given. Since it is established that shear forces on a flexural member are defined as those force components which act normal to the member axis, Figs. b to e in Example 5.8 present FBDs for each gable frame member and provide the necessary tools of statics and trigonometry to determine the shear forces at member ends. The internal resisting moment at points B and D are also evaluated by statics in Example 5.8, Figs. b to d. Note that the internal forces at point B of Fig. b are shown equal in magnitude and opposite in direction at point B of Fig. c in accordance with Newton's laws.

EXAMPLE 5.8

Draw the V and M diagrams for the gable frame under wind-load forces illustrated below. Reaction components are given. Interior hinge at C. Hinge supports at A and E.

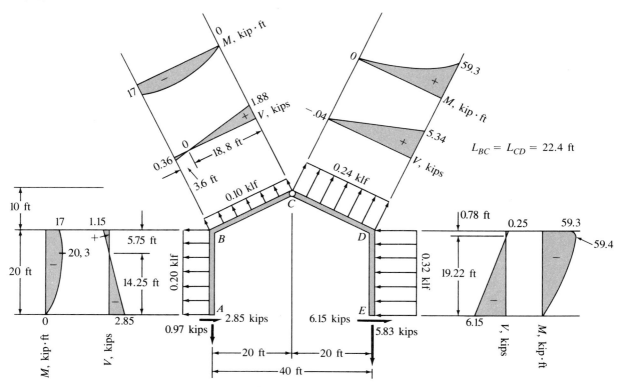

Free-body diagram (FBD) support data for V and M diagrams:

$+\uparrow \Sigma F_y = 0 = S_{BA} - 0.97 \text{ kips}; \qquad \therefore \quad S_{BA} = +0.97 \text{ kips } (T)$

$+ \rightarrow \Sigma F_x = 0 = F_{Bx} + 2.85 \text{ kips} - 0.20 \text{ klf}(20 \text{ ft}); \qquad \therefore \quad F_{Bx} = 1.15 \text{ kips}$

$\zeta + \Sigma M_B = 0 = M_B - 2.85 \text{ kips}(20 \text{ ft}) + 0.20 \text{ klf}(20 \text{ ft})\left(\dfrac{20 \text{ ft}}{2}\right);$

$\therefore \quad M_B = 17 \text{ kip} \cdot \text{ft}$

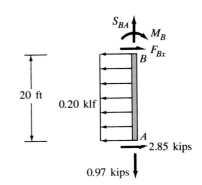

From point A, locate point of zero shear:

$-2.85 \text{ kips} + 0.20x = 0$

$x = 2.85/0.20 = 14.25 \text{ ft}$

$L_{BC} = 22.4 \text{ ft}$

$V_B = \dfrac{1}{\sqrt{5}} (1.15 \text{ kips}) - \dfrac{2}{\sqrt{5}} (0.97 \text{ kips})$

$V_B = 0.51 - 0.87 = -0.36 \text{ kips} \searrow$

$V_C = -0.36 \text{ kips} + 0.10(22.4)$

$V_C = -0.36 + 2.24 = +1.88 \text{ kips} \searrow$

$S_{CB} = 1.03 + 0.43 = +1.46 \text{ kips } (T)$

$\zeta + \Sigma M_C = 0 = -M_B + (0.51 - 0.87)(22.4) + 0.10(22.4)\left(\dfrac{22.4}{2}\right)$

$0 = -17 - 8.06 + 25.09 \approx 0 \quad \checkmark$

$L_{CD} = 22.4 \text{ ft}$

$\zeta + \Sigma M_D = 0 = M_D + (0.96 \text{ kips} - 0.92 \text{ kips})(22.4 \text{ ft}) - 0.24 \text{ klf}(22.4 \text{ ft})(11.2 \text{ ft})$

$0 = M_D + 0.90 - 60.2$

$M_D = 59.3 \text{ kip} \cdot \text{ft}$

An FBD of frame section ABC yields the following force set at C:

$\Sigma F_x = 0 = 2.85 \text{ kips} - 0.20 \text{ klf}(20 \text{ ft}) - 0.10 \text{ klf}(10) + F_{cx}$

$F_{cx} = 2.15 \text{ kips}$

$\Sigma F_y = 0 = -0.97 \text{ kips} + 0.10 \text{ klf}(20 \text{ ft}) + F_{cy}$

$F_{cy} = 1.03 \text{ kips}$

$V_C = 1.03 \text{ kips} \cdot \dfrac{2}{\sqrt{5}} - 2.15 \text{ kips} \cdot \dfrac{1}{\sqrt{5}} = 0.92 \text{ kips} - 0.96 \text{ kips}$

$V_C = 0.04 \text{ kips} \approx 0$

$S_{DC} = 1.03 \text{ kips} \cdot \dfrac{1}{\sqrt{5}} + 2.15 \text{ kips} \cdot \dfrac{2}{\sqrt{5}} = 2.38 \text{ kips}$

$V_D = 0.92 \text{ kips} - 0.96 \text{ kips} + 0.24 \text{ klf}(22.4 \text{ ft}) = 5.34 \text{ kips}$

$+ \rightarrow \Sigma F_x = 0 = V_D - 0.32 \text{ klf}(20 \text{ ft}) + 6.15 \text{ kips}$

$V_D = 0.25 \text{ kips} \rightarrow$

$\uparrow + \Sigma F_y = 0 = S_{DE} - 5.83 \text{ kips}$

$S_{DE} = +5.83 \text{ kips } (T)$

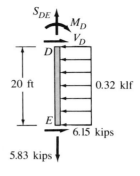

See Fig. *a* above for final V and M diagrams.

5.8. V AND M DIAGRAMS BY METHOD OF SUPERPOSITION

When a structure supports many loads and various load types, it may be difficult to develop the shear and bending-moment diagrams, especially if it is desirable to use known area properties for integration. If a system of applied loads results in linear elastic behavior of a structure, resultant shears (or moments) can be determined by summation of the shears (or moments) due to each individual load. This is known as the *principle of superposition*. This principle is an important and frequently used tool in structural analysis which, in this application, can be used to draw the shear and bending-moment diagrams. In later chapters, the principle of superposition is shown to be particularly useful to compute displacements by geometric methods and for solutions of statically indeterminate linear elastic structures.

In a linear elastic system:

1 Stress is proportional to strain (Hooke's law is valid).
2 Strain values are at or below the elastic limit strain which usually results in insignificant member deformations.
3 For a stable structure, elastic-load deformations usually cause minor change to the original structure geometry.
4 Accurate analysis is thus obtained by use of the undeformed geometry (neglecting second-order deformation effects).

Therefore, the shear (V) and bending-moment (M) diagrams can be initially drawn for each load type followed by summation into final V and M diagrams. Figure 5.10 presents a simple beam under the action of both uniform and concentrated loads; the load, V, and M diagrams are determined using the method of superposition of separate diagrams associated with each load type. Figure 5.11 shows the application of the method of superposition to

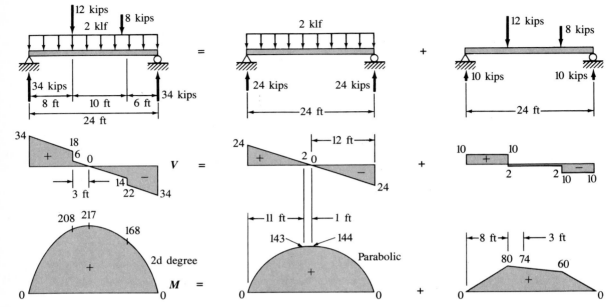

Figure 5.10

Figure 5.11

determine the V and M diagrams for a cantilever beam supporting two uniform loads. The moment diagrams for each separated load type in Figs. 5.10 and 5.11 have readily identified area properties whereas the area of the combined moment diagram is not easily determined. Figure 5.12 shows a fixed-fixed beam under a uniform load w with known fixed-end reactions. Separation of the vertical forces and the moment forces are shown to provide two simple-moment diagram areas that, by superposition, are equivalent to the final-moment diagram. In Chaps. 9 and 11, the method of superposition is used to develop separate moment diagrams of known areas that ease the task of solving displacements by geometric methods.

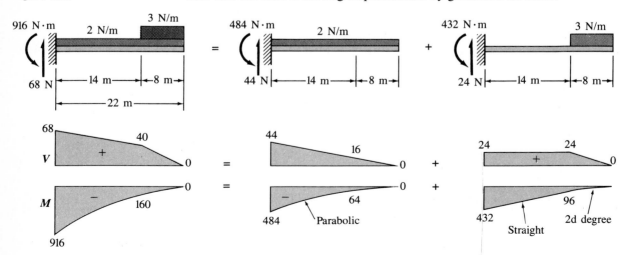

Figure 5.12

5.9. COMPUTER APPLICATIONS

Example Problem 6

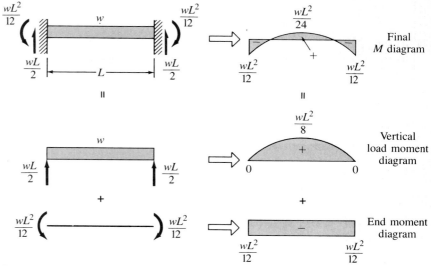

Figure 5.13

Node 3 Hinge support
Node 8 Roller support

Element	Nodes i	j
1	1	2
2	2	3
3	3	4
4	4	5
5	5	6
6	6	7
7	7	8

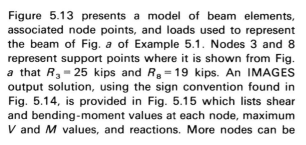

Figure 5.13 presents a model of beam elements, associated node points, and loads used to represent the beam of Fig. *a* of Example 5.1. Nodes 3 and 8 represent support points where it is shown from Fig. *a* that $R_3 = 25$ kips and $R_8 = 19$ kips. An IMAGES output solution, using the sign convention found in Fig. 5.14, is provided in Fig. 5.15 which lists shear and bending-moment values at each node, maximum V and M values, and reactions. More nodes can be generated between the existing nodes if intermediate V and M values are desired.

Figure 5.14

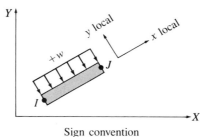

Sign convention

$w =$ Distributed load (force/unit length)

Note: Positive value of w results in a load in the negative local y direction

Figure 5.15 SOLVE BEAM LOADS AND STRESSES   Version 3.0   08/15/85

LOAD CASE 1

LOCAL LOADS

Element	Node	Axial	Shear	Bending
1	1	0.00000E+00	9.79426E-06	1.95885E-05
1	2	0.00000E+00	-9.79426E-06	1.95885E-05
2	2	0.00000E+00	-1.00000E+01	-3.50540E-06
2	3	0.00000E+00	1.00000E+01	-6.00000E+01
3	3	0.00000E+00	1.50000E+01	6.00000E+01
3	4	0.00000E+00	-1.50000E+01	-3.27216E-05
4	4	0.00000E+00	1.50000E+01	1.71661E-05
4	5	0.00000E+00	1.00001E+00	5.59999E+01
5	5	0.00000E+00	-1.00000E+00	-5.59999E+01
5	6	0.00000E+00	1.00000E+00	4.79999E+01
6	6	0.00000E+00	-1.00001E+00	-7.39999E+01
6	7	0.00000E+00	1.00001E+00	6.29999E+01
7	7	0.00000E+00	-1.00001E+00	-6.29998E+01
7	8	0.00000E+00	1.90000E+01	3.05176E-05

Maximum Load Summary

Beam	End	Comp.	Load
7	I	Axial	0.000E+00
4	I	Shear	1.500E+01
6	I	Bend.	-7.400E+01
7	J	Axial	0.000E+00
7	J	Shear	1.900E+01
6	J	Bend.	6.300E+01

RESTRAINT REACTIONS

Node	Direction	Reaction
3	Fx	0.00000E+00
3	Fy	2.50000E+01
8	Fy	1.90000E+01

Example Problem 7

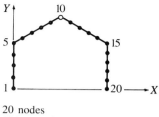

Figure 5.16a

20 nodes
19 beam elements

Figure 5.16a shows a computer model of the wind-loaded gable frame of Fig. a in Example 5.8 with numerous node points (20) selected for better definition of the V and M diagram distributions. Obviously, hinge restraints exist at the base of both vertical members and an interior hinge at node 10 joins the crown members. Local shears and moments found for each frame element by the IMAGES computer solution of Fig. 5.16b compare well with the hand-calculation results shown in Example 5.8, Fig. a. The nodes between 1 and 5, 5 and 10, 10 and 15, and 15 and 20 were established by use of the "Generate" command discussed in the tutorial instructions of the appendix.

The student is encouraged to attempt to reproduce the two computer solutions above by the IMAGES program to gain experience and confidence in its use. The ability to use IMAGES properly will provide a student with a means to verify hand generated V and M diagrams. However, the student should note that the mastery of V- and M-diagram construction is most important to the structural engineer and can be achieved only by practice through numerous hand solutions.

Figure 5.16b SOLVE BEAM LOADS AND STRESSES   Version 3.0   08/15/85

LOAD CASE 1

LOCAL LOADS

Element	Node	Axial	Shear	Bending
1	1	-9.75001E-01	-2.85001E+00	-1.09673E-05
1	2	9.75001E-01	1.85001E+00	-1.17500E+01
2	2	-9.75001E-01	-1.85002E+00	1.17500E+01
2	3	9.75001E-01	8.50019E-01	-1.85000E+01
3	3	-9.75001E-01	-8.49998E-01	1.85000E+01
3	4	9.75001E-01	-1.50003E-01	-2.02501E+01
4	4	-9.75001E-01	1.49989E-01	2.02501E+01
4	5	9.75001E-01	-1.14999E+00	-1.70001E+01
5	5	-1.46462E+00	-3.57775E-01	1.70001E+01
5	6	1.46462E+00	-8.94381E-02	-1.76000E+01
6	6	-1.46462E+00	8.94442E-02	1.76000E+01
6	7	1.46462E+00	-5.36658E-01	-1.62000E+01
7	7	-1.46462E+00	5.36657E-01	1.62000E+01
7	8	1.46462E+00	-9.83870E-01	-1.28000E+01
8	8	-1.46462E+00	9.83364E-01	1.28000E+01
8	9	1.46462E+00	-1.43108E+00	-7.39999E+00

Figure 5.16b (continued)

Element	Node	Axial	Shear	Bending
9	9	-1.46462E+00	1.43108E+00	7.40000E+00
9	10	1.46462E+00	-1.87829E+00	1.87755E-06
10	10	-2.38141E+00	-4.47201E-02	0.00000E+00
10	11	2.38141E+00	-1.02859E+00	2.19997E+00
11	11	-2.38140E+00	1.02859E+00	-2.19997E+00
11	12	2.38140E+00	-2.10190E+00	9.19994E+00
12	12	-2.38141E+00	2.10191E+00	-9.19995E+00
12	13	2.38141E+00	-3.17522E+00	2.09999E+01
13	13	-2.38141E+00	3.17522E+00	-2.09999E+01
13	14	2.38141E+00	-4.24853E+00	3.75999E+01
14	14	-2.38141E+00	4.24853E+00	-3.75999E+01
14	15	2.38141E+00	-5.32184E+00	5.89999E+01
15	15	-5.82500E+00	2.49999E-01	-5.89999E+01
15	16	5.82500E+00	1.03000E+00	5.74399E+01
16	16	-5.82500E+00	-1.03000E+00	-5.74399E+01
16	17	5.82500E+00	2.31000E+00	5.07600E+01
17	17	-5.82500E+00	-2.30998E+00	-5.07599E+01
17	18	5.82500E+00	3.58998E+00	3.89600E+01
18	18	-5.82500E+00	-3.58999E+00	-3.89600E+01
18	19	5.82500E+00	4.86999E+00	2.20400E+01
19	19	-5.82500E+00	-4.87000E+00	-2.20400E+01
19	20	5.82500E+00	6.15000E+00	-1.50800E-05

Maximum Load Summary

Beam	End	Comp.	Load
19	I	Axial	-5.825E+00
19	I	Shear	-4.870E+00
15	I	Bend.	-5.900E+01
19	J	Axial	5.825E+00
19	J	Shear	6.150E+00
14	J	Bend.	5.900E+01

Figure 5.16b (continued)

```
          RESTRAINT REACTIONS

Node    Direction       Reaction

  1        Fx         2.85001E+00
  1        Fy        -9.75001E-01
 20        Fx         6.15000E+00
 20        Fy        -5.82500E+00
```

Problems

Draw V and M diagrams for each structure.

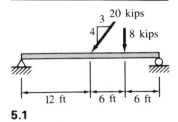

5.1

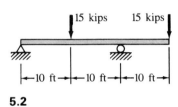

5.2

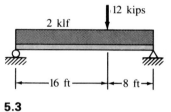

5.3

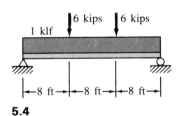

5.4

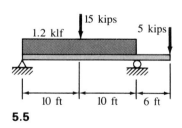

5.5

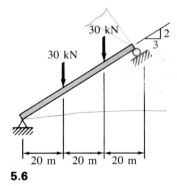

5.6

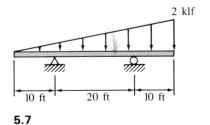

5.7

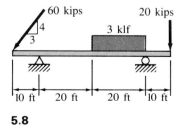

5.8

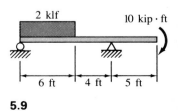

5.9

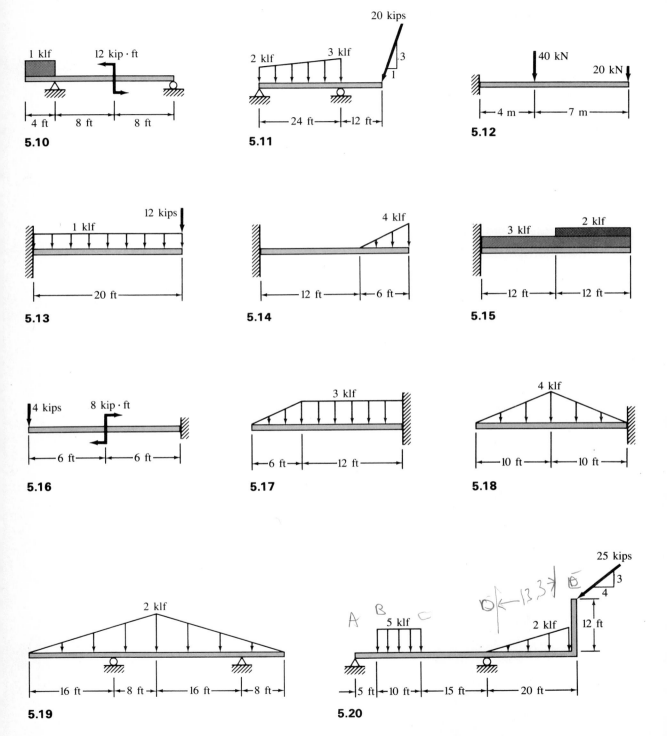

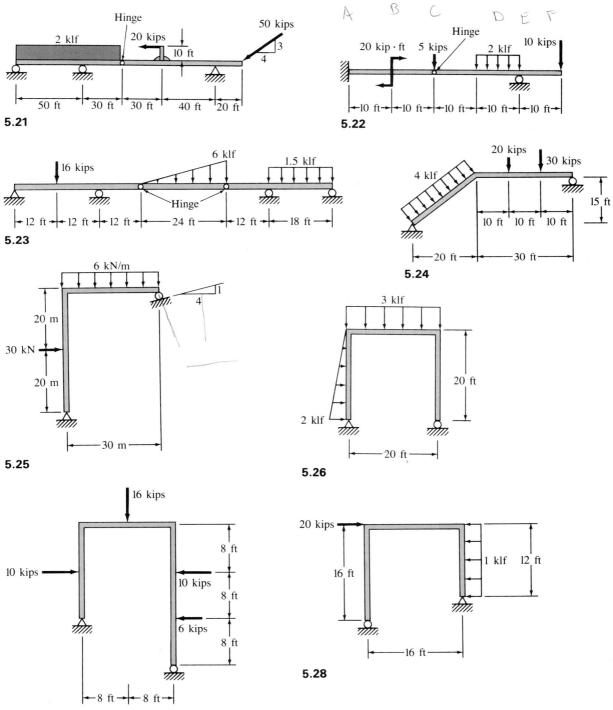

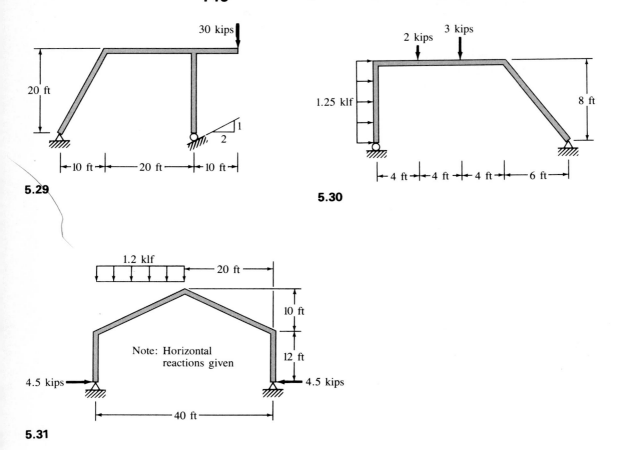

5.29

5.30

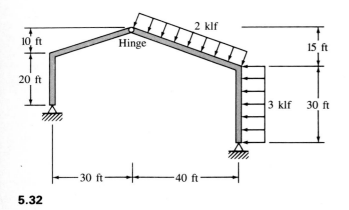

5.31

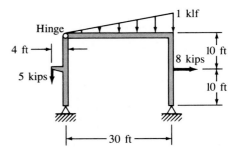

5.32

5.33

CHAPTER SIX

Influence Lines—Beams

6.1. THE INFLUENCE LINE—A DESIGN AID

6.1.1. Introduction   Civil engineering structures are subjected to various dead- and live-load conditions. The specific design of each member on a structure must consider various design-load combinations and establish where to position the movable live loads for maximum effect. An influence line is a design aid that can be used to study the structural effects of moving loads, and to decide where to position live loads for maximum effect.

Golden Gate bridge, San Francisco. Main span is 4200 ft long. Bridge completed in 1937. The principal designer was Joseph Strauss. (*Courtesy of International Structural Slides, Berkeley.*)

The live loads can be classified into two categories:

Category A includes natural live loads such as wind, snow, rain, and earthquake forces. The forces and data associated with these natural events are recorded at many geographical regions, and studied and statistically evaluated for design implementation. Therefore, category A includes prescribed design live loads that are recommended for specific design usage and based on probability studies (see Sec. 1.5.3). Influence lines are not developed for category A live loads.

Category B includes predictable moving live loads such as railway trains on a railroad bridge, trucks and automobiles on a highway bridge, or passenger chair lifts at a ski lodge. Minimum design live loads on floors of buildings represent another form of category B live loading (see Table 1.2). Consequently, the variable position of live loads on a bridge or a building-floor system can change axial member forces and cause variations in beam shear, bending moment, and displacements. The influence line is frequently used to study the effects of category B-type live loads.

6.1.2. The Influence Line

An influence line *is defined as a diagram which describes the variation of a design function as a concentrated live load of unit value moves across the structure. The design function may be the force in a truss member, shear or bending moment at a specific location on a beam, a beam support reaction, or the displacement at a specified point on a structure.* Therefore, *influence lines are visual aids* that help the designer to understand structural behavior under moving-load conditions. Since the influence line plots the variation of a design function due to a *unit load* moving across a structure, it represents a general curve that can be applied for any magnitude of elastic concentrated load.

All influence lines developed in this chapter are readily determined from the principles of statics and the ability to compute values of shear and bending moment at any point on a beam. Furthermore, influence lines for statically determinate beams will be shown to be linear (straight-line) functions.

6.2. BEAM SUPPORT REACTIONS

Mobile crane girders are used in mill and industrial buildings to lift and transport heavy objects within the building interior (see Fig. 6.1). As the lifting hoist moves its load along the girder, the magnitudes of the support reactions change. Consider a 100-ft-long simple beam which supports a motor driven hoist with a lift capacity of 1000 lb (1 kip) as shown in Fig. 6.2. The hoist is seen to be movable along the beam span. Figure 6.3 shows the value of the left support reaction (R_L), determined from statics, for various moving positions of the unit (1 kip) load along the beam. The diagram of Fig. 6.4a is obtained by plotting each value of the left reaction at the location associated with the unit-load position on the beam. Figure

Figure 6.1 Movable crane girder showing motor-driven lifting hoist. Photo taken by author.

6.4b is a general FBD from which it is evident, by use of statics, that $R_L = 1$ kip$(L - x)/L$ and $R_R = 1$ kip(x/L); that is, each reaction varies linearly due to the unit moving load. Therefore, all reaction values shown on Fig. 6.4a are joined by a straight line. Since Fig. 6.4a shows how the position of the unit load "influences" the value of the left reaction, Fig. 6.4a is called the influence line for the left reaction of the beam. In a similar manner, the influence line for the right reaction (R_R) can easily be determined with ordinate values as shown in Fig. 6.4c. The positive sense shown on the influence line of Fig. 6.4c indicates that the direction of R_R is upward. Since these reactions are the shears at the beam support ends, the influence lines of Figs. 6.4a and 6.4c are influence lines for shear. One can now continue to study influence lines for shear at interior beam locations.

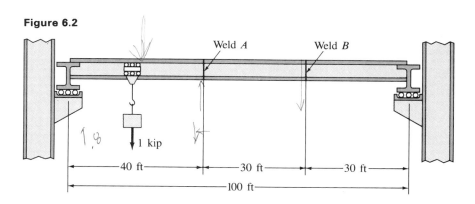

Figure 6.2

150 Influence Lines—Beams

Figure 6.3

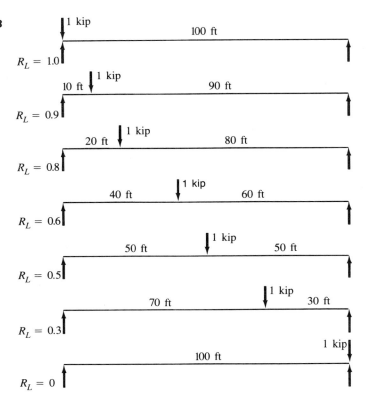

Figure 6.4

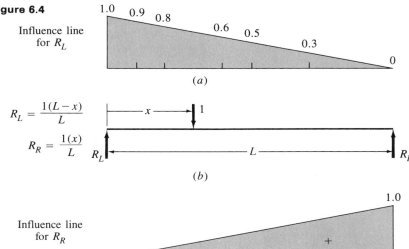

Influence line for R_L

(a)

$R_L = \dfrac{1(L-x)}{L}$

$R_R = \dfrac{1(x)}{L}$

(b)

Influence line for R_R

(c)

6.3. BEAM SHEARS

A study of Fig. 6.2 shows a 100-ft beam made by welding sections together at A and B. Since the welds must be able to transmit shear between joining sections, it is important to determine the shear value at each weld location as the hoist moves across the beam. Therefore, influence lines are to be developed for the shear at points A and B. Figure 6.5a is an FBD of the left section of the beam terminating at point A. The sign convention presented

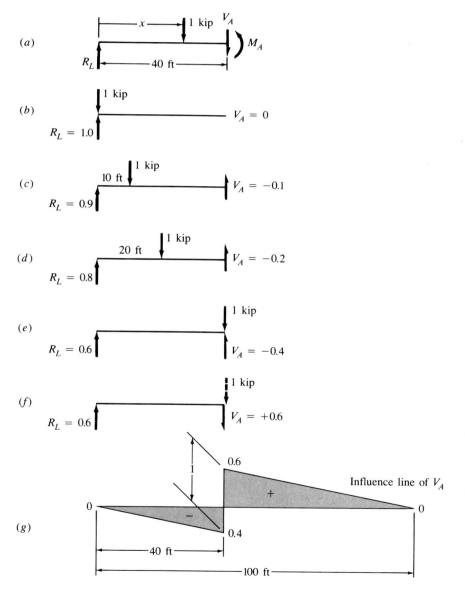

Figure 6.5

for shear in Chap. 5 is used for shear influence lines (i.e., shear at a section is defined as positive if the summation of normal forces from the left end of a beam is directed upward). Figures 6.5b to e show the shear value at A required to maintain vertical equilibrium for different positions of the unit load. The shear (V_A) shown in Fig. 6.5 is the equilibrant shear force at point A. For each position of the unit load shown in Figs. 6.5b to e, values of R_L can be obtained from the influence line of Fig. 6.4a and $V_A = (R_L - 1)$. If the unit load at A, as shown in Fig. 6.5e, moves an infinitesimal distance to the right of A, as shown in Fig. 6.5f, the left reaction remains virtually unchanged. However, the shear at point A required to maintain vertical equilibrium suddenly changes since the unit load is no longer on the beam section; thus, $V_A = R_L$. From Fig. 6.5f and the influence line of Fig. 6.4a, it

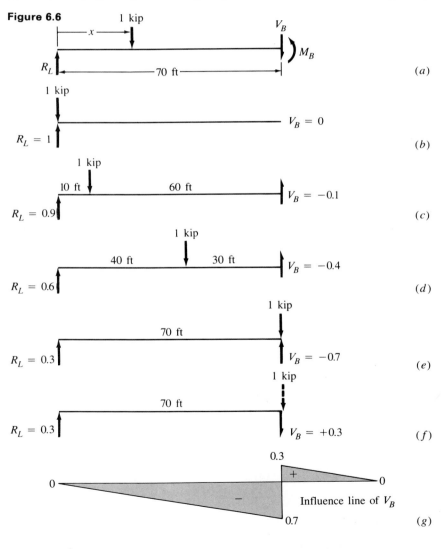

can be reasoned that the shear at point A will linearly decrease to a value of zero as the unit load continues to move toward the right support. The influence line for the shear at A is shown in Fig. 6.5g with a noted unit step change at point A. Positive and negative values shown in the influence line of Fig. 6.5g reflect the sign convention used for shear.

An FBD of the left section of the beam terminating at point B is shown in Fig. 6.6a. Again, influence line values of R_L are found from Fig. 6.4a. Figures 6.6b to e easily identify the value of shear at B required to maintain equilibrium for various positions of the unit load on the beam. Figure 6.6f shows the sudden change in shear value at B as the unit load moves an infinitesimal distance to the right of point B. The influence line of shear at point B is shown as Fig. 6.6g.

The diving board of Fig. 6.7a is readily identified as a simple beam with an overhang. The moving position of a diver from the supported end to the free end of the board must be studied to ensure a safe and functional design. Let us consider that the shear at point A and the support reactions are parameters of interest. Note again that the shear at a beam section is the summation of all normal forces at the section measured from either end of the beam. Therefore, the influence line for shear at $A(V_A)$ can be developed

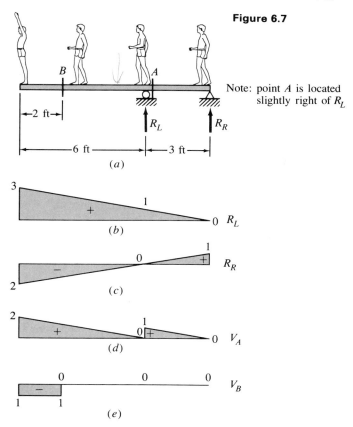

Figure 6.7

easily *after* the influence line for either reaction is known. Figures 6.7b to d show the influence lines of R_L, R_R, and V_A, respectively. In addition, the influence line for shear at point $B(V_B)$ is given in Fig. 6.7e.

Consider the statically determinate continuous beam of Fig. 6.8a which has an interior hinge at C. As before, by the use of static equilibrium equations, a student will be able to determine each reaction value associated with the placement of a unit (1 kip) concentrated load anywhere on the beam. Likewise, with known reactions for each particular position of the unit load, shear at a specific location on the beam can be found. Repetition of these steps for unit load placement at various positions across the beam provides the needed data to draw the influence lines. Figures 6.8b to e show the influence lines for the beam reactions R_D, R_B, and R_A, and V_C—the shear at hinge C. All the influence lines drawn thus far are called quantitative influence lines since numerical values are determined at pertinent locations along the beam length. Qualitative influence lines, which provide the shape of an influence line without ordinate values, will be studied in Sec. 6.5 of this chapter and in Chap. 11. Before we continue our studies of influence lines, let us reflect upon the following comparison:

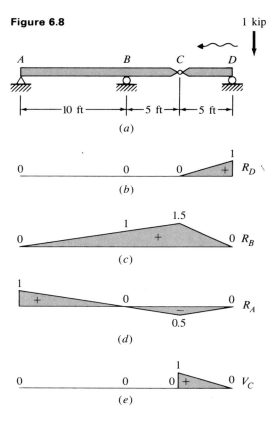

Figure 6.8

For a system of loads, fixed in position on a beam, a unique set of reactions are found. Also, the shear and bending-moment diagrams can be drawn which provide specific V and M values at *all* locations along the beam structure.

In contrast, when an influence line is to be drawn, the load is not fixed in position, but is a moving concentrated unit load; moreover, the function to be drawn, such as reaction, shear, or bending moment, is determined at a *specific location* on the structure and may vary in magnitude as the unit load moves from one position to another.

6.4. BEAM MOMENTS

Since beams are generally expected to support transverse loads and behave as flexural support members, their shape selections are usually controlled by the bending stress. Therefore, the influence line for beam bending moment is particularly helpful for beams subjected to moving design loads. The moment sign convention from Chap. 5 (which states that positive moment produces compression above and tension below the beam neutral axis, respectively) will be used to develop moment influence lines.

Consider again, the 100-ft beam of Fig. 6.2 where the moment influence lines at weld joints A and B are of interest. The moment influence line at midspan is also desirable since the midspan bending moment often proves to be the maximum moment for typical simple beam loads. For each location of the unit load on the beam, the left reaction (R_L) is known from Fig.

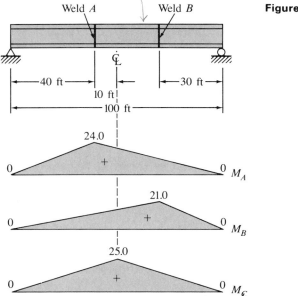

Figure 6.9

Figure 6.10

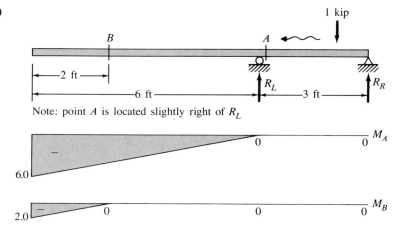

6.4a. From the FBD of Fig. 6.5a, a value of M_A can be found for each position of the unit load by use of $\Sigma M_A = 0$. For example, when the unit kip load is at weld A, $R_L = 0.6$ kip and $M_A = 0.6$ kip(40 ft) = 24.0 kip·ft. When the unit load is 10 ft from the left end (see Fig. 6.5c), $M_A = 0.9$ kip(40 ft) $- 1$ kip(30 ft) = 6.0 kip·ft. Thus, the influence line for M_A can be easily found. Similarly, influence lines for moment at B and at midspan can be determined and are presented in Fig. 6.9.

The influence lines for bending moment at points A and B on the diving board of Fig. 6.7 are presented in Fig. 6.10 and are easily determined from statics as demonstrated before.

Influence lines for the bending moment at points B and E of the beam in Fig. 6.11a are presented in Figs. 6.11b to c. The influence lines of the beam reactions available in Fig. 6.8 are used to determine the moment influence lines. For example, when the unit load is at E, R_A can be found equal to 0.60 from Fig. 6.8d, and $M_E = 0.60$ kip(4 ft) = 2.4 kip·ft.

Reviewing our study of influence lines for statically determinate beams

Figure 6.11

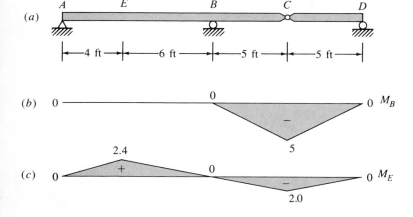

thus far, suggests the following procedure:

1. Establish a parameter of interest at a *specific location* on the beam (e.g., R_L, V_A, and M_B).
2. Using principles of statics, determine the quantitative value of the parameter for a position of the unit load on the beam.
3. Move the unit load to another position on the beam and recompute the parameter of interest. This step may be repeated for numerous positions of the unit load on the beam. Consider placing the unit load immediately before and after the location of interest for possible step changes.
4. Establish a horizontal axis to represent the beam length on which an influence line is to be plotted.
5. At each location where the unit load is placed on the beam, plot the corresponding parameter value found in 2 and 3 above.
6. Construct the influence line of the parameter by drawing straight lines between appropriate values associated with the motion of the unit load.

6.5. QUALITATIVE INFLUENCE LINES FOR BEAMS

A qualitative influence line is easy to draw and provides the student with an independent approach to verify its respective quantitative influence line. This is beneficial since students often experience difficulty in the construction of quantitative influence lines. This difficulty may arise partly due to a lack of clarity to distinguish between the graphical presentation of V and M diagrams, which represent V and M values at all points on a beam for fixed loads, and influence lines, which focus on a function at one point in the beam for a moving unit load.

In 1886, Professor Heinrich Müller-Breslau presented a simple approach to determine the correct shape of influence lines for functions such as reactions, shears, and moments in beams. This approach is called the *Müller-Breslau principle*. The principle may be stated as follows:

> *If a function at a point on a beam, such as reaction, or shear, or moment, is allowed to act without restraint, the deflected shape of the beam, to some scale, represents the influence line of the function.*

Although the emphasis in this chapter is on beams, the principle can be applied to linear elastic trusses and frames and for statically indeterminate structures (see Chap. 11).

Examples In Fig. 6.12*a*, the left reaction restraint is removed and replaced with a guided roller to permit the R_L function to displace the beam in the direction of the reaction as shown. The guided roller can resist moment and axial load, but not shear. If the motion at R_L is of unit value, the resulting

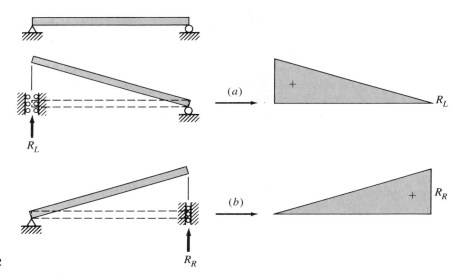

Figure 6.12

deflected beam shape, shown in Fig. 6.12a, proves to be the quantitative influence line of R_L. Figure 6.12b provides a similar action for the right reaction (R_R).

In Figure 6.13, the influence line for shear at point x is desired. Therefore, an assumed system of plates and rollers is placed at beam-point x in order to remove shear restraint only; the shape of the influence line for shear at x is developed by a unit displacement resulting from a set of opposite shear forces applied at x.

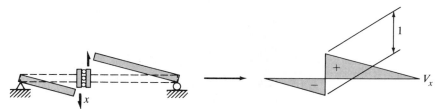

Figure 6.13

Figures 6.14 and 6.15 present qualitative influence lines for beam moments. In Fig. 6.14, an interior hinge is placed at point C; then, a positive moment is applied as shown. The ensuing deformation draws the shape of the desired influence line for moment at C since the fictitious hinge removes moment restraint at C and permits the function M_c (drawn positively) to displace the beam as shown. In a similar manner, qualitative influence lines for moment at points E and B are shown in Figs. 6.15b

Figure 6.14

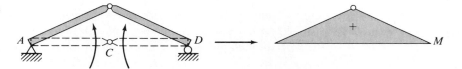

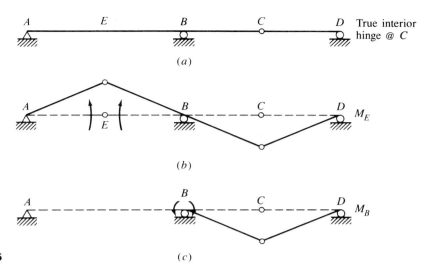

Figure 6.15

through 6.15c. In Fig. 6.15c, the freedom for M_B to rotate at B is provided but the vertical support restraint is maintained at A and B so that the left section does not displace (the curvature between A and B is neglected). However, a moment applied at B causes point B to rotate freely between B and D, since rotational freedom exists at C and D, and C is also free to displace. Figures 6.14 and 6.15 clearly show the resemblance to the influence lines drawn in Figs. 6.9 and 6.11, respectively.

Figure 6.16 considers the beam of Fig. 6.10. By introducing a fictitious interior hinge at point A, the left section is free to rotate as shown when a moment is applied at A. The displaced shape that results from the free rotation at A is comparable to the influence line shape of Fig. 6.10. The same approach can be used to find the influence-line shape for the moment at point B.

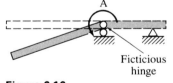

Figure 6.16

Computer application The student may choose to attempt to use the IMAGES-2D program to establish qualitative influence lines of beam functions. For example, the influence lines for the reactions of Fig. 6.15a can be obtained as follows: After the Geometry Menu data is defined in IMAGES, enter the Static Menu and press 1, Create/Edit Loads. Then, press 4, Define Enforced Displacements. By entering a nominal displacement in the direction of a vertical beam reaction at its position on the beam, the plot of the deflected shape will reflect the qualitative influence line. If the plotted shape is not clearly visible, retry the plot using a scale factor of 5, 10, 20, or greater.

6.6. BRIDGE DECK AND BUILDING FLOOR SYSTEMS

In addition to studying isolated beams, it must be realized that live loads typically are applied to systems of beams and girders that comprise either bridge decks or building floor systems. In order to facilitate the task of developing the influence lines for these floor members, it is appropriate to first study how the moving loads are transferred and supported in each system.

Figures 6.17a and b show a simple arrangement typically found in bridge decks that is comprised of a deck slab, stringers, floor beams, and girders.

Figure 6.17

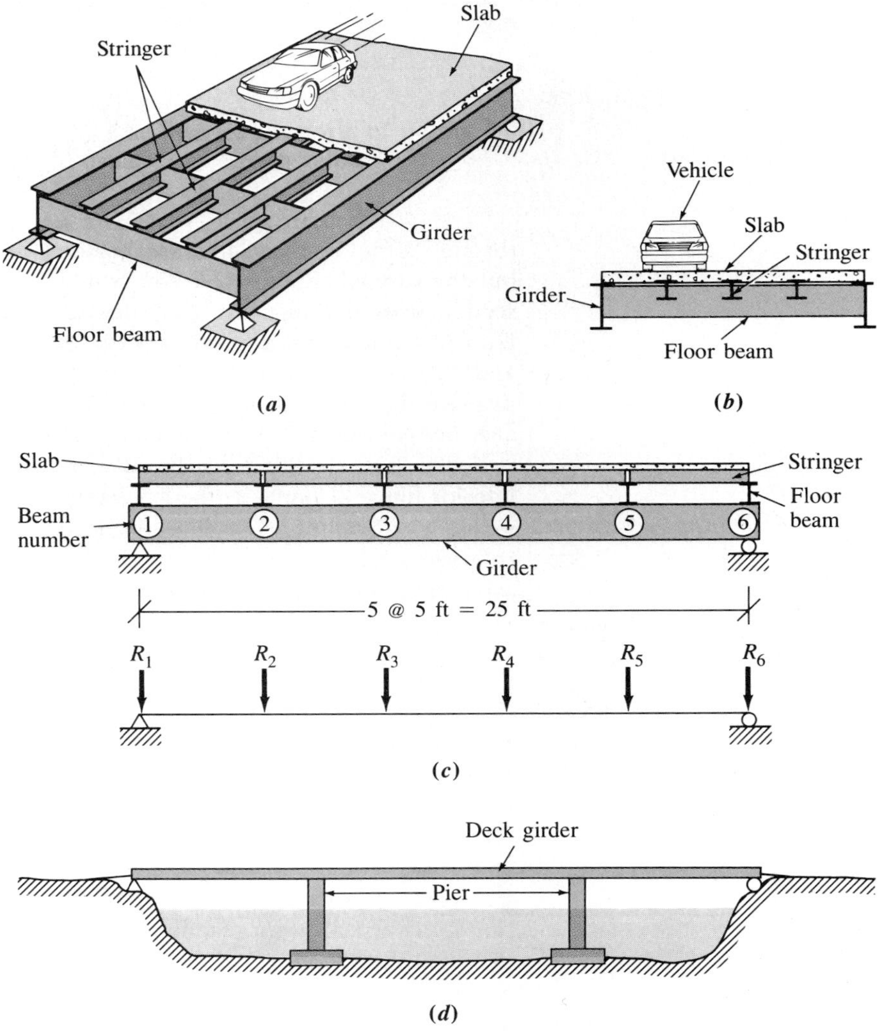

The load path is as follows:

1 The vehicle weight rests on the concrete slab which is supported by longitudinal beam members known as *stringers*.

2 Each stringer is assumed to support a proportional share of the concrete slab weight. In addition, the concentrated wheel loads of a vehicle, at any location on the deck, are "beamed out" (transferred) to adjacent stringers for support.

3 The stringers are assumed to be pin-end connected to floor beams. Thus, the stringer reactions act as concentrated forces on the floor beams. When two stringers are attached at opposite sides of a common floor-beam joint, both stringer reactions are summed to form a resultant concentrated force on the beam.

4 The floor beam end reactions, in turn, act as concentrated loads on each of its attaching girders (see schematic on Fig. 6.17c). Each girder may support several floor beam loads and transmit them to either ground supports or bridge piers (see Fig. 6.17d). All member joints are usually assumed as simple support connections.

Aside: In general, girders, stringers, and floor beams alike, basically behave as flexural members. The name *girder* is usually given to those flexural members that support many beams and represent a primary load path for transmitting loads directly to main structural supports.

Figure 6.18a shows a typical building schematic displaying a floor system composed of slab, beams, girders, columns, and walls. Exterior-wall flexural members are either referred to as *spandrels*, or spandrel beams, or girders. Note that all girders are directly supported by vertical building columns whereas beams are usually attached at girder locations beneath which no direct support exists. Figure 6.18b shows a common floor framing plan and Fig. 6.18c displays a schematic model that can be used to determine influence lines for a typical girder. Similar to bridge decks, loads on building floors are transmitted from the flooring to beams, from beams to girders, from girders to columns, and from columns to the foundation supports.

6.6.1. Influence Lines for Bridge Deck Girders

Reactions The schematics of Fig. 6.17c show a deck girder with six floor beams spaced 5 ft c–c. R_1 through R_6 represent the six beam reactions on the girder that result from dead and live loads on the slab. In order to develop the influence line for the left support reaction of the girder, we place a unit concentrated live load on the slab; then, it is apparent that either four or five of the reactions are zero depending on the location of the unit load. For example, when the unit load is directly over beam 1, $R_1 = 1.0$ and all other beam reactions are zero. When the unit load is applied to the stringer halfway between beams 1 and 2, $R_1 = R_2 = \frac{1}{2}$, and the remaining four beam reactions are zero. Similar results occur when the unit load is directly over or between the remaining beams.

162 Influence Lines—Beams

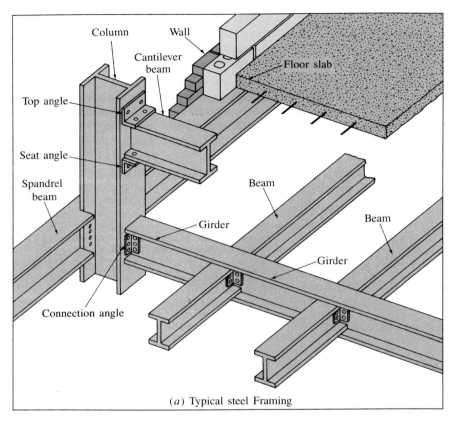

(*a*) Typical steel Framing

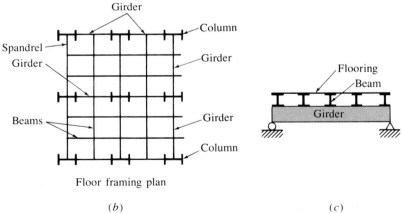

Floor framing plan

Figure 6.18   (*b*)   (*c*)

The influence line for the left reaction of the girder is determined in a manner similar to that for simple beams, namely, by use of the equation $\Sigma M = 0$ about the right support. In Figure 6.19a, the upper section shows the unit load placed directly over and reacted by beam 1; the lower beam schematic shows the beam 1 reaction on the girder from which the left girder reaction, R_L, is equal to 1.0 from statics. In Fig. 6.19b, the unit load is located at the midspan of stringer 1–2 so that $R_1 = R_2 = \frac{1}{2}$ and $R_L = 0.90$ from statics. In Fig. 6.19c, the unit load is directly over beam 2 and the left reaction is computed as shown to equal 0.80. Continuation of this simple procedure leads to Fig. 6.19d which is the influence line for R_L of the girder.

Panel shear The influence line for shear in the deck girder reveals two interesting features: (1) For any position of the unit load on the deck, the shear in any girder panel between floor beams is constant. This is true since the load path onto the girder is directly through the floor beams. (2) As the unit load moves along the deck, it causes gradual changes in the neighbor-

Figure 6.19

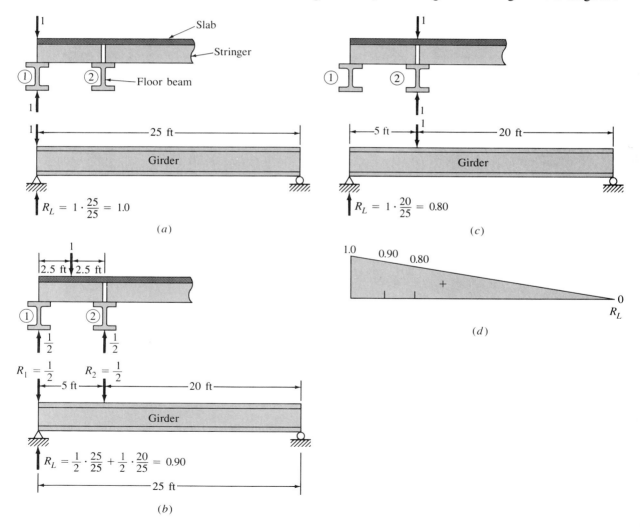

ing floor beam reactions such that the influence line for shear in any girder panel will change gradually and not exhibit the unit step change associated with beam shear. For example, Fig. 6.20a shows a section of the deck girder used to compute the influence line for the shear in panel 3–4. Figures 6.20b to g show the load path and resulting shear in panel 3–4 due to moving positions of the unit load. Observe that as the unit load moves to the right onto the stringer between beams 3 and 4, the beam reaction at 3 decreases. From Fig. 6.20f, the beam reaction at 3 is 0.50 when the load is equidistant from 3 and 4 and results in zero shear in panel 3–4. Figure 6.20h shows the complete influence line for shear in panel 3–4. In a similar manner, influence lines for shear in panels 1–2 and 2–3, as shown in Fig. 6.21, can be determined. A study of the influence lines of Figs. 6.20 and 6.21 reveal that the sense of the shear is constant in end girder panels but changes in interior panels.

Moment The influence line for moment at a point on a deck girder can also be established by principles of statics. Figure 6.22a shows a 7-ft length of section measured from the left end of the girder of Fig. 6.17c which is used to draw the influence line of the bending moment at point Z.

When the unit load (assume 1 kip) is over beam 1, $R_L = 1$ kip and $M_z = 0$. When the unit load is halfway between beams 1 and 2, the beam reactions at 1 and 2 on the girder are both equal to 0.5 kip and $R_L = 0.90$ kip; then $M_z = 0.90$ kip(7 ft) $- \frac{1}{2}$ kip(7 ft) $- \frac{1}{2}$ kip(2 ft) = 1.8 kip·ft. When the unit load is over beam 2, $R_L = 0.80$ kip, the beam reaction at 2 is 1.0 kip, and $M_z = 0.80$ kip(7 ft) $- 1.0$ kip(2 ft) = 3.6 kip·ft. When the unit load is directly over point Z, $R_L = 18/25$ kip, the beam reaction at point 2 is 0.60 kip, and $M_z = 18/25$ kips(7 ft) $- 0.60$ kip(2 ft) = 3.84 kip·ft. When the unit load is over beam 3, $R_L = 0.60$ kip, the beam reaction at 2 is zero, and $M_z = 0.60$ kip(7 ft) = 4.2 kip·ft.

The complete influence line for moment at point Z on the girder is shown in Fig. 6.22b.

6.6.2. Influence Lines for Building Girders

Influence lines for girders in building floor systems resemble those found for bridge decks and are determined by using the same principles of statics.

6.7. USE OF INFLUENCE LINES FOR CONCENTRATED LOADS

Once the ability to construct influence lines is established, a student can use the influence line of a design parameter to determine its maximum value for various load conditions. Since the ordinate values on any influence line (y) are based on a moving concentrated load of unit magnitude, the parameter value for any concentrated load (P) at a point on the beam is equal to $P \cdot y$. For example, the influence line for the left reaction of the crane girder shown in Fig. 6.2 was developed in Fig. 6.3 and presented in Fig. 6.4a. If the

165 Use of Influence Lines for Concentrated Loads

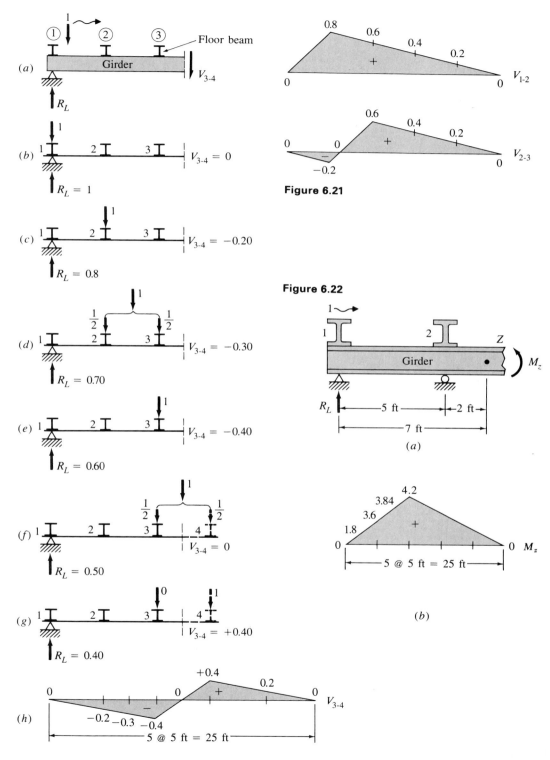

Figure 6.20

Figure 6.21

Figure 6.22

hoist supports a load of 800 lb at 20 ft from the left end of the beam, $R_L = P \cdot y = 800(0.8) = 640$ lb. If the hoist is moved to 40 ft from the left end, $R_L = P \cdot y = 800(0.6) = 480$ lb.

Using the influence lines of Fig. 6.7 If a 150-lb diver is at the left end of the board:

$$R_L = 150(3) = 450 \text{ lb} \qquad R_R = 150(-2) = -300 \text{ lb}$$

and $\qquad V_A = 150(2) = 300$ lb

If a 100-lb diver steps onto the board at the right end, with the first diver still on the left end:

$$R_L = 150(3) + 100(0) = 450 \text{ lb}$$
$$R_R = 150(-2) + 100(1) = -300 + 100 = -200 \text{ lb}$$

and $\qquad V_A = 150(2) + 100(0) = 300$ lb

Now consider that the 100-lb diver moves to point A on the board while the 150-lb diver is still at the left end. With both divers in place on the board:

$$R_L = 150(3) + 100(1) = 450 + 100 = 550 \text{ lb}$$
$$R_R = 150(-2) + 100(0) = -300 \text{ lb}$$

and $\qquad V_A = 150(2) + 100(1) = 300 + 100 = 400$ lb

It should also be observed that if the 100-lb diver moves slightly to the left of R_L, both R_L and R_R remain unchanged whereas the magnitude of V_A will suddenly change to 300 lb.

Assume that the 100-ft girder of Fig. 6.9 is used as part of a highway deck system to support the moving concentrated loads of an HS 20–44 semitrailer shown in Fig. 1.23. Let the distance between the two rear axles of the semitrailer (V) be 20 ft. Figure 6.23 shows the girder with the middle axle of the semitrailer located over weld A. Using the influence lines to compute the moment values at the two welds, we get

$$M_A = 8(15.6) + 32(24) + 32(16) \approx 1405 \text{ kip} \cdot \text{ft}$$

and $\qquad M_B = 8(7.8) + 32(12) + 32(18) \approx 1022$ kip·ft

6.8. USE OF INFLUENCE LINES FOR UNIFORM LOADS

Since uniform distributed live loads are common to highway, railway, and building design code specifications, it is also beneficial to establish how to use influence lines for these loads.

167 Use of Influence Lines for Uniform Loads

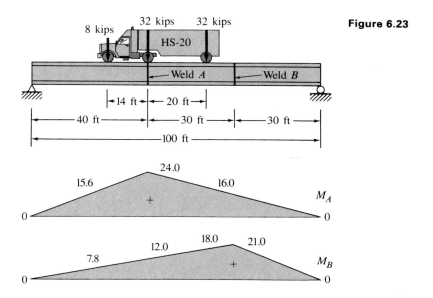

Figure 6.23

Let the function $F(x)$ on a typical xy plot shown in Fig. 6.24 represent the influence line of a design function of a beam, where x is measured along the beam axis, the ordinate value of the design function is designated by y, and w represents a uniformly distributed load acting on the beam. At a distance x (measured from the left end of the beam), integration of the uniform distributed load over a differential length dx is equal to $w(dx)$ and is equivalent to a local concentrated load. Therefore, a uniform distributed loading can be viewed as a series of closely spaced concentrated loads spaced dx apart, with each load equal in intensity to $w(dx)$. For each local concentrated load equivalent, the design function, $F(x) = w(dx) \cdot y$. The value of the function for the entire uniform load is found by integration to be

$$F(x) = \int_1^2 w(dx)y = w \int_1^2 y(dx)$$

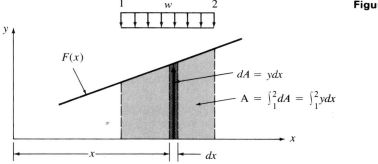

Figure 6.24

where $\int_1^2 y(dx)$ is the area of the influence line over the range from 1 to 2 of the uniform load on the beam.

Therefore, the value of a function of a beam subjected to uniform distributed loading is equal to the product of the area under the influence line function where the load acts, multiplied by the intensity of the uniform load.

EXAMPLE 6.1

The figure shows a 12-ft simple beam supporting a uniform distributed load of 2 klf. It is easily verified from statics that the bending moment at midspan is $wL^2/8 = 36$ kip·ft. The influence line for moment at midspan is given in Fig. b which is used to show that the moment due to the uniform load over the entire beam is equal to 36 kip·ft.

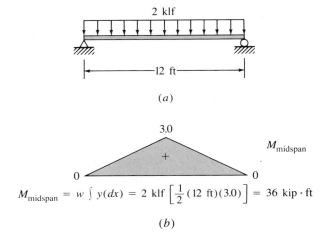

$$M_{midspan} = w \int y(dx) = 2 \text{ klf} \left[\frac{1}{2}(12 \text{ ft})(3.0)\right] = 36 \text{ kip}\cdot\text{ft}$$

(b)

6.9. APPLICATIONS FOR DESIGN-LOAD COMBINATIONS

The examples presented herein demonstrate applications dealing with design live loads only. However, when using influence lines to study structural behavior, the student is reminded of two facts: (1) the locations of the dead loads are fixed in position and cannot be moved about at the option of the analyst, and (2) impact factors, such as those associated with auto wheels bouncing along a highway (see Sec. 1.5.10) or the dynamics of railroad locomotive motion on a railway bridge, must be applied to that portion of a function affected by live-load motion.

6.9.1. Railroad-Bridge Live Loads The Cooper loadings of Fig. 1.25 is a moving load system that includes nine concentrated wheel loads for each of two locomotive engines plus a uniformly distributed loading to represent the trailing freight and passenger cars. A Cooper train-load system can proceed to cross a railroad structure from either end and assume many possible positions along its path. Consequently, many load cases (with many axle loads) may be required to determine how each affects one or more design functions. Therefore, the rationale for using uniform distributed loading in place of numerous passenger and freight-car axle loads is to minimize the tedium related to dealing with too many concentrated loads.

EXAMPLE 6.2

Railroad Deck Girder under Cooper Loading

The influence line for the shear in panel 2–3 of an 80-ft railway deck girder is given here. The girder is made by welding two sections together at joint A and supports five floor beams equally spaced 20 ft apart. The girder is designed to support an E–70 Cooper loading moving in the direction shown.

Part A: By use of the influence line given, determine the shear at joint A when the 36-kip axle of engine no. 1 is over the left support.

Part B: Reevaluate the shear at A at the instant when the 36-kip axle of engine no. 2 is over the left support.

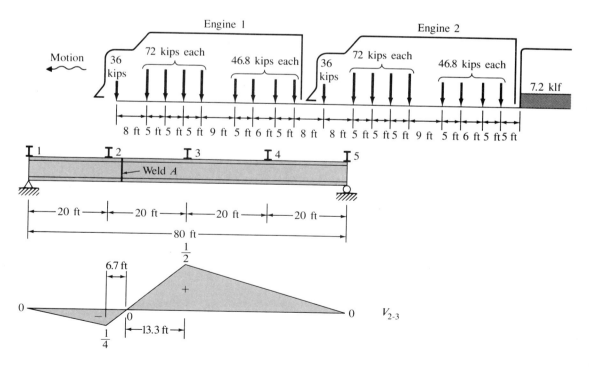

Solution

Part A

$$V_A = V_{2-3} = 36 \text{ kips}(0) + 72 \text{ kips}\left[\frac{8}{20} + \frac{13}{20} + \frac{18}{20}\right]\left(-\frac{1}{4}\right) + 72 \text{ kips}\left[\frac{3.7}{6.7}\right]\left(-\frac{1}{4}\right)$$
$$+ 46.8 \text{ kips}\left[\frac{5.3}{13.3} + \frac{10.3}{13.3}\right]\left(+\frac{1}{2}\right) + 46.8 \text{ kips}\left[\frac{37}{40} + \frac{32}{40}\right]\left(+\frac{1}{2}\right)$$
$$+ 36 \text{ kips}\left[\frac{24}{40}\right]\left(+\frac{1}{2}\right) + 72 \text{ kips}\left[\frac{16}{40} + \frac{11}{40} + \frac{6}{40} + \frac{1}{40}\right]\left(+\frac{1}{2}\right)\right]$$

$\}$ Engine no. 1

$\}$ Engine no. 2

$V_A = V_{2-3} = 0 - 35.1 - 9.9 + 27.4 + 40.3 + 10.8 + 30.6 \cong +64 \text{ kips}$

(Note: Only a portion of engine no. 2 is on the member.)

Part B

$$V_A = V_{2-3} = 36 \text{ kips}(0) + 72 \text{ kips}\left[\frac{8}{20} + \frac{13}{20} + \frac{18}{20}\right]\left(-\frac{1}{4}\right) + 72 \text{ kips}\left[\frac{3.7}{6.7}\right]\left(-\frac{1}{4}\right)$$
$$+ 46.8 \text{ kips}\left[\frac{5.3}{13.3} + \frac{10.3}{13.3}\right]\left(+\frac{1}{2}\right) + 46.8 \text{ kips}\left[\frac{37}{40} + \frac{32}{40}\right]\left(+\frac{1}{2}\right)$$
$$+ 7.2 \text{ klf}\left[\frac{1}{2}\left(\frac{27}{40}\right)\left(+\frac{1}{2}\right)(27 \text{ ft})\right]$$

$\}$ Engine no. 2

$\}$ Uniform load

$V_A = V_{2-3} = 0 - 35.1 - 9.9 + 27.4 + 40.3 + 32.8 \cong +56 \text{ kips}$

6.9.2. Highway-Bridge Live Loads

Currently, the American Association of State Highway and Transportation Officials (AASHTO) requires that primary highway bridges be designed to support (in each lane) either the HS20–44 truck load or the uniform lane loading described in Sec. 1.5.8. The 100-ft beam of Fig. 6.23 uses the HS20–44 truck loading to determine the moments at welds A and B with the variable spacing of the rear axles chosen as 20 ft. However, the student is reminded that the variable spacing range is from 14 ft to 30 ft and should be selected to produce maximum stresses.

EXAMPLE 6.3

Determine the moments at welds A and B of the 100-ft deck girder shown resulting from the AASHTO uniform-lane design loading (see Sec. 1.5.8); use the available

171 Applications for Design-Load Combinations

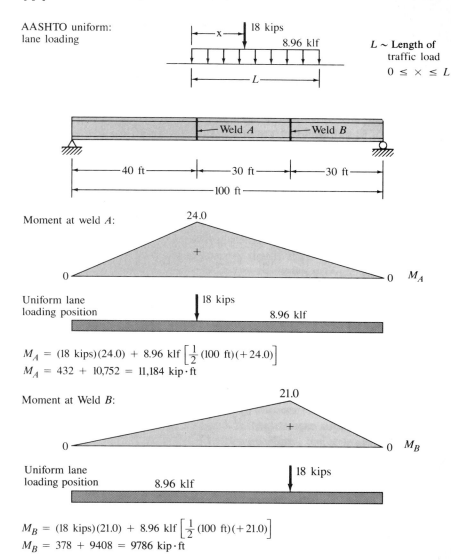

$M_A = (18 \text{ kips})(24.0) + 8.96 \text{ klf} \left[\frac{1}{2}(100 \text{ ft})(+24.0)\right]$
$M_A = 432 + 10{,}752 = 11{,}184 \text{ kip} \cdot \text{ft}$

$M_B = (18 \text{ kips})(21.0) + 8.96 \text{ klf} \left[\frac{1}{2}(100 \text{ ft})(+21.0)\right]$
$M_B = 378 + 9408 = 9786 \text{ kip} \cdot \text{ft}$

bending-moment influence lines from Fig. 6.23. The member is assumed to support one 14-ft lane load; that is, it must support a uniform load of 0.64 klf/ft(14 ft) = 8.96 klf. For moment, an added concentrated load of 18 kips is also included to account for an extra heavy truck somewhere in the assumed line of traffic. For maximum effect, the 18-kip concentrated load is placed to produce maximum moment. The results are shown above.

172 Influence Lines—Beams

6.10. MOVING-LOAD PLACEMENT FOR MAXIMUM EFFECT

6.10.1. Singular Concentrated Load

Figure 6.25 shows four familiar influence line shapes for simply supported members. If only one concentrated moving load (P) is to be placed on the members, positioning the load at the maximum positive ordinate of each respective influence line (see arrow designations in Fig. 6.25) will affect maximum positive values of reaction, shear, or moment. Placing the load at the maximum negative ordinate of Fig. 6.25b and 6.25d will yield maximum negative shear. Locating a cluster of closely spaced concentrated loads in the vicinity of the peak amplitude of the influence line will approximate the maximum function value. The actual location of the cluster loading to produce a maximum effect, called the *critical position*, can be accomplished by trial and error. Section 6.10.3 offers a method which uses influence lines to determine the critical position of a series of fixed-spaced concentrated moving loads.

Figure 6.25

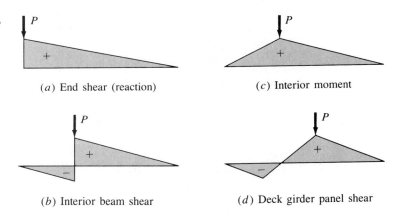

(*a*) End shear (reaction)

(*b*) Interior beam shear

(*c*) Interior moment

(*d*) Deck girder panel shear

6.10.2. Uniform Distributed Load

With reference to the influence lines of Fig. 6.25, maximum function values can easily be obtained for uniformly distributed live loads. As in the following examples:

1. A maximum positive end reaction will occur by placing the uniform load over the entire beam length of Fig. 6.25a.
2. Maximum positive beam shear or panel shear is found by placing the uniform load over the positive regions only for the influence lines of Figs. 6.25b and 6.25d, respectively; vice versa for negative shear.
3. Maximum positive interior moment will occur by placing the uniform load over the entire beam length of Fig. 6.25c.

6.10.3. Series of Concentrated Loads—Shear

Increase-decrease method Railroad and highway design loads include the motion of a series of fixed-spaced concentrated axle loads. Since the distance moved from one position to another is the same for all axles, a

173 Moving-Load Placement for Maximum Effect

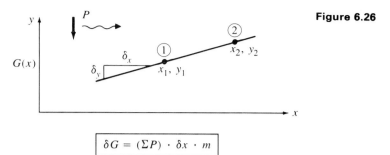

Figure 6.26

$$\delta G = (\Sigma P) \cdot \delta x \cdot m$$

ΣP Sum of moving loads on the member
$\delta x = (x_2 - x_1)$
$m = \delta_y/\delta_x$
$\delta G = (G_2 - G_1)$

systematic approach is available to find the critical position of a load series which will maximize a function. A segment of an influence line is given for a function, $G(x)$, in Fig. 6.26 where a moving concentrated load (P) is assumed to move from left to right.

When P is at point 1, $G(x) = G_1 = P \cdot y_1$

When P is at point 2, $G(x) = G_2 = P \cdot y_2$

The change in the function, $\delta G = P \cdot y_2 - P \cdot y_1 = P(y_2 - y_1)$

The slope of the influence line, $\delta y/\delta x = (y_2 - y_1)/(x_2 - x_1)$

Therefore, $\delta G = G_2 - G_1 = P \cdot (x_2 - x_1) \cdot \delta y/\delta x$ which is the product of three factors, namely, the load intensity, the distance moved, and the slope of the influence line.

Substituting δx for $(x_2 - x_1)$ and m for $\delta y/\delta x$, we can rewrite $\delta G = P \cdot \delta x \cdot m$. Furthermore, if instead of one load P, a series of concentrated loads, defined as ΣP, undergo the same motion along the member, it can be generalized that $\delta G = (\Sigma P) \cdot \delta x \cdot m$. When using this equation, the analyst should realize that during the live-load motion, ΣP may change as some of the series loads arrive onto the structure and others depart.

In general, for each position of the load series, δG is computed. If δG is found to be positive, the value of G is larger than it was for the last position, the largest value of G is yet to be found and the process is continued. However, if δG is found to be negative, it signifies that the last position is a critical position from which a maximum value of G can be found. This procedure is called the *increase-decrease method*.

EXAMPLE 6.4

The influence line of the left reaction of a simple beam is given. The beam is subjected to the right-to-left motion of a series of seven wheel loads of a multiaxle truck. By use of the influence line (IL) and the increase-decrease method, determine the maximum value of the left reaction.

Maximum Beam End Shear

End shear maxima due to a series of moving concentrated loads.

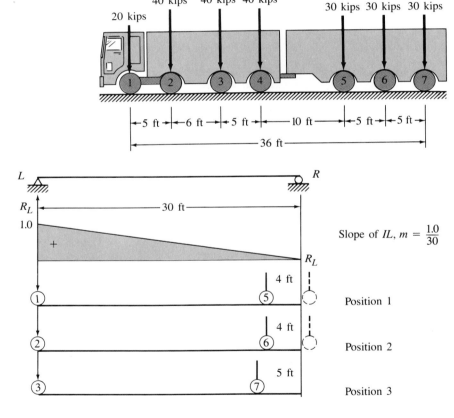

Remarks

1 Although the direction of vehicle motion is specified in this example, the effects on the structure due to vehicle motion from both directions should be studied.

2 It is logical to assume that placement of one of the heavier 40-kip wheels over the maximum ordinate of the IL will yield a maximum reaction value.

Three load positions are considered:

Position 1: Wheel 1 is over the left support.
Position 2: Wheel 2 is over the left support.
Position 3: Wheel 3 is over the left support.

When the loads are in any position, the function R_L is equal to the sum of the products of the loads multiplied by their respective ordinate value on the IL; that is, $R_L = \Sigma P_n \cdot y_n$. However, the maximum value is found by studying the relative change in R_L as the loads move from one position to another. Let the change in the end shear (R_L) from one position (a) to another position (b) be expressed as δV_{a-b}. Then, motion from position 1 to position 2 yields

$$\delta V_{1-2} = [40 + 40 + 40 + 30] \cdot 5 \text{ ft} \cdot (1/30) + 30 \cdot 4 \text{ ft} \cdot (1/30) - 20$$
$$= 25 + 4 - 20 = +9 \text{ kips} \ldots \text{increase}$$

Note: As the load series moves from position 1 to 2, the 20-kip load at the left support leaves the beam; this represents a *decrease* in the function of $20(-1) = -20$ as shown above.

$$\delta V_{2-3} = [40 + 40 + 30 + 30] \cdot 6 \text{ ft} \cdot (1/30) + 30 \cdot 5 \text{ ft} \cdot (1/30) - 40$$
$$= 33 - 40 = -7 \text{ kips} \ldots \text{decrease}$$

Since a decrease in the left reaction value occurs when the load series moves from position 2 to 3, the maximum value is found by placing the loads in position 2. Then,

$$\text{Maximum } R_L = \Sigma P_n \cdot y_n = 40 \text{ kips} \cdot [(30/30) + (24/30) + (19/30)]$$
$$+ 30 \text{ kips} \cdot [(9/30) + (4/30)] = 110.3 \text{ kips}$$

EXAMPLE 6.5

Maximum Beam Interior Shear

The seven axle semitrailer of Example 6.4 is to pass from right to left over the same 30-ft simple beam shown. Using the influence line in this example for shear at point P (located 10 ft from the left end), determine the maximum shear at P for the specified motion of the trailer.

Assume the maximum shear at P is positive and occurs when one of the wheel loads is located at the peak value on the positive region of the shear IL. Since more loads can occur over the positive region as compared to the negative region of the IL, the assumption seems logical but remains to be proven. In the calculations that follow, observe that during the initiation of motion from one position to another, the shear at point P suddenly decreases by the value of the load at P due to a unit step change (from $+2/3$ to $-1/3$).

$$\delta V_{1-2} = [20 + 40 + 40 + 40] \times 5 \text{ ft}(1/30) - 20 = 23.3 - 20 = +3.3 \text{ kips}$$

$$\delta V_{2-3} = 20 \times 5 \text{ ft}(1/30) + [40 + 40 + 40] \times 6 \text{ ft}(1/30)$$
$$+ 30 \times 5 \text{ ft}(1/30) - 40$$
$$= 3.3 + 24 + 5 - 40 = -7.7 \text{ kips}$$

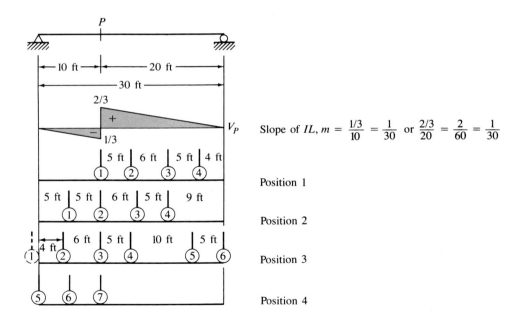

Since motion from position 2 to 3 results in a decrease of the shear function at P, the critical position is position 2 and the maximum positive shear at P is

$$\text{Maximum } V_P = \Sigma P_n \cdot y_n = 20[-(1/6)] + 40 \cdot (2/3) + 40(28/60)$$
$$+ 40(18/60) = -3.3 + 26.7 + 18.7 + 12$$
$$= +54.1 \text{ kips}$$

If the maximum negative shear at P is of interest, position 4 of Example 6.5 represents one possible critical load location since none of the wheels are on the positive portion of the IL to negate the contribution of the wheels at the leftmost negative region.

For load position 4

$$V_P = 30[0 - (1/6) - (1/3)] = -15 \text{ kips}$$

It remains for the student to verify if position no. 4 is the critical position for maximum negative shear at point P. It does appear that the magnitude of maximum positive shear at P is greater than the maximum negative shear at P. Although the increase-decrease method has been demonstrated using shear influence lines, the method is suitable for all types of ILs.

6.10.4. Series of Concentrated Loads—Bending Moment

Average load method Consider a series of fixed-spaced moving loads on a simple beam. The shape of the IL for the moment at an interior point x on a simple beam is always triangular with the peak ordinate occurring at x. The maximum moment at point x on the beam generally occurs when one of the large loads is at the peak ordinate of the moment IL and as many of the large loads are clustered around it (see Fig. 6.27a). When the loads are positioned on the beam, the moment at x is readily found equal to $\Sigma P_n \cdot y_n$ by use of the IL. A variation of the increase-decrease method that was developed specifically for triangular shaped ILs is known as the *average load method*. This method locates critical load positions by evaluating relative changes of the moment at x (M_x) due to unit left and right virtual motions of the load set at the IL peak. The moment change, δM_x, is equal to $\Sigma P_n \cdot \delta y_n$, where δy_n is the product of the unit distance moved multiplied by the slope of the IL. Both positive and negative changes of M_x may occur as the motion of the live loads appears to ride up and down the slopes of the IL on both sides of x. The analysis may proceed as follows:

1 Position the second load (P_2) of the series at the peak ordinate of the IL (see Fig. 6.27b).

2 For a virtual movement of the load series to the left, say 1 ft, a moment increase, $\delta Mi = (\Sigma P_r) \times 1 \text{ ft} \times (y/b)$ occurs, where ΣP_r is the load sum to the right of x that remains on the beam with a sense of moving up the IL (see Fig. 6.27c).

3 During the same 1-ft motion of the load series to the left, a moment decrease, $\delta Md = (\Sigma P_l) \times 1 \text{ ft} \times (y/a)$ occurs, where ΣP_l is the load sum remaining on the beam to the left of x with a sense of moving down the IL (see Fig. 6.27c). In this first load position, $\Sigma P_l = P_1 + P_2$.

A comparison of the moment increase on the right of x to the moment decrease on the left of x due to the unit motion of the load series yields

Figure 6.27

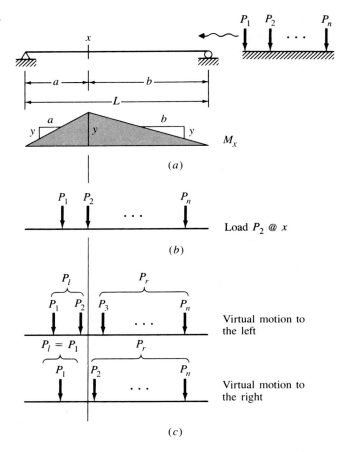

Moment decrease ≈ moment increase

$$(\Sigma P_l) \times 1 \times \frac{y}{a} \approx (\Sigma P_r) \times 1 \times \frac{y}{b}$$

Division of both terms by y results in $\Sigma P_l/a \approx \Sigma P_r/b$ which appears to compare the average load on the right side of the IL to the average load on the left side of the IL.

4 Now, move the load series a 1-ft distance to the right. This may result in a change to both ΣP_l and ΣP_r. Compare the average load on the left side to the average load on the right side.

5 For a given load position, if a virtual motion of the load series to the left side of x results in $\Sigma P_r/b < \Sigma P_l/a$ and a virtual motion to the right results in $\Sigma P_r/b > \Sigma P_l/a$, or vice versa, a critical position exists for which the maximum moment at x can be found.

6 In step 5 above, if the average load on the left > the average load on the right for *both* left and right virtual motions of the load series, a critical location has not been reached. The same is true if the average load on the left < the average load on the right for *both* left and right virtual motion of the load series.

7 If step 6 proves true, then advance the load series on the beam until the next load (P_3) is at the peak ordinate of the IL and repeat the procedure; continue to move the next load to the peak ordinate position until the rate of change is zero or the condition of step 5 exists.

The absolute maximum moment at x occurs at the load position which demonstrates that the average load on the left is equal to the average load on the right for both virtual motions of the load series. The following example provides a convenient tabular form to demonstrate the average load method.

EXAMPLE 6.6

Average Load Method—Beam Moment

The influence line for bending moment at point A on a 60-ft simple beam is given here. A series of nine fixed-spaced wheel loads are assumed to move from right to left as shown. Determine the maximum value of M_A.

Solution

For a long series of variable loads, there may be more than one critical position to be investigated as demonstrated in the tabular solution below.

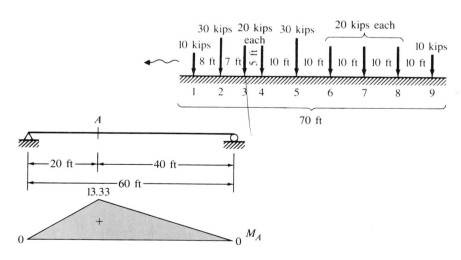

Wheel @ A	Virtual motion	Average load—left side $\Sigma P_l/a$	$><$	Average load—right side $\Sigma P_r/b$
2	Right	$\dfrac{10}{20} = 0.5$	$<$	$\dfrac{30+20+20+30+20}{40} = \dfrac{120}{40} = 3.0$
	Left	$\dfrac{10+30}{20} = \dfrac{40}{20} = 2.0$	$<$	$\dfrac{20+20+30+20}{40} = \dfrac{90}{40} = 2.25$
3*	Right	$\dfrac{10+30}{20} = \dfrac{40}{20} = 2.0$	$<$	$\dfrac{20+20+30+20+20}{40} = \dfrac{110}{40} = 2.75$
	Left	$\dfrac{10+30+20}{20} = \dfrac{60}{20} = 3.0$	$>$	$\dfrac{20+30+20+20}{40} = \dfrac{90}{40} = 2.25$
4	Right	$10 + \dfrac{30+20}{20} = \dfrac{60}{20} = 3.0$	$>$	$\dfrac{20+30+20+20}{40} = \dfrac{90}{40} = 2.25$
	Left	$\dfrac{30+20+20}{20} = \dfrac{70}{20} = 3.5$	$>$	$\dfrac{30+20+20+20}{40} = \dfrac{90}{40} = 2.25$
5*	Right	$\dfrac{20+20}{20} = \dfrac{40}{20} = 2.0$	$<$	$\dfrac{30+20+20+20}{40} = \dfrac{90}{40} = 2.25$
	Left	$\dfrac{20+20+30}{20} = \dfrac{70}{20} = 3.5$	$>$	$\dfrac{20+20+20+10}{40} = \dfrac{70}{40} = 1.75$

* A critical position of the load series.

For wheel 3 @ A:
Left reaction (using reaction IL):

$$R_L = [10(55) + 30(47) + 20(40 + 35) + 30(25) + 20(15 + 5)]/60$$

$$= 76.8 \text{ kips}$$

$$M_A = 20R_L - 10(15) - 30(7) = 20(76.8) - 150 - 210$$

$$= 1176 \text{ kip} \cdot \text{ft}$$

For wheel 5 @ A:

$$R_L = [20(55) + 20(50) + 30(40) + 20(30 + 20 + 10)]/60$$

$$= 75 \text{ kips}$$

$$M_A = 20(75) - 20(15) - 20(10) = 1500 - 300 - 200$$

$$= 1000 \text{ kip} \cdot \text{ft}$$

$\therefore \; M_{A_{\max}} = 1176 \text{ kip} \cdot \text{ft}$ when wheel 3 is @ A

The methods previously presented demonstrate a trial-and-error approach to determine which load of a series of concentrated live loads should be placed at the maximum ordinate of the IL to find the maximum moment. References [6.1] and [6.2] at the end of this chapter describe the use of a moment chart that may prove helpful to expidite the numbers of computations.

6.10.5. Absolute Maximum Moment

The design of a beam is usually governed by the maximum bending-moment force. Therefore, it is imperative that the analyst determine the maximum moment the beam may experience for all design load conditions. When a simple beam supports a uniformly distributed load over its entire length, it is obvious that the maximum moment will occur at the midspan; for a single concentrated live load, the maximum beam moment will also occur at midspan when the load is placed at midspan. For a moving concentrated load series, it is generally found that the maximum moment in a long span beam will occur at or near the beam centerline when the load series is clustered about it. Yet, placement of one of the loads in the series at the beam centerline will not guarantee maximum beam moment, particularly for shorter span beams.

The classical derivation which follows establishes where to position a series of fixed-spaced concentrated live loads on a simple beam to produce the absolute maximum moment in the beam. The bending-moment diagram of a simple beam under a set of concentrated loads always consists of straight-line segments. These line segments join at the load locations with the maximum moment occurring at one of the loads. As loads move across the beam (even if new loads enter on or existing loads leave the beam), the moment diagram still retains a straight-line segmented shape and the maximum moment will occur under one of the loads. In fact, the maximum moment will usually occur at the load nearest the resultant of the loads that remain on the beam or at the largest load near the resultant of the load system on the beam.

Consider the simple beam of Fig. 6.28 which shows six of eight truck wheels on the beam. From statics, the resultant of the loads on the beam (Q) is located a distance x from the left support. The maximum moment is considered to occur at load P_3 which is closest to the resultant Q.

$$\Sigma M_B = 0: R_A = Q(L - x)/L$$

Maximum moment, $M_3 = [Q(L - x)/L] \cdot (x + s)$
$$- P_1(d_1 + d_2) - P_2 \cdot d_2$$

Since the maximum moment will occur at shear equal to zero,

$$dM_3/dx = V = 0 = Q(L - 2x - s)/L$$

Figure 6.28

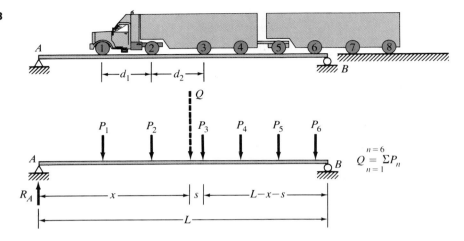

Since Q/L is nonzero

$$L - 2x - s = 0 \quad \text{or} \quad x = L/2 - s/2$$

From this derivation, it can be stated that:

> The maximum moment in a simple beam subjected to a series of moving concentrated loads occurs under the load closest to the resultant of the load system on the beam when the resultant and the load closest to the resultant are spaced equidistant from the beam centerline.

Since the number and magnitudes of loads on the beam can vary, it may be necessary to evaluate the moment for several load-set positions to establish which one will actually produce the absolute maximum beam moment.

EXAMPLE 6.7

Absolute Maximum Moment

Find the absolute maximum moment on the 20-ft beam for the given loading.

First condition: Assume all wheel loads on the beam.

From ④: $\bar{x} = \dfrac{\Sigma Px}{\Sigma P} = \dfrac{20(4) + 16(10) + 10(16)}{10 + 16 + 20 + 4} = 8$ ft

Thus Q is closest to wheel ②. Thus straddle beam centerline with wheel ② and Q as shown, from which $R_L = 22.5$ kips and

183 Moving-Load Placement for Maximum Effect

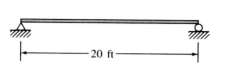
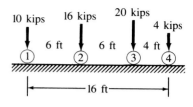

First Condition:

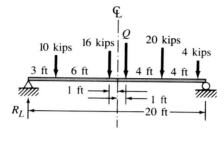

Second Condition:

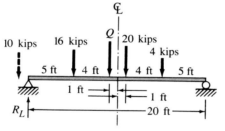

$M_{②} = 22.5(9) - 10(6) = 202.5 - 60 = 142.5$ kip·ft

$M_{ℂ} = 22.5(10) - 10(7) - 16(1) = 139$ kip·ft

Second Condition: Assume only wheels ②, ③, and ④ are on the beam.

From ④: $\bar{x} = \dfrac{\Sigma Px}{\Sigma P} = \dfrac{20(4) + 16(10)}{16 + 20 + 4} = 6$ ft

Although the resultant Q is equidistant from both wheels ② and ③, $P_3 > P_2$. Thus, straddle the beam centerline with the greater load, P_3 and Q to solve for absolute maximum moment at ③.

$R_L = 22$ kips

$M_{③} = 22(11) - 16(6) = 242 - 96 = 146$ kip·ft

$M_{ℂ} = 22(10) - 16(5) = 220 - 80 = 140$ kip·ft

Note: $M_{ℂ}$ in both cases is less than absolute maximum moment.

Absolute maximum moment = 146 kip·ft for the loading.

References

6.1 Charles H. Norris, John B. Wilbur, and Senoi Utku, *Elementary Structural Analysis*, 3d ed., McGraw-Hill, New York, 1976, sec. 5.7.

6.2 Tung Au and Paul Christiano, *Structural Analysis*, Prentice-Hall, Englewood Cliffs, N.J., 1987, sec. 8.3.

Problems

Draw quantitative influence lines for vertical reaction components and for shear (V) and moment (M) at specified beam locations.

6.1 Both reactions and V and M at B.

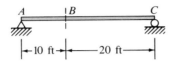

6.2 Both reactions plus V and M at B and C.

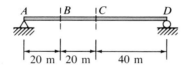

6.3 Both reactions, shear just to left of B, shear to right of B, and M at B.

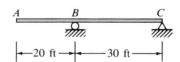

6.4 Both reactions plus V and M at C and D.

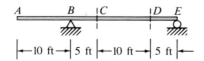

6.5 Both reactions plus V and M at C, D, and F.

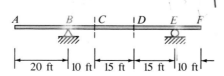

185 Problems

6.6 Both reactions plus V and M at C.

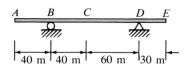

6.7 V and M at A, B, and C.

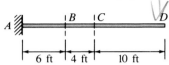

6.8 V and M at A, B, and C.

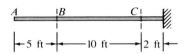

6.9 V and M at A, B, and C (V left of C).

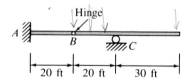

6.10 All reactions and V and M at C (V right of C).

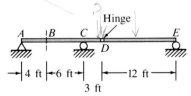

6.11 All reactions, shear to right of C, shear at D, and moment at D.

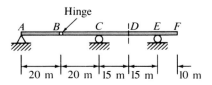

6.12 Reactions at A and B plus V at C.

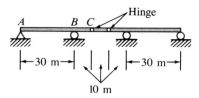

6.13 Draw the influence lines for reaction at A and V and M on lower beam at B as unit load passes from A to C.

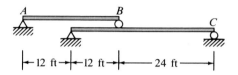

For Probs. 6.14 to 6.17, using the Müller-Breslau principle, draw the qualitative influence lines specified to a scale.

6.14 Shear at B and C, and moment at C.

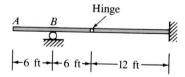

6.15 Shear and moment at B (shear to right of B).

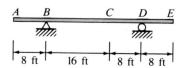

6.16 Shear to left of B, shear to right of B, and moment at C.

6.17 Shear at C and moment at D.

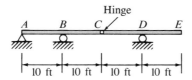

6.18 Draw the influence lines for the moment at C and the shear in panel BC.

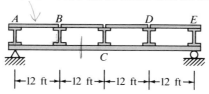

6.19 Shear in panel BC and moment at B of lower beam.

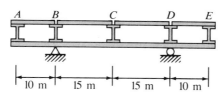

6.20 Shear in panel AB and moment at C in lower beam.

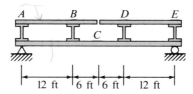

6.21 to 27: Using the influence lines, determine the value(s) of the function(s) requested for a uniform dead load of 0.60 klf, a moving uniform live load of 2 klf, and a concentrated live load of 10 kips. Assume an impact factor of 20 percent for each problem.

6.21 Maximum positive shear at B and moment at B for beam of Prob. 6.1.

6.22 Maximum negative shear and maximum negative moment at C for beam of Prob. 6.4.

6.23 Maximum positive reaction at B, and maximum positive shear and moment at C for beam of Prob. 6.5.

6.24 Maximum positive shear at C, and maximum negative moment at B for the beam of Prob. 6.7.

188 Influence Lines—Beams

6.25 Maximum positive shear at *A* and at *E* for beam of Prob. 6.10.

6.26 Maximum moment at *C*, and maximum positive shear in panel *BC* for the beam of Prob. 6.18.

6.27 Maximum moment at *C*, and maximum negative shear in panel *AB* for the beam of Prob. 6.20.

6.28 to 6.33 A truck shown below can travel in either direction.

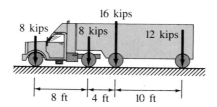

6.28 Determine the maximum positive shear in panel *BC* of the beam of Prob. 6.18 due to the moving truck loads, using the increase-decrease method.

6.29 Repeat Prob. 6.28 for the beam of Prob. 6.20.

6.30 Determine the maximum positive moment at point *C* of the beam of Prob. 6.18 due to the moving truck wheel loads using the average load method.

6.31 Repeat Prob. 6.30 for the beam of Prob. 6.4.

6.32 Determine the absolute maximum bending moment in a 100-ft simple end supported beam for the truck loading above.

6.33 Repeat Prob. 6.32 if the span of the simple beam is 70 ft and the 12-kip rear axle is reduced to 6 kips.

6.34 A movable object is supported by a crane girder. The girder is 80-ft long with a roller support at one end and a pin support at the other end. The wheel forces on the girder are shown below. Determine the absolute maximum bending moment in the 80-ft girder due to the loads.

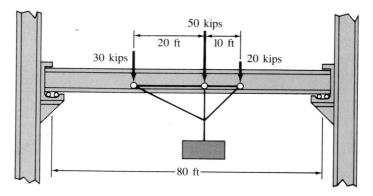

CHAPTER SEVEN

Influence Lines—Plane Trusses

7.1. LOAD-PATH IDEALIZATION FOR AXIALLY LOADED TRUSS MEMBERS

Trusses are described in Sec. 3.1 as structures composed of straight members that are usually arranged in triangular geometric form and joined together by frictionless pins. The members of a truss are primarily designed

Michigan Avenue bascule bridge, Chicago. This double-deck bridge handles both pedestrian and vehicular traffic. (*Courtesy of International Slides, Berkeley*.)

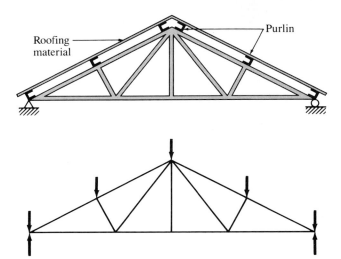

Figure 7.1

as two-force members which provide axial-force resistance only. Since shear and bending forces are to be avoided, the design loads must be applied and resisted at member joints. In a gable-truss frame building, roof loads are usually transmitted to the top chord joints through roof purlins (see Fig. 7.1). On highway and railway bridge-truss structures, the moving loads that occur on a floor deck system are transferred from the floor beams directly to truss joints (see Fig. 3.14 and the load path description given in Sec. 3.3). The direct path of live loads to the truss joints provides us with reasonable assurance that the primary resistance in the truss members is in the form of axial force. Consequently, the influence line (IL) for the axial force in a truss member is developed by placing a unit load at each joint on the chord that supports load. The influence lines for plane truss members are developed using the principles of statics and the methods of truss analysis presented in Chap. 3.

7.2. PARALLEL CHORD TRUSS MEMBERS

Example 7.1 provides a detailed presentation of IL development for truss members. The process may be started by first drawing the left reaction (A_y) influence line since A_y is needed for three of the ILs to be found. For a diagonal truss member, it is easier to construct the IL for a component of the axial force and later factor each IL value, as shown, to obtain the IL for the resultant member force. The author recommends the use of a free-body diagram for visual assistance in the computation of member force ILs as the unit load moves across a truss span. Example 7.2 shows some member influence lines for another parallel chord truss.

EXAMPLE 7.1

Determine influence lines for members *BH*, *IJ*, *AB*, and *ID* of the bridge truss below.

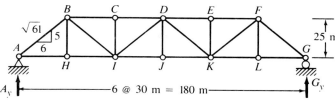

Solution

1. It should be obvious that the live loads are placed on a floor deck connected to the lower chord of the bridge truss.
2. It will be observed that ILs of members are obtained following construction of IL of the left reaction (A_y).

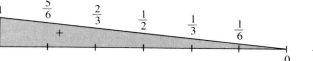

Member *BH* (using method of joints):

$$AH \leftarrow \overset{BH}{\underset{H}{\circ}} \rightarrow HI$$

$\Sigma F_y = 0$:

1. When the unit load is at H, $BH = 1$.
2. When the unit load is at A, I, J, K, L, or G, $BH = 0$.
3. When the unit load is located between A and H or H and I, $0 < BH < 1$.

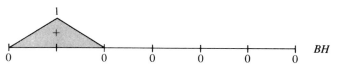

Member *IJ* (method of sections—moments):

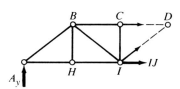

$\Sigma M_D = 0$:

1. For unit load at A, $IJ = 0$
2. For unit load at H, $\Sigma M_D = 0 = 5/6[90] - 1[60] - IJ[25]$, $IJ = +0.60$
3. For unit load at I, $\Sigma M_D = 0 = 2/3[90] - 1[30] - IJ[25]$, $IJ = +1.20$.
4. For unit load right of point I, $\Sigma M_D = 0 = 90A_y - 25IJ$ or $IJ = (90/25)A_y$, Thus, $IJ = 3.6A_y$ for unit load in path J to G.

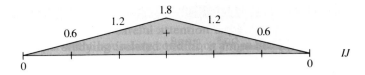

IJ

Member *AB* (method of joints):

1. By $\Sigma F_y = 0$ and use of IL for A_y, AB_y is easily found for each position of the unit load along the truss span.
2. For unit load at A, $\Sigma F_y = 0 = 1 + AB_y - 1$ $\therefore$ $AB_y = 0$.
3. For unit load at H, $\Sigma Fy = 0 = 5/6 + AB_y$ $\therefore$ $AB_y = -5/6$.
4. As unit load moves to right of H, $AB_y = -A_y$.
5. IL below is drawn for AB_y, $AB = AB_y \cdot \sqrt{61}/5$.

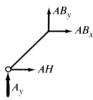

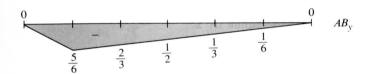

AB_y

Member *ID* (method of sections—shear):

For the FBD section shown and using $\Sigma F_y = 0$:

1. For unit load at A, $1 + ID_y - 1 = 0$ $\therefore$ $ID_y = 0$.
2. For unit load at H, $5/6 - 1 + ID_y = 0$ $\therefore$ $ID_y = +1/6$.
3. For unit load at I, $2/3 - 1 + ID_y = 0$ $\therefore$ $ID_y = +1/3$.
4. For unit load at J, $1/2 + ID_y = 0$ $\therefore$ $ID_y = -1/2$.
5. For unit load right of J, $ID_y = -A_y$.
6. IL below is drawn for ID_y knowing that the resultant $ID = ID_y \cdot \sqrt{61}/5$.

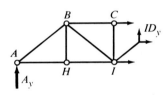

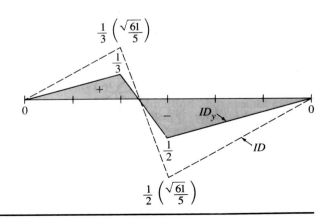

EXAMPLE 7.2

Influence Lines—Parallel Chord Truss

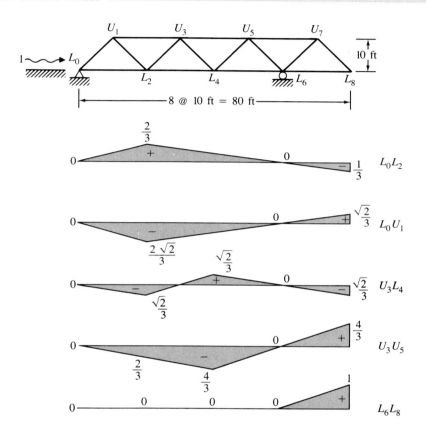

7.3. NONPARALLEL CHORD TRUSS MEMBERS

Influence lines for nonparallel chord truss members are developed using the method of sections (moments) presented in Chap. 3. In Fig. 7.2a, a section is taken which contains member U; the section FBD is shown as Fig. 7.2b from which $\Sigma M_o = 0$ is expressed to find the force in member U for varied joint placements of the unit load on the truss span. When the unit load is

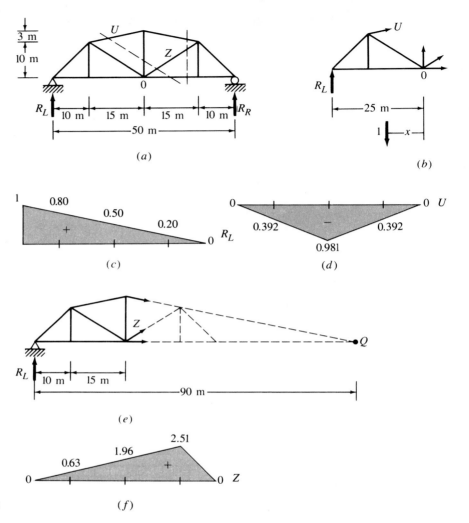

Figure 7.2

over or to the right of point 0, it has no direct contribution in the moment summation equation *but* its influence is present in the value of the left reaction, R_L. Figure 7.2c shows the IL of the left reaction, and Fig. 7.2d presents the IL of member U. In Fig. 7.2a, a vertical section line is passed through the panel which includes member Z. By the method of sections, a free-body diagram of the left section is drawn in Fig. 7.2e which shows that two of the three members "cut" by the vertical section intersect at a point Q in two-dimensional space. Thus, moment equilibrium can be taken about point Q, $\Sigma M_Q = 0$, to determine the force in member Z for each joint position of the unit load on the lower chord. The IL of member Z is shown in Fig. 7.2f. Influence lines for other truss members can be found using the same basic principles of statics used above.

7.4. APPLICATIONS FOR DESIGN-LOAD COMBINATIONS

Once an influence line for a truss member is established, the same methods developed in Chap. 6 (increase-decrease and average load methods) can be applied to determine the maximum member force due to dead, live, and impact load combinations.

Example 7.3 presents an influence line for a truss bridge diagonal member (S) found by the method of sections. The slope of each line segment of the influence line is shown on the IL. A series of six concentrated live loads, 20 N each and equally spaced 5 m apart, are assumed to represent passing vehicles moving across the bridge. A tabular form of the increase-decrease method is used to determine the critical position of the loading that produces maximum member force. The solution is assumed to be self-explanatory.

EXAMPLE 7.3

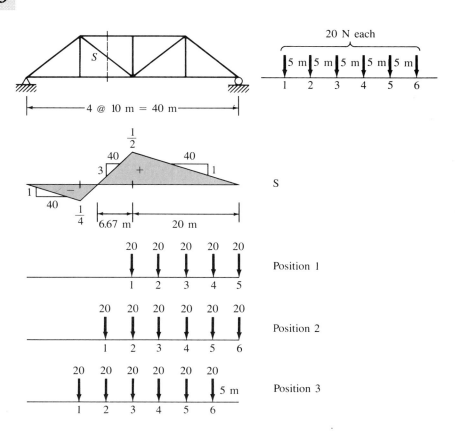

		Change in Force S		
Motion		Decrease		Increase
From	To			
① →	②	$20 \text{ N} \cdot 5 \cdot \dfrac{3}{40} = \dfrac{300}{40}$	<	$[4 \cdot 20 \text{ N}] \cdot 5 \cdot \dfrac{1}{40} = \dfrac{400}{40}$
② →	③	$[2 \cdot 20 \text{ N}] \cdot 5 \cdot \dfrac{3}{40} = \dfrac{600}{40}$	>	$[4 \cdot 20 \text{ N}] \cdot 5 \cdot \dfrac{1}{40} = \dfrac{400}{40}$

Therefore, position ② is critical and

$$S_{max} = 20 \text{ N}\left(\dfrac{0 + 5 + 10 + 15 + 20}{40}\right) + 20 \text{ N} \cdot \dfrac{5}{40} = +27.5 \text{ N (tension)}$$

Influence-line analysis of structures subjected to heavy traffic load conditions may involve numerous wheel loads and can prove to be very tedious. Quite often, maximum forces in bridge truss members due to highway-design live loads are found with minimum tedium by use of the AASHTO uniform lane load (see Sec. 1.5.8); the AASHTO uniform lane load approach uses a uniformly distributed live load in place of the many moving wheel loads plus one or more concentrated live loads per lane to account for unusually heavy vehicles in the line of traffic.

Example 7.4 demonstrates the ease of evaluating the maximum tensile and/or compressive force in a highway bridge-truss member due to a combination of dead and live loads including a dynamic impact factor (see Sec. 1.5.10). In addition, this example establishes that member U_1L_2 will experience stress reversal when subjected to constant moving traffic load conditions. The fatigue strength and endurance limit of the member material are used to help establish the structural adequacy and reliability of the member to support repetitive live-load motions. The examples in this chapter are equally suited to determine maximum frame or beam shear and bending forces under live loading by use of influence lines.

EXAMPLE 7.4

Given the influence lines for members L_0L_1 and U_1L_2 (vertical component), determine maximum tensile and compressive member forces if the loading consists of

1. Uniform dead load of 1.2 klf

2. Uniform live load of 1.8 klf
3. Moving concentrated load of 15 kips
4. Impact factor defined by $I = 50/(L + 125) \leq 0.30$ where L = length in feet < 100 ft

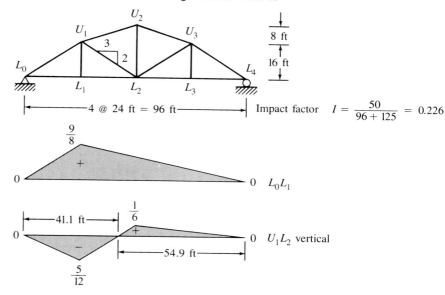

Solution

Member $L_0 L_1$ only experiences tensile load (see IL above). Thus maximum $L_0 L_1$ is found as follows:

$$
\begin{aligned}
\text{UDL} &= 1.2[1/2(96)(9/8)] &&= +64.8 \text{ kips} \\
\text{ULL} &= 1.8[1/2(96)(9/8)] &&= +97.2 \text{ kips} \\
\text{CLL} &= 15[+9/8] &&= +16.9 \text{ kips} \\
I &= 0.226[97.2 + 16.9] &&= +25.8 \text{ kips} \\
L_0 L_1 \text{ maximum} & &&= \overline{204.7 \text{ kips}} \text{ (tensile)}
\end{aligned}
$$

Note 15-kip live load placed at maximum positive ordinate.

Member $U_1 L_2$:

Maximum tensile force:

$$
\begin{aligned}
\text{UDL} &= 1.2[1/2(41.1)(-5/12) + 1/2(54.9)(1/6)] = -4.8 \text{ kips} \\
\text{ULL} &= 1.8[1/2(54.9)(1/6)] &&= +8.2 \text{ kips} \\
\text{CLL} &= 15[1/6] &&= +2.5 \text{ kips} \\
I &= 0.226[8.2 + 2.5] &&= +2.4 \text{ kips} \\
\text{Maximum tension in } U_1 L_2 & &&= +8.3 \text{ kips}\left[\frac{\sqrt{13}}{2}\right] = +15 \text{ kips}
\end{aligned}
$$

Maximum compressive force:

UDL = (remains unchanged) $\quad = -4.8$ kips

ULL = $1.8[1/2(41.1)(-5/12)]$ $\quad = -15.4$ kips

CLL = $15[-5/12]$ $\quad = -6.3$ kips

$I = 0.226[-15.4 - 6.3]$ $\quad = \underline{-4.9 \text{ kips}}$

$$\text{Maximum compression in } U_1L_2 = -31.4\left[\frac{\sqrt{13}}{2}\right] = -56.6 \text{ kips}$$

Remarks
1. Stress reversal occurs in member U_1L_2.
2. ULL and CLL are placed for maximum effect.
3. UDL is placed over entire span.

7.5. CLOSING REMARKS ON INFLUENCE LINES

The exercise of constructing and using influence lines is an excellent way to enhance a student's understanding of structural behavior under moving loads. It also provides a reinforcement of the student's knowledge and ability to analyze truss members, beam shear, beam bending moment, and so forth. A major benefit of drawing influence lines is realized by the engineer who learns where to place design live loads to effect maximum forces, stresses, or displacements.

Quantitative influence lines can be developed by the hand calculation methods presented above or by the use of microcomputer structural analysis programs. The IL hand-calculation approach is a multistep procedure, namely,

1 Construct an FBD that permits member force solution.
2 Place the unit load at each joint along the loaded chord.
3 Compute the member force for each unit load position.
4 Draw the IL of the truss member.

Many structural analysis software packages, such as IMAGES-2D can evaluate a truss structure for multiple load conditions. Consequently, the ILs of *all* truss members can be developed simultaneously from the computer output of multiple load cases where each load case relates to placement of the unit load at a joint on the loaded chord. In addition, multiple load case computer analyses can provide an alternate way of studying the effects of multiple moving wheel load positions on highway and railway bridge structures in lieu of hand methods, such as the increase-decrease method. However, it is worth repeating that the fundamentals of structural analysis are reinforced in a student who has developed the skills of drawing influence lines.

Problems

7.1 to 7.6 Draw influence lines for the specified members (loading moves at support level).

7.1

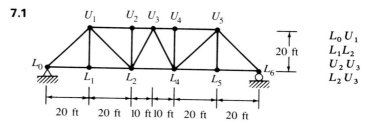

$L_0 U_1$
$L_1 L_2$
$U_2 U_3$
$L_2 U_3$

7.2

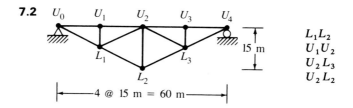

$L_1 L_2$
$U_1 U_2$
$U_2 L_3$
$U_2 L_2$

7.3

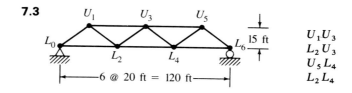

$U_1 U_3$
$L_2 U_3$
$U_5 L_4$
$L_2 L_4$

7.4

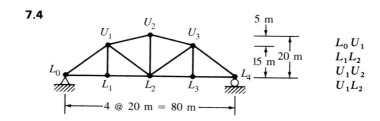

$L_0 U_1$
$L_1 L_2$
$U_1 U_2$
$U_1 L_2$

7.5

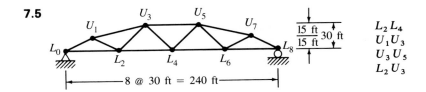

L_2L_4
U_1U_3
U_3U_5
L_2U_3

7.6

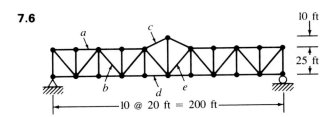

7.7 to 7.17 Draw influence lines for specified members (unit live-load motion is shown).

7.7

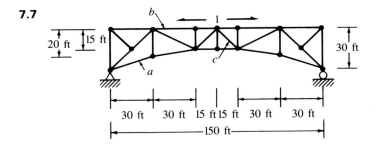

7.8

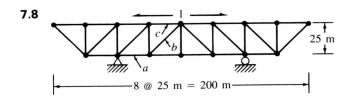

7.9

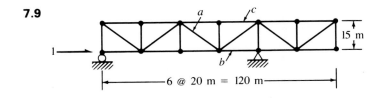

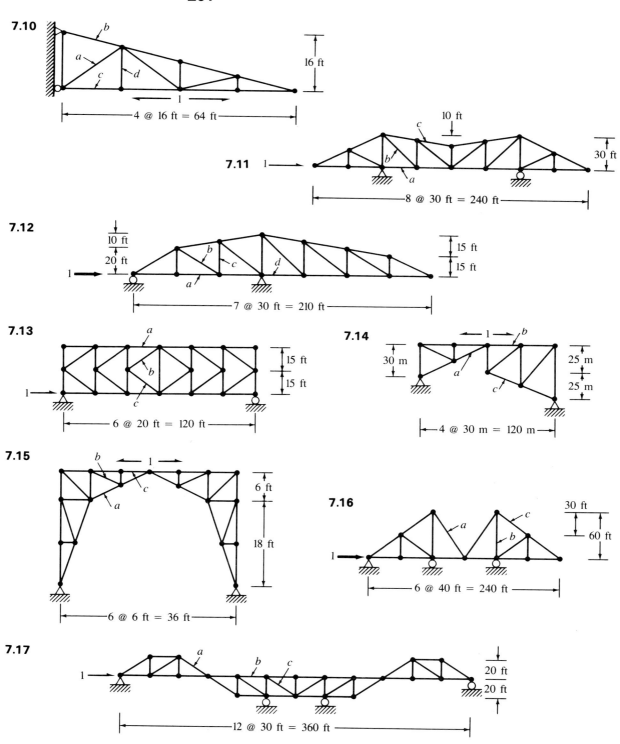

7.18 The influence line for a lower chord truss member of a 320-ft span truss bridge is given below. The bridge is designed to receive one-way traffic loads from right to left. By use of the increase-decrease method and given influence line, determine the maximum force in the truss member due to the series of moving concentrated loads shown below.

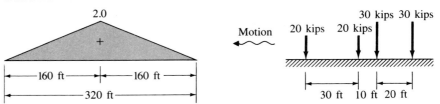

7.19 Repeat Prob. 7.18 using the average load method described in Chap. 6.

7.20 The influence line for a truss diagonal member of a 240-ft simple truss is given below. Determine the maximum tensile force in the member if the loads on the truss include a uniform dead load of 3 klf, a uniform live load of 2 klf, and a series of concentrated loads shown below which can move from right to left only. Also, use an impact factor of 20 percent.

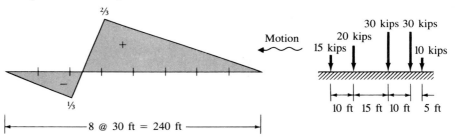

7.21 Repeat Prob. 7.20 to establish if stress reversal can occur (that is, if the loads can be positioned to cause compressive force in the member).

7.22 The influence line of truss member U_1L_2 is given below. Determine if stress reversal in member U_1L_2 can occur if the truss is subjected to a uniform dead load of 3 klf, a uniform live load of 5 klf, a floating concentrated live load of 20 kips, and an impact factor of 15 percent.

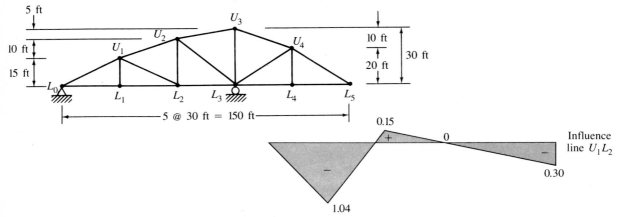

7.23 The influence line of truss member U_4L_5 is given below. For the combination of a uniform dead load of 1 klf, a uniform live load of 4 klf, the series of concentrated loads shown below, and an impact factor of 20 percent, determine (a) the maximum compressive force in member U_4L_5, and (b) if stress reversal can occur.

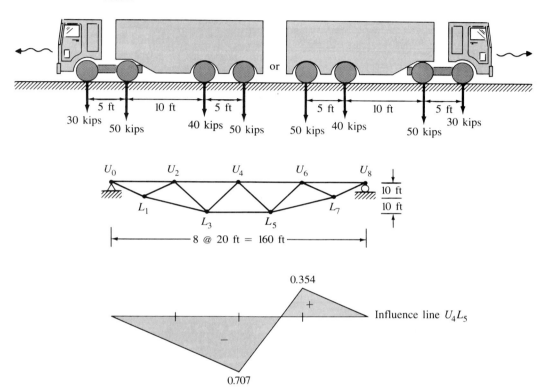

CHAPTER EIGHT

Approximate Methods for Statically Indeterminate Structures

8.1. INTRODUCTION TO STATICALLY INDETERMINATE STRUCTURES

Most civil engineering structures are probably statically indeterminate even though they are frequently analyzed as statically determinate. Concrete buildings are classic examples of indeterminate structures since they usually consist of beams, floor slabs, and column sections that are formed into a continuous rigid-frame assembly from integrated concrete pours. Steel rigid-frame buildings and composite bridge decks represent other forms of structures that are highly indeterminate.

Very often, apparent statically determinate structures are actually statically indeterminate. Consider the statically determinate simple beam which is viewed as having a roller support at one end, a frictionless pin support at the other end, and freedom to rotate, without restraint, at both ends. Yet, a steel-beam end support may actually consist of a weld pattern or a series of bolts for connection to steel columns, either of which provide some degree of moment resistance. In fact, if a beam is initially supported by true pin joints, the joints may, after a time, rust in place and develop rotational restraint. Therefore, the simple beam, that initially behaves as a statically determinate structure, may become statically indeterminate.

Modern design actually favors statically indeterminate structures for several reasons: (1) smaller forces are often generated in its members which result in smaller member size and potential cost savings, (2) the redundant members provide alternate load paths for redistribution of inelastic forces

Pratt Truss, Exhibition Hall, Las Vegas. (*Courtesy of International Structural Slides, Berkeley.*)

that may result from overload conditions, and (3) the greater rigidity associated with statically indeterminate structures usually proves to be beneficial for highway and railway systems where vibration and impact forces are ever present.

In the past, statically indeterminate structures were viewed as complex structures which were difficult to analyze and design. However, this position has been softened by the advancements of computer technological capabilities in both the analysis and design of structures.

Exact analyses of statically indeterminate structures are presented in Chaps. 11 and 12 using the principles to be developed in Chaps. 9 and 10. These methods are considered as "exact" since they are based on assumed ideal or exact conditions. Some of these traditional assumptions include:

1 Members contain homogeneous material properties
2 Members are usually assumed as perfectly straight
3 Linear elastic material behavior of structure under load
4 Knowledge of magnitude, location, and direction of loads
5 Supports provide pin or fully fixed boundary restraints

However, if the ideal conditions are not fully realized, a so-called exact method is, at best, a good approximation of the true structural behavior. Also, the performance of an exact analysis may be complex, require voluminous computations, be prone to error, time consuming, and expensive. Therefore, an approximate analysis may be viewed as an expedient option for the preliminary analysis of an indeterminate structure.

8.2. USE OF APPROXIMATE ANALYSES

This chapter presents approximate methods of analysis that attempt to simplify the solution of statically indeterminate structures. These methods are based on engineering judgment and may provide results that are accurate within 20 percent of those produced by "exact" methods. Furthermore, these approximate methods are based on elastic behavior and require that the stuctures under investigation are geometrically stable and in equilibrium. Some reasons for performing approximate analyses of statically indeterminate structures include:

1. Limited time and monies may demand a quick and simple solution to an otherwise complex problem
2. To assist in performing several preliminary designs in the process of arriving at a final design selection
3. To verify the accuracy of an exact analysis
4. To establish geometric design input data that is required to perform a computer program analysis or an exact analysis
5. Available individuals may lack the ability to perform complex exact analyses

The approximate methods presented in the following sections reduce a statically indeterminate structure to one that is statically determinate. This is generally achieved by using engineering judgment to estimate the force distribution between the redundant members and to decide where points of inflection may occur. The ability to make proper engineering estimates should improve after deflection methods and stiffness effects of structural behavior are mastered.

8.3. INDUSTRIAL (MILL) BUILDING ANALYSIS

Light industrial buildings, often called mill buildings, are commonly assembled from a series of rigid frames which are attached to small isolated or wall footings. Each building frame, also called a mill bent, is commonly constructed of a flat or pitched-truss roof assembly that is rigidly attached to steel columns. Each frame is properly braced and attached to adjacent frames to form a complete building as shown in Fig. 8.1. However, mill building walls are usually chosen from light homogeneous or matrix materials that do not offer significant lateral force resistance. Small aircraft hangers and warehouses, which exemplify this type of construction, are often found with walls consisting of corrugated-steel sheet panels attached to the frames. Thus, mill building frames must be able to withstand design lateral wind forces. In addition, these buildings often support crane girders

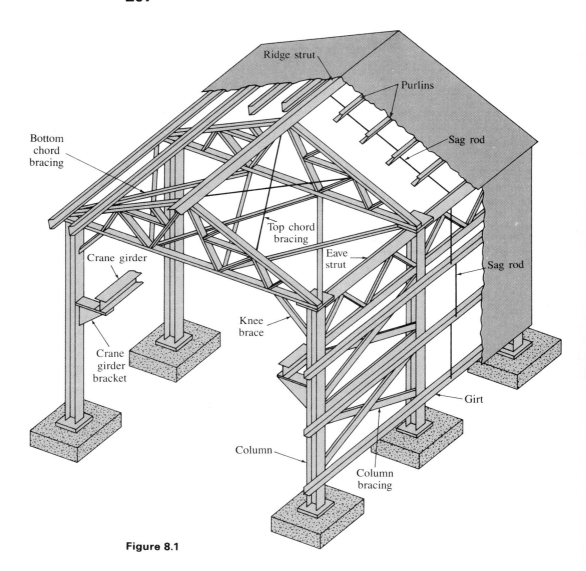

Figure 8.1

which can impart added lateral forces on the frames while moving loads within the building.

A flat-roof mill building frame subjected to a lateral wind loading is shown in Fig. 8.2a. If both vertical columns of the frame are assumed as fixed at their base supports, the frame is statically indeterminate to the third degree (six unknown reaction components and three static equilibrium equations). If both frame columns are assumed to have pin supports, the frame is statically indeterminate to the first degree (four unknown reaction components and three static equilibrium equations available). Consequently, three assumptions must be made to perform an approximate analysis for

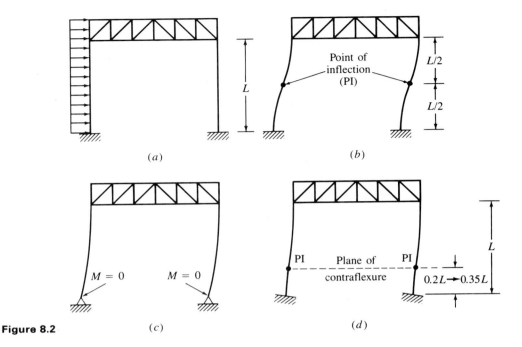

Figure 8.2

the fixed-support condition; likewise, one assumption must be made if the frame is to be studied for the pin-support condition.

Mill bent with fixed supports A mill bent under lateral wind load will deform as shown in Fig. 8.2b if the column ends are rigidly attached at the truss and fixed at the support base. That is, if the rotation at both ends of the frame columns is prevented, a point of inflection (PI), shown as a black dot, will exist at column midheight (between the base support and the truss connection). *A point of inflection occurs at the transition point of zero moment where the curvature changes from positive to negative bending, or vice versa.* If instead, the columns are pin connected at the base supports (where $M = 0$), they will deform in single curvature under lateral load, as shown in Fig. 8.2c. Since these mill bent columns are often attached to small concrete footings, some degree of rotation is likely to occur, but not without some moment restraint. Therefore, it is probable that a point of inflection (PI) will occur somewhere between the base and the midheight of the column as shown in Fig. 8.2d. The dashed line drawn between the PIs of Fig. 8.2d is often referred to as the plane of contraflexure. A common choice is to assume that the PI in each column occurs in the range of 20 to 35 percent of the unsupported column length above the base. This choice satisfies two of the three assumptions required for analysis.

The lateral wind forces must be resisted by horizontal shear in the frame columns. Therefore, a third assumption is made to select the horizontal shear distribution between the frame columns. It is convenient to compute

209 Industrial (Mill) Building Analysis

the shears at the plane of contraflexure where the shears must resist the lateral forces above the plane. If the columns are of equal size, we can assume the shear to be equally resisted by the columns. If the columns are of different sizes, the stiffer column is apt to resist more shear than the less stiff (flexible) column. One common practice is to distribute the shears in proportion to their relative stiffnesses (I/L^3) for dissimilar column sizes. After the PI locations are selected and the column shears at the PIs are computed, the lateral shear forces at the base supports are determined from simple cantilever action below the plane of contraflexure. Example 8.1 provides a clear demonstration of an approximate analysis of a mill bent under lateral loading.

EXAMPLE 8.1

Mill Frame Analysis—Fixed Supports

1. Assume PIs occur at 25 percent of free column length above base
2. Assume equal shear resistance between columns ($H_E = H_F$)

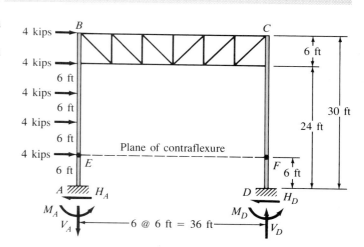

$\Sigma H = 0 = 4 \text{ kips}(5) - H_E - H_F$

since $H_E = H_F$,

$H_E = H_F = 10 \text{ kips}$ ◀

$\curvearrowleft + \Sigma M_F = 0 = 4 \text{ kips}(0 + 6 \text{ ft} + 12 \text{ ft} + 18 \text{ ft} + 24 \text{ ft}) - 36 V_E$

$V_E = 6.67 \text{ kips} \downarrow$ ◀

$\Sigma V = 0 = V_E - V_F$

$V_F = V_E = 6.67 \text{ kips} \uparrow$ ◀

FBD (EBCF):

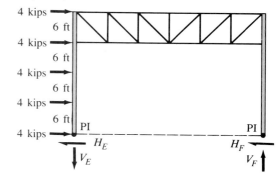

210 Approximate Methods for Statically Indeterminate Structures

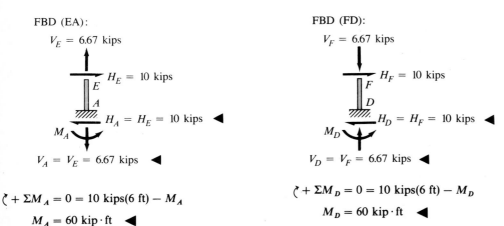

FBD (EA):
$V_E = 6.67$ kips
$H_E = 10$ kips
$H_A = H_E = 10$ kips ◄
$V_A = V_E = 6.67$ kips ◄

$\circlearrowleft + \Sigma M_A = 0 = 10$ kips(6 ft) $- M_A$
$M_A = 60$ kip·ft ◄

FBD (FD):
$V_F = 6.67$ kips
$H_F = 10$ kips
$H_D = H_F = 10$ kips ◄
$V_D = V_F = 6.67$ kips ◄

$\circlearrowleft + \Sigma M_D = 0 = 10$ kips(6 ft) $- M_D$
$M_D = 60$ kip·ft ◄

Mill bent with simple supports A mill bent with simple column supports is statically indeterminate to the first degree. Therefore, an approximate analysis can easily be performed after a distribution of the lateral-load shear resistance at the PIs of the columns is selected. Example 8.2 demonstrates the direct manner of solving support reaction forces for mill building frames by means of an approximate analysis. Once the reaction forces are found, the truss member forces can readily be determined from the principles of statics.

EXAMPLE 8.2

Mill Frame Analysis—Simple Supports

Determine support reactions and forces in truss members L, M, and N.

$\circlearrowleft + \Sigma M_D = 0 = 4$ kips(6 ft + 12 ft + 24 ft + 30 ft) + 8 kips(18 ft) $- 36 V_A$
$\therefore V_A = 12$ kips↓ ◄

$\Sigma V = 0 = V_A - V_D \quad \therefore V_D = V_A = 12$ kips↑ ◄

$\Sigma H = 0 = 4$ kips + 4 kips + 8 kips + 4 kips + 4 kips $- H_A - H_D = 24 - H_A - H_D$

211 Industrial (Mill) Building Analysis

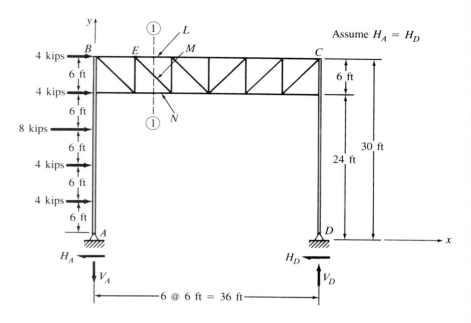

since $H_A = H_D$ $H_A = H_D = 12$ kips ← ◀

Consider section ①–① (left FBD): Assume L, M, and N act in tension. By method of section (shear):

$+ \uparrow \Sigma V = 0 = -M_y - V_A$ or $M_y = -V_A = -12$ kips. Also, $M_x = -12$ kips

Force in member $M = \sqrt{2}(M_y) = -17$ kips (C) ◀

By method of section (moment):

$\zeta + \Sigma M_E = 0 = V_A(6 \text{ ft}) - H_A(30 \text{ ft}) + 4 \text{ kips}(0 \text{ ft} + 6 \text{ ft} + 18 \text{ ft} + 24 \text{ ft})$
$\qquad + 8 \text{ kips}(12 \text{ ft}) + N(6 \text{ ft})$

$0 = 12(6 \text{ ft}) - 12(30 \text{ ft}) + 192 + 96 + 6N$

$\therefore$ Force in member $N = 0$ kips ◀

$+ \rightarrow \Sigma H = 0 = L + M_x + N - H_A + 4 + 4 + 8 + 4 + 4$

$0 = L - 12 + 0 - 12 + 24$

$\therefore$ Force in member $L = 0$ kips ◀

8.4. TRUSSES WITH DOUBLE DIAGONALS

8.4.1. Truss with Slender Double Diagonals in Each Panel

Truss systems for roofs, bridges, and building walls often contain double diagonals in each panel which make the truss statically indeterminate. If the diagonals are slender, it may be assumed that diagonals are only able to resist tensile force and that the diagonals subjected to compressive force

Figure 8.3

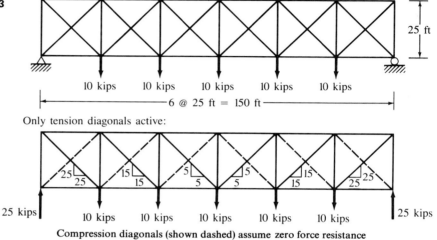

Compression diagonals (shown dashed) assume zero force resistance

(a)

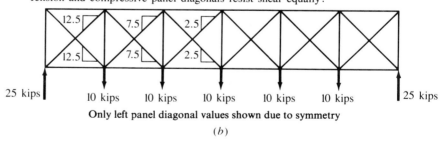

Only left panel diagonal values shown due to symmetry

(b)

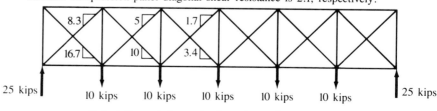

Only left panel diagonal values shown due to symmetry

(c)

become ineffective due to potential buckling behavior. In a parallel chord truss with double diagonals, as shown in Fig. 8.3a, the method of sections (shear) is used to evaluate the shear force component in the tension diagonal of each panel. The slope of the diagonal determines whether a panel diagonal provides tensile or compressive resistance to the panel shear. The diagonal in each panel that acts in compression is assumed to carry zero force. Therefore, the compressive diagonals are essentially removed and the truss is basically reduced to one that is statically determinate. Circular-rod and small light angle shapes typify slender diagonal members.

8.4.2. Truss with Stiff Double Diagonals in Each Panel

If the double diagonals in truss panels are able to resist compression as well as tensile forces, the vertical shear resistance can be divided between the diagonals in an assumed proportion. Then, the remaining member forces are found by usual truss analysis methods. The shear resistance may be proportioned equally between the two diagonals in each panel. Another choice is to assume a greater proportion of shear resistance by the tensile diagonal of a panel. Some analysts elect to proportion the relative vertical shears with respect to the geometric area or the stiffness parameters of the panel diagonals. Figure 8.3b shows the results of an approximate analysis based on a 1 : 1 proportion of vertical shear between the diagonals in each truss panel. Figure 8.3c displays the results of an approximate analysis which assumes that the tension diagonal will resist twice as much vertical shear as the compression diagonal in each panel.

8.5. CONTINUOUS BEAMS UNDER GRAVITY LOADS

Guideline aids for approximate analysis The traditional approach used to perform approximate analyses of statically indeterminate structures supplements the existing equations of statics with equations of condition at assumed PI locations. Once the PI locations are selected, many structures become statically determinate. The student should realize that subsequent chapters will present exact methods that provide the means to locate the PIs under ideal and specified conditions. It is also worth noting that numerous engineering aids are readily available to lighten the task of performing approximate analyses. Chapter 2 of the AISC *Manual of Steel Construction* provides exact solutions for beam shear, deflection, and bending moment. These AISC solutions pertain to single-span or continuous equal-span beams of constant E (Young's modulus) and I (moment of inertia) that are subjected to uniform gravity loading. In addition, from these AISC data points of inflection along the beam can be located [8.1]. Figure 8.4 provides moment diagram shapes and locations of points of inflection for several individual and multispan beams of equal span and constant EI under uniform loading. Epstein [8.2] has presented quick and easy procedures to approximate, with reasonable accuracy, the location of points of inflections for continuous beams. The ACI Building Code Requirements also provide

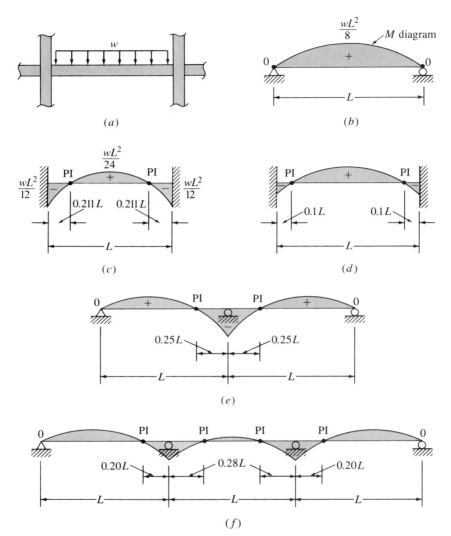

Figure 8.4

approximate moment and shear expressions for beams of continuous concrete spans under uniform gravity loading that may be used for preliminary design if certain conditions are satisfied [8.3].

Simplified approach for beams in buildings Figure 8.4a shows a standard interior arrangement in a building frame consisting of adjacent beams and columns. Furthermore, the beams are typically designed to support specified uniform dead and live loads. If a beam is either joined to members that do not offer substantial resistance to rotation or is attached at its ends by simple support connections, the beam may experience single curvature deflection. Figure 8.4b reflects the classical simple-beam bending-moment diagram with zero end moments. However, if a beam is attached at both ends by moment-resisting connections to stiff members, fixed-end supports may be assumed and the beam is likely to experience both concave and convex deflection (contraflexure) under uniform gravity load; for this condition, the moment diagram of Fig. 8.4c can be used to locate the PIs. However, both Figs. 8.4b and 8.4c represent ideal conditions. In reality, it is most probable that the beam-end supports are neither completely free or completely restrained from rotation when the beam deforms under load. Thus, it is logical to assume that the true PIs are located closer to the beam ends than shown in Fig. 8.4c; a traditional assumed PI location is 0.10L as shown in Fig. 8.4d. Example 8.3 demonstrates the solution of a typical interior floor beam under uniform loading with PIs assumed at 0.10L from each moment-resisting end. The interior beam section between PIs behaves as a simple beam experiencing positive bending moment whereas the end sections are treated as cantilevers which undergo negative bending moment due to gravity loading. The elastic deflection curve of the beam due to the gravity loading is also presented in Example 8.3.

EXAMPLE 8.3

Fixed-Fixed Beam under Gravity Loads

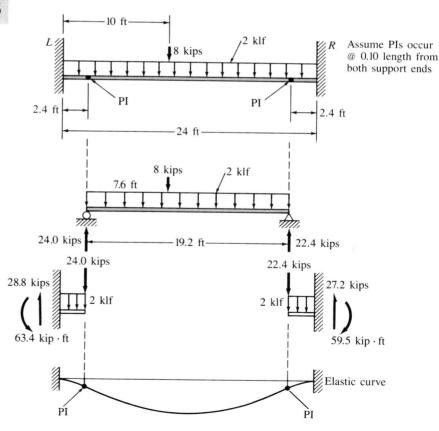

8.6. BUILDING FRAMES SUBJECTED TO LATERAL LOADS

In addition to dead and live gravity loading, building frames must also sustain lateral wind forces. This is especially noted in tall buildings where trussed bracing systems, shear walls, moment resisting connections, and tuned mass dampers, are commonly used to minimize occupant discomfort and prevent excessive lateral drift under high wind forces. When a building is properly secured at its base, it behaves like a vertical cantilever beam that is able to safely transmit the lateral shear from floor to floor and resist the overturning moments produced by the design wind forces.

Two-dimensional analyses of building structures under lateral loads are usually simplified by assuming that the skeleton frame of a structure completely resists the lateral loads without contributions from floors, partitions and exterior walls, and so forth. However, building-frame members are usually assumed to be joined by rigid connections where shear, axial, and bending-moment forces are unknown. Therefore, building-frame analyses prove to be highly indeterminate for solution by exact methods. If a building frame is viewed as being composed of many individual portals with fixed supports, the degree of indeterminacy of the total frame is equal to three times the number of portals. On this basis, a five-story building that is four bays wide is statically indeterminate to the 60th degree; a building, ten stories high by five bays wide, is statically indeterminate to the 150th degree. Consequently, a simplified approximate method which provides reasonable results when compared to an otherwise complex and tedious exact method will prove to be attractive. Two of the most popular approximate methods of analyzing building frames under lateral loads are the portal and

Merchant Exchange Building, Chicago. Outside building trusses consist of x-braced 50-ft-square panels. Clear span between support columns is 100 ft and the end of the building has a 50-ft overhang. (*Courtesy of International Structural Slides, Berkeley.*)

cantilever methods which are presented in the paragraphs that follow. Both methods have been used with success for many decades since they are very simple and expedient. Since these methods determine member forces from statics alone, without consideration of member stiffness variations and elastic deformations, their results are, at best, approximate.

A particular benefit of the portal and cantilever methods is the ability to quickly estimate member forces for subsequent selection of preliminary member sizes of the structure. Once the preliminary member sizes are selected, the analyst can proceed to perform an exact structural analysis.

8.6.1. Portal Method[1]

Five basic assumptions are made in order to analyze a building frame subjected to lateral loads as follows:

1 All frame member joints are rigid.

2 All lateral loads are applied at joints.

3 Each column is deformed under load so that a PI occurs at midheight (see Fig. 8.5).

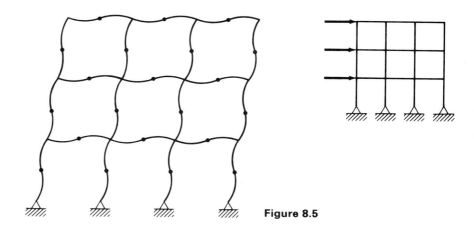

Figure 8.5

4 Each girder is deformed under load so that a PI occurs at midspan (see Fig. 8.5).

5 At each midstory level (plane of contraflexure) a distribution of resisting shear between columns is selected. One choice is to assume that the frame is composed of equal individual portals (see Fig. 8.6) so that the interior frame columns consist of two portal verticals whereas the exterior frame columns consist of one portal vertical. Therefore, at each level, the interior columns may be considered to resist twice as much shear compared to the exterior columns. Other selections of shear distribution between columns can be made.

[1] From Ref. 8.4.

Portal method:

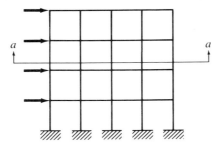
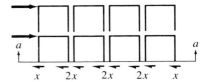

x Shear resistance—exterior column
$2x$ Shear resistance—interior column

Figure 8.6

Using these five assumptions, the multistory frame is stable and statically determinate and the analysis proceeds in the following order:

1 *Column shears:* At each midstory height of the frame, the total lateral load above is resisted by shear in the columns ($\Sigma H = 0$). Compute column shears in the proportion of 2:1 for interior and exterior columns, respectively.

2 *Column moments:* Since PIs exist at column midheights, the maximum moment occurs at the top and bottom of each column and is equal to the column shear times half of the column height. Compute all column moments.

3 *Girder moments:* Girder moments are found by enforcing moment equilibrium at each frame joint starting at the uppermost windward joint of the frame and progressing across to the next frame joint in successive order. The author prefers to construct an FBD at each joint showing the adjacent column and girder sections drawn to their respective PIs and including the unknown shear and axial forces at the PIs. From an FBD at a joint, it is evident that the sum of the moments in the columns is equal to the sum of the moments in the girders.

4 *Girder shears:* After a girder moment is known, the girder shear is determined by dividing the girder moment by half of the girder length. For frames of equal bay widths, girder shears on the same level will be found equal in value.

5 *Axial member forces:* The axial forces in the members at each joint are found by using joint equilibrium equations ($\Sigma H = 0$ and $\Sigma V = 0$). Steps 3, 4, and 5 are executed in consecutive order at each joint before proceeding to the next joint. The axial forces in the girders are usually not computed since the preliminary girder selections are usually based on bending and shear.

For frames of equal bay widths, the portal method will result in zero axial force for all interior columns. Preliminary sizes of interior columns are likely to be selected based on axial compression transmitted from the columns above plus the design gravity loads on a portion of the floor above.

Example 8.4 demonstrates the simple procedure of the portal method. This example also remarks on the fact that this method does not inspire confidence in the column axial-force values computed. The results, which are shown in Fig. 8.7, provide logical sense for the exterior columns (i.e., tension on the windward side and compression on the leeward side); yet, sense of the interior-column axial forces is opposite to that of their respective exterior columns which is illogical.

Figure 8.7   Portal method

```
18 kips   A   V = 2.5    B   V = 1.88   C   V = 2.5    D
     ───▶ ├──── M = 15 ────┼──── M = 15 ────┼──── M = 15 ────┤
          │ V = 3          │ V = 6          │ V = 6          │ V = 3
          │ M = 15         │ M = 30         │ M = 30         │ M = 15     10 ft
          │ S = +2.5       │ S = −.62       │ S = +.62       │ S = −2.5
 6 kips   E   V = 9.17    F   V = 6.88    G   V = 9.17    H
     ───▶ ├──── M = 55 ────┼──── M = 55 ────┼──── M = 55 ────┤
          │ V = 4          │ V = 8          │ V = 8          │ V = 4
          │ M = 40         │ M = 80         │ M = 80         │ M = 40     20 ft
          │ S = +11.7      │ S = −2.92      │ S = +2.92      │ S = −11.7
          │                │                │                │
         I▓               J▓               K▓               L▓
         ├──── 12 ft ─────┼──── 16 ft ─────┼──── 12 ft ─────┤
```

Remarks:
1. If bay spacings are equal, interior columns have zero axial force.
2. If interior bay length is greater than outer bay lengths, the axial forces will change in magnitude and sign for interior columns.
3. Column axial forces do not exhibit reliable data for design.

EXAMPLE 8.4

Portal Method

Determine all member shears and moments in beams and shears, and moments and axial force in columns of the building frame. Summary of results are presented in Fig. 8.7 above.

220 Approximate Methods for Statically Indeterminate Structures

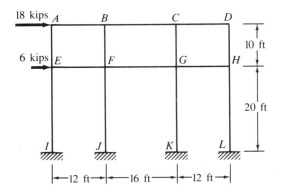

1. Column shears:

   $AE = DH = x = 3$ kips
   $BF = CG = 2x = 6$ kips
   $EI = HL = x' = 4$ kips
   $FJ = GK = 2x' = 8$ kips

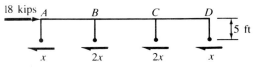

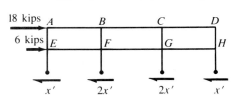

2. Column moments:

   $AE = DH = 3 \text{ kips} \cdot 5 \text{ ft} = 15 \text{ kip} \cdot \text{ft}$
   $BF = CG = 6 \text{ kips} \cdot 5 \text{ ft} = 30 \text{ kip} \cdot \text{ft}$
   $EI = HL = 4 \text{ kips} \cdot 10 \text{ ft} = 40 \text{ kip} \cdot \text{ft}$
   $FJ = GK = 8 \text{ kips} \cdot 10 \text{ ft} = 80 \text{ kip} \cdot \text{ft}$

3. Girder moments and shears plus column axial force found from joint FBDs:

   $\Sigma M_{\text{girders}} = \Sigma M_{\text{columns}}$
   $\therefore \ M_{AB} = M_{AE} = 15 \text{ kip} \cdot \text{ft}$

   $V_{AB} = \dfrac{M_{AE}}{6} = 2.5 \text{ kips}$

   $\Sigma V = 0 = S_{AE} - V_{AB} \quad \therefore \ S_{AE} = +2.5 \text{ kips}$

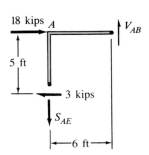

221 Building Frames Subjected to Lateral Loads

$\Sigma M_{\text{girders}} = \Sigma M_{\text{columns}}$

2.5 kips · 6 ft + V_{BC} · 8 ft = 6 kips · 5 ft

$\therefore V_{BC} = (30 - 15)/8 = 15/8$ kips ≈ 1.88 kips

$M_{BC} = 15$ kip · ft (that is, 1.88×8)

$\Sigma V = 0 = 1.88 - 2.5 - S_{BF}$ $\therefore S_{BF} = -0.62$ kips

$\Sigma M_{\text{girders}} = \Sigma M_{\text{columns}}$

15 kip · ft + V_{CD} · 6 ft = 30 kip · ft $\therefore V_{CD} = \frac{15}{6} = 2.5$ kips

Thus, $M_{CD} = V_{CD} \cdot 6 = 2.5(6) = 15$ kip · ft

$\Sigma V = 0 = 1.88 + S_{CG} - 2.5$ $\therefore S_{CG} = +0.62$ kips

$\Sigma M_{\text{girders}} = \Sigma M_{\text{columns}}$

2.5 kip · 6 ft = 3 kip · 5 ft   or   15 = 15 ✓

$\Sigma V = 0 = 2.5 + S_{DH}$ $\therefore S_{DH} = -2.5$ kips

$\Sigma M_{\text{girders}} = \Sigma M_{\text{columns}}$

$V_{EF} \cdot 6 = 15 + 40 = 55$ $\therefore V_{EF} = 9.17$ kips

$M_{EF} = 55$ kip · ft

$\Sigma V = 0 = 2.5 + 9.17 - S_{EI}$ $\therefore S_{EI} = +11.67$ kips

$\Sigma M_{\text{girders}} = \Sigma M_{\text{columns}}$

$55 + V_{FG} \cdot 8$ ft $= 30 + 80$

$M_{FG} = 55$ kip · ft   and $V_{FG} = 6.88$ kips

$\Sigma V = 0 = 6.88 - 9.17 - 0.62 - S_{FJ}$

$S_{FJ} = -2.92$ kips

The complete results are summarized in Fig. 8.7

8.6.2. Cantilever Method[1]

The cantilever method provides an alternate concept for the approximate analysis of tall building frames under lateral loads. The method presumes that a building frame under lateral wind force behaves like a vertical cantilever beam which must resist the overturning moment produced by the wind. Therefore, axial tension forces are expected to develop in the columns on the windward side of the frame, while compression is developed in the columns on the leeward side (see Fig. 8.8). Moreover, the magnitude of axial force in each column is assumed to increase in linear proportion to its respective distance from the centroid of the column cross-section areas at each floor level. The cantilever method also assumes that all frame member joints are rigid, all loads are applied at joints, and points of inflection exist at the midheights of all columns and at the midspans of all girders.

The magnitudes of axial tension and compression forces in the columns are proportioned in the following manner:

Determine the overturning moment produced by the applied lateral forces about each midstory plane of contraflexure. Moment equilibrium requires that the overturning moment must be resisted by the internal moment produced from the column axial forces. Thus, the column axial forces are found from moment equilibrium applied at each plane of contraflexure; that is, by setting the overturning moment due to the external loads equal to the sum of the internal resisting moments of the columns about the column centroid. Each column axial force is usually of nonzero value unless a column is located at the centroid of the column group.

This statically determinate analysis proceeds in the following order:

1. *Column axial forces:* Determine the overturning moment at each plane of contraflexure due to the lateral forces above the plane. Equate the overturning moment to the resisting moment produced by the axial column forces at this level as described above.

2. *Girder shears:* Starting at the upper windward joint and proceeding across from joint to joint, compute each girder shear by the $\Sigma F_y = 0$ equilibrium equation. Use of FBDs, as employed in the portal method, is recommended.

Cantilever method:

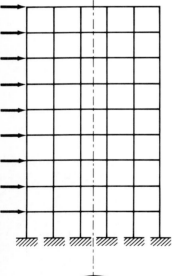

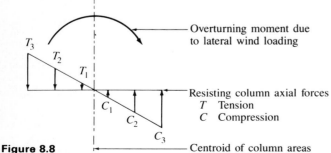

Figure 8.8

[1] From Ref. 8.5.

3 *Girder moments:* Each girder moment is equal to the girder shear times the girder half-length.

4 *Column moments:* Starting at the upper windward joint and proceeding across from joint to joint, column moments are determined from the $\Sigma M = 0$ equation about each joint.

5 *Column shears:* Each column shear is found by dividing the column moment by half the column height.

Example 8.5 illustrates the use of the cantilever method; the results of this example are summarized in Fig. 8.9.

EXAMPLE 8.5

Cantilever Method

Repeat Example 8.4 using the cantilever method.

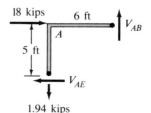

(a)

1. Column axial forces:
   Assume all column cross-section areas are equal. Let axial force in columns at B and C from centroid = Y. Then, the axial force in columns located at A and D from centroid = $20/8(Y) = 2.5Y$

$\Sigma M_{\text{plane of contraflexure}} = 0 = 18 \text{ kips} \cdot 5 \text{ ft} - 16Y - 40(2.5Y)$

$\therefore \quad Y \cong 0.78$ kips, and $2.5Y \cong 1.94$ kips

2. Girder shears and moments:

$\Sigma V = 0 = V_{AB} - 1.94 \text{ kip} \quad \therefore \quad V_{AB} = 1.94 \text{ kips}$
$M_{AB} = (V_{AB})(6 \text{ ft}) = 11.64 \text{ kip} \cdot \text{ft}$

3. Column shears and moments:

$M_{AE} = M_{AB} = 11.64 \text{ kip} \cdot \text{ft}$

and $\quad V_{AE} = M_{AE} \div 5 \cong 2.33 \text{ kip} \cdot \text{ft}$

(b)

Moving in succession to the FBD of joints B, C, D, E, F, G, and H, the girder and column shears and moments are readily found by statics.

The column axial forces at the first floor level are as follows:

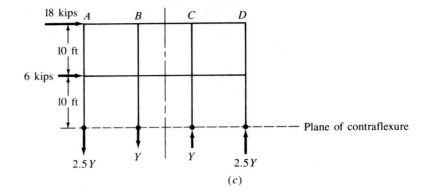

$\Sigma M_{\text{plane of contraflexure}} = 0$:

$0 = 18 \text{ kips} \cdot 20 \text{ ft} + 6 \text{ kips} \cdot 10 \text{ ft} - 16Y - 40(2.5Y)$

$\therefore \quad Y \cong 3.62 \text{ kips}$

and $2.5Y \cong 9.05 \text{ kips}$

Cantilever method

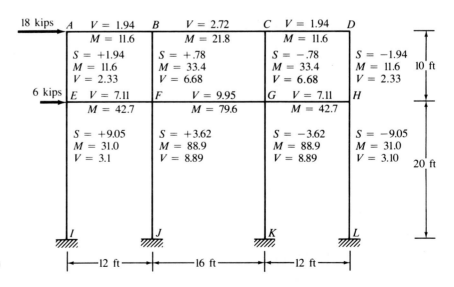

Figure 8.9

8.7. COMPARISON WITH COMPUTER RESULTS

A frame analysis by the IMAGES-2D finite element computer program was made to compare with the results of the portal and cantilever analyses shown in Figs. 8.7 and 8.9. Since both the portal and cantilever methods are independent of member area properties, the IMAGES analysis used constant area, moment of inertia, and material properties for all frame members. A comparison of shear and moment values are tabulated in Fig. 8.10. Where the PIs are assumed at midlengths of all frame members in both the portal and the cantilever methods, PIs will not actually occur at member midlengths. However, the shear and moment values by all three analyses are shown to be within the same order of magnitude.

8.8. VIERENDEEL TRUSS

The portal method can also be used to analyze a structure known as a Vierendeel truss. This structure has become very popular in the United States for use as an enclosed pedestrian crossover. The Vierendeel truss is actually not a truss since it lacks diagonal members and consists of vertical and horizontal members which must resist bending moment and shear forces. Consequently, it is a rigid frame structure. The following assumptions are used for an approximate analysis of a Vierendeel truss:

1 PIs occur at midlengths of all members.

2 The vertical shear in each panel is resisted between the upper and lower horizontal chord members in a selected proportion. A common choice is

Figure 8.10 Comparison with finite element analysis by IMAGES

Member	Shear (V) ~ kips			Moment (M) ~ kip·ft		
	Portal	Cantilever	IMAGES	Portal	Cantilever	IMAGES
AB	2.5	1.94	3.6	15	11.6	22.6
BC	1.88	2.72	1.87	15	21.8	15.0
CD	2.5	1.94	3.55	15	11.6	21.1
EF	9.17	7.11	8.98	55	42.7	53.3
FG	6.88	9.95	4.21	55	79.6	33.7
GH	9.17	7.11	8.94	55	42.7	53.2
AE	3	2.33	2.96	15	11.6	14.8
BF	6	6.68	6.10	30	33.4	30.5
CG	6	6.68	6.06	30	33.4	30.2
DH	3	2.33	2.89	15	11.6	14.5
EI	4	3.10	5.69	40	31.0	56.9
FJ	8	8.89	6.31	80	88.9	63.1
GK	8	8.89	6.31	80	88.9	63.1
HL	4	3.10	5.69	40	31.0	56.9

to divide the shear resistance equally between the top and bottom panel chord members since they often have the same cross-section area.

Remaining member shear and moment values are determined by statics.

An approximate analysis of a Vierendeel truss by the portal method is displayed in Example 8.6 with the results summarized in Fig. 8.11a. Shown

Vierendeel girder frame used in the construction of Cobo Hall, Detroit. *(Courtesy of BEI Associates, Inc.)*

Vierendeel girder rigid frame serves as an overpass footbridge in Garden Grove, California. The bridge is composed of 17 panels (6 ft × 7.5 ft high and 4 ft wide). Ten WF steel beams are used throughout. *(Courtesy of International Structural Slides, Berkeley.)*

227 Vierendeel Truss

in Fig. 8.11b are the results of an exact analysis by the IMAGES computer program which used constant area and moment of inertia properties for all members of the Vierendeel truss. Example 8.6 provides a sufficient number of FBDs needed to solve the member shears and moments by statics.

EXAMPLE 8.6

Vierendeel Truss by Portal Method

Assumptions:

1. PIs occur at midlengths of all members.
2. Shear resistance per panel equally divided between horizontal members.

Note
Axial forces in horizontal members are not shown in FBDs.

$M_{AB} = 5 \text{ kip} \cdot 5 \text{ ft} = 25 \text{ kip} \cdot \text{ft} = M_{GH}$

$M_{AG} = 25 \text{ kip} \cdot \text{ft}$

$V_{AG} = \frac{25}{5} = 5 \text{ kips}$

$S_{AG} = -5 \text{ kips}$

$V_{BH} = \frac{25 + 2.5(5)}{5} = 7.5 \text{ kips}$

$M_{BH} = 7.5(5) = 37.5 \text{ kip} \cdot \text{ft}$

$S_{BH} = 5 - 2.5 = +2.5 \text{ kips}$

$M_{CI} = 2.5(5) = 12.5 \text{ kips}$
$V_{CI} = 2.5 \text{ kips}$

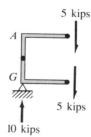

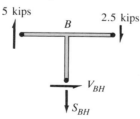

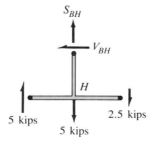

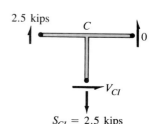

228 Approximate Methods for Statically Indeterminate Structures

$V_{CD} = V_{IJ} = 0$
$M_{CD} = M_{IJ} = 0$

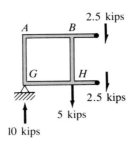
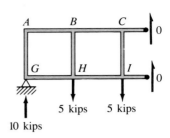

By portal method

V = 5 kips	V = 2.5	V = 0	V = 2.5	V = 5	
M = 25	M = 12.5	M = 0	M = 12.5	M = 25	
M = 25 V = 5 S = −5	M = 37.5 V = 7.5 S = +2.5	M = 12.5 V = 2.5 S = +2.5	M = 12.5 V = 2.5 S = +2.5	M = 37.5 V = 7.5 S = +2.5	M = 25 V = 5 S = −5
V = 5	V = 2.5	V = 0	V = 2.5	V = 5	
M = 25	M = 12.5	M = 0	M = 12.5	M = 25	
	↓5 kips	↓5 kips	↓5 kips	↓5 kips	

⟵ 5 @ 10 ft = 50 ft ⟶   10 ft

(a)

By IMAGES computer program   *Average M values shown for all members

V = 5	V = 2.5	V = 0	V = 2.5	V = 5	
*M = 25	M = 12.5	M = 0	M = 12.5	M = 25	
M = 26.1 V = 5.2 S = −4.99	M = 32.8 V = 6.56 S = 2.49	M = 11.96 V = 2.39 S = 2.50	M = 11.96 V = 2.39 S = 2.50	M = 32.8 V = 6.56 S = 2.49	M = 26.1 V = 5.2 S = −4.99
V = 5	V = 2.5	V = 0	V = 2.50	V = 5	
M = 25	M = 12.5	M = 0	M = 12.5	M = 25	
	↓5 kips	↓5 kips	↓5 kips	↓5 kips	

Figure 8.11   (b)

Problems

8.1 The industrial frame shown has fixed column supports. Assume that each column provides equal shear resistance and that points of inflection occur 6 ft above the supports. Determine support forces and the force in each truss member.

8.2 Repeat Prob. 8.1 if the frame columns are secured by small isolated footings and the points of inflection are assumed to occur 4 ft above the supports.

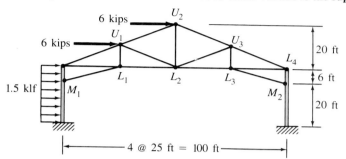

8.3 The flat-roof industrial frame is fixed-supported with assumed points of inflection at (a) 10 ft above the base, and (b) 5 ft above the base. Also, assume that the columns provide equal shear resistance to lateral force at the plane of contraflexure. Compute all support forces and the force in each truss member.

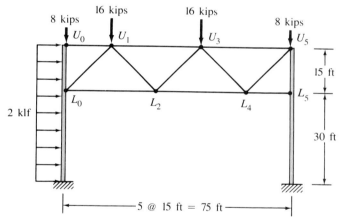

8.4 and 8.5 Determine the force in each interior *diagonal* truss member for each of the following approximations: (a) interior diagonals are unable to support compressive force, (b) the compressive diagonals can support 50 percent of the panel shear, and (c) the compression diagonals support 25 percent of the panel shear.

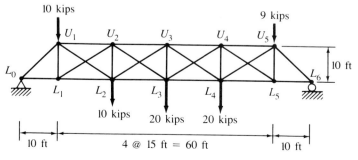

8.5

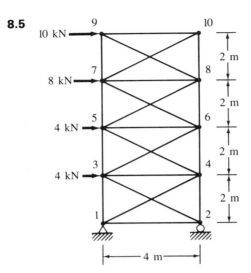

8.6 Determine the support reactions and draw the shear and moment diagrams if $L = 60$ ft and $k = 0.10$.

8.7 Repeat Prob. 8.6 if $k = 0.20$.

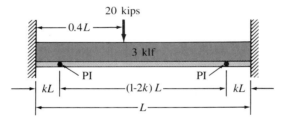

8.8 to 8.13 Using the portal method of approximate analysis determine the shear, moment, and axial force in all members of the frame under lateral loading.

8.8

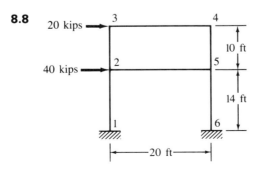

231 Problems

8.9

8.10

8.11

8.12

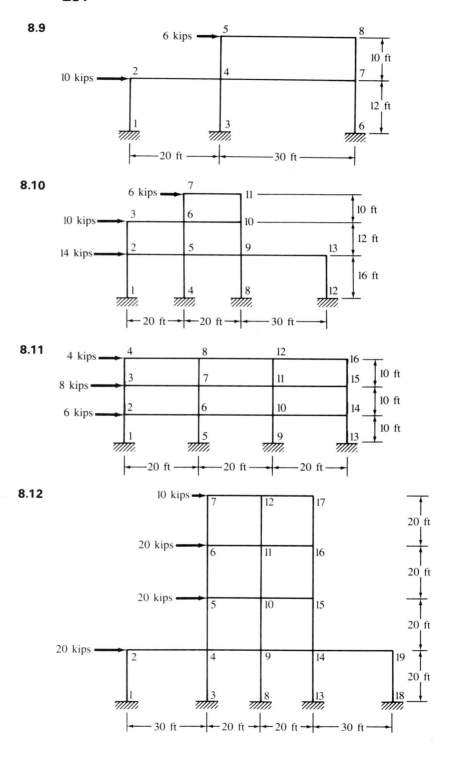

8.13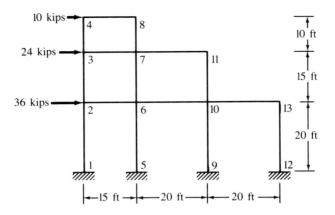

8.14 to 8.19 By the cantilever method, determine shears, moments, and axial forces in all frame members. Assume equal cross-section areas for all columns. (Hint: Locate centroid of column group at each plane of contraflexure.)

8.14 The frame of Prob. 8.8 by the cantilever method.
8.15 The frame of Prob. 8.9 by the cantilever method.
8.16 The frame of Prob. 8.10 by the cantilever method.
8.17 The frame of Prob. 8.11 by the cantilever method.
8.18 The frame of Prob. 8.12 by the cantilever method.
8.19 The frame of Prob. 8.13 by the cantilever method.

8.20 to 8.25 Perform a rigid frame analysis by use of the IMAGES-2D computer program for each frame. Assume $E = 30 \times 10^6$ psi, $A = 10$ in^2, and $I = 120$ in^4 for all members.

8.20 The frame of Prob. 8.8 by IMAGES-2D.
8.21 The frame of Prob. 8.9 by IMAGES-2D.
8.22 The frame of Prob. 8.10 by IMAGES-2D.
8.23 The frame of Prob. 8.11 by IMAGES-2D.
8.24 The frame of Prob. 8.12 by IMAGES-2D.
8.25 The frame of Prob. 8.13 by IMAGES-2D.

8.26 Compute shears, moments, and axial forces for all members of the Vierendeel truss using the portal method. (Reminder: All joints are rigid.)

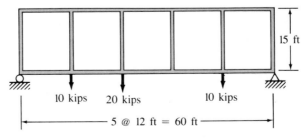

8.27 Repeat Prob. 8.26 by performing an IMAGES-2D computer analysis using $E = 30 \times 10^6$ psi, $A = 10$ in^2, and $I = 80$ in^4 for all members.

References

8.1 *Manual of Steel Construction*, 8th ed., American Institute of Steel Construction, Inc., 1980, pp. 2-112 to 2-125.

8.2 Howard I. Epstein, "Approximate Location of Inflection Points," *Journal of Structural Engineering*, vol. 114, no. 6, June 1988, pp. 1403–1413.

8.3 American Concrete Institute, "Building Code Requirements for Reinforced Concrete," ACI 318-83, chap. 8.

8.4 Albert Smith, "Wind Stresses in the Frames of Office Buildings," *Journal of the Western Society of Engineers*, April 1915, p. 341.

8.5 A. C. Wilson, "Wind Bracing with Knee Braces or Gusset Plates," *Engineering Record*, September 5, 1908, pp. 272–274.

CHAPTER NINE

Deformations: Beams

9.1. INTRODUCTION—WHY COMPUTE DEFLECTIONS?

In general, when a structure supports load or experiences temperature change, the individual members will undergo strain deformations and the overall structure will deflect. Yet, the deflections of civil engineering structures under the action of usual design loads are known to be small in relation to both the overall dimensions and the member lengths. The logical question arises: Why bother to compute deflections? Clearly, one proper answer is to establish that the predicted design loads will not result in large deflections which can cause structural failure, impede serviceability, or result in an aesthetically displeasing and distorted structure. In addition, the skills derived from a mastery of deflection methods form a vital part of an engineer's training to both understand the behavior of various structures and develop the needed insights to evaluate the structural adequacy and performance of a design system.

Several examples which clearly demonstrate the value of deflection analysis are listed as follows:

1 Wind forces on tall buildings have been known to produce excessive lateral deflections which have resulted in cracked windows and walls, as well as discomfort to occupants.

2 Large floor deflections in a building are aesthetically unattractive, do not inspire confidence, may crack brittle finishes or cause other damage, and can be unsafe.

235 Introduction—Why Compute Deflections?

Historic Craigellachie Iron Bridge over the River Spey, Scotland. Built in 1815. (*Courtesy of International Structural Slides, Berkeley.*)

3 Floor systems are often designed to support motor-driven machines or sensitive equipment which will run satisfactorily only if the support system undergoes limited deflections.

4 Large deflections on a railway or highway structural support system may impair ride quality, cause passenger discomfort, and be unsafe.

5 Deflection control and camber behavior of prestressed concrete beams during the various stages of construction and loading, are vital for a successful design.

6 Deflection computations serve to establish the vibration and dynamic characteristics of structures that must withstand moving loads, vibration, and shock environments—inclusive of seismic design loads.

7 One of the most immediate and important benefits derived from deflection analysis is that statically indeterminate structures are solvable by deflection methods (see Chaps. 11 to 14).

Many design codes specify or recommend limited deflections to ensure serviceability without damage. For example, Sec. 1.13.1 of the AISC *Manual of Steel Construction* requires beams and girders which support floors or roofs with plastered ceilings to be proportioned so that the maximum live-load deflection does not exceed 1/360 of the span. Also, Sec. 9.5 of the ACI 318-83 Building Code Requirements provides deflection control recommendations for reinforced and prestressed concrete design. The American Institute of Timber Construction (AITC) and many building codes stipulate maximum recommended deflections for structural wood members based on parameters such as use group, seasoned or unseasoned lumber, live load, or live-plus dead-load combinations, and so forth.

Civil engineering structures are generally manufactured from construction materials such as steel, concrete, and timber, with the general expectation that the original geometry of the undeformed structure will be reasonably maintained both before and after design loads are applied. In general, it can be stated that the original geometric form is maintained if small elastic deformations of the structural system exist; that is, if the elastic limit of the structural materials used are not exceeded. For example, the elongation of a 15-ft structural steel truss member with an allowable tensile strength of 22 ksi and an elastic modulus of 29,000 ksi is $L \cdot \varepsilon = L \cdot (\sigma/E) = 15 \cdot (22/29,000) = 0.0114$ ft $= 0.137$ in. Thus, deflections of typical civil engineering structures are usually conducted by using the original undeformed geometry of the structure. All the deflection methods presented in this text are based on linear elastic material behavior.

Several beam-deflection methods are presented in this book. An engineer may favor some methods more than others or decide one is more suitable for solution or easier to use. The ability to solve a deflection problem by several methods will enhance an engineer's versatility and provide the means to perform independent check solutions.

9.2. DOUBLE INTEGRATION METHOD

The double integration method for computing slopes and displacements at various points along a beam axis has its usual introduction in mechanics of materials. The method is based on pure bending of a prismatic beam section into a circular arc under elastic material behavior (i.e., strain varies linearly from the neutral axis, stress is proportional to strain, Young's modulus is constant, etc.).

In general, the transverse displacements and slope changes along a beam length are primarily the result of bending-moment forces. The axial and torsion beam forces, as a rule, provide secondary effects which are deemed negligible when compared to bending deflection values. Although, normal beam loading produces both internal bending and shear forces, the shear forces usually provide secondary beam deflections which, in general, are also neglected. However, significant transverse shear deformations can occur, especially for beams that have thin webs, a deep cross section, a very short span, or are made from a low modulus material; in these cases, shear deformations should be investigated. In general, the double integration method will provide accurate computations for linear elastic prismatic beam deflections, especially when the shear deformations are negligible.

The curvature of the neutral axis of a prismatic beam under pure elastic bending is expressed as

$$\frac{1}{p} = \frac{M(x)}{EI} \tag{9.1}$$

where $M(x)$ = the bending moment with the beam oriented along the x axis
E = Young's modulus
I = the moment of inertia about the neutral axis
EI = the *flexural rigidity* of the beam

For a prismatic beam, the curvature is known to vary in direct relation to the change of $M(x)$ along the beam axis.

From elementary calculus, the curvature of a plane curve at a point is expressed as

$$\frac{1}{p} = \frac{d^2y/dx^2}{[1 + (dy/dx)^2]^{3/2}} \tag{9.2}$$

where y = the beam displacement normal to the x axis
$\frac{dy}{dx}$ = the slope of the curve of the beam

and x is directed along the beam axis. When beam stresses are within the proportional limit, the slope of the elastic curve of a beam (dy/dx) is a small quantity; thus, $(dy/dx)^2$ is considered negligible when compared to unity in the denominator of Eq. (9.2). Therefore, Eq. (9.2) may be rewritten as

$$\frac{1}{p} = \frac{d^2y}{dx^2} \tag{9.3}$$

By equating Eq. (9.1) to Eq. (9.3), we have

$$\frac{d^2y}{dx^2} = \frac{M(x)}{EI} \tag{9.4}$$

which is a second-order linear differential equation. If a single integration is performed on Eq. (9.4), we have the equation of the slope change at a point on the elastic curve of the beam; double integration of Eq. (9.4) provides the equation for the transverse displacement (deflection) along the beam. Obviously, a constant of integration is identified with each integration performed.

238 Deformations: Beams

For a prismatic beam (constant EI), the integration process is conducted after the bending moment is algebraically expressed as a function of x. However, if the flexural rigidity (EI) varies along the x axis, it also must be expressed as a function of x before integration can occur. Examples 9.1, 9.2, and 9.3 provide slope and deflection relationships of statically determinate prismatic beams by the double integration method.

EXAMPLE 9.1

Determine the deflection equation of the elastic curve of the prismatic beam with simple end supports for a uniform distributed load of w per unit length. Also, determine the maximum deflection of the beam.

$$\frac{M(x)}{EI} = \frac{d^2y}{dx^2} = \frac{1}{EI}\left(\frac{wL}{2}x - \frac{wx^2}{2}\right)$$

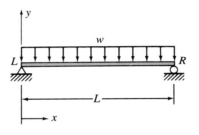
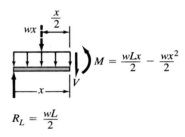

Integrating once, we have

$$\frac{dy}{dx} = \frac{1}{EI}\left(\frac{wLx^2}{4} - \frac{wx^3}{6} + C_1\right)$$

Integrating again, we have

$$y = \frac{1}{EI}\left(\frac{wLx^3}{12} - \frac{wx^4}{24} + C_1x + C_2\right)$$

Boundary condition: at $x = 0$, $y = 0$. Therefore,

$$y = 0 = C_2.$$

Boundary condition: at $x = L$, $y = 0$. Therefore,

$$y = 0 = \frac{1}{EI}\left(\frac{wL^4}{12} - \frac{wL^4}{24} + C_1L\right)$$

$$C_1 = -\frac{wL^3}{24}$$

Thus, equation of elastic curve deflection is

$$y = \frac{1}{EI}\left(\frac{wLx^3}{12} - \frac{wx^4}{24} - \frac{wL^3}{24}x\right)$$

$$= \frac{w}{24EI}(-x^4 + 2Lx^3 - L^3x)$$

239 Double Integration Method

Because of symmetry of geometry and load, it is apparent that maximum deflection occurs at midspan. Also, the beam's maximum deflection occurs where slope (dy/dx) is zero. Thus, by setting the slope equation above equal to zero, x is found equal to $L/2$. At $x = L/2$,

$$y_{max} = \frac{w}{24EI}\left(-\frac{L^4}{16} + \frac{L^4}{8} - \frac{L^4}{2}\right)$$

$$= -\frac{5wL^4}{384EI} \quad \text{Answer}$$

EXAMPLE 9.2

Write the deflection equation of the elastic curve of the prismatic cantilever beam which carries a linear varying distributed load. Also, determine the rotation at the free end of the beam.

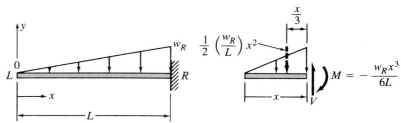

$$\frac{M(x)}{EI} = \frac{d^2y}{dx^2} = \frac{1}{EI}\left(-\frac{w_R x^3}{6L}\right)$$

By integrating twice, we have

$$\frac{dy}{dx} = \frac{1}{EI}\left(-\frac{w_R x^4}{24L} + C_1\right) \quad \text{and}$$

$$y = \frac{1}{EI}\left(-\frac{w_R x^5}{120L} + C_1 x + C_2\right)$$

The boundary conditions to be satisfied are:

at $x = L$: $y = 0$ and $\frac{dy}{dx} = 0$

Thus,

$$\frac{dy}{dx} = 0 = \frac{1}{EI}\left(-\frac{w_R L^4}{24L} + C_1\right) \quad \therefore \quad C_1 = \frac{w_R L^3}{24}$$

and

$$y = 0 = \frac{1}{EI}\left[-\frac{w_R L^5}{120L} + \left(\frac{w_R L^3}{24}\right)L + C_2\right]$$

$$\therefore\ C_2 = \frac{w_R L^4}{120} - \frac{w_R L^4}{24} = \frac{w_R L^4}{240}(2-10) = -\frac{w_R L^4}{30}$$

$$\therefore\ y = \frac{1}{EI}\left(-\frac{w_R x^5}{120L} + \frac{w_R L^3 x}{24} - \frac{w_R L^4}{30}\right)$$

or

$$y = \frac{w_R}{240EI}\left(-\frac{2x^5}{L} + 10L^3 x - 8L^4\right) \quad \text{Answer}$$

Rotation at free end:

$$\theta = \frac{dy}{dx} = \frac{1}{EI}\left(-\frac{w_R(0)^4}{24L} + \frac{w_R L^3}{24}\right) = \frac{w_R L^3}{24EI} \quad \text{Answer}$$

EXAMPLE 9.3

Determine the deflection equation of the elastic curve of the prismatic simple beam with a single concentrated load of P at point O.

$$\frac{M(x)}{EI} = \frac{d^2y}{dx^2}$$

For $0 \leqslant x \leqslant L/3$:

$$\frac{M(x)}{EI} = \frac{d^2y}{dx^2} = \frac{1}{EI}\left(\frac{2}{3}Px\right)$$

Integrating yields

$$\left(\frac{dy}{dx}\right)_L = \frac{1}{EI}\left(\frac{Px^2}{3} + C_1\right) \quad \text{and} \quad y_L = \frac{1}{EI}\left(\frac{Px^3}{9} + C_1 x + C_2\right)$$

For $L/3 \leqslant x \leqslant L$:

$$\frac{M(x)}{EI} = \frac{d^2y}{dx^2} = \frac{1}{EI}\left(-\frac{Px}{3} + \frac{PL}{3}\right)$$

$M = \frac{2}{3}Px \quad \text{for } 0 \leq x \leq \frac{L}{3}$

$M = \frac{2}{3}Px - P\left(x - \frac{L}{3}\right) \quad \text{or}$

$M = -\frac{Px}{3} + \frac{PL}{3} \quad \text{for } \frac{L}{3} \leq x \leq L$

Double Integration Method

Integration yields

$$\left(\frac{dy}{dx}\right)_R = \frac{1}{EI}\left(-\frac{Px^2}{6} + \frac{PLx}{3} + C_3\right) \quad \text{and}$$

$$y_R = \frac{1}{EI}\left[-\frac{Px^3}{18} + \frac{PLx^2}{6} + C_3 x + C_4\right]$$

Boundary conditions to be satisfied:

At $x = 0$, $y = 0$

At $x = L$, $y = 0$

Compatibility equations to be satisfied:

At point O, $(x = L/3)$:

$$y_L = y_R$$

$$\left(\frac{dy}{dx}\right)_L = \left(\frac{dy}{dx}\right)_R$$

Thus, using these four condition equations, the four constants of integration can be found.

At $x = 0$, $y = 0$:

$$\therefore \quad y_L = 0 = C_2$$

At $x = L$, $y = 0$:

$$y_R = 0 = \frac{1}{EI}\left(-\frac{PL^3}{18} + \frac{PL^3}{6} + C_3 L + C_4\right)$$

At $x = L/3$, $(dy/dx)_L = (dy/dx)_R$:

$$\frac{1}{EI}\left(\frac{PL^2}{27} + C_1\right) = \frac{1}{EI}\left(-\frac{PL^2}{54} + \frac{PL^2}{9} + C_3\right)$$

At $x = L/3$, $y_L = y_R$:

$$\frac{1}{EI}\left(\frac{PL^3}{243} + \frac{C_1 L}{3}\right) = \frac{1}{EI}\left(-\frac{PL^3}{486} + \frac{PL^3}{54} + \frac{C_3 L}{3} + C_4\right)$$

Solving these three equations for the three constants reveals that

$$C_1 = -10PL^2/162$$

$$C_3 = -19PL^2/162$$

and $C_4 = +PL^3/162$

At $x = \dfrac{L}{3}$: $\left(\dfrac{dy}{dx}\right)_{L/3} = -\dfrac{2PL^2}{81EI}$

At $x = \dfrac{L}{3}$: $y_{L/3} = -\dfrac{4PL^3}{243EI}$

Notes
1. The negative sign of $y_{L/3}$ indicates that the displacement is directed downward.
2. The negative sign of $(dy/dx)_{L/3}$ indicates a negative slope change at $x = L/3$.
3. The maximum displacement (y_{max}) occurs at the location where $dy/dx = 0$.

The limits of integration for prismatic beams are set as the range over which the bending moment can be expressed by one continuous function. The algebraic expression for bending moment will suddenly change at interior points along the beam where a reaction, concentrated load, or local couple is introduced, or where a distributed force begins or ends. Therefore, several integrals may be required (with an equal number of integration constants) to describe displacements along a beam. The constants of integration are usually found by substitution of known slope and deflection values at specific boundary locations; this process is also referred to as satisfying *boundary conditions*. For example, an ideal-pin simple support experiences zero deflection, whereas both zero slope and zero deflection are presumed to occur at fixed-end supports. In addition to boundary conditions, *compatibility conditions* are often introduced and needed to resolve some integration constants. In Example 9.3, four constants of integration must be evaluated. However, only two boundary conditions are known, namely, the deflection at both simple end supports is zero. The remaining two unknown constants are found by the use of *compatibility equations* that are written to express the continuity of deflection on both sides of the beam at point O; that is, only one value of slope (dy/dx) and one value of displacement exists at point O.

An alternative mathematical approach that may be of interest for solving beam deflection problems is to write bending-moment expressions by use of singularity functions [9.1]. The double integration method can also be used to determine deflections of statically indeterminate beams.

9.3. METHOD OF MOMENT AREA

Theorems In 1873, Charles E. Greene of the University of Michigan presented his moment-area theorems which are useful to determine beam deflections. This deflection method is based on linear elastic relationships between the shape of the elastic curve, the bending moment, and the flexural rigidity of the beam. The method is particularly simple to use if the area properties of the moment diagram of a loaded prismatic beam are known. It is also beneficial for the solution of deflections for beams with sudden changes in moment of inertia (e.g., beams with cover plates).

In the double integration method presented in Sec. 9.2 above, it was found that $d^2y/dx^2 = M/EI$, where y is the transverse beam displacement and dy/dx is the slope change at any point on the elastic curve. Letting $\theta = dy/dx$, the expression above can be rewritten as

$$\frac{d^2y}{dx^2} = \frac{d\theta}{dx} = \frac{M}{EI} \tag{9.5}$$

Equation (9.5) can be rewritten as

$$d\theta = \frac{M \cdot dx}{EI} \tag{9.6}$$

Now, observe the simple prismatic beam under an arbitrary load in Fig. 9.1a. Two points, A and B, are a dx distance apart and located at a distance x measured from the left support. An exaggerated version of the deformed elastic curve is presented in Fig. 9.1b where tangents to the curve are drawn at A and B. In the unloaded state, the beam is assumed to be perfectly straight and the tangents at A and B are concurrent with the beam axis. When the beam is subjected to elastic loads, a change in slope between the tangents at A and B occurs, namely, $d\theta$, which is defined by Eq. (9.6). If the moment diagram of the beam is divided by the EI value of the beam, the resulting construction is called the M/EI diagram as shown in Fig. 9.1c. A study of Eq. (9.6) and the shaded area of the M/EI diagram reveals that the change in slope between tangents drawn at A and B, $d\theta$, is equivalent to the *area* under the M/EI diagram between A and B. Therefore, the change in slope between tangents drawn at two points on the elastic curve, such as at the left support (L) and the right support (R), can be expressed as

$$\theta_{lr} = \int_L^R \frac{M \cdot dx}{EI} \tag{9.7}$$

This expression is valid if the elastic curve is a continuous function between L and R. If a discontinuity in the elastic curve exists (e.g., at an interior hinge in a beam) and the slope at a point to one side of the discontinuity is desired, Eq. (9.7) can be applied to the continuous side of the discontinuity that includes the point. The first moment-area theorem can be defined as follows:

First theorem: *The change in slope between two points on a continuous elastic curve is equal to the area of the M/EI diagram between the two points.*

The second moment-area theorem provides a way to compute the transverse beam displacement at any point on the elastic curve by use of the M/EI diagram area properties and relative distances measured from

244 Deformations: Beams

Figure 9.1

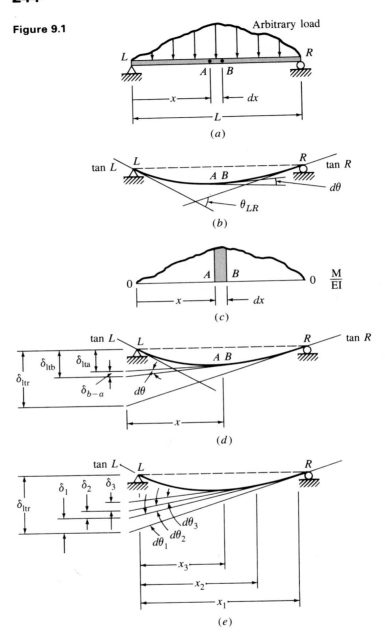

tangents to the elastic curve. The deformed elastic beam curve is repeated in Fig. 9.1d showing a tangent drawn at R. The vertical distance measured to the elastic curve at L from the tangent at R is identified as δ_{ltr}. Figure 9.1d also shows tangents drawn at points A and B extended below the left support. The relative distance to point L, as measured from the tangents drawn from A and B, is equal to $\delta_{ltb} - \delta_{lta}$ and redefined as δ_{b-a} in Fig. 9.1d. This relative distance approximates the arc length of a circle drawn by a radius x sweeping through an angle of $d\theta$; since small deformations are characteristic of linear elastic behavior, it is reasonable to assume that δ_{b-a} is equivalent to a circular arc length by the expression

$$\delta_{b-a} = x \cdot d\theta = x \cdot \left(\frac{M \cdot dx}{EI}\right) \tag{9.8}$$

Furthermore, the vertical distance to the elastic curve at L from the tangent drawn from R (δ_{ltr}) is equal to the sum of the component arc segments between adjacent tangents; that is, $\delta_{ltr} = x_1 \cdot d\theta_1 + x_2 \cdot d\theta_2 + x_3 \cdot d\theta_3 + \cdots + x_n \cdot d\theta_n$ as seen in Fig. 9.1e. In integral form, we have

$$\delta_{ltr} = \int_L^R x \cdot \left(\frac{M \cdot dx}{EI}\right) = \bar{x} \cdot A_{M/EI} \tag{9.9}$$

where $\bar{x}$ is the distance measured about L from the centroid of the M/EI area between L and R, and $A_{M/EI}$ represents the area of the M/EI diagram between L and R. Equation (9.9) applies to any two points on a continuous elastic curve such that the second moment-area theorem can be stated as follows:

Second theorem: Consider points A and B on a straight undeformed beam element in a horizontal orientation. Upon loading, A and B represent two points on a continuous elastic curve. *The vertical distance to point* **A** *on the elastic curve, as measured from a tangent at point* **B** *on the elastic curve, is equal to the product of the area of the* M/EI *diagram between points* **A** *and* **B** *and the distance from the centroid of the* M/EI *area about point* **A**.

Both moment-area theorems can be used to determine slopes and vertical displacements of beams under elastic loading. The recommended procedure used in subsequent example problems is presented as follows:

1 Draw the bending-moment diagram of the beam.
2 Draw an accurate sketch of the deformed elastic curve. The elastic curve is concave upward in regions of positive bending moment and concave downward in regions of negative bending moment.
3 Convert the M diagram to an M/EI diagram.
4 On the elastic curve, draw a tangent at a point of known slope or displacement.
5 Establish a geometric strategy to determine slopes and displacements at points of interest on the elastic curve by the moment-area theorems (other tangents to the elastic curve may be required). Figure 9.2 presents geometric approaches to determine slopes and deflections on a cantilever beam, a simple beam, and a simple beam with an overhanging segment by use of the moment-area theorems.

1. Draw elastic curve due to load P.
2. Draw tangent at a which is coincident with unloaded beam axis.
3. $\theta_b = \theta_b - \theta_a = \theta_b - 0 = A_{M/EI}\big|_a^b$
4. $\theta_c = \theta_c - \theta_a = \theta_c - 0 = A_{M/EI}\big|_a^c$
5. $\delta_c = cc' = \delta_{cta} = \bar{x} \cdot A_{M/EI}\big|_a^c$ about c
6. $\delta_b = bb' = \delta_{bta} = \bar{x} \cdot A_{M/EI}\big|_a^b$ about b

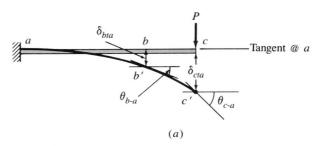

(a)

---

1. Draw elastic curve due to load P.
2. Draw tangent at support a.
3. For elastic deformations,

$$\tan \theta_a \approx \theta_a = \frac{cc'}{L_{ac}} = \frac{\delta_{cta}}{L_{ac}}$$

4. Find δ_{cta} from second theorem.
5. Evaluate θ_a from step 3.
6. From similar triangles,

$$\frac{bb'}{L_{ab}} = \frac{cc'}{L_{ac}} = \frac{\delta_{cta}}{L_{ac}}$$

7. Deflection at $b = bb' - \delta_{bta}$.
8. With θ_a known, $\theta_b + \theta_a = A_{M/EI}\big|_a^b$

thus, $\theta_b = \left(A_{M/EI}\big|_a^b\right) - \theta_a$

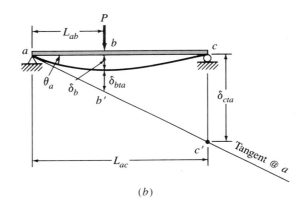

(b)

---

1. Draw elastic curve due to P loads.
2. Draw tangent at support a.
3. $\delta_{bta} = bd$ found by second theorem.
4. By similar triangles,

$$\frac{bd}{L_{ab}} = \frac{\delta_{bta}}{L_{ab}} = \frac{ce}{L_{ac}}$$

5. By second theorem, determine δ_{cta}.
6. $\delta_c = \delta_{cta} - ce$
7. $\theta_a \approx \tan \theta_a = \dfrac{\delta_{bta}}{L_{ab}}$

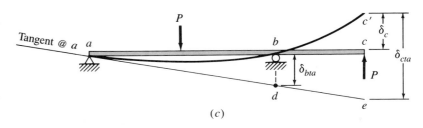

(c)

Figure 9.2 Geometric strategies for moment-area solutions of beam deflections. (a) Cantilever beam; (b) simple beam; (c) simple beam with overhang.

The moment-area method is particularly easy to use when the area properties of the M/EI *diagram are readily known.* A beam that only supports concentrated loads has a moment diagram formed by a series of straight chords which can easily be divided into rectangles and triangles. A simple beam that supports a uniform distributed loading over its length has a moment diagram of known parabolic shape. These and other geometric area properties that may be useful for *M/EI* areas appear in Fig. 9.3, App. 2, and on the inner front cover of the text.

When a beam is subjected to several types of load, the resultant moment diagram can be complex and not easily defined from conventional geometric area properties. One option is to solve the problem by another method such as the double integration method. Another option is offered as follows: Construct separate moment diagrams for each type of load. If the area of the *M/EI* diagrams for each type of load is readily known, solve the problem by the moment-area method using the principle of superposition (see Sec. 2.11).

Applications of the moment-area method Examples 9.4 to 9.7 demonstrate that the moment-area method is mathematically simple and

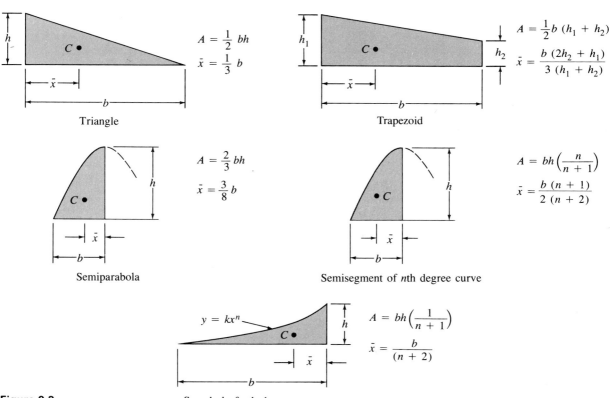

Figure 9.3

uncomplicated when the geometric area properties of the M/EI diagram are readily known. The careful sketch of the elastic curve and a logical geometric strategy (see Fig. 9.2) are essential steps for the successful application of this method. Examples 9.4, 9.5, and 9.6 employ the tangent at the fixed support since it coincides with the unloaded cantilever-beam axis and leads to direct solutions of slopes and displacements along the beam. Example 9.6 also shows the importance of developing separate moment diagrams for concentrated and uniform loads which lead to correct results by use of the principle of superposition; a combined moment diagram would require a significant effort to arrive at the geometric area properties and thereby defeat the intent of this method. In Example 9.7, the second moment-area theorem is used to compute the maximum deflection on a simple beam; this approach can be used to determine the deflection at any point on the simple beam and can be applied to beams of other supports.

EXAMPLE 9.4

Determine the slopes and deflections at each load on the prismatic cantilever beam. $E = 29 \times 10^6$ psi and $I = 900$ in^4.

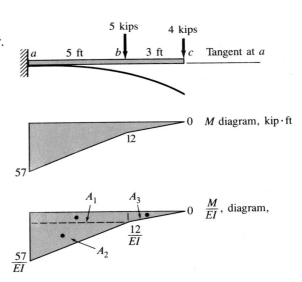

Since EI is constant over the entire beam length, the shape of the M/EI diagram is identical to the M diagram. The M/EI is subdivided into triangular and rectangular areas where dots identify the centroid locations.

$A_1 = 5 \cdot 12/EI = 60$ kip $\cdot$ ft$^2/EI$

$A_2 = \dfrac{1}{2} \cdot 5 \cdot \dfrac{45}{EI} = \dfrac{112.5 \text{ kip} \cdot \text{ft}^2}{EI}$

$A_3 = \dfrac{1}{2} \cdot 3 \cdot \dfrac{12}{EI} = \dfrac{18 \text{ kip} \cdot \text{ft}^2}{EI}$

Slope Solutions

Draw tangents at a, b, and c.

$$\theta_c - \theta_a = \theta_c - 0 = A_{M/EI}\Big|_a^c$$

$$\theta_c = \frac{1}{EI}(60 + 112.5 + 18) = 190.5 \text{ kip} \cdot \text{ft}^2/EI$$

$$\theta_c = \frac{190.5 \text{ kip} \cdot \text{ft}^2 \times 144 \text{ in}^2/\text{ft}^2 \times 1000 \text{ lb/kips}}{29 \times 10^6 \times 900 \text{ kip} \cdot \text{in}^2} = 1.05 \times 10^{-3} \text{ radians} \blacktriangleleft$$

$$\theta_b - \theta_a = \theta_b - 0 = A_{M/EI}\Big|_a^b$$

$$\theta_b = A_1 + A_2 = \frac{172.5 \text{ kip} \cdot \text{ft}^2}{EI} = 0.95 \times 10^{-3} \text{ radians} \blacktriangleleft$$

Deflection Solutions

From the tangent drawn at a,

$$\delta_c = \delta_{cta} \quad \text{and} \quad \delta_b = \delta_{bta}$$

$$\delta_c = A_1\left(\frac{5}{2} + 3\right) + A_2\left[\frac{2(5)}{3} + 3\right]$$
$$\quad + A_3\left[\frac{2(3)}{3}\right]$$

$$\delta_c = \frac{60}{EI}(5.5) + \frac{112.5}{EI}(6.33) + \frac{18}{EI}(2)$$

$$\delta_c = \frac{1078 \text{ kip} \cdot \text{ft}^3}{EI} = \frac{1078 \times 1728 \text{ in}^3/\text{ft}^3 \times 1000 \text{ lb/kip}}{29 \times 10^6 \times 900 \text{ kip} \cdot \text{in}^2}$$

$$\delta_c = 0.071 \text{ in} \blacktriangleleft$$

$$\delta_b = \delta_{bta} = A_1\left(\frac{5}{2}\right) + A_2\left[\frac{2(5)}{3}\right]$$

$$\delta_b = \frac{60}{EI}(2.5) + \frac{112.5}{EI}(3.33) = \frac{525 \text{ kip} \cdot \text{ft}^3}{EI}$$

$$\delta_b = \frac{525 \text{ kip} \cdot \text{ft}^3 \times 1728 \text{ in}^3/\text{ft}^3 \times 1000 \text{ lb/kip}}{29 \times 10^6 \times 900 \text{ kip} \cdot \text{in}^2}$$

$$\delta_b = 0.035 \text{ in} \blacktriangleleft$$

EXAMPLE 9.5

Determine the slope and deflection at the free end of the cantilever beam. $E = 29 \times 10^6$ psi, $I_1 = 1000$ in^4, $I_2 = 2000$ in^4.

The substitution of the EI value is postponed until the end of a solution since it represents a large quantity. If EI values are substituted in the M/EI diagram ordinates, very small quantities arise which are cumbersome. Note that the M/EI diagram is given in terms of I_1 only. Thus, in the I_2 region:

At b: $\dfrac{20}{EI_2} = \dfrac{20}{E(2I_1)} = \dfrac{10}{EI_1}$

At a: $\dfrac{50}{EI_2} = \dfrac{50}{E(2I_1)} = \dfrac{25}{EI_1}$

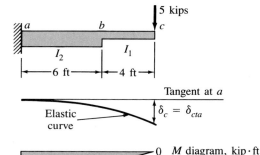

From a tangent at support a,

$\theta_c = \theta_c - \theta_a = \theta_c - 0 = A_{M/EI}\Big|_a^c$

$\theta_c = \left(\dfrac{1}{2} \cdot \dfrac{20}{EI_1} \cdot 4\right) + \left(\dfrac{1}{2} \cdot \dfrac{10}{EI_1} \cdot 6\right) + \left(\dfrac{1}{2} \cdot \dfrac{25}{EI_1} \cdot 6\right)$

$\theta_c = \dfrac{40}{EI_1} + \dfrac{30}{EI_1} + \dfrac{75}{EI_1} = \dfrac{145 \text{ kip} \cdot \text{ft}^2}{EI_1}$

$\theta_c = \dfrac{145 \text{ kip} \cdot \text{ft}^2 \cdot 144 \text{ in}^2/\text{ft}^2}{29{,}000 \text{ ksi} \cdot 1000 \text{ in}^4}$

$\theta_c = 7.2 \times 10^{-4}$ radians ◀

$\delta_c = \delta_{ctA}$

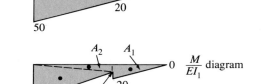

$\delta_c = A_1 \cdot \dfrac{8}{3} + A_2 \cdot (2 + 4) + A_3 \cdot (4 + 4)$

$\delta_c = \left(\dfrac{40}{EI_1} \cdot \dfrac{8}{3}\right) + \left(\dfrac{30}{EI_1} \cdot 6\right) + \left(\dfrac{75}{EI_1} \cdot 8\right)$

$\delta_c = \dfrac{106.7}{EI_1} + \dfrac{180}{EI_1} + \dfrac{600}{EI_1} = \dfrac{886.7 \text{ kip} \cdot \text{ft}^3}{EI_1}$

$\delta_c = \dfrac{886.7 \text{ kip} \cdot \text{ft}^3 \cdot 1728 \text{ in}^3/\text{ft}^3}{29{,}000 \text{ ksi} \cdot 1000 \text{ in}^4}$

$\delta_c = 0.053$ in ◀

EXAMPLE 9.6

Determine the deflection at the free end of the prismatic cantilever beam. Thus, EI is constant.

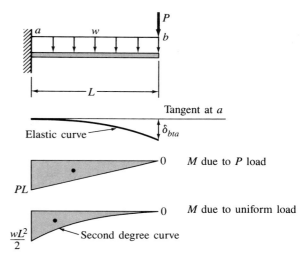

Solution

Using the principle of superposition, bending-moment diagrams for each load are found. M/EI diagrams are of identical shape of M diagrams. See Fig. 9.3 for area properties.

$$\delta_b = \delta_{bta} = \delta_{bp} + \delta_{bw}$$

where δ_{bp} is deflection due to P and δ_{bw} is deflection due to w.

$$\delta_{bp} = \frac{1}{2}\left(\frac{PL}{EI} \cdot L\right)\frac{2L}{3} = \frac{PL^3}{3EI}$$

$$\delta_{bw} = \frac{1}{3}\left(\frac{wL^2}{2EI} \cdot L\right)\frac{3L}{4} = \frac{wL^4}{8EI}$$

$$\therefore \quad \delta_b = \frac{PL^3}{3EI} + \frac{wL^4}{8EI} \quad \text{Answer}$$

EXAMPLE 9.7

Find the maximum deflection for the prismatic simple beam under uniform distributed loading.

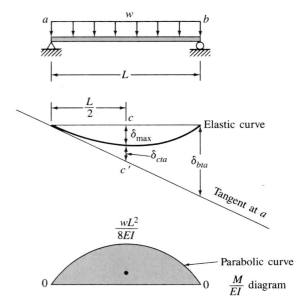

Due to symmetry, δ_{max} occurs at midspan,

$$\delta_{bta} = \left(\frac{2}{3} \cdot \frac{wL^2}{8EI} \cdot L\right) \cdot \frac{L}{2} = \frac{wL^4}{24EI}$$

by similar triangles,

$$\frac{cc'}{L/2} = \frac{\delta_{bta}}{L} \quad \therefore \quad cc' = \frac{\delta_{bta}}{2} = \frac{wL^4}{48EI}$$

$$\delta_{cta} = \left[\frac{2}{3} \cdot \frac{wL^2}{8EI} \cdot \frac{L}{2}\right] \cdot \frac{3}{8}\left(\frac{L}{2}\right) = \frac{wL^4}{128EI}$$

$$\delta_{max} = cc' - \delta_{cta} = \frac{wL^4}{48EI} - \frac{wL^4}{128EI}$$

$$= \frac{5wL^4}{384EI} \quad \text{Answer}$$

Student exercise: Using the example data above and a moment-area theorem, determine the slope at support *a* and verify your answer independently.

A sign convention is noticeably absent in this presentation of the moment-area method. The author feels that directions of slopes and deflections are evident from a properly drawn sketch of the elastic curve for the beam under load.

9.4. METHOD OF CONJUGATE BEAM

The conjugate-beam method was originally introduced as the elastic-weights method by Otto Mohr in 1860, and is closely related to the moment-area method. In Sec. 5.5, elastic relationships are derived between load, shear, and moment so that the beam load, $w = dV/dx$, and $V = dM/dx$. From Sec. 9.2, we note that y = beam deflection, $dy/dx = \theta$ = slope, and $d^2y/dx^2 = M/EI$. Successive derivatives of y with respect to x for a section of constant EI yields

$$\frac{d^3y}{dx^3} = \left(\frac{dM}{dx}\right)\frac{1}{EI} = \frac{V}{EI}$$

and

$$\frac{d^4y}{dx^4} = \left(\frac{dV}{dx}\right)\frac{1}{EI} = \frac{w}{EI}$$

where

$$y = \text{deflection}$$

$$\frac{dy}{dx} = \theta = \text{slope}$$

$$EI\left(\frac{d^2y}{dx^2}\right) = M = \text{moment} \qquad (9.10)$$

$$EI\left(\frac{d^3y}{dx^3}\right) = V = \text{shear}$$

$$EI\left(\frac{d^4y}{dx^4}\right) = w = \text{load}$$

Therefore, successive derivatives of y lead from deflection to load. In reverse fashion, it is shown in Chap. 5 that shear and moment diagrams are determined by integration of load to obtain shear (V), and integration of shear to obtain moment (M). Also, by way of analogy, it is seen from Eq. (9.10) that integration of the M/EI diagram of an elastic beam will result in a slope diagram and integration of a slope diagram will yield a deflection diagram. That is, slope and deflection values on the real beam are found from analogous shear and moment diagrams of the conjugate beam under M/EI loading, respectively. Moreover, the conjugate beam V and M diagrams are developed using the same sign convention as presented in Sec. 5.3 for beam shear and bending moment.

Conjugate-beam theorems The conjugate-beam theorems are defined as follows:

Consider the M/EI diagram of an elastic beam as the load on a fictitious beam called a conjugate beam. The conjugate beam contains a span length equal to the real beam and has substitute supports (called *conjugate supports*) that satisfy certain boundary conditions.

Theorem 1: *The shear at any point on a conjugate beam that is loaded with the M/EI diagram of the real beam, is equal to the slope at the identical point on the real beam.*

Theorem 2: *The moment at any point on a conjugate beam that is loaded with the M/EI diagram of the real beam, is equal to the deflection at the identical point on the real beam.*

The conjugate-beam theorems at any point on an elastic beam can be expressed as

$$V_{CB} = \theta_{RB}$$

and

$$M_{CB} = y_{RB} \tag{9.11}$$

where the subscripts CB and RB are symbolic for conjugate beam and real beam, respectively.

Sign convention Consider a beam in a horizontal orientation.
1 Positive segments of the *M/EI* curve on the real beam are placed on the conjugate beam as a load in a downward direction (negative *M/EI* load acts upward).
2 Positive shear on the conjugate beam corresponds to clockwise slope (rotation) on the real beam and negative shear corresponds to counterclockwise slope (rotation).
3 Positive bending moment on the conjugate beam corresponds to downward displacement on the real beam and negative bending moment corresponds to upward displacement.

Applications of conjugate-beam method to simple beams
Examples 9.8 to 9.10 demonstrate the mathematical ease of using the conjugate-beam method to determine beam slopes and displacements, especially if the area properties of the *M/EI* diagram loading are readily known. The method is also known as the elastic weights method when it is applied to simple beams. In simple-beam applications, correct results are obtained since simple end supports rotate and have zero displacement which is analogous to nonzero shear and zero bending moment on the simply supported conjugate beam, respectively. However, when the conjugate-beam method is applied to beams other than simple end supported beams, the real supports must be replaced by substitute supports.

EXAMPLE 9.8

Determine the rotation (slope change) at *a* and *b* and both slope and displacement at midspan in terms of *P*, *L*, and *EI* for the simple prismatic beam.

$$\curvearrowleft + M_b = 0 = R_a \cdot L - \frac{1}{2}\left[\frac{PL}{4EI}\right](L) \cdot \frac{L}{2}$$

$$\therefore R_a = PL^2/16EI \uparrow$$

$$\theta_a = V_{a_{CB}} = +\frac{PL^2}{16EI}$$

$$\theta_b = V_{b_{CB}} = -\frac{PL^2}{16EI}$$

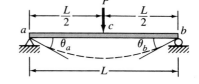

From symmetry, it is apparent that $\theta = 0$ at midspan; this is verified by the conjugate beam method as follows:

$$\theta_c = V_{c_{CB}} = \frac{PL^2}{16EI} - \frac{1}{2}\left(\frac{PL}{4EI}\right)\left(\frac{L}{2}\right) = 0$$

Displacement at midspan, δ_c:

$$\delta_c = M_{c_{CB}} = \frac{PL^2}{16EI}\left(\frac{L}{2}\right) - \left[\frac{1}{2}\left(\frac{PL}{4EI}\right)\frac{L}{2}\right]\left(\frac{L}{6}\right)$$

$$\delta_c = \frac{PL^3}{32EI} - \frac{PL^3}{96EI} = +\frac{PL^3}{48EI} \downarrow$$

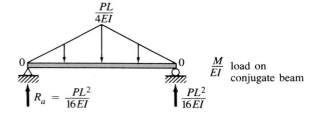

Since θ at $c = 0$, $\delta_c = \delta_{max}$.

EXAMPLE 9.9

For the simple beam under uniform loading, find θ_a, θ_b, and δ_{max} for $w = 2$ klf, $L = 30$ ft, $E = 30 \times 10^3$ ksi, and $I = 1000$ in^4.

$$\curvearrowleft + M_b = 0 = R_a \cdot L - \frac{2}{3}\left(\frac{wL^2}{8EI}\right)(L) \cdot \frac{L}{2}$$

$$\therefore R_a = \frac{wL^3}{24EI} \uparrow$$

By symmetry, $\theta_a = -\theta_b = V_{\text{conjugate beam}}$

$$\theta_a = +\frac{wL^3}{24EI} = \frac{2(30)^3 \cdot 144}{24(30 \times 10^3)1000} = 10.8 \times 10^{-3} \text{ rads}$$

$$\theta_b = 10.8 \times 10^{-3} \text{ radians}$$

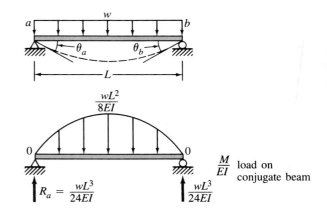

256 Deformations: Beams

By symmetry, $\delta_{max} = \delta_{midspan}$

$\delta_{max} = M_{midspan\ on\ conjugate\ beam}$

$$= \frac{wL^3}{24EI} \cdot \frac{L}{2} - \left[\frac{2}{3}\left(\frac{wL^2}{8EI}\right)\frac{L}{2}\right] \cdot \frac{3L}{8} = +\frac{5wL^4}{384EI}$$

$\delta_{max} = +\frac{5}{384} \frac{2\ klf(30)^4 \cdot 1728}{30 \times 10^3 \cdot 1000} = +1.22\ in\downarrow$

EXAMPLE 9.10

Determine the rotation and displacement at a and b by the conjugate beam method.
$E = 29 \times 10^3$ ksi, $I_1 = 400$ in^4, and $I_2 = 800$ in^4.

$A_1 = \frac{1}{2}\left(\frac{80}{EI_1}\right)12 = \frac{480}{EI_1}$

$A_2 = \frac{1}{2}\left(\frac{20}{EI_1}\right)12 = \frac{120}{EI_1}$

$A_3 = \left(\frac{20}{EI_1}\right)12 = \frac{240}{EI_1}$

$A_4 = \frac{1}{2}\left(\frac{40}{EI_1}\right)12 = \frac{240}{EI_1}$

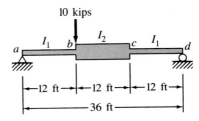

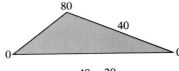

M diagram kip·ft

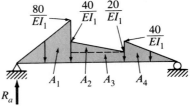

$\frac{M}{EI_1}$ load on conjugate beam

$\zeta + \Sigma M_d = 0 = 36R_a - A_1 \cdot 28\ ft - A_2 \cdot 20\ ft - A_3 \cdot 18\ ft - A_4 \cdot 8\ ft$

$0 = 36R_a - \frac{1}{EI_1}(13,440 + 2400 + 4320 + 1920)$

$= \frac{22,080}{EI}$

$R_a = +\frac{613.3}{EI_1}\uparrow$

$\theta_a = V_{a_{CB}} = +\frac{613.3\ kip \cdot ft^2}{EI_1} = \frac{613.3 \cdot 144}{29 \times 10^3 \cdot 400} = 0.0076\ radians$

$$\theta_b = V_{b_{CB}} = R_a - A_1 = \frac{613.3}{EI_1} - \frac{480}{EI_1}$$

$$= +\frac{133.3 \text{ kip} \cdot \text{ft}^2}{EI_1} = \frac{133.3 \cdot 144}{29 \times 10^3 \cdot 400} = +0.0017 \text{ radians}$$

$$\delta_b = M_{b_{CB}} = R_a \cdot 12 \text{ ft} - A_1 \cdot 4 \text{ ft} = \frac{613.3}{EI_1} \cdot 12 \text{ ft} - \frac{480}{EI_1} \cdot 4 \text{ ft}$$

$$= \frac{7360 - 1920}{EI_1} = +\frac{5440 \text{ kip} \cdot \text{ft}^3}{EI_1}$$

$$\delta_b = \frac{5440 \text{ kip} \cdot \text{ft}^3 \cdot 1728}{29 \times 10^3 \cdot 400} = 0.81 \text{ in} \downarrow$$

$$\delta_c = M_{c_{CB}} = R_a \cdot 24 \text{ ft} - A_1 \cdot 16 \text{ ft} - A_2 \cdot 8 \text{ ft} - A_3 \cdot 6 \text{ ft}$$

$$\delta_c = \frac{613.3}{EI_1} \cdot 24 - \frac{480}{EI_1} \cdot 16 - \frac{120}{EI_1} \cdot 8 - \frac{240}{EI_1} \cdot 6$$

$$\delta_c = \frac{14{,}720 - 7680 - 960 - 1440}{EI_1}$$

$$= +\frac{4640 \text{ kip} \cdot \text{ft}^3}{EI_1} = \frac{4640 \times 1728}{29 \times 10^3 \times 400} = 0.69 \text{ in} \downarrow$$

Conjugate supports The conjugate-beam method will provide correct elastic deflection results if the real supports are replaced by substitute supports that satisfy the analogous boundary conditions of Eqs. (9.11). Explanations are provided in the following paragraphs to justify each conjugate support type.

1. *Simple end support (hinge or roller)* The substitute (conjugate) support for a hinge or roller at a simple end of a beam remains a hinge or roller support since it satisfies the analogous boundary conditions of Eqs. (9.11) (see explanation provided above).

2. *Fixed-end support* A beam under load does not deflect or rotate at a fixed-end support. Therefore, at that point on the conjugate beam, no shear or bending moment must occur. This is accomplished by replacing the fixed support with a free end.

3. *Free end* A beam under load deflects and rotates at its free end. Therefore, the corresponding point on the conjugate beam must have both shear and bending moment which is provided by a fixed support.

4. *Simple interior support* A beam under load rotates but does not deflect at both simple interior and simple end supports. However, the slope is continuous across a simple interior support as compared with a simple end support where the slope begins abruptly. In order to satisfy the slope continuity over a simple interior support, the substitute support must

experience zero moment and be able to transmit shear without change. However, any type of external support will cause a change in shear; the conditions of Eqs. (9.11) are satisfied by replacing a simple interior support with an unsupported interior hinge.

5 *Interior hinge* Both slope change and deflection will occur at an unsupported interior hinge (pin) on a beam under load; therefore, its conjugate support must have both shear and bending moment. This is achieved by replacing an unsupported interior hinge by a simple interior support.

Figure 9.4 displays many examples of substitute (conjugate) supports for real-beam supports. Examples 9.11 and 9.12 show deflection solutions for a cantilever and simple beam with an overhang, respectively, by the method of conjugate beam.

Figure 9.4

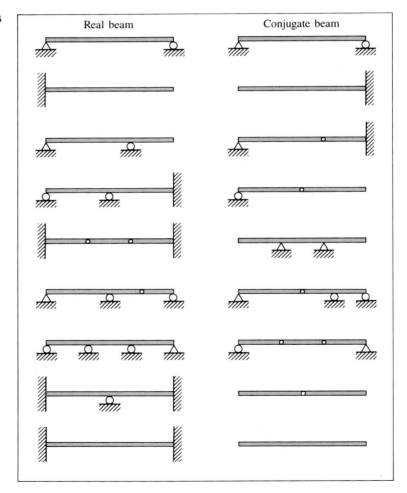

EXAMPLE 9.11

Find θ_a and δ_a by the method of conjugate beam in terms of P, w, E, I, and L for the prismatic beam.

$$\theta_a = V_{a_{CB}} = +\frac{1}{2}\left(\frac{PL}{2EI}\right)\left(\frac{L}{2}\right) = \frac{PL^2 \text{ kip} \cdot \text{ft}^2}{8EI}$$

$$\delta_a = M_{a_{CB}} = +\left[\frac{1}{2}\left(\frac{PL}{2EI}\right)\left(\frac{L}{2}\right)\right]\left(\frac{L}{2} + \frac{2}{3} \cdot \frac{L}{2}\right)$$

$$= \frac{5PL^3 \text{ kip} \cdot \text{ft}^3}{48EI} \downarrow$$

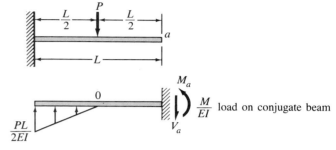

load on conjugate beam

EXAMPLE 9.12

Compute the slope and deflection 8 ft from the left support by the conjugate beam method. $E = 29{,}000$ ksi, $I = 1400$ in^4.

Reaction R_a on the conjugate beam is found by moment summation of forces from a to b about $b = 0$:

For uniform load M/EI:

$$16R_a - \left(\frac{2}{3} \cdot \frac{64}{EI} \cdot 16\right) \cdot 8 = 0 \qquad \therefore \quad R_a = 341.3/EI \uparrow$$

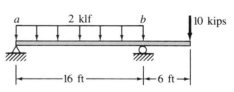

For concentrated load M/EI:

$$16R_a + \left(\frac{1}{2} \cdot \frac{60}{EI} \cdot 16\right) \cdot \frac{16}{3} = 0 \qquad \therefore \quad R_a = -\frac{160}{EI} \downarrow$$

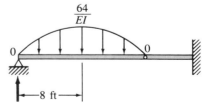

Uniform load $\frac{M}{EI}$ on conjugate beam

By the method of superposition,

$$\theta_{8\,\text{ft}} = \frac{341}{EI} - \frac{2}{3} \cdot \frac{64}{EI} \cdot 8 - \frac{160}{EI} + \frac{1}{2} \cdot \frac{30}{EI} \cdot 8$$

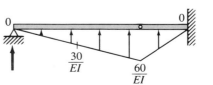

Concentrated load $\frac{M}{EI}$ on conjugate beam

$$\theta_{8\,\text{ft}} = \frac{1}{EI}(341 - 341 - 160 + 120) = -\frac{40}{EI}$$

$$\theta_{8\,\text{ft}} = -\frac{40 \cdot 144}{29{,}000 \cdot 1400} = 1.42 \times 10^{-4} \text{ radians}$$

$$\delta_{8\,\text{ft}} = \frac{341}{EI} \cdot 8\,\text{ft} - \frac{2}{3} \cdot \frac{64}{EI} \cdot 8\left(\frac{3}{8} \cdot 8\right)\text{ft}$$

$$\quad - \frac{160}{EI} \cdot 8\,\text{ft} + \frac{1}{2} \cdot \frac{30}{EI} \cdot 8\left(\frac{8}{3}\right)\text{ft}$$

$$\delta_{8\,\text{ft}} = \frac{1}{EI}(2730 - 1024 - 1280 + 320)$$

$$\delta_{8\,\text{ft}} = +\frac{746\,\text{kip}\cdot\text{ft}^3}{EI} = \frac{746 \cdot 1728}{29{,}000 \cdot 1400}$$

$$\delta_{8\,\text{ft}} = 0.032\,\text{in}\downarrow$$

Closing remarks on the conjugate-beam method

1 The conjugate beam is *always* statically determinate even if the real beam is statically indeterminate.

2 The conjugate beam for a statically indeterminate beam may appear to be unstable; however, it is in equilibrium and is stabilized by the M/EI elastic loading. Consider the fixed-fixed beam of Fig. 9.5 which has an unsupported (free-free) conjugate beam; the opposing M/EI loads on the conjugate beam satisfy equilibrium.

Figure 9.5

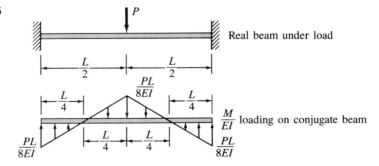

3 A comparison of the moment area and conjugate beam methods reveals that (*a*) they both use M/EI diagrams to compute beam deflections and slopes, (*b*) a sign convention is easier to establish and use correctly for the conjugate beam method, (*c*) the conjugate beam method is usually easier to apply, and (*d*) individual sets of computations are required to determine slope and deflection values at each point on a beam by the moment-area method, whereas construction of the V and M diagrams of a conjugate beam under M/EI loading provides the slope and displacement at *every* point along the beam.

4 The moment-area and conjugate-beam methods can be used, in a practical manner, to furnish approximate deflection analyses for beam loads that result in complex M/EI diagrams. This is accomplished by careful construction of the M/EI diagram followed by careful subdivision and replacement of high-order curves with straight chords as shown in Fig. 9.6. This simplification will result in M/EI areas comprised of triangles and rectangles. The author does not recommend either method for application to frame structures since both procedures usually prove to be confusing; instead, frame deflections are more easily found by energy methods such as virtual work or the theorems of Castigliano.

Figure 9.6

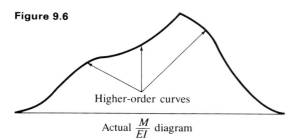

Higher-order curves

Actual $\frac{M}{EI}$ diagram

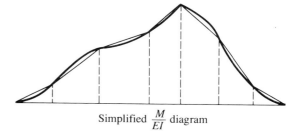

Simplified $\frac{M}{EI}$ diagram

9.5. SPECIAL APPLICATION FOR FIXED-END BEAMS

The moment-area and conjugate-beam methods can be used to determine the fixed-end moments of a prismatic beam under load. The approach is a first-order elastic analysis which neglects axial force effects. Example 9.13 shows a uniformly loaded fixed-fixed beam with no slope change or dis-

Oakland Colisium Arena, built in 1966. The roof is 420 ft in circular diameter and has an inside height of 106 ft. A composite structure of steel cables, reinforced concrete, and structural steel were used to form this impressive arena with a seating capacity of 15,000. (*Courtesy of International Structural Slides, Berkeley.*)

placement at its support ends. By use of superposition, the beam can be viewed as a simple beam with each end moment and the uniform load treated as three separate load cases as shown in Example 9.13. The M/EI diagram for each load case is also provided in Example 9.13 realizing that M_A and M_B are unknown.

EXAMPLE 9.13

Determine the fixed-end moments for the uniformly loaded beam by the moment-area or conjugate-beam method. EI is constant.

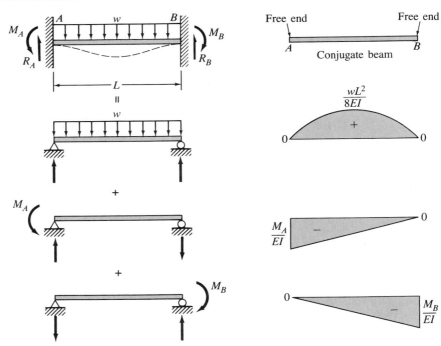

Solution by moment area

Theorem 1: $\Sigma A_{M/EI}$ between A and $B = 0$ since tangents drawn at A and B coincide.

$$\frac{2}{3}\left(\frac{wL^2}{8EI} \cdot L\right) - \frac{1}{2}\left(\frac{M_A}{EI} \cdot L\right) - \frac{1}{2}\left(\frac{M_B}{EI} \cdot L\right) = 0$$

$$\frac{wL^3}{12EI} - \frac{M_A L}{2EI} - \frac{M_B L}{2EI} = 0 \qquad (1)$$

Theorem 2: Moments of areas of M/EI between A and B about B:

$$\frac{wL^3}{12EI} \cdot \frac{L}{2} - \frac{M_A L}{2EI} \cdot \frac{2L}{3} - \frac{M_B L}{2EI} \cdot \frac{L}{3} = 0$$

$$\frac{wL^4}{24EI} - \frac{M_A L^2}{3EI} - \frac{M_B L^2}{6EI} = 0 \qquad (2)$$

Solving equations (1) and (2) simultaneously for M_A and M_B yields

$$M_A = M_B = \frac{wL^2}{12}$$

Moment area The beam elastic curve reveals no slope change or deflection from a tangent at A to a tangent at B (or vice versa). Thus, by the moment-area theorems, the area of the M/EI diagram between A and B, and the moment of the M/EI area about either A or B, are equal to zero. Both moment-area theorem equations shown in Example 9.13 contain the unknown end moments M_A and M_B which are solved by simultaneous equation solution.

Conjugate beam In the conjugate-beam method, a fixed support is replaced by a "free" support. Thus, the fixed-fixed beam of Example 9.13 is replaced by a free-free conjugate beam, and it is evident that the M/EI loads on the conjugate beam must be in equilibrium. A vertical equilibrium equation can be written to state that the sum of the three M/EI areas equal zero; the result is the same Eq. (1) in Example 9.13 found by moment area. By inspection and use of superposition, the shear at each beam end is zero which satisfies the first conjugate-beam theorem (shear at a point on the conjugate beam equals slope at the point on the real beam). Since the displacement is zero at both ends of the real beam, the second conjugate-beam theorem requires that the moment at each end of the loaded conjugate beam is zero. Therefore, an equation can be written where the moments of the three M/EI loads (in terms of w, M_A, and M_B) are taken about either beam end and their sum is set equal to zero. This equation will be identical to Eq. (2) as developed by the moment-area method in Example 9.13. Fixed-end moments for other loads may be determined by either of these area methods.

Problems

9.1 to 9.6 Derive the equation of the elastic curve for each prismatic beam (EI constant over entire length) in terms of L, E, and I. Also determine specified rotation and/or displacement.

9.1

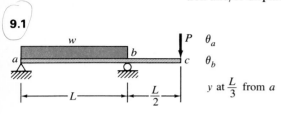

y at $\frac{L}{3}$ from a

(Define elastic curve between a and b only)

9.2

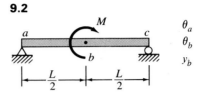

9.3

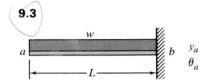

9.4

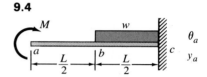

9.5

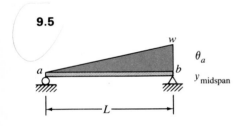

9.6

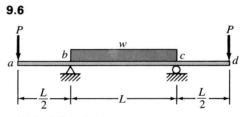

y_{max} between b and c

(Only define elastic curve between b and c)

9.7 to 9.20 Moment-area method

9.7 Determine θ_a, θ_b, δ_a, δ_b. $E = 29 \times 10^6$ psi, $I = 800$ in⁴.

9.8 Compute θ_a, θ_b, δ_a, δ_b. $E = 30 \times 10^6$ psi, $I = 1000$ in⁴.

9.9 Compute θ_a, θ_b, and δ_a. $E = 1.4 \times 10^6$ psi, $I = 1200$ in⁴.

9.10 Find θ_b and δ_b. $E = 29 \times 10^6$ psi, $I = 2000$ in⁴.

9.11 Compute θ_a, δ_a. $E = 200$ GPa, $I = 80 \times 10^8$ mm⁴.

9.12 Determine θ_a, θ_b, δ_a, δ_b. $E = 29 \times 10^3$ ksi, $I = 3000$ in⁴.

9.13 Determine θ_a and δ_b. $E = 30 \times 10^3$ ksi, $I = 600$ in⁴.

9.7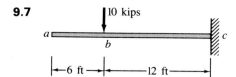

9.8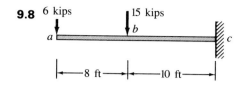

9.9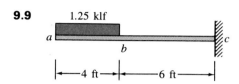

9.10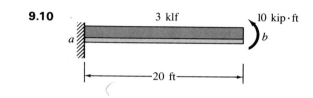

9.11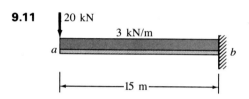

9.12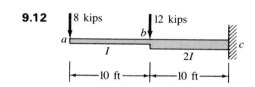

9.13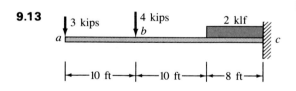

9.14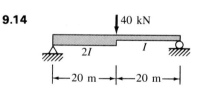

9.14 Compute the midspan rotation and displacement. $E = 200$ GPa, $I = 200 \times 10^7$ mm^4.

9.15 Determine θ_A and δ_c. $E = 30 \times 10^3$ ksi, $I = 800$ in^4.

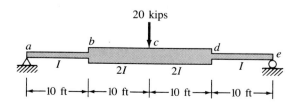

266 Deformations: Beams

9.16 Compute δ_b and δ_d in terms of EI (E constant).

9.17 Compute θ_c and δ_c in terms of EI.

9.18 Determine the displacement at hinge b. $E = 29 \times 10^6$ psi, $I = 2000$ in^4.

9.16
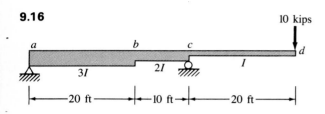

9.17
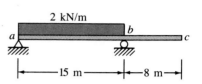

9.18
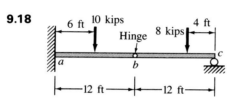

9.19 and 9.20 Determine the fixed-end moments by the moment-area method. $E = 30 \times 10^3$ ksi, $I = 1000$ in^4.

9.19
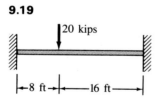

9.20
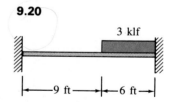

9.21 to 9.39 Conjugate-beam method
Repeat the following problems by the conjugate-beam method.
9.21 Redo 9.7.
9.22 Redo 9.8.
9.23 Redo 9.9.
9.24 Redo 9.10.
9.25 Redo 9.11.
9.26 Redo 9.12.
9.27 Redo 9.14.
9.28 Redo 9.15.

9.29 Compute θ_a, δ_b, θ_c. $E = 29 \times 10^3$ ksi, $I = 2400$ in^4.

9.29
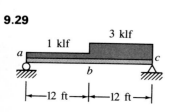

9.30

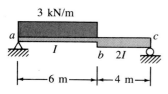

9.30 Determine θ_a, θ_b, θ_c, and δ_b. $E = 20$ GPa, $I = 4 \times 10^7$ mm^4.
9.31 Repeat Prob. 9.16 by conjugate beam.
9.32 Repeat Prob. 9.17 by conjugate beam.
9.33 Repeat Prob. 9.18 by conjugate beam.
9.34 Repeat Prob. 9.19 by conjugate beam.
9.35 Determine θ_a, δ_a, δ_c, θ_c, θ_e, and δ_e. $E = 29 \times 10^3$ ksi, $I = 1500$ in^4.

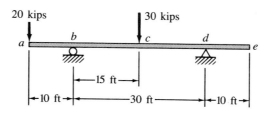

9.36 Determine θ_b, θ_d, and δ_d. $E = 29 \times 10^6$ psi, $I = 1500$ in^4.

9.37 Compute θ_b, δ_b, θ_d, and δ_d in terms of EI.

9.38 Determine θ_a, θ_c, and δ_c. $E = 30 \times 10^3$ ksi, $I = 4000$ in^4.

9.39 Determine θ_a, δ_a, and δ_c. $E = 2 \times 10^6$ psi, $I = 1800$ in^4.

9.36

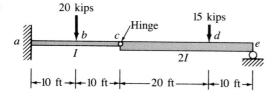

9.37

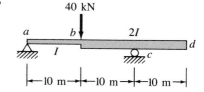

9.38

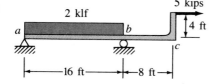

9.39

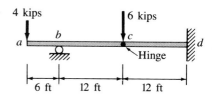

Reference

9.1 Ferdinand P. Beer and E. Russell Johnston, Jr., *Mechanics of Materials*, McGraw-Hill, New York, 1981, chap. 8.

CHAPTER TEN

Deformations: Virtual Work

10.1. INTRODUCTION TO VIRTUAL WORK AND ENERGY METHODS

The method of virtual work was developed in 1717 by John Bernoulli and remains as one of the most versatile methods used to compute deflections of truss, beam, and frame structures. In concept, the virtual work method is based on the principle of conservation of energy. It considers the work that results from placing an imaginary system of forces in equilibrium on a deformable structure. The work due to the imaginary forces is called virtual work and the imaginary forces are called virtual forces. It will become evident to the reader that virtual work is accomplished by a real force moving through an imaginary (virtual) displacement, or by a virtual force moving through a real displacement.

In this text, the application of the virtual-work method is limited to elastically deformable plane structures where (1) the real and virtual force systems, apart and independent of each other, are in equilibrium both before and after their actions produce structural deformations, and (2) the deformations of the structural elements are small and compatible with each other and with their support constraints so that the original structural geometry remains unchanged.

Deflections by virtual work can also be found for three-dimensional structures and can be used to evaluate the effects of inelastic material behavior if the conditions of equilibrium and geometric stability, as stated above, are satisfied. However, the virtual-work method does not provide

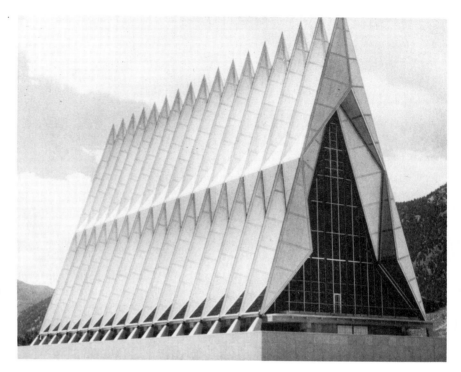

Air Force Academy Chapel, Colorado Springs. Built in 1962. The chapel is a space frame covered with aluminum sheeting. The basic structure is an A frame made of identical tetrahedron space frame units. (*Courtesy of International Structural Slides, Berkeley.*)

accurate results for cable structures since cables generally undergo notable deflections when loaded which significantly alters their geometry.

In Sec. 10.3, truss-joint deflections are determined from virtual work due to axial deformations of the truss members. Beam deflections (Secs. 10.4 and 10.5) and frame deflections (Sec. 10.6) are based on the virtual work associated with the internal bending-moment forces of plane structures; shear and torsional deformation computations are not included in this text.

The following concepts of work and energy are developed which serve to introduce the principle of virtual work.

External Work Consider a uniform rod with one end fixed and the other end free. The action of an axial force P, gradually applied to the free end of the rod, results in elastic behavior and an elongation of amount y in the direction of P. Thus, the linear increase of force from 0 to P is proportional to a linear increase in displacement from 0 to y as shown in Fig. 10.1a. The external work, W_e, can be expressed as the area under the P–y curve, that is,

$$W_e = \tfrac{1}{2} P \cdot y \tag{10.1}$$

Suppose that after force P is fully developed, an additional force P' is applied at the member end which results in an added displacement y'. The

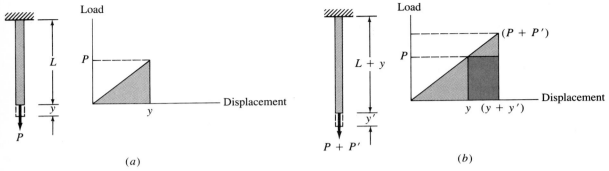

Figure 10.1

external work that is done by force P (not $P + P'$), when the member undergoes an added displacement y', can be expressed as

$$W_e = P \cdot y' \qquad (10.2)$$

as shown by the dark shaded area of Fig. 10.1b.

In a similar manner, the external work of a moment that is gradually applied to a structure from 0 to M is equal to the product of the average moment and the angle θ through which it rotates; that is,

$$W_e = \tfrac{1}{2} M \cdot \theta \qquad (10.3)$$

Presume that after M is fully developed at a point on the structure, other forces are gradually applied which cause an added rotation of θ' at M. The added external work produced by M can be expressed as

$$W_e = M \cdot \theta' \qquad (10.4)$$

Internal Work **Axial-force structures (trusses)** In Chap. 3, methods are present to determine the internal axial member forces that develop when a plane truss is subjected to static loading. From mechanics of materials, the axial deformation δ of an elastic rod under a constant axial force S, of Young's modulus E, length L, and constant cross-sectional area A, is found as follows:

$$E = \frac{\sigma}{\varepsilon} = \frac{(S/A)}{(\delta/L)} \quad \text{from which} \quad \delta = \frac{SL}{AE} \qquad (10.5)$$

Since the internal force in a truss member gradually builds from 0 to S, the internal work of the member is equal to the product of the average force $(S/2)$ and its axial deformation. That is,

$$W_i = \tfrac{1}{2} S \left(\frac{SL}{AE} \right) = \tfrac{1}{2} S^2 \frac{L}{AE} \qquad (10.6)$$

which is also called the *internal strain energy*. Since a truss consists of many members, the total internal work on a truss under a system of one or more loads can be defined as

$$W_i = \Sigma \tfrac{1}{2} S^2 \frac{L}{AE} \qquad (10.7)$$

Consider a truss subjected to only one concentrated load P that undergoes a displacement y at P, in the direction of the load P (see Fig. 10.2). In accordance with the principle of conservation of energy and Eqs. (10.1) and (10.7), we have

$$\tfrac{1}{2} P y = \Sigma \tfrac{1}{2} S^2 \frac{L}{AE} \qquad (10.8)$$

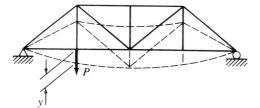

Figure 10.2

Thus, the external work produced by the load on the deformable truss is converted to internal strain energy. Equation (10.8) contains only one unknown, namely y, which can therefore be found. However, Eq. (10.8) cannot be used to compute other displacement components at P and cannot be used to compute displacements at other truss joints. If a truss is subjected to more than one load force, Eq. (10.1) can be modified to $\Sigma \tfrac{1}{2} P y$ (with an unknown y value at each P load). For a multiload condition, Eq. (10.8) can then be modified as

$$\Sigma \tfrac{1}{2} P y = \Sigma \tfrac{1}{2} S^2 \frac{L}{AE} \qquad (10.8a)$$

where y represents the unknown joint displacement at each applied load in the respective direction of each load. Thus, the unknown y's cannot be found from the singular equation of Eq. (10.8a) and this expression of real work proves to be of limited use.

Moment-force structures (beams and frames) When elastic loads are gradually applied to a beam or frame, the internal bending moment at a point will increase linearly from 0 to M as the loads are developed on the structure. Since the external work is converted to internal strain energy as the structure deforms under load, the internal work can be derived from the differential form of Eq. (10.3). That is, $dW_i = dW_e = \tfrac{1}{2} M \cdot d\theta$, where dW represents the energy developed in a differential element and $d\theta$ is the rela-

Figure 10.3

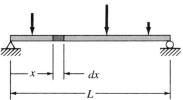
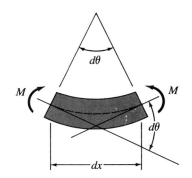

tive rotation over an element length dx (see Fig. 10.3). Substitution of Eq. (9.6) for $d\theta$ into dW_i above yields

$$dW_i = \tfrac{1}{2} M \cdot d\theta = \tfrac{1}{2} M \left(\frac{M\,dx}{EI} \right) = \frac{M^2 dx}{2EI} \qquad (10.9)$$

The internal strain energy stored over the entire length L of a flexural member is

$$W_i = \int_0^L dW_i = \int_0^L \frac{M^2 dx}{2EI} \qquad (10.10)$$

In accordance with the principle of conservation of energy, Eqs. (10.1) and (10.10) are equated to yield

$$\tfrac{1}{2} P \cdot y = \int_0^L \frac{M^2 dx}{2EI} \qquad (10.11)$$

Equation (10.11) is an expression of real work which is limited in use to a structure subjected to only one applied concentrated load. Furthermore, only the displacement component in the direction of the load, at the point where the load is applied, can be determined from Eq. (10.11).

10.2. BERNOULLI'S PRINCIPLE OF VIRTUAL WORK

Suppose that a system of virtual (imaginary) forces Q is applied to a structure in equilibrium, such that the structure remains in equilibrium while it undergoes small virtual displacements. Since energy must be conserved, the external virtual work, δW_e, and the internal virtual work, δW_i, must be equal. Therefore, the energy change must remain constant when either a real or a virtual system of loads in equilibrium are placed on a structure.

Reconsider the system of virtual (imaginary) forces Q placed on the deformable structural body. If the force system is reduced to one force and given a unit value, a technique for computing a specific displacement (or rotation) at any point on the structure becomes evident. The principle of virtual work can be stated as follows:

> *Consider the external work done by a unit virtual load applied to a structure in equilibrium that moves due to deformations associated with a real-load system. By the principle of conservation of energy, the external work is equal to the internal strain energy done by the internal virtual forces undergoing real deformations.*

The virtual-work method is frequently called the *unit-load method* or *dummy unit-load method*.

The examples that follow demonstrate that the virtual-work method provides a direct and clear approach to compute any directed displacement component due to one or more loads on a structure. The method can also be used to determine deflections due to temperature change, support settlement, and fabrication errors (misfits).

10.3. TRUSS DEFLECTIONS BY VIRTUAL WORK

When the truss of Fig. 10.4a is acted upon by P_1, P_2, and P_3, axial internal forces develop in each member. This action causes the truss members to either elongate or shorten in length which results in overall deflection of the structure. Assume that the vertical deflection at joint x, y_x, is desired and the original truss geometry remains basically unchanged by the load deformations. By the principle of virtual work, the following method is presented:

1. Consider all the P loads removed from the truss.

2. Place a unit vertical force at joint x in an assumed direction (either up or down). By standard methods of truss analysis, determine the internal force in each truss member; the forces are designated as $u_1 \ldots u_n$ in Fig. 10.4b.

3. Now, place the external forces (P loads) on the truss. The member forces due to the P loads can be found by truss analysis and are designated as $S_1 \ldots S_n$ in Fig. 10.4c.

Figure 10.4

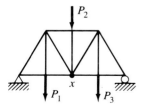

(a)

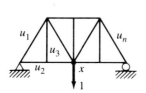

(b)

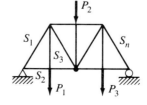

(c)

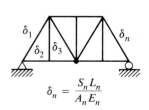

$$\delta_n = \frac{S_n L_n}{A_n E_n}$$

(d)

4 With the S forces known, the axial deformation of each truss member can be found by use of Eq. (10.5) which is shown as $\delta_1 \ldots \delta_n$ in Fig. 10.4d.

5 With the unit load fully in place, the virtual external work, δW_e, is attributed to the motion of the unit load at x through the unknown real elastic displacement, y_x. That is,

$$\delta W_e = 1 \cdot y_x \qquad (10.12)$$

6 The virtual internal work (virtual strain energy), δW_i, occurs as the fully developed internal virtual u force of each member undergoes a δ strain deformation due to the real P loads. That is,

$$\delta W_i = \Sigma u_n \cdot \delta_n \qquad (10.13)$$

7 Since $\delta W_e = \delta W_i$, substitution of Eq. (10.12) and Eq. (10.13) leads to

$$1 \cdot y_x = \Sigma u_n \cdot \delta_n \qquad (10.14)$$

Since all product terms on the right side of the equation are readily known, y_x can be determined.

The following examples prepare the essential computations in a presentable tabular format. Sign conventions used in these examples are:

1 Truss member u and S forces are positive $(+)$ for tension and negative $(-)$ for compression.

2 If the value of a computed displacement is found to be positive $(+)$, it is in the same direction as the virtual unit load; if negative, it is in the opposite direction of the virtual unit load.

In Example 10.1, the unit load is placed at joint L_2 in an assumed downward direction. Individual sketches show the reactions and truss member forces due to the external P loads and the unit load at L_2, respectively. It is observed that several members have a zero S or u value. Since all members have the same values of A and E, the product AE is intentionally absent from the table and factored into the energy equation following the summation, ΣSuL. In Example 10.2, all truss members have the same modulus of elasticity; thus, tabular computations are simplified by treating E as a constant factor which is applied after the summation, $\Sigma SuL/A$. The tabular approach is simple but demands careful accounting of all members and attention to signs of S and u forces and the product Su. Example 10.3 shows a table that is expanded to include data for two unknown displacement components. It is observed that δ_{V_A} is found to be of negative value; thus, it is opposite to the assumed direction of the vertical unit load placed at joint A. Example 10.4 shows the versatility of virtual work as it is used to determine a truss-joint displacement that occurs due to a uniform temperature change on several of the truss members.

EXAMPLE 10.1

Determine the vertical deflection of joint L_2 by the method of virtual work. The cross-section area of all bars, A, = 0.60 in² and $E = 29 \times 10^6$ psi.

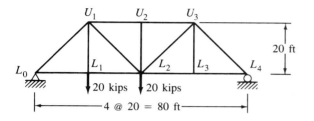

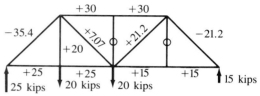

S forces due to external loads

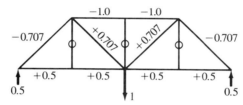

u forces due to unit load at L_2

Member	L (in)	S (kips)	u	SuL
$L_0 L_1$	240	+25	+0.5	+3000
$L_1 L_2$	240	+25	+0.5	+3000
$L_2 L_3$	240	+15	+0.5	+1800
$L_3 L_4$	240	+15	+0.5	+1800
$U_1 U_2$	240	+30	−1.0	−7200
$U_2 U_3$	240	+30	−1.0	−7200
$L_0 U_1$	339	−35.4	−0.707	+8484
$L_1 U_1$	240	+20	0	0
$U_1 L_2$	339	+7.07	+0.707	+1694
$L_2 U_2$	240	0	0	0
$L_2 U_3$	339	+21.2	+0.707	+5081
$L_3 U_3$	240	0	0	0
$U_3 L_4$	339	−21.2	−0.707	+5081
Σ				+15,540

Let δ_{VL_2} = vertical deflection at L_2. Equation (10.14):

$$1 \cdot \delta_{VL_2} = \Sigma u_n \cdot \delta_n = \Sigma u \cdot \frac{SL}{AE}$$

$$1 \cdot \delta_{VL_2} = \Sigma \frac{SuL}{AE} = +\frac{15{,}540}{AE}$$

$$1 \cdot \delta_{VL_2} = +\frac{15{,}540(1000)}{0.60(29 \times 10^6)}$$

$$\delta_{VL_2} = +0.89 \text{ in} \downarrow$$

EXAMPLE 10.2

By the virtual-work method, determine the motion of the roller support at L_4.
$E = 29 \times 10^3$ ksi.

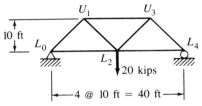

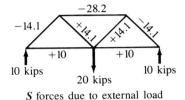

S forces due to external load

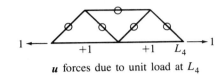

u forces due to unit load at L_4

A all diagonals = 1.5 in^2
A all horizontal members = 0.70 in^2

Member	L (in)	A (in^2)	L/A	S (kips)	u	SuL/A
L_0L_2	240	0.70	343	+10	+1	+3430
L_2L_4	240	0.70	343	+10	+1	+3430
U_1U_3	240	0.70	343	−28.2	0	0
L_0U_1	170	1.50	113	−14.1	0	0
U_1L_2	170	1.50	113	+14.1	0	0
L_2U_3	170	1.50	113	+14.1	0	0
U_3L_4	170	1.50	113	−14.1	0	0
Σ						+6860

Let δ_{HL_4} = the horizontal motion of roller at L_4.

$$1 \cdot \delta_{HL_4} = \Sigma u_n \cdot \delta_N = \Sigma u \cdot \frac{SL}{AE}$$

$$1 \cdot \delta_{HL_4} = +6860/E = +\frac{6860}{29 \times 10^3} = +0.237 \text{ in}$$

$\therefore \quad \delta_{HL_4} = 0.24$ in →

EXAMPLE 10.3

Find the horizontal and vertical deflection component of joint A by the method of virtual work. $E = 20 \times 10^6$ psi. Area of bars AB, AC, and $BC = 1.0$ in^2. All other bar areas are 2.0 in^2.

277 Truss Deflections by Virtual Work

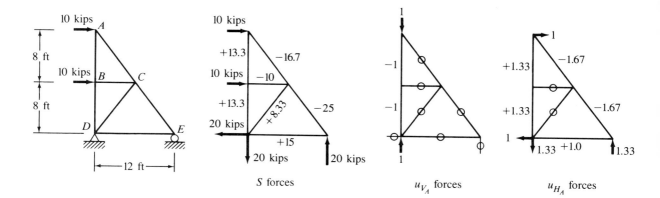

Member	L (in)	A (in²)	L/A	S	u_{V_A}	u_{H_A}	$Su_{V_A} \cdot \dfrac{L}{A}$	$Su_{H_A} \cdot \dfrac{L}{A}$
AB	96	1.0	96	+13.3	−1	+1.33	−1280	+1700
AC	120	1.0	120	−16.7	0	−1.67	0	+3350
BC	72	1.0	72	−10	0	0	0	0
BD	96	2.0	48	+13.3	−1	+1.33	−640	+850
CD	120	2.0	60	+8.33	0	0	0	0
CE	120	2.0	60	−25	0	−1.67	0	+2500
DE	144	2.0	72	+15	0	+1	0	+1080
Σ							−1920	+9480

$$1 \cdot \delta_{HA} = \Sigma u_{HA} \cdot \frac{SL}{AE} = +\frac{9480}{E} = +\frac{9480(1000)}{20 \times 10^6} = +0.474$$

$$\therefore \quad \delta_{HA} = 0.47 \text{ in} \rightarrow$$

$$1 \cdot \delta_{VA} = \Sigma u_{VA} \cdot \frac{SL}{AE} = \frac{-1920}{E} = \frac{-1920(1000)}{20 \times 10^6} = -0.096$$

$$\therefore \quad \delta_{VA} = 0.096 \approx 0.10 \text{ in} \uparrow$$

EXAMPLE 10.4

A steel roof truss experiences a 50°F temperature rise on all upper chord members. While the verticals and other diagonals rise +30°F, other members remain unchanged. By the virtual-work method, determine the vertical motion at joint G. Linear coefficient of thermal expansion $\alpha = 6 \times 10^{-6}/°F$ for all members.

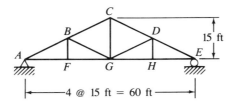
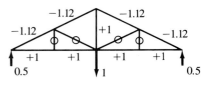

u forces due to unit load at G

Solution

$1 \cdot \delta_{VG} = \Sigma u_n \cdot \delta_n = \Sigma u \cdot L\alpha(\Delta T)$

Member*	L (in)	$a \times 10^{-6}/°F$	$\Delta T(°F)$	u	$uL\alpha(\Delta T)$
AB	202	6	+50	−1.12	−0.068
BC	202	6	+50	−1.12	−0.068
CD	202	6	+50	−1.12	−0.068
DE	202	6	+50	−1.12	−0.068
BF	90	6	+30	0	0
BG	202	6	+30	0	0
CG	180	6	+30	+1	+0.032
DH	90	6	+30	0	0
DG	202	6	+30	0	0
Σ					−0.24

* Other members omitted since $(\Delta T) = 0$.

$1 \cdot \delta_{VG} = -0.24$

$\therefore \quad \delta_{VG} = 0.24 \text{ in} \uparrow$

10.4. BEAM DISPLACEMENTS BY VIRTUAL WORK

A beam subjected to a system of real loads is shown in Fig. 10.5a where the vertical displacement at point O, y_o, is desired. For simplicity, the beam cross section is assumed to be of uniform rectangular shape. By the virtual-work method, let us assume that the beam is initially subjected to a unit vertical load placed at point O (see Fig. 10.5b). Let us then presume that elastic beam deformations occur due to placement of the real-load system on the beam. The resulting external virtual work on the beam is equal to the product of the unit virtual load times the real unknown displacement at location O; that is, $W_e = 1 \cdot y_o$.

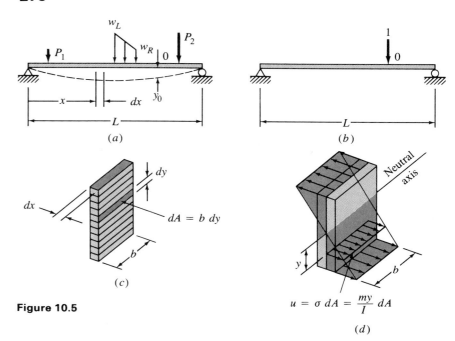

Figure 10.5

The internal virtual work done in a beam and a truss are determined in a similar way. In a truss, internal virtual work was previously found as the sum of the products of each member u force times the axial deformation due to real loading. The internal virtual work in a beam can be found as follows: Assume that significant elastic beam deflections are due to internal bending forces alone; that is, axial, shear, and torsional forces are assumed to produce negligible beam deflections. In Fig. 10.5c, a differential slice of beam of length dx is taken at an arbitrary location x along the beam. The composition of the cross section is presented as a stack of rectangular strips, each having an area of $dA = b \cdot dy$.

Let M be the bending moment at location x on the beam due to the real external load system.

Let m be the bending moment at location x on the beam due to the unit virtual load applied at point O.

Plane bending, due to the imaginary (*virtual*) unit load at O, results in a u force at each differential strip in the beam cross section as shown in Fig. 10.5d. It can be established that

$$u = \left(\frac{my}{I}\right)(dA) = my\frac{b \cdot dy}{I} \tag{10.15}$$

where I is the moment of inertia about the beam neutral axis.

Plane bending, due to the *real* load system on the beam, causes a longitudinal deformation to occur at each differential strip in the beam cross section which can be expressed as

$$\delta = \varepsilon \cdot dx = \left(\frac{\sigma}{E}\right) \cdot dx = \left(\frac{My}{EI}\right) \cdot dx \tag{10.16}$$

The internal virtual work at each differential strip in the beam cross section is found from Eqs. (10.15) and (10.16) above as

$$\delta W_i = u \cdot \delta = \left(my \cdot \frac{dA}{I}\right) \frac{(My \cdot dx)}{EI}$$

The virtual work for the entire cross section is found by area integration to be

$$\int Mmy^2 dA \cdot \frac{dx}{EI^2} = \frac{Mm}{EI^2} \int y^2 dA \cdot dx$$

Since $I = \int y^2 dA$, substitution in the above expression shows the virtual work in $dA = Mm \cdot dx/EI$.

Finally, the internal virtual work over the entire beam length from 0 to L is

$$W_i = \int_0^L \frac{Mm \cdot dx}{EI}$$

By the principle of conservation of energy,

$$W_e = W_i: \qquad 1 \cdot y_o = \int_0^L \frac{Mm \cdot dx}{EI} \tag{10.17}$$

Equation (10.17) is an expression by which the deflection at any point in a beam can be found. (Note: In general, y_o represents a displacement in any unrestrained direction.)

Applications Several notable comments relative to beam applications of virtual work are offered as follows:

1 Algebraic expressions must be written for M and m along the entire beam length. A consistent sign convention is used for both the M and m expressions; a positive moment (+) produces tension in the bottom fibers of a beam.

2 Application of Eq. (10.17) above requires that *all* terms under the integral are constant or expressed as a continuous function over the range of the integral. In Sec. 5.4, expressions for beam moment are shown to be discontinuous at points where concentrated forces occur and at regional ends of distributed loads. Although EI is usually constant over a beam's

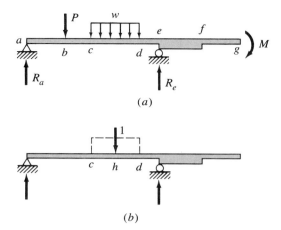

Figure 10.6

entire length, a sudden change in *I* on a beam is not uncommon (e.g., a simple *WF* beam with cover plates attached to the middle third of the beam length). *Therefore, a new integral is required when a discontinuity in M, m, E, or I occurs along a beam.*

Figure 10.6a shows a beam that extends from *a* to *g* under a system of three real loads. The beam has constant *E* and *I* values along its length except for the region between points *e* and *f* where *I* is double in value. Figure 10.6b shows the beam with a unit virtual load applied at point *h* where the vertical displacement of the beam is desired. A careful study of both load conditions indicates that a separate integral of Eq. (10.17) must be written from *a* to *b*, *b* to *c*, *c* to *h*, *h* to *d*, *d* to *e*, *e* to *f*, and *f* to *g*, since a discontinuity of *M*, *m*, or *I* is shown to occur.

3 The internal work integral(s) of Eq. (10.17) can be written in any order of integration (i.e., left to right, right to left, etc.).

4 If the result of integration is positive, the direction of deflection is the same as the applied virtual unit load.

5 Beam weight is assumed negligible in comparison to loads.

Examples 10.5, 10.6, and 10.7 attempt to demonstrate various beam problem solutions by virtual work. Example 10.5 is a simple and direct solution that only requires one integral of internal work since continuity of *M*, *m*, and *EI* exists over the entire beam length. In Example 10.6, the discontinuity of moment of inertia at point *B* requires that the internal work be found from the sum of two integrals. Examples 10.5 and 10.6 can easily be solved by integration from right to left after the values of the fixed-end support reactions are determined. Example 10.7 demonstrates that (1) any order of beam integration of internal virtual work is acceptable (note: left to right from *A* to *C*, and right to left from *D* to *C*). (2) Zero virtual work is done between *C* and *D* on the beam. (3) Does point *B* move up or down? The 4-kips load alone will move point *B* down whereas the

282 Deformations: Virtual Work

30-kips load alone will move point B upward. The virtual work method offers a direct approach to resolve both the magnitude and direction of a beam displacement.

EXAMPLE 10.5

Determine the deflection at point A on the beam. EI is constant and equal to 30×10^6 kip·in^2; $w = 3$ klf; $L = 8$ ft.

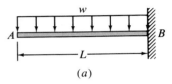

(a)

Solution

Assume a virtual 1-kip load placed at A as shown in (b).

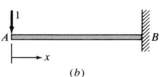

(b)

Both M and m are expressed by continuous functions of x originating from A to B as follows:

$$M = -wx\left(\frac{x}{2}\right) = -wx^2/2 \quad \text{and}$$

$$m = -1 \cdot x = -x$$

From Eq. (10.17):

$$1 \cdot \delta_A = \int_A^B \frac{Mmd}{EI} x = \int_0^L [-wx^2/2] \cdot [-x] \frac{dx}{EI} = \int_0^L \frac{wx^3}{2EI} dx$$

$$\delta_A = \frac{wx^4}{8EI}\bigg|_0^8 = \frac{3(8)^4(1728)}{8(30 \times 10^6)} = +0.09 \text{ in}\downarrow \quad \text{Answer}$$

EXAMPLE 10.6

Determine the deflection at point A on the beam in terms of EI.

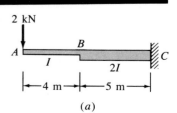

(a)

Solution

Assume a virtual 1 kN load placed at A as shown in (b).

Continuous expressions are written for M and m as follows:

$M = -2x$ and $m = -1x$

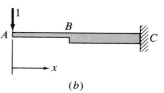

(b)

Since I is discontinuous at B, separate integrals are required from A to B, and B to C, that is,

$$\int_{A=0}^{B=4} (-2x)(-1x) \frac{dx}{EI} \quad \text{and} \quad \int_{B=4}^{C=9} (-2x)(-1x) \frac{dx}{E(2I)}$$

From Eq. (10.17):

$$1 \text{ kN} \cdot \delta_A = \int_0^4 \frac{2x^2 dx}{EI} + \int_4^9 \frac{2x^2 dx}{2EI} = \frac{2x^3}{3EI}\bigg|_0^4 + \frac{2x^3}{6EI}\bigg|_4^9$$

$$\delta_A = \frac{42.7}{EI} + \left(\frac{243 - 21.3}{EI}\right) = +\frac{264.4}{EI} \downarrow \quad \text{Answer}$$

EXAMPLE 10.7

Determine the deflection at B if $EI = 29 \times 10^6$ kip·in².

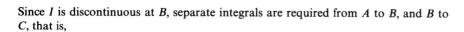

Solution

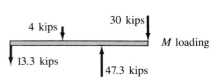

M loading

From A to B (0 ft → 8 ft):

$\left.\begin{array}{l} M = 13.3x \\ m = +0.43 \end{array}\right\} Mm = -5.72x^2$

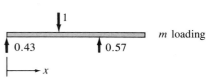

m loading

From B to C (8 ft → 14 ft):

$\left.\begin{array}{l} M = -13.3x - 4(x-8) = -17.3x + 32 \\ m = +0.43x - 1(x-8) = -0.57x + 8 \end{array}\right\} Mm = 9.86x^2 - 156.6x + 256$

From D to C (0 ft → 7 ft):

$$M = -30x \atop m = 0 \Big\} Mm = 0$$

Therefore,

$$1 \cdot \delta_B = \int_A^D \frac{Mm\,dx}{EI} = \int_0^8 (-5.72) \frac{dx}{EI}$$

$$+ \int_8^{14} (9.86x^2 - 156.6x + 256) \frac{dx}{EI}$$

$$= -\frac{5.72x^3}{3EI}\Big|_0^8 + \left(\frac{9.86x^3}{3EI} - \frac{78.3x^2}{EI} + \frac{256x}{EI}\right)\Big|_8^{14}$$

$$\delta_B = -\frac{976}{EI} + \frac{1}{EI}(-2744 + 1280)$$

$$= -\frac{2440}{EI} = -\frac{2440(1728)}{29 \times 10^6} = 0.145 \text{ in}\uparrow \quad \text{Answer}$$

10.5. BEAM ROTATIONS BY VIRTUAL WORK

The slope change (rotation) at any point in a beam can be determined by virtual work. In Fig. 10.7, a unit couple is applied at a point O on a beam where the rotation θ due to a real system of loads is to be found.

Let m' be the bending moment at location x on the beam due to the unit virtual load applied at point 0.

Now, consider placing the real-load system on the beam. The external work done is the product of the unit couple times the real rotation θ, that is,

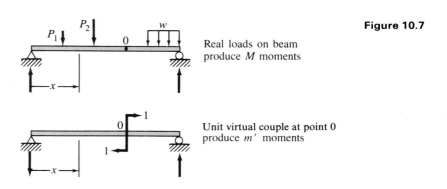

Figure 10.7

Real loads on beam produce M moments

Unit virtual couple at point 0 produce m' moments

285 Beam Rotations by Virtual Work

$W_e = 1 \cdot \theta$. The internal virtual work is $\int_0^L Mm' \cdot dx/EI$, and by the principle of conservation of energy,

$$1 \cdot \theta = \int_0^L Mm' \cdot \frac{dx}{EI} \qquad (10.18)$$

where θ is expressed in radians. If the integration of Eq. (10.18) is positive, θ rotates in the direction of the unit couple. Examples 10.8 and 10.9 illustrate beam applications of virtual work.

EXAMPLE 10.8

Find the rotation at the free end of the beam. $E = 29 \times 10^3$ ksi, $I = 600$ in^4.

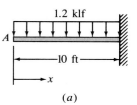

(a)

Solution

Apply a unit virtual couple at A.

$$1 \cdot \theta_A = \int_0^L \frac{Mm'dx}{EI} = \int_0^{10} -0.6x^2 \frac{dx}{EI} = -\frac{0.6x^3}{3EI}\Big|_0^{10}$$

$$1 \cdot \theta_A = -\frac{200}{EI} = -\frac{200(144)}{29 \times 10^3(600)} = -1.66 \times 10^{-3}$$

$$\therefore \theta_A = 1.66 \times 10^{-3} \text{ radians} \qquad \text{Answer}$$

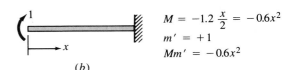

$M = -1.2 \frac{x}{2} = -0.6x^2$
$m' = +1$
$Mm' = -0.6x^2$

(b)

EXAMPLE 10.9

Determine the slope θ at point D on the beam. EI is constant.

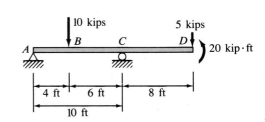

Solution

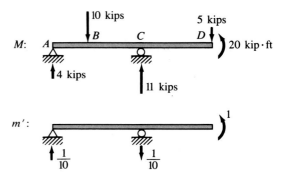

From A to B (0 ft → 4 ft):

$M = 4x$

$m' = x/10$

$Mm' = 0.4x^2$

From B to C (4 ft → 10 ft):

$M = 4x - 10(x - 4)$

$M = -6x + 40$

$m' = x/10$

$Mm' = -0.6x^2 + 4x$

From D to C (0 ft → 8 ft):

$M = 20 - 5x$

$m' = 1$

$Mm' = 20 - 5x$

$$1 \cdot \theta_D = \int_0^4 0.4x^2 \frac{dx}{EI} + \int_4^{10} (-0.6x^2 + 4x) \frac{dx}{EI} + \int_0^8 (20 - 5x) \frac{dx}{EI}$$

$$1 \cdot \theta_D = \frac{0.4x^3}{3EI}\Big|_0^4 + \left(-\frac{0.2x^3}{EI} + \frac{2x^2}{EI}\right)\Big|_4^{10} + \left(\frac{20x}{EI} - \frac{5x^2}{2EI}\right)\Big|_0^8$$

$$\theta_D = \frac{1}{EI}(8.5 - 19.2 + 0) = -10.7/EI \text{ radians} \quad \text{Answer}$$

10.6. FRAME DEFLECTIONS

Displacements and rotations of points on elastic frame structures can be determined by the same virtual-work relationships of Eqs. (10.17) and (10.18) as used for beams. It is imperative that a consistent sign convention is adapted for M, m, and m' of each element of a frame structure. Example 10.10 demonstrates a simple application of virtual work to determine frame deformations.

EXAMPLE 10.10

Determine the horizontal displacement and rotation at point *B* of the frame. $E = 30 \times 10^3$ ksi. $I = 500$ in^4.

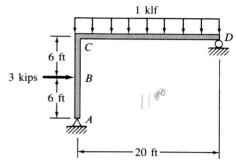

Solution

A separate schematic is given for *M*, *m*, and *m'* loading.

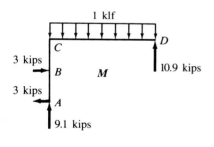
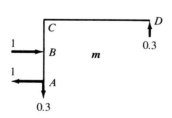
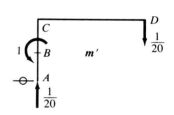

From A to B (0 ft → 6 ft):

$M = 3x$

$m = 1x$

$m' = 0$

$Mm = 3x^2$

$Mm' = 0$

From B to C (6 ft → 12 ft):

$M = 3x - 3(x - 6) = +18$

$m = 1x - 1(x - 6) = +6$

$m' = -1$

$Mm = +108$

$Mm' = -18$

From D to C (0 ft → 20 ft):

$M = 10.9x - 0.5x^2$

$m = 0.3x$

$m' = -x/20$

$Mm = 3.27x^2 - 0.15x^3$

$Mm' = -0.545x^2 + 0.025x^3$

$$1 \cdot \delta_B = \int_0^6 3x^2 \frac{dx}{EI} + \int_6^{12} 108 \frac{dx}{EI} + \int_0^{20} (3.27x^2 - 0.15x^3) \frac{dx}{EI}$$

$$= \frac{1}{EI}(216 + 648 + 2720) = +\frac{3580}{EI}$$

$$\delta_B = \frac{3580(1728)}{30 \times 10^3(500)} = 0.41 \text{ in} \rightarrow \text{Answer}$$

$$1 \cdot \theta_B = \int_0^6 (0) \frac{dx}{EI} + \int_6^{12} (-18) \frac{dx}{EI} + \int_0^{20} (-0.545x^2 + 0.025x^3) \frac{dx}{EI}$$

$$= \frac{1}{EI}(-108 - 453) = -\frac{561}{EI}$$

$$\theta_B = -\frac{561(144)}{30 \times 10^3 (500)} = -5.3 \times 10^{-3} \text{ radians} \quad \text{Answer}$$

10.7. COMPUTER APPLICATIONS AND COMPARISONS

Computer solutions can be used to verify the deflection results of structures determined by the virtual-work method. It is presumed that the reader has successfully used the IMAGES-2D computer program to solve truss, beam, and frame structure problems with a reasonable degree of success. Comparative analyses have been performed by IMAGES-2D for the structures of Examples 10.3, 10.5, and 10.10 above.

The computer output deflection data for the truss of Example 10.3 is given in Fig. 10.8 with a magnified *Print Screen* plot of the deformed truss (scale of 25 : 1); the load vectors, axes, supports, and joint numbers are inserted for clarity and are not plot data. Excellent numerical correlation is observed between the virtual-work and computer-deflection computations.

Figure 10.8

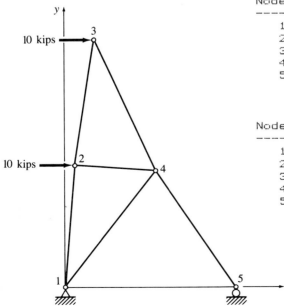

```
                    Applied Load Vector

     Node         Fx              Fy              Mz
     ----         --              --              --
      1       0.00000E+00     0.00000E+00     0.00000E+00
      2       1.00000E+01     0.00000E+00     0.00000E+00
      3       1.00000E+01     0.00000E+00     0.00000E+00
      4       0.00000E+00     0.00000E+00     0.00000E+00
      5       0.00000E+00     0.00000E+00     0.00000E+00

                       Displacements

     Node        X-Trans.        Y-Trans.        Z-Rot.
     ----        --------        --------        ------
      1       0.00000E+00     0.00000E+00     -9.93055E-04
      2       1.46333E-01     3.20000E-02     -1.15972E-03
      3       4.73667E-01     9.60000E-02     -3.15972E-03
      4       1.10333E-01    -5.15000E-02     -1.15972E-03
      5       5.40000E-02     0.00000E+00      2.01320E-21

                  Maximum Displacement Summary

            Node     Dir.     Displacement
            ----     ----     ------------
             3        Fx       4.73667E-01
             3        Fy       9.60000E-02
             3        Mz      -3.15972E-03
```

```
                       Applied Load Vector

        Node         Fx              Fy              Mz

         1       0.00000E+00    -1.50000E+00    -3.00000E+00
         2       0.00000E+00    -3.00000E+00     0.00000E+00
         3       0.00000E+00    -3.00000E+00     0.00000E+00
         4       0.00000E+00    -3.00000E+00     0.00000E+00
         5       0.00000E+00    -3.00000E+00     0.00000E+00
         6       0.00000E+00    -3.00000E+00     0.00000E+00
         7       0.00000E+00    -3.00000E+00     0.00000E+00
         8       0.00000E+00    -3.00000E+00     0.00000E+00
         9       0.00000E+00    -1.50000E+00     3.00000E+00


                          Displacements

        Node       X-Trans.         Y-Trans.         Z-Rot.

         1       0.00000E+00    -8.84723E-02     1.22878E-03
         2       0.00000E+00    -7.37341E-02     1.22638E-03
         3       0.00000E+00    -5.90968E-02     1.20958E-03
         4       0.00000E+00    -4.48194E-02     1.16398E-03
         5       0.00000E+00    -3.13340E-02     1.07518E-03
         6       0.00000E+00    -1.92454E-02     9.28788E-04
         7       0.00000E+00    -9.33110E-03     7.10392E-04
         8       0.00000E+00    -2.54157E-03     4.05596E-04
         9       0.00000E+00     0.00000E+00     0.00000E+00


                   Maximum Displacement Summary

             Node      Dir.        Displacement

              9        Fx         0.00000E+00
              1        Fy        -8.84723E-02
              1        Mz         1.22878E-03
```

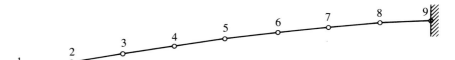

Figure 10.9

Figure 10.9 shows a computer model of the cantilever beam of Example 10.5 which contains nine nodes and eight beam elements. Deformation results by IMAGES is given at each node. Again, excellent deflection data correlation is found between virtual work and IMAGES. A Print Screen plot of the beam deformation is presented as seen on the PC monitor with node numbers and support added for clarity. Figure 10.10 shows an IMAGES computer model of the frame structure of Example 10.10 along with nodal deformation output data. Excellent agreement with the virtual-work hand solution is demonstrated.

290 Deformations: Virtual Work

```
                      Displacements

    Node      X-Trans.            Y-Trans.           Z-Rot.
    ----    --------------    --------------    --------------
     1       0.00000E+00       0.00000E+00       -5.90359E-03
     2       4.12617E-01      -4.36799E-04       -5.38519E-03
     3       7.63025E-01      -8.73599E-04       -4.34838E-03
     4       7.63025E-01       0.00000E+00        3.77965E-03

              Maximum Displacement Summary

              Node      Dir.       Displacement
              ----     ------    --------------
               4         Fx        7.63025E-01
               3         Fy       -8.73599E-04
               1         Mz       -5.90359E-03
```

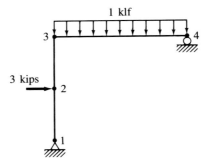

Figure 10.10

10.8. MAXWELL-BETTI THEOREM OF RECIPROCAL DEFLECTIONS

A useful theorem in the analysis of linear elastic statically indeterminate structures was presented by James Clerk Maxwell in 1864. The theorem is developed as follows with the aid of the virtual-work method:

Consider an undeformed elastic beam resting on simple and unyielding supports. Figure 10.11a shows the beam deflected by an external vertical load P_A applied at A which causes a deflection at point B, namely, δ_{BA} (subscript notation BA denotes the displacement at B due to a load at A). Let M_A define the moment at any point on the beam due to the load applied at A. Figure 10.11b shows the beam deflected by a vertical load P_B applied at B which causes a deflection at point A, namely, δ_{AB}. Let M_B define the moment at any point on the beam due to the load applied at B.

To find the displacement at A by virtual work, a unit load is placed at A; let m_A define the moment at any point on the beam due to the unit load at A. Likewise, to determine the displacement at point B, a unit load is placed

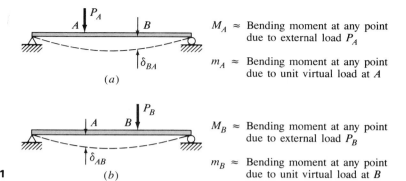

Figure 10.11

at B; let m_B define the moment at any point on the beam due to the unit load applied at B. Then, the deflection at B due to the load at A is found from

$$1 \cdot \delta_{BA} = \int \frac{M_A m_B \cdot dx}{EI} \tag{10.19}$$

and the deflection at A due to the load at B is found from

$$1 \cdot \delta_{AB} = \int \frac{M_B m_A \cdot dx}{EI} \tag{10.20}$$

Let $P_A = P_B = P$. Then, $M_A = P \cdot m_A$ and $M_B = P \cdot m_B$ which can be substituted into Eqs. (10.19) and (10.20) to yield

$$\delta_{AB} = \int \frac{(P \cdot m_A) m_B \cdot dx}{EI} \tag{10.19a}$$

and

$$\delta_{BA} = \int \frac{(P \cdot m_B) m_A \cdot dx}{EI} \tag{10.20a}$$

Since the integrals above are equal, $\delta_{AB} = \delta_{BA}$.

Maxwell's theorem of reciprocal deflection can be stated as follows:

The deflection at A due to a load applied at B is equal to the deflection at B if the same load is applied at A.

Maxwell's reciprocal theorem can be shown to be valid for rotations as well as displacements, and for truss and frame structures by the approach presented above. A general form of this theorem is often credited to E. Betti (1872) and called *Betti's law* which can be expressed as

$$\frac{\delta_{BA}}{P_B} = \frac{\delta_{AB}}{P_A} \tag{10.21}$$

10.9. CLOSURE ON VIRTUAL WORK

The sections above provide significant attention to truss deflections due to internal axial bar forces, and to beam and frame deflections due to internal bending forces. The virtual work method can also be applied to determine beam deflections due to shear and torsion. Conservation of energy relationships are given as follows:

For shear:
$$1 \cdot \delta = \alpha \int Vv \cdot \frac{dx}{AG}$$

where 1 = the unit virtual load applied at the point of unknown displacement on a beam in the direction of the displacement

δ = the unknown beam displacement

α = a form factor (usually 1.2 for WF and rectangular beams)

V = the shear at a section due to the real external loads

v = the shear at a section due to the unit virtual load

A and G = area and shear modulus of the beam, respectively

For torsion of circular members:

A displacement is found from

$$1 \cdot \delta = \int Tt \cdot \frac{dx}{JG}$$

where T = the torque at a section due to the real external loads

t = the torque at a section due to the unit virtual load

J = the polar moment of inertia of the member

If rotations instead of displacements are desired, the equations above can be modified as follows:

"1" represents a unit virtual couple applied at a point where the rotation of the structure is desired, δ is replaced by θ (the rotation to be determined), v is replaced by v' (the shear at a section due to the unit virtual couple), and t is replaced by t' (the torque at a section due to the unit virtual couple).

10.10. HISTORICAL REFERENCE—CASTIGLIANO'S THEOREMS

In 1879, Carlo Alberto Castigliano published a book that contained his two classical theorems which were originally presented as his "theorems of the differential coefficients on internal work." Castigliano's second theorem is presented first since it is similar to the virtual-work method for the determination of deflections of elastic structures. An extension of the second theorem, called the theorem of least work, provides a useful method for the analysis of statically indeterminate structures (see Chap. 11). Castigliano's first theorem is not a deflection method but is discussed for the sake of completeness at the end of this section.

The second theorem is developed using the same approach shown by Kinney[1] and considers the following assumptions:

1 The structure material is elastic, remains at a constant temperature, and follows Hooke's law under gradually applied loading.

2 The structural supports are unyielding and the geometry does not change appreciably due to loading.

3 By the principle of conservation of energy, for an elastic structure in equilibrium and at rest under a system of gradually applied loads, the external work is equal to the internal strain energy stored in the structure. Since the structure is elastic and follows Hooke's law, the total internal strain energy is constant and independent of the order in which the loads are applied; therefore, the law of superposition is valid.

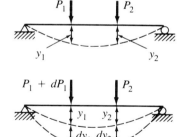

Figure 10.12

Figure 10.12 shows a beam deflected by two loads, P_1 and P_2, which are applied slowly and simultaneously. Since the external work and internal work are equal, we have

$$W_i = W_e = \tfrac{1}{2} P_1 \cdot y_1 + \tfrac{1}{2} P_2 \cdot y_2 \tag{10.22}$$

If an infinitesimal increase in P_1 is added to P_1, it causes added deflections of dy_1 and dy_2 as shown in Fig. 10.12. The additional increase in strain energy can be written as

$$dW_i = (P_1 + dP_1) \cdot dy_1 + P_2 \cdot dy_2$$

If the product of the differentials is neglected, the above change in work becomes

$$dW_i = P_1 \cdot dy_1 + P_2 \cdot dy_2 \tag{10.23}$$

---

[1] J. S. Kinney, *Indeterminate Structural Analysis*, Addison-Wesley, Reading, Mass., 1957, pp. 84–86.

The differential change in internal work can also be found by applying P_1, dP_1, and P_2 gradually and simultaneously. Let W'_i represent the total strain energy in the beam that develops when all loads are applied at the same time. This results in

$$W'_i = \tfrac{1}{2}(P_1 + dP_1)\cdot(y_1 + dy_1) + \tfrac{1}{2}P_2 \cdot (y_2 + dy_2)$$

which by neglecting the product of the differentials becomes

$$W'_i = \tfrac{1}{2}[(P_1 \cdot y_1) + (P_2 \cdot y_2) + (dP_1 \cdot y_1) + (P_1 \cdot dy_1) + (P_2 \cdot dy_2)] \quad (10.24)$$

Since $dW_i = W'_i - W_i$, it is found from Eqs. (10.22) and (10.24) that

$$dW_i = \tfrac{1}{2}[(dP_1 \cdot y_1) + (P_1 \cdot dy_1) + (P_2 \cdot dy_2)]$$

The last two terms in the bracketed expression above can be replaced by dW_i from Eq. (10.23) which results in

$$dW_i = \tfrac{1}{2}dP_1 \cdot y_1 + \tfrac{1}{2}dW_i$$

from which

$$y_1 = \frac{dW_i}{dP_1}$$

In general, Castigliano's second theorem is expressed by a partial derivative since more than one force is usually applied to a structure. The deflection components can be written in terms of a Q force as

$$y_x = \frac{\partial W_i}{\partial Q_x} \quad \text{or} \quad \theta_x = \frac{\partial W_i}{\partial Q_x} \quad (10.25)$$

where x is a location on a structure, and Q represents a real or imaginary force on the structure.

Castigliano's second theorem Consider a linear elastic structure of constant temperature and unyielding supports subjected to a set of loads. The partial derivative of the internal strain energy with respect to a force is equal to the deflection at the point where the force is applied and directed along the action of the force. If the force is a couple, the deflection at the point is the associated rotation; if the force is vertical, the associated deflection at the point is a vertical displacement, and so forth.

In Examples 10.11 and 10.12, it is noted that a positive answer indicates that the deflection component is in the direction of the force to which the derivative is taken. In addition, the second theorem can be used to determine a deflection component at a point on a structure where no force is applied. This is demonstrated in Example 10.12 (part c), where an imaginary force Q is placed at the point where a deflection component is desired; reactions and moments in terms of Q are used to evaluate the internal strain energy after which the Q force is set to zero.

EXAMPLE 10.11

Compute the vertical displacement at joint A due to the 5-kip load at A. Bar areas (in^2) are shown circled. $E = 30 \times 10^3$ ksi.

Solution

Replace load by Q.

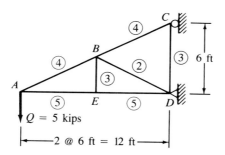
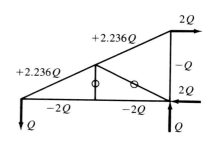

Member	$\dfrac{L}{A} \sim \dfrac{1}{\text{in}}$	S	$\dfrac{\partial S}{\partial Q}$	$S\left(\dfrac{\partial S}{\partial Q}\right)\dfrac{L}{A}$
AE	14.4	$-2Q$	-2	$+57.6Q$
ED	14.4	$-2Q$	-2	$+57.6Q$
BE	12.0	0	0	0
CD	24.0	$-Q$	-1	$+24.0Q$
AB	20.1	$2.236Q$	2.236	$+100.5Q$
BC	20.1	$2.236Q$	2.236	$+100.5Q$
BD	40.2	0	0	0
Σ				$+340.2Q$

$$1 \cdot y_A = \Sigma S\left(\dfrac{\partial S}{\partial Q}\right)\dfrac{L}{AE}$$

$$y_A = \dfrac{+340.2Q}{E} = +\dfrac{340.2(5)}{30 \times 10^3} = 0.0567 \text{ in}\downarrow \quad \text{Answer}$$

EXAMPLE 10.12

By Castigliano's second theorem, find
a. Vertical displacement at point B
b. Rotation at point C
c. Vertical displacement at point C
EI is constant.

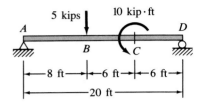

Solution

Part a: Replace load at B by Q:

From A to B (0 ft → 8 ft):

$$M = 0.6Qx + 0.50x$$

$$\frac{\partial M}{\partial Q} = 0.6x$$

$$M\left(\frac{\partial M}{\partial Q}\right) = 0.36Qx^2 + 0.30x^2$$

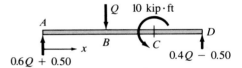

From B to C (8 ft → 14 ft):

$$M = 0.6Qx + 0.50x - Q(x - 8)$$
$$M = -0.4Qx + 0.50x + 8Q$$

$$\frac{\partial M}{\partial Q} = -0.4x + 8$$

$$M\left(\frac{\partial M}{\partial Q}\right) = 0.16Qx^2 - 0.2x^2 - 6.4Qx + 4x + 64Q$$

From D to C (0 ft → 6 ft):

$$M = 0.4Qx - 0.50x$$

$$\frac{\partial M}{\partial Q} = 0.4x$$

$$M\left(\frac{\partial M}{\partial Q}\right) = 0.16Qx^2 - 0.20x^2$$

$$1 \cdot y_B = \int M(\partial M/\partial Q)\frac{dx}{EI}$$

$$y_B = \int_0^8 (0.36Qx^2 + 0.3x^2)\frac{dx}{EI} + \int_8^{14}(0.16Qx^2 - 0.2x^2 - 6.4Qx + 4x + 64Q)\frac{dx}{EI}$$

$$+ \int_0^6 (0.16Qx^2 - 0.2x^2)\frac{dx}{EI}$$

$$y_B = \left(\frac{0.12Qx^3 + 0.1x^3}{EI}\right)_0^8 + \left(\frac{0.053Qx^3 - 0.067x^3 - 3.2Qx^2 + 2x^2 + 64Qx}{EI}\right)_8^{14}$$

$$+ \left(\frac{0.053Qx^3 - 0.067x^3}{EI}\right)_0^6$$

$$y_B = \frac{61.4Q + 51.2 + 118Q - 150 - 422Q + 264 + 384Q + 11.4Q - 14.5}{EI}$$

$$y_B = \frac{153Q + 154}{EI} = \frac{153(5) + 154}{EI} = +\frac{919}{EI}\downarrow \quad \text{Answer}$$

Part b: Replace couple at C by Q:

From A to B (0 ft → 8 ft):

$$M = 3x + 0.05Qx$$

$$\frac{\partial M}{\partial Q} = 0.05x$$

$$M\left(\frac{\partial M}{\partial Q}\right) = 0.15x^2 + 0.0025Qx^2$$

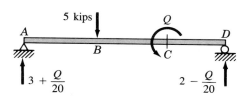

From B to C (8 ft → 14 ft):

$$M = 3x + 0.05Qx - 5(x-8)$$

$$M = -2x + 0.05Qx + 40$$

$$\frac{\partial M}{\partial Q} = +0.05x$$

$$M\left(\frac{\partial M}{\partial Q}\right) = -0.1x^2 + 0.0025Qx^2 + 2x$$

From D to C (0 ft → 6 ft):

$$M = 2x - 0.05Qx$$

$$\frac{\partial M}{\partial Q} = -0.05x$$

$$M\left(\frac{\partial M}{\partial Q}\right) = -0.1x^2 + 0.0025Qx^2$$

$$1 \cdot \theta_C = \int M\left(\frac{\partial M}{\partial Q}\right)\frac{dx}{EI}$$

$$\theta_c = \int_0^8 (0.15x^2 + 0.0025Qx^2)\frac{dx}{EI}$$

$$+ \int_8^{14}(-0.1x^2 + 0.0025x^2 + 2x)\frac{dx}{EI} + \int_0^6 (-0.1x^2 + 0.0025Qx^2)\frac{dx}{EI}$$

$$\theta_c = \left(\frac{0.5x^3 + 0.0083Qx^3}{EI}\right)\Big|_0^8 + \left[\frac{-0.333x^3 + (8.3 \times 10^{-4})x^3 + x^2}{EI}\right]\Big|_8^{14}$$

$$+ \left[\frac{-0.033x^3 + (8.3 \times 10^{-4})Qx^3}{EI}\right]\Big|_0^6$$

$$\theta_c = \frac{25.6 + 0.425Q - 74 + 1.86Q + 132 - 7 + 0.18Q}{EI} = \frac{+76 + 2.47Q}{EI}$$

$$\theta_c = \frac{76 + 2.47(10)}{EI} = -\frac{100}{EI}\bigg) \quad \text{Answer}$$

Part c: Add a vertical force Q at point C:

From A to B (0 ft → 8 ft):

$$M = 3.5x + 0.3Qx$$

$$\frac{\partial M}{\partial Q} = 0.3x$$

$$M\left(\frac{\partial M}{\partial Q}\right) = 1.05x^2 + 0.09Qx^2$$

From B to C (8 ft → 14 ft):

$$M = 3.5x + 0.3Qx - 5(x - 8)$$

$$M = -1.5x + 0.3Qx + 40$$

$$\frac{\partial M}{\partial Q} = 0.3x$$

$$M\left(\frac{\partial M}{\partial Q}\right) = -0.45x^2 + 0.09Qx^2 + 12x$$

From D to C (0 ft → 6 ft):

$$M = 1.5x + 0.7Qx$$

$$\frac{\partial M}{\partial Q} = 0.7x$$

$$M\left(\frac{\partial M}{\partial Q}\right) = 1.05x^2 + 0.49Qx^2$$

$$1 \cdot y_c = \int_0^8 (1.05x^2) \frac{dx}{EI} + \int_8^{14} (-0.45x^2 + 12x) \frac{dx}{EI} + \int_0^6 (1.05x^2) \frac{dx}{EI}$$

Note

Since $Q = 0$, all integral terms above factored by Q are set equal to zero. Thus,

$$y_c = \left(\frac{0.350x^3}{EI}\right)_0^8 + \left(\frac{-0.15x^3 + 6x^2}{EI}\right)_8^{14} + \left(\frac{0.35x^3}{EI}\right)_0^6$$

$$y_c = \frac{179 - 335 + 792 + 76}{EI} = +\frac{712}{EI} \downarrow \quad \text{Answer}$$

In terms of a linear displacement component, the internal strain energy can be written as follows:

For axial forces (in summation form):

$$y = \frac{\partial W_i}{\partial Q} = \Sigma S \cdot \left(\frac{\partial S}{\partial Q}\right) \cdot \frac{L}{AE}$$

For bending forces:

$$y = \frac{\partial W_i}{\partial Q} = \int M \cdot \left(\frac{\partial M}{\partial Q}\right) \cdot \frac{dx}{EI}$$

For torsion forces:

$$y = \frac{\partial W_i}{\partial Q} = \int T \cdot \left(\frac{\partial T}{\partial Q}\right) \cdot \frac{dx}{JG}$$

For shear forces:

$$y = \frac{\partial W_i}{\partial Q} = \int V \cdot \left(\frac{\partial V}{\partial Q}\right) \cdot \frac{dx}{AG}$$

In general, the unit-load method has been more popular than Castigliano's second theorem, probably since it is slightly easier to apply. However, they both serve the engineer with independent methods to compute structural deformations.

Castigliano's first theorem The first theorem of Castigliano can be derived in a similar manner to that of the second theorem and is expressed as

$$\frac{\partial W_i}{\partial y} = Q$$

300 Deformations: Virtual Work

which indicates that the partial derivative of the internal work of a structure as a function of a displacement is equal to the value of the corresponding force. This theorem provides a displacement method (see Sec. 14.2.1) which can be used for the analysis of statically indeterminate structures. Applications of Castigliano's first theorem are not included in this text.

Problems

Problems 10.1 through 10.26 by virtual-work method.

10.1 The truss crane is designed to support 20 kips. Determine the horizontal and vertical displacements of pickup point E due to the 20 kip maximum loading. $E = 30 \times 10^3$ ksi and $A = 4.0$ in² for all truss members.

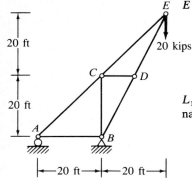

10.2 Determine the horizontal and vertical displacement components of joint L_1. $E = 200$ GPa for all members. Cross-section areas are 1000 mm² for all diagonal members; other member areas = 600 mm⁴.

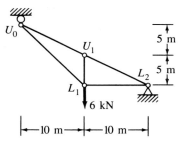

10.3 Compute horizontal and vertical displacement components at joint F. $E = 30 \times 10^3$ ksi. Member areas are given in circles in square inch.

10.4

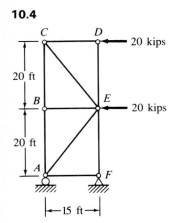

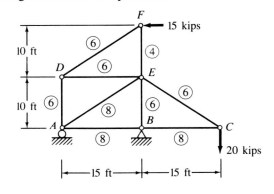

10.4 Determine the horizontal and vertical displacement components at joint C. $E = 29 \times 10^3$ ksi. Areas of horizontal members = 2 in². Areas of vertical members = 4 in². Areas of diagonal members = 5 in².

10.5 Compute the vertical displacement at joint A due to the loads shown if (a) all members are at room temperature and $E = 30 \times 10^3$ ksi and $A = 2.0$ in^2 for all members, and (b) in addition to the applied loads at A, member CE experiences a uniform temperature decrease of $80°F$ ($\alpha = 6 \times 10^{-6}$ in/in/°F).

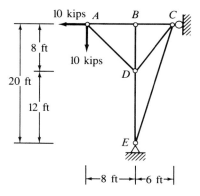

10.6 (a) Determine the vertical displacement at joint L_2 of the roof truss shown below due to gravity roof loads.
(b) Determine the horizontal motion of the roller support at L_4 if all upper chord truss members undergo a $+100°F$ temperature increase.

$$E = 29 \times 10^6 \text{ psi}, \qquad A = 4.0 \text{ in}^2$$
$$\alpha = 6 \times 10^{-6} \text{ in/in/°F for all members}$$

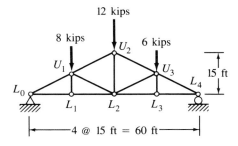

10.7 to 10.26 Use the virtual-work method for all problem solutions. Also, use $E = 29 \times 10^3$ ksi unless specified otherwise.

10.7 Rework Prob. 9.7 by virtual work.
10.8 Rework Prob. 9.8 by virtual work.
10.9 Rework Prob. 9.9 by virtual work.
10.10 Rework Prob. 9.10 by virtual work.

10.11 Determine θ_a and δ_a. $I = 2 \times 10^8$ mm^4, $E = 200$ MPa.

302 Deformations: Virtual Work

10.12 Compute θ_a, θ_b and the vertical displacement at midspan. $I = 800$ in^4.
10.13 Rework Prob. 9.12 by virtual work.
10.14 Determine θ_c and δ_c. $I = 1400$ in^4.

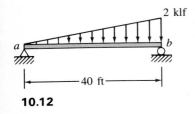

10.12

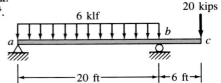

10.14

10.15 Compute the displacement at B and the rotation at D. $I = 900$ in^4.

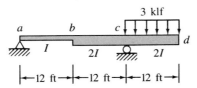

10.16 Determine deflection and rotation at point C. $I = 600$ in^4.

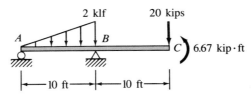

10.17 Rework Prob. 9.15 by virtual work.
10.18 Rework Prob. 9.18 by virtual work.
10.19 Determine θ_b and δ_b if $E = 10 \times 10^6$ psi and $I = 1000$ in^4.

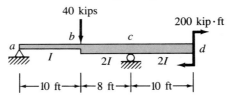

10.20 Determine the rotation and displacement under the 40-kip load on the beam. $I = 9000$ in^4, $E = 30 \times 10^3$ ksi.

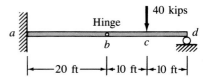

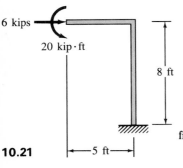

10.21

10.21 Determine the horizontal displacement and rotation at the free end of the frame. $E = 10 \times 10^3$ ksi, $I = 1000$ in^4.

10.22 Compute θ_b and the horizontal displacement at d. $I = 400$ in^4.

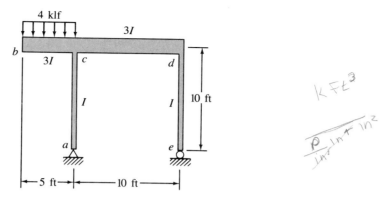

10.23 Determine the motion at the roller support due to the loading. $E = 200$ GPa, $I = 6 \times 10^6$ mm^4.

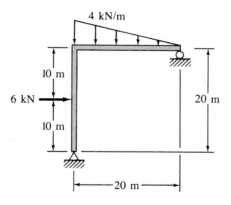

10.24 Compute the horizontal motion at E and the vertical displacement at C. $E = 30 \times 10^3$ ksi, $I = 2000$ in^4.

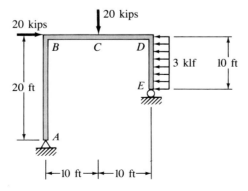

10.25 Determine the horizontal and vertical motion at C. $I = 600$ in^4.

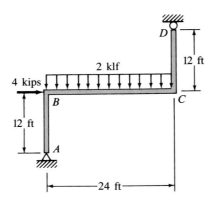

10.26 Determine the vertical displacement at B. $E = 2 \times 10^3$ ksi. $I = 4000$ in^4.

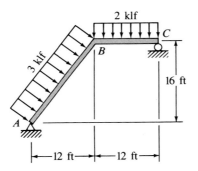

10.27 to 10.31 Determine the vertical displacement component requested in each truss structure by use of Castigliano's theorem.

10.27 The truss of Prob. 10.1
10.28 The truss of Prob. 10.2
10.29 The truss of Prob. 10.3
10.30 The truss of Prob. 10.4
10.31 The vertical displacement at joint L_3 by Castigliano's theorem.

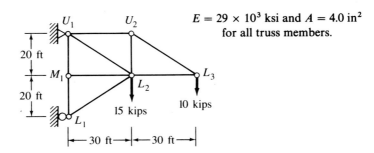

10.32 Compute θ_c and δ_c by Castigliano's theorem. $E = 30 \times 10^3$ ksi, $I = 300$ in^4.

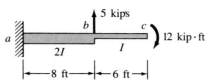

10.33 to 10.36 Solve the following beam displacements by Castigliano's theorem.
- **10.33** θ_b and δ_b of Prob. 9.10
- **10.34** θ_a and δ_a of Prob. 9.11
- **10.35** Midspan rotation and displacement of Prob. 9.14
- **10.36** θ_c and δ_c of Prob. 9.17
- **10.37** Find δ_a in terms of EI using Castigliano's theorem.

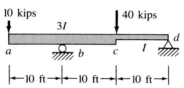

10.38 Vertical displacement at free end by Castigliano's theorem

$$E = 30 \times 10^3 \text{ ksi}$$
$$I = 5000 \text{ in}^4$$

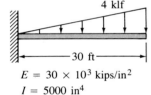

Analysis of frames by Castigliano's theorem

10.39 Determine the vertical displacement at e by the Castigliano theorem. $E = 29 \times 10^3$ ksi, $I = 300$ in^4.

10.40 Compute the roller support displacement by use of Castigliano's theorem. $E = 29 \times 10^3$ ksi, $I = 900$ in^4.

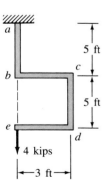

10.39

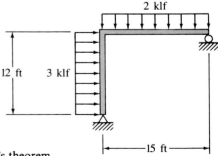

- **10.41** Solve Prob. 10.21 by Castigliano's theorem.
- **10.42** Solve Prob. 10.23 by Castigliano's theorem.
- **10.43** Solve Prob. 10.24 by Castigliano's theorem.
- **10.44** Solve Prob. 10.25 by Castigliano's theorem.

CHAPTER ELEVEN
Statically Indeterminate Structures

11.1. INTRODUCTION

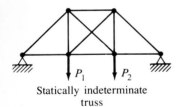
Statically indeterminate truss

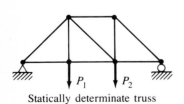
Statically determinate truss (primary structure)

Figure 11.1

The analyses of statically indeterminate structures are generally performed by force and displacement methods. A general discussion of these classical approaches is presented below with the aid of the truss model shown in Fig. 11.1.

The statically indeterminate truss of Fig. 11.1 contains both a redundant member and a redundant reaction component. In a force-method analysis, a sufficient number of truss bars and reaction components are assumed to be removed to form a *primary structure* that is statically determinate and geometrically stable. Obviously, removal of these members and support constraints will allow the truss to deform more freely under load. Thus, we are able to compute "free" displacements in the directions of the released members and the released reactions due to the applied loads on the primary truss. Corresponding displacements can also be found due to a unit force applied at and in the direction of a released force; similar "unit" force displacements are found for each released force. Then, equations can be written in terms of free displacements due to applied loads and the free displacements due to the unknown redundant forces. These equations are used to determine the unknown forces which will restore each released support to its known location, and will ensure consistent deformation and force compatibility between the released members and the primary truss. The force method is also called the *method of consistent deformations* and is known as the *compatibility* or *flexibility method*.

George Washington Bridge, New York (after the lower roadway was added in 1962). The designers were Ammann and Strauss. (*Courtesy of International Structural Slides, Berkeley.*)

The displacement method differs from the force method in that a set of simultaneous equations are written in terms of unknown joint displacements and rotations instead of redundant forces. After joint displacements are determined, the elongation or contraction of each member is computed. Then, the force in each truss bar (S) is solved from the familiar relationship, $\delta = SL/AE$ or $S = \delta \cdot AE/L$, where AE/L represents the relative stiffness of an axially loaded member. Thus, the displacement method is also known as the *stiffness method*. The analyses of structures by use of displacement methods are presented in Chap. 12 (Slope Deflection), Chap. 13 (Moment Distribution), and Chap. 14 (Matrix Structural Analysis).

The force methods were the first to appear in technical print and have been more widely used for civil engineering structures. However, the displacement methods may be regarded as more systematic and attractive for the analysis of most structural systems; this attraction relates to the ability to perform determinate and indeterminate structural analysis without the need to identify redundant forces. Many factors influence the choice of method by an analyst, namely, the nature of the structure, the degree of indeterminacy, the type of data required, and so forth. Although the choice of method is beyond the scope of this text, it is worth noting that recent advances in computer technology have encouraged significant expansion and refinement to both methods, particularly in the field of finite element analysis. The contents of this chapter deal mainly with classical analyses by the force method. Chapter 14 studies the matrix displacement method.

Force method In 1864, James Clerk Maxwell published the first consistent force method for the analysis of statically indeterminate structures

[11.1]. In 1874, Otto Mohr, unaware of Maxwell's work, independently presented the same force approach and demonstrated applications for the solution of statically indeterminate trusses [11.2]. This classical method is known as the method of consistent deformations and is also referred to as the Maxwell-Mohr method (see Secs. 11.2 to 11.5).

Figure 11.2

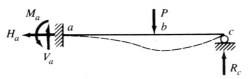

(*a*) Indeterminate beam; first degree

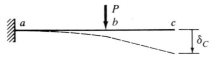

(*b*) Primary beam under load *P*

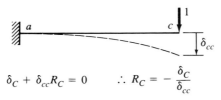

$$\delta_C + \delta_{cc} R_C = 0 \qquad \therefore R_C = -\frac{\delta_C}{\delta_{cc}}$$

(*c*) Primary beam under unit load at *c*

Procedure of the force method of consistent deformations The propped cantilever beam shown in Fig. 11.2 is used to introduce a detailed procedure of a force method analysis. In the steps that follow, added commentary is provided in parentheses:

1 Establish the degree of indeterminacy of the structure. (The propped cantilever has four unknown reactions and three equations of statics; thus, the beam is statically indeterminate to the first degree.)

2 Remove a sufficient number of unknown forces (reactions and/or member forces) from the indeterminate structure so that the structure is statically determinate and remains geometrically stable. As stated earlier, the stable and determinate structure is called the *primary structure* (or released structure) and the extra forces that are removed are referred to as *redundants*. (The beam of Fig. 11.2*a* is statically indeterminate to the first degree; therefore, it is elected to remove the redundant roller force at point *c*.)

3. Compute the unrestrained (released) deformation of the primary structure at each redundant location due to the applied loads on the structure. (δ_C—the vertical displacement at c in the direction of the redundant roller reaction at c; see Fig. 11.2b.)

4. The structure has one redundant. Therefore, place a unit force at the redundant location in the direction of the redundant force and compute the deflection of the primary structure at the redundant location, in the direction of the redundant force. This unit load deformation to the primary structure is referred to as a *flexibility coefficient* or *influence coefficient*. (δ_{cc}—subscript denotes the deflection at c due to a unit load at c; see Fig. 11.2c.)

5. Point c represents an unyielding support; that is, the total vertical deflection at c is zero. Therefore, the redundant force R_C must have a magnitude necessary to restore the unrestrained deflection of step 3 above. Using the principle of superposition, the deflection is formulated for the primary structure in terms of the applied loads, the influence coefficient, and the unknown redundant force, and set equal to zero (see Fig. 11.2c). Then, the redundant force at c is found directly from the deflection equation. (The product $\delta_{cc} \cdot R_C$ represents the deflection that is necessary to restore the unrestrained displacement δ_C; thus, $R_C = -\delta_C/\delta_{cc}$. The minus sign indicates that R_C is opposite in direction to the unit redundant load.)

This procedure is demonstrated for beam structures with more than one redundant in Sec. 11.2.

11.2. METHOD OF CONSISTENT DEFORMATIONS—BEAMS

11.2.1. Beam with One Redundant

Examples 11.1, 11.2, and 11.3 demonstrate beam applications of the method of consistent deformations using the moment area, conjugate beam, and virtual-work methods, respectively. In these examples, a stable primary structure is established by limiting the choice of redundant to a transverse or a rotational restraint; therefore, secondary effects due to axial deformations of members are neglected.

EXAMPLE 11.1

By the methods of consistent distortions and moment area, determine all support reactions. EI is constant.

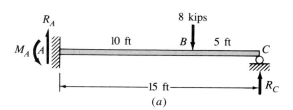

(a)

Solution

Remove support at C.

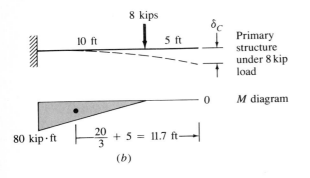

(b)

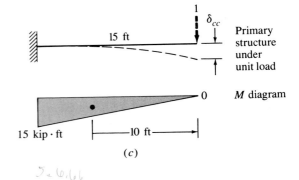

(c)

By second moment area theorem,

$$\delta_C = \frac{1}{2}\left(\frac{80}{EI}\right)10 \cdot 11.7 = \frac{4680}{EI}; \qquad \delta_{cc} = \frac{1}{2}\left(\frac{15}{EI}\right)15 \cdot 10 = \frac{1125}{EI}$$

$$\delta_C + \delta_{cc} R_C = 0 \qquad \therefore \quad R_C = -\frac{\delta_C}{\delta_{cc}} = -\frac{4680/EI}{1125/EI} = -4.16 \text{ kips} = 4.16 \text{ kips}\uparrow$$

$$\uparrow + \Sigma F_y = 0 = R_A + R_C - 8 \text{ kips} \qquad \therefore \quad R_A = 8 - R_C = 8 - 4.16 = 3.84 \text{ kips}\uparrow$$

$$\circlearrowleft + \Sigma M_C = 0 = R_A \cdot 15 \text{ ft} - M_A - 8 \text{ kips} \cdot 5 \text{ ft} \qquad \text{or} \qquad M_A = 15 R_A - 40$$

$$\therefore \quad M_A = 3.84(15) - 40 = 17.6 \text{ ft} \cdot \text{kips} \,\circlearrowright$$

EXAMPLE 11.2

By the methods of consistent distortion and conjugate beam, determine all support reactions. EI is constant.

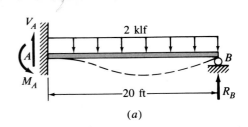

(a)

Solution

Remove moment restraint at A, see Figure b:

$$\theta_A + \theta_{aa} M_A = 0$$

$$\therefore \quad M_A = -\frac{\theta_A}{\theta_{aa}} = -\frac{6670/EI}{6.67/EI} = -100 \text{ kip} \cdot \text{ft}$$

311 Method of Consistent Deformations—Beams

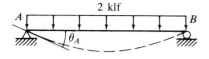

Primary structure for 2 klf load

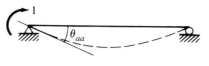

Primary structure for unit couple

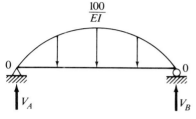

$\dfrac{M}{EI}$ load on conjugate beam

$\theta_{A_{\text{real beam}}} = V_{A_{\text{conj bm}}} = \dfrac{6670}{EI}$

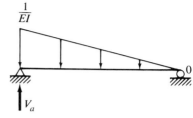

$\dfrac{M}{EI}$ load on conjugate beam

$\theta_{aa} = V_{a_{\text{conj bm}}} = \dfrac{6.67}{EI}$

(b)

$\therefore \quad M_A \equiv 100 \text{ kip} \cdot \text{ft} \circlearrowright$

$\zeta + \Sigma M_B = 0 = V_A \times 20 - M_A - 2(20) \times 10$

$\therefore \quad V_A = \dfrac{1}{20}[100 + 400] = 25 \text{ kips}\uparrow$

$+\uparrow \Sigma F_y = 0 = V_A + R_B - 2(20)$

$\therefore \quad R_B = 40 - V_A = 40 - 25 = 15 \text{ kips}\uparrow$

EXAMPLE 11.3

By the methods of consistent distortion and virtual work, find all support reactions. EI is constant.

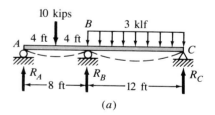

(a)

Solution

Remove support at B.

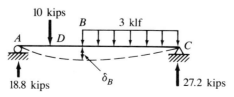

Primary structure; real loading

(b)

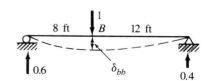

Primary structure; unit load at B

(c)

A to D (0 ft → 4 ft) D to B (0 ft → 4 ft) C to D (0 ft → 12 ft)

$M = 18.8x$

$M = 18.8(4 + x) - 10x$

$M = 27.2x - 1.5x^2$

$m = 0.6x$

$M = 8.8x + 75.2$

$m = 0.4x$

$Mm = 11.3x^2$

$m = 0.6(4 + x)$

$Mm = 10.9x^2 - 0.6x^3$

$m^2 = 0.36x^2$

$m = 2.4 + 0.6x$

$m^2 = 0.16x^2$

$Mm = 5.28x^2 + 66.2x + 181$

$m^2 = 0.36x^2 + 2.88x + 5.76$

$$1 \cdot \delta_B = \int_0^L \frac{Mm\,dx}{EI} = \int_0^4 (11.3x^2)\frac{dx}{EI}$$

$$+ \int_0^4 (5.28x^2 + 66.2x + 181)\frac{dx}{EI} + \int_0^{12}(10.9x^2 - 0.6x^3)\frac{dx}{EI}$$

$$\delta_B = \frac{241 + 1367 + 3168}{EI} = \frac{4776}{EI}$$

$$\delta_{bb} = \int_0^L \frac{m^2\,dx}{EI} = \int_0^8 0.36x^2 \frac{dx}{EI}$$

$$+ \int_0^{12} 0.16x^2 \frac{dx}{EI} = \frac{61.5 + 92.1}{EI} \cong \frac{154}{EI}$$

$$\delta_B + \delta_{bb} R_B = 0$$

$$R_B = -\frac{\delta_B}{\delta_{bb}} = -\frac{4776/EI}{154/EI} = -31.0 \text{ kips}; \quad R_B = 31 \text{ kips}\uparrow$$

From equations of statics,

$R_A = 0.2$ kips↑ and $R_C = 14.8$ kips↑

11.2.2. Beam with Two or More Redundants

When a structure has more than one redundant, the deflection of the primary structure is determined at each redundant location due to the applied loads. In addition, flexibility coefficients are established at each redundant location for each unit redundant force. Then, a deflection equation is written for each directed redundant on the primary structure in terms of the applied loads, the flexibility coefficients, and the unknown redundant forces. Simultaneous solution of these equations determine the redundant forces needed to satisfy the support boundary conditions and/or to restore the compatibility between structure members.

Figure 11.3 shows a continuous beam over several supports that is statically indeterminate to the third degree. Three options of a suitable primary (released) structure are shown in Figs. 11.3a, 11.3b, and 11.3c for the solu-

Figure 11.3

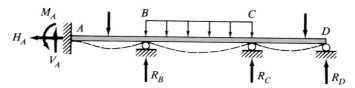

Primary structure options

1. Remove M_A, R_B, and R_C

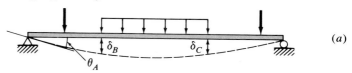

(a)

2. Remove R_B, R_C, and R_D

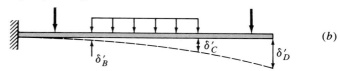

(b)

3. Remove M_A, R_B, and R_D

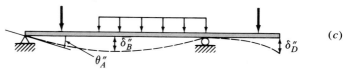

(c)

Flexibility coefficients for primary structure—(Option 1)

(d)

(e)

(f)

Rotation at point A: $\theta_A + \theta_{aa}M_A + \theta_{ab}R_B + \theta_{ac}R_C = 0$ (11.1)

Displacement at point B: $\delta_B + \delta_{ba}M_A + \delta_{bb}R_B + \delta_{bc}R_C = 0$ (11.2)

Displacement at point C: $\delta_C + \delta_{ca}M_A + \delta_{cb}R_B + \delta_{cc}R_C = 0$ (11.3)

tion of beam-support reactions by consistent deformation. Use of the primary structure of Fig. 11.3a shows released deformations of θ_A, δ_B, and δ_C due to the beam loading; the associated flexibility coefficients are illustrated in Figs. 11.3d, 11.3e, and 11.3f due to each unit redundant force. Since the slope change at A and the vertical displacements at B and C are known to be zero, deflection Eqs. (11.1), (11.2), and (11.3) are formulated and presented in Fig. 11.3. Simultaneous solution of these equations is performed to determine the redundant forces M_A, R_B, and R_C. In these equations, the flexibility coefficients and the released deformations associated with the applied loads can be found by use of the deflection methods described in Chap. 10. It is also evident, from Maxwell's law of reciprocal deflections, that $\delta_{bc} = \delta_{cb}$.

11.3. SUPPORT SETTLEMENT

In the preceding sections, it has been assumed that the beam supports are stationary. However, a settlement of one or more supports may occur if a beam is resting against a weak foundation. It is also possible for a gap to occur between a beam end and its support due to temperature changes, faulty erection, improper manufacturing, and so forth. Moreover, a support movement will alter the magnitudes of the reactions, shears, moments, and stresses in the beam structure. In Secs. 11.1 and 11.2, deflection equations at redundant locations are set equal to zero since the redundant forces are in the direction of unyielding supports. However, if a redundant support settles or undergoes creep displacement under sustained load, the deflection equation written at that redundant is set equal to the value of the displacement.

As an example, consider the continuous beam shown in Fig. 11.4a, to rest on three *unyielding* supports. If the support force at B is released to form a primary structure, the deflection equation can be expressed as

$$\delta_B + \delta_{bb} \cdot R_B = 0$$

where δ_B is the displacement of the primary structure at B, and δ_{bb} is the displacement at B due to a unit load at B.

In Fig. 11.4b, it is shown that the support at B undergoes a 2-in settlement. By the method of consistent deformations, the deflection equation at B can now be written as

$$\delta_B + \delta_{bb} \cdot R_B = 2.0 \text{ in}$$

or

$$R_B = \frac{2.0 \text{ in} - \delta_B}{\delta_{bb}}$$

Figure 11.4

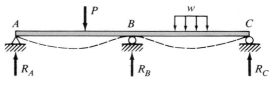

With R_B removed: $\delta_B + \delta_{bb} R_B = 0$

(a)

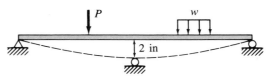

With R_B removed: $\delta_B + \delta_{bb} R_B = 2$ in

(b)

Common support settlement 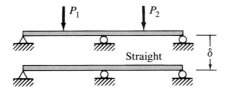 Linear support settlement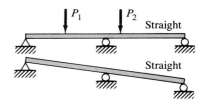

(c)

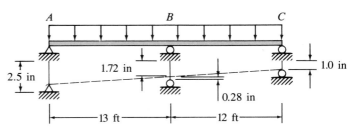

Relative settlement with R_B removed: $\delta_B + \delta_{bb} R_B = 0.28$ in

(d)

If all supports experience the same settlement, or displace different amounts but remain in a straight line as shown in Fig. 11.4c, the results are the same as if the beam supports are unyielding.

When support settlements occur that do not lie in a straight line, the beam analysis can be performed on the basis of relative displacements. For example, Fig. 11.4d shows settlements of 2.5 in, 1.72 in, and 1.0 in, at points

316 Statically Indeterminate Structures

A, B, and C, respectively. A straight line drawn from A to C shows a relative settlement at B of 0.28-in such that the deflection equation at B can be written as

$$\delta_B + \delta_{bb} \cdot R_B = 0.28 \text{ in}$$

Example 11.4 determines the reactions of a continuous beam under uniform loading by first considering unyielding supports, and then by assuming that a 0.50-in settlement occurs at support B. The resulting shear and bending-moment diagrams show the significant changes that occur due to the 0.50-in settlement of support B.

EXAMPLE 11.4

Compute the reactions and draw shear and moment diagrams for the beam shown below if (a) supports are unyielding, and (b) if support B settles $\frac{1}{2}$ in.

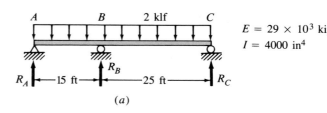

(a)

Solution

Part a: Remove R_B and use virtual work to find redundant at B.

From A to B (0 ft → 15 ft)

$M = 40x - 2x^2/2 = 40x - x^2$

$m = \dfrac{5x}{8};\ m^2 = \dfrac{25x^2}{64}$

$Mm = 25x^2 - \dfrac{5x^3}{8}$

From C to B (0 ft → 25 ft)

$M = 40x - x^2$

$m = \dfrac{3x}{8};\ m^2 = \dfrac{9x^2}{64}$

$Mm = 15x^2 - \dfrac{3x^3}{8}$

Primary structure:

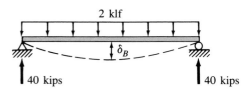

Unit load at B:

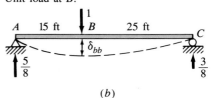

(b)

$1 \cdot \delta_B = \displaystyle\int_0^L \dfrac{(Mm)dx}{EI} = \int_0^{15}(25x^2 - 5x^3/8)\dfrac{dx}{EI} + \int_0^{25}(15x^2 - 3x^3/8)\dfrac{dx}{EI}$

$\delta_B = 61720/EI$

$1 \cdot \delta_{bb} = \displaystyle\int_0^L \dfrac{m^2 dx}{EI} = \int_0^{15} \dfrac{25x^2 dx}{64EI} + \int_0^{25} \dfrac{9x^2 dx}{64EI} = \dfrac{1172}{EI}$

$\delta_B + \delta_{bb} R_B = 0$

$\therefore\ R_B = -\dfrac{\delta_B}{\delta_{bb}} = -\dfrac{61720/EI}{1172/EI} = -52.7 \text{ kips}\uparrow$

By superposition,

$$R_A = 40 + \frac{5}{8}(-52.7) = 7.06 \text{ kips}\uparrow$$

$$R_C = 40 + \frac{3}{8}(-52.7) = 20.24 \text{ kips}\uparrow$$

If support B settles $\frac{1}{2}$ in:

$$\delta_B + \delta_{bb} R_B = 0.50 \text{ in}$$

$$\therefore R_B = \frac{0.50 - \delta_B}{\delta_{bb}} = \frac{0.50 - 61720/EI}{1172/EI}$$

$$R_B = \frac{0.50 - [61720(1728)]/[29 \times 10^3(4000)]}{[1172(1728)]/[29 \times 10^3(4000)]} = \frac{0.50 - 0.919}{0.0175} = -23.9 \text{ kips}\uparrow$$

By superposition,

$$R_A = 40 + \tfrac{5}{8}(-23.9) = 25.1 \text{ kips}\uparrow$$
$$R_C = 40 + \tfrac{3}{8}(-23.9) = 31.0 \text{ kips}\uparrow$$

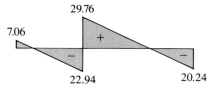

V (kips)   No settlement

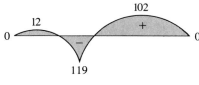

M, (ft·kips)   No settlement

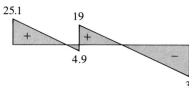

V, (kips) $\frac{1}{2}$-in settlement at B

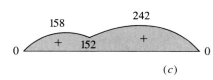

M, (ft·kips) $\frac{1}{2}$-in settlement at B.

(c)

It is worth noting that the effects of support settlement are computed using the original orientation of both the structure geometry and the applied forces. Truss and frame structures that experience support settlements, erection misfits, manufacturing misalignments, and so on, can also be evaluated by the approach presented above.

11.4. METHOD OF CONSISTENT DEFORMATIONS—FRAMES

The method of consistent deformations can be used to perform plane indeterminate frame analysis by the same procedure described for beams in the preceding sections. Example 11.5 applies the method for the analysis of a two-member frame using the principle of virtual work.

EXAMPLE 11.5

Compute the reactions for the frame shown below. EI is constant.

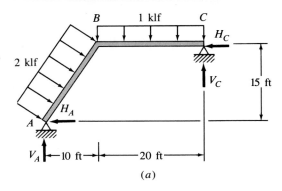

(a)

Solution

Remove H_A to form a primary structure.

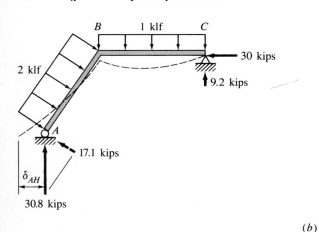
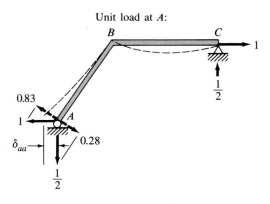

(b)

From A to B (0 ft → 18 ft)

$M = 17.1x - x^2$

$m = 0.55x;\ m^2 = 0.303x^2$

$Mm = 9.4x^2 - 0.55x^3$

From C to B (0 ft → 20 ft)

$M = 9.2x - x^2/2$

$m = 0.5x;\ m^2 = 0.25x^2$

$Mm = 4.6x^2 - x^3/4$

$$1 \cdot \delta_{AH} = \int_0^L \frac{Mm\,dx}{EI} = \int_0^{18} (9.4x^2 - 0.55x^3)\frac{dx}{EI} + \int_0^{20}(4.6x^2 - x^3/4)\frac{dx}{EI}$$

$$\delta_{AH} = \frac{3840}{EI} + \frac{2267}{EI} \cong \frac{6100}{EI}$$

$$1 \cdot \delta_{aa} = \int_0^L \frac{m^2\,dx}{EI} = \int_0^{18}(0.308x^2)\frac{dx}{EI} + \int_0^{20}(0.25x^2)\frac{dx}{EI}$$

$$\delta_{aa} \cong \frac{1270}{EI}$$

$$\delta_{AH} - \delta_{aa}H_A = 0$$

$$H_A = +\frac{\delta_{AH}}{\delta_{aa}} = +\frac{6100/EI}{1270/EI} = +4.8 \text{ kips} \leftarrow$$

By use of superposition,

$V_A = 30.8 + \frac{1}{2}(4.8) = 33.2$ kips↑

$H_C = 30 + 1(4.8) = 34.8$ kips ←

$V_C = 9.2 - \frac{1}{2}(4.8) = 6.8$ kips↑

Check by IMAGES-2D
$H_A = 4.2$ kips, $V_A = 32.9$ kips, $H_C = 34.1$ kips, $V_C = 7.1$ kips (reasonable accuracy)

11.5. METHOD OF CONSISTENT DEFORMATIONS—TRUSSES

Truss structures are often statically indeterminate. In Sec. 3.8, it is observed that trusses may be statically indeterminate externally (redundant reactions), internally (redundant members), or statically indeterminate due to a combination of both redundant reactions and members. Following are truss analyses for each of these conditions of indeterminacy which use the force method of consistent deformations and the principle of virtual work.

11.5.1. Externally Indeterminate Truss with One Redundant Reaction Component

The plane truss of Example 11.6 is statically indeterminate to the first degree since it has four reaction components and three available equations of statics. It is assumed that all members have the same modulus of elasticity (E) and a common cross-section area of 2.0 in². A primary structure is formed by releasing the horizontal restraint at D. Then, the free horizontal motion at D due to the applied loading is computed by virtual work (δ_D); horizontal motion at D on the primary structure is also determined for a

320 Statically Indeterminate Structures

unit horizontal load applied at D (δ_{dd}). The horizontal force at joint D that is needed to move joint D back to its original location is found from the deflection equation

$$\delta_D + \delta_{dd} \cdot H_D = 0 \quad \text{or} \quad H_D = -\delta_D/\delta_{dd}$$

Observe that after H_D is found, the total force in each truss member is tabulated using the principle of superposition.

EXAMPLE 11.6

Determine all reaction components and member forces.

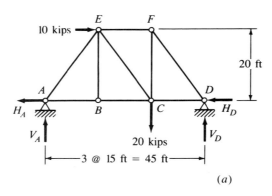

(a)

Solution

Remove horizontal restraint at D.

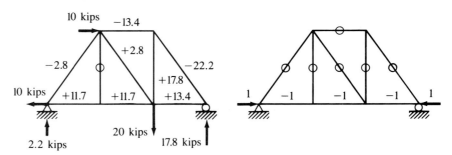

S forces due to applied loads u forces due to unit load at D

(b)

Member	L (in)	$\dfrac{L}{A}$	S	u_D	$Su_D \dfrac{L}{A}$	$u_D^2 \dfrac{L}{A}$	$S_{\text{total}} = S + u_D H_D$		
AB	180	90	11.7	−1	−1053	90	$11.7 + [(-1)(12.2)]$	=	−0.5
AE	300	150	−2.8	0	0	0	$-2.8 + 0$	=	−2.8
BC	180	90	11.7	−1	−1053	90	$11.7 + [(-1)(12.2)]$	=	−0.5
BE	240	120	0	0	0	0	$0 + 0$	=	0
CE	300	150	2.8	0	0	0	$2.8 + 0$	=	2.8
CD	180	90	13.4	−1	−1206	90	$13.4 + [(-1)(12.2)]$	=	1.2
CF	240	120	17.8	0	0	0	$17.8 + 0$	=	17.8
DF	300	150	−22.2	0	0	0	$-22.2 + 0$	=	−22.2
EF	180	90	−13.4	0	0	0	$-13.4 + 0$	=	−13.4
Σ					−3312	270			

$\delta_D + \delta_{dd} H_D = 0$ or

$$H_D = -\dfrac{\delta_D}{\delta_{dd}} = -\dfrac{\Sigma S u_D L/AE}{\Sigma u_D^2 L/AE} = -\dfrac{(-3312/E)}{(270/E)} = +12.2 \text{ kips} \leftarrow$$

$+ \rightarrow \Sigma H = 0 = 10 \text{ kips} - H_A - H_D$ or

$H_A = 10 - H_D = 10 - 12.2 = -2.2$ $\therefore$ $H_A = 2.2 \text{ kips} \rightarrow$

$\circlearrowleft + \Sigma M_D = 0 = 45 V_A + 10 \text{ kips}(20 \text{ ft}) - 20 \text{ kips}(15 \text{ ft})$ $\therefore$ $V_A = 2.2 \text{ kips} \uparrow$

$+ \uparrow \Sigma V = 0 = V_A + V_D - 20 \text{ kips}$ or

$V_D = 20 \text{ kips} - V_A = 20 - 2.2 = +17.8 \text{ kips} \uparrow$

11.5.2. Internally Indeterminate Truss with One Redundant Member

The truss shown in Example 11.7 is similar to the truss of Example 11.6 except that the hinge support at joint D is replaced with a roller and a diagonal member is added to the truss bar arrangement and joined at B and F. In the truss of Example 11.7, the reactions are easily obtained by statics. However, a comparison of the number of static joint equations ($2j$) versus the number of unknown member forces and reactions ($m + r$) reveals that the truss is statically indeterminate internally and has one redundant member. It is observed that removal of diagonal BF as the redundant member provides the same *primary* structure used in Example 11.6 and permits joints B and F to pull apart freely due to the absence of the restraint of member BF. Let δ_{BF} represent the *free* motion of B relative to F on the primary structure due to the applied loads, in the direction of BF. Also, let δ_{bf} denote the movement of B relative to F on the primary structure due to a unit tensile force in member BF. The actual force in member BF, needed to restore the unrestrained motion δ_{BF} and maintain compatible

deformations with the truss at joints B and F, is determined as follows:
From virtual work,

$$1 \cdot \delta_{BF} = \Sigma S \cdot u_{bf} \cdot \frac{L}{AE} \quad \text{and} \quad 1 \cdot \delta_{bf} = \Sigma(u_{bf})^2 \cdot \frac{L}{AE}$$

The force in member BF, namely S_{BF}, is found from the consistent deformation expression

$$\delta_{BF} + S_{BF} \cdot \delta_{bf} = 0$$

which can be rewritten as

$$\Sigma S \cdot u_{bf} \cdot \frac{L}{AE} + S_{BF} \cdot \Sigma(u_{bf})^2 \cdot \frac{L}{AE} = 0$$

Thus, $\quad S_{BF} = -\left(\Sigma S \cdot u_{bf} \cdot \frac{L}{AE}\right) \Big/ \left[\Sigma(u_{bf})^2 \cdot \frac{L}{AE}\right]$

After S_{BF} is determined, the forces in all truss members are tabulated using the principle of superposition. A positive value of S_{BF} indicates tensile force; a negative value implies a compressive force in S_{BF}.

EXAMPLE 11.7

Determine all reaction components and member forces.

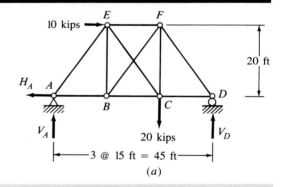

(a)

Solution

Release member BF. Truss support reactions found by statics.

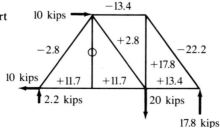

S forces due to applied loads

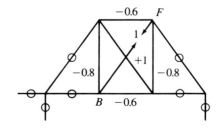

u forces due to unit load along BF

(b)

323 Method of Consistent Deformations—Trusses

Member	L (in)	$\frac{L}{A}$	S	u_{bf}	u_{bf}^2	$Su_{bf}\frac{L}{A}$	$u_{bf}^2\frac{L}{A}$	$S_{total} = S + u_{bf}S_{BF}$	
AB	180	90	11.7	0	0	0	0	11.7 + 0	= 11.7
AE	300	150	−2.8	0	0	0	0	−2.8 + 0	= −2.8
BC	180	90	11.7	−0.6	0.36	−632	32.4	11.7 + (−0.6)(2.3)	= 10.3
BE	240	120	0	−0.8	0.64	0	76.8	0 + (−0.8)(2.3)	= −1.8
CE	300	150	2.8	1	1	420	150	2.8 + 1(2.3)	= 5.1
CD	180	90	13.4	0	0	0	0	13.4 + 0	= 13.4
CF	240	120	17.8	−0.8	0.64	−1709	76.8	17.8 + (−0.8)(2.3)	= 16.0
DF	300	150	−22.2	0	0	0	0	−22.2 + 0	= −22.2
EF	180	90	−13.4	−0.6	0.36	724	32.4	−13.4 + (−0.6)(2.3)	= −14.8
BF	300	150	0	1	1	0	150	0 + 1(2.3)	= 2.3
Σ						−1197	518		

$\delta_{BF} + S_{BF}\delta_{bf} = 0$

$\Sigma Su_{bf}\dfrac{L}{AE} + S_{BF}\Sigma u_{bf}^2\dfrac{L}{AE} = 0$ or $S_{BF} = -\dfrac{\Sigma Su_{bf}L/AE}{\Sigma u_{bf}^2 L/AE} = -\dfrac{(-1197/E)}{518/E}$

$S_{BF} = +2.3$ kips

Note
Truss reactions are shown on the *primary* truss above and are not influenced by the value of S_{BF}.

11.5.3. Indeterminate Truss with Two or More Redundants

A truss can also be solved by the force method of consistent deformations if it is statically indeterminate to the second degree or higher.

Truss with two redundant reactions The truss of Fig. 11.5 has five support reaction components and is therefore statically indeterminate externally to the second degree (two redundant reactions). If redundants are removed at A and B, two deflection equations can be written in terms of the two unknown redundant forces as follows:

$$\delta_A + V_A \cdot \delta_{aa} + V_B \cdot \delta_{ab} = 0$$
$$\delta_B + V_A \cdot \delta_{ba} + V_B \cdot \delta_{bb} = 0$$

Figure 11.5

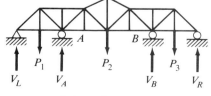

Indeterminate truss

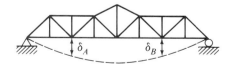

Determinate (primary) truss in deformed state

Using virtual work, the displacement terms above can be written as

$$\delta_A = \Sigma S \cdot u_a \cdot \frac{L}{AE} \qquad \delta_B = \Sigma S \cdot u_b \cdot \frac{L}{AE}$$

$$\delta_{aa} = \Sigma(u_a)^2 \cdot \frac{L}{AE} \qquad \delta_{bb} = \Sigma(u_b)^2 \cdot \frac{L}{AE}$$

and

$$\delta_{ab} = \delta_{ba} = \Sigma u_a \cdot u_b \cdot \frac{L}{AE}$$

where the member forces of the primary structure are represented by

S due to the applied loads
u_a due to a unit load at A
u_b due to a unit load at B

The two redundant reactions, V_A and V_B, are determined by simultaneous solution of the two equations above.

Truss with two redundant members A statically indeterminate truss with two redundant members is shown in Fig. 11.6, where the redundant member forces are noted as S_1 and S_2. The method of virtual work can be used to determine displacements of the primary truss from the real loads. In addition, virtual work can be used to establish the flexibility coefficients associated with each redundant member force. Then, deformation equations are written in terms of the unknown redundant forces, S_1 and S_2, which are required to maintain compatible deformations with the remaining truss at their respective member ends (truss joints). These displacement relationships are expressed as

$$\Sigma S \cdot u_1 \cdot \frac{L}{AE} + S_1 \Sigma (u_1)^2 \cdot \frac{L}{AE} + S_2 \Sigma u_1 \cdot u_2 \cdot \frac{L}{AE} = 0$$

$$\Sigma S \cdot u_2 \cdot \frac{L}{AE} + S_1 \Sigma u_2 \cdot u_1 \cdot \frac{L}{AE} + S_2 \Sigma (u_2)^2 \cdot \frac{L}{AE} = 0$$

where u_1 = the member forces due to an applied unit tensile force in member 1
u_2 = the member forces due to an applied unit tensile force in member 2

The two redundant member forces, S_1 and S_2, are found by simultaneous solution of the equations above.

Truss with one redundant reaction and one redundant member The truss shown in Example 11.8 combines the indeterminate features of Examples 11.6 and 11.7; that is, the truss has the identical

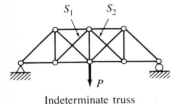
Indeterminate truss

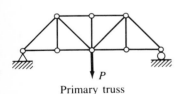
Primary truss

Figure 11.6

325 Method of Consistent Deformations—Trusses

geometry of the truss of Example 11.7 and contains identical hinge supports as shown in Example 11.6. Therefore, the truss is statically indeterminate to the first degree—both externally and internally! By releasing the horizontal restraint at D and member BF, the truss is, once again, reduced to the primary truss used in Examples 11.6 and 11.7, above.

EXAMPLE 11.8

Determine all reactions and truss member forces.

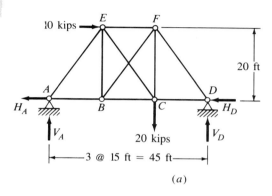

(a)

Solution

Remove horizontal restraint at D and release member BF.

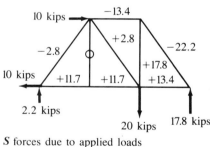

S forces due to applied loads

u_D forces due to unit load at D

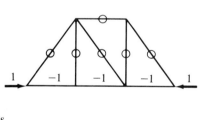

u_{bf} forces due to unit load along BF

(b)

$$\delta_D + \delta_{dd} \cdot H_D + \delta_{d(bf)} \cdot S_{BF} = 0$$

$$= \Sigma \frac{S u_D L}{AE} + H_D \Sigma \frac{u_D^2 L}{AE} + S_{BF} \Sigma \frac{u_{bf} \cdot u_d L}{AE}$$

$$\delta_{BF} + \delta_{(bf)d} \cdot H_D + \delta_{(bf)bf} \cdot S_{BF} = 0$$

$$= \Sigma \frac{S u_{bf} L}{AE} + H_D \Sigma \frac{u_d \cdot u_{bf} L}{AE} + S_{BF} \Sigma (u_{bf})^2 \frac{L}{AE}$$

Member	L (in)	L/A	S	u_D	u_{bf}	$u_D u_{bf}$	$\frac{L}{A}(u_D u_{bf})$	u_D^2	$u_D^2 \cdot \frac{L}{A}$	u_{bf}^2	$u_{bf}^2 \cdot \frac{L}{A}$	$Su_D \frac{L}{A}$	$Su_{bf} \frac{L}{A}$	S_{total}*
AB	180	90	11.7	−1	0	0	0	1	90	0	0	−1053	0	−0.4
AE	300	150	−2.8	0	0	0	0	0	0	0	0	0	0	−2.8
BC	180	90	11.7	−1	−0.6	0.6	54	1	90	0.36	32.4	−1053	−632	−1.0
BE	240	120	0	0	−0.8	0	0	0	0	0.64	76.8	0	0	−0.8
CE	300	150	2.8	0	1	0	0	0	0	1	150	0	420	3.9
CD	180	90	13.4	−1	0	0	0	1	90	0	0	−1206	0	1.3
CF	240	120	17.8	0	−0.8	0	0	0	0	0.64	76.8	0	−1709	17.0
DF	300	150	−22.2	0	0	0	0	0	0	0	0	0	0	−22.2
EF	180	90	−13.4	0	−0.6	0	0	0	0	0.36	32.4	0	724	−14.0
BF	300	150	0	0	1	0	0	0	0	1	150	0	0	1.05
Σ							54		270		518	−3312	−1197	

* $S_{total} = S + u_D H_D + u_{bf} S_{BF}$

Substituting displacements and flexibility coefficients from the tabular summations into the consistent deformation equations yields

$$\frac{1}{E}[-3312 + 270 H_D + 54 S_{BF}] = 0$$

and

$$\frac{1}{E}[-1197 + 54 H_D + 518 S_{BF}] = 0$$

Simultaneous solution of these equations results in

$$H_D = 12.1 \text{ kips} \leftarrow \quad \text{and} \quad S_{BC} = 1.05 \text{ kips } (T)$$

Total bar forces (S_{total}) are presented in the last column of the table above.

As before, the influence coefficients for each redundant force must be determined in order to formulate deformation equations that will lead to solution of the redundant forces. It is observed that the applied loads on the primary truss result in both a horizontal displacement at support D and a relative displacement of point B with respect to point F. Likewise, a unit force applied horizontally at D, or a unit tensile force applied to the primary truss along BF will result in unit displacements (called influence coefficients) both at D and along BF. The method of virtual work is used to compute these displacements. After all displacements are found, the method of consistent deformations is exercised to write the displacement equations from which the redundant forces are solved.

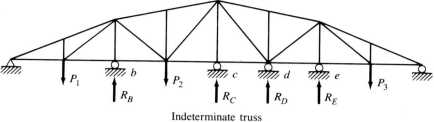

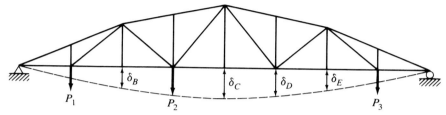

$$\delta_B + \delta_{bb}R_B + \delta_{bc}R_C + \delta_{bd}R_D + \delta_{be}R_E = 0$$
$$\delta_C + \delta_{cb}R_B + \delta_{cc}R_C + \delta_{cd}R_D + \delta_{ce}R_E = 0$$
$$\delta_D + \delta_{db}R_B + \delta_{dc}R_C + \delta_{dd}R_D + \delta_{de}R_E = 0$$
$$\delta_E + \delta_{eb}R_B + \delta_{ec}R_C + \delta_{ed}R_D + \delta_{ee}R_E = 0$$

Figure 11.7

Indeterminate truss with more than two redundants A truss (or a continuous beam) placed over many unyielding supports is found to be highly indeterminate, as shown in Fig. 11.7. By the method of consistent deformations, the selection of redundant supports at b, c, d, and e can be removed to provide a primary truss that is both statically determinate and stable. Thereafter, a set of consistent deflection equations are written at each redundant location (see Fig. 11.7); R_B, R_C, R_D, and R_E represent the redundant reaction forces for which an equal number of equations exist that easily lead to their solution. In these four equations, the flexibility coefficients are expressed in terms of lowercase subscript notation (e.g., δ_{bc} which is interpreted as the displacement at b due to a unit load applied at c). Initially, it may appear that 20 δ values must be determined in order to solve these four deflection equations. However, by Maxwell's law of reciprocal deflections, it is found that

$$\delta_{bc} = \delta_{cb} \qquad \delta_{bd} = \delta_{db} \qquad \delta_{be} = \delta_{eb}$$
$$\delta_{cd} = \delta_{dc} \qquad \delta_{ce} = \delta_{ec} \qquad \delta_{de} = \delta_{ed}$$

which results in 14 independent constants. This represents a 30 percent reduction of computations for evaluation of equation constants.

11.6. CASTIGLIANO'S THEOREM OF LEAST WORK

The theorem of least work is a special application of Castigliano's second theorem that can be used for the analysis of statically indeterminate structures. The term *theorem of least work* was first suggested in a paper by Ménabréa in 1858 [11.3]. However, a satisfactory proof of the theorem was first presented in Castigliano's thesis for a diploma in engineering at Turin (1875); the theorem was later published in 1879 [11.4]. The method, as presented below, applies to elastic structures which have unyielding supports and do not experience temperature changes or fabrication errors. That is, the theorem will consider structures which experience deflections due to loads alone.

Consider the indeterminate beam of Fig. 11.8a resting on supports at A,

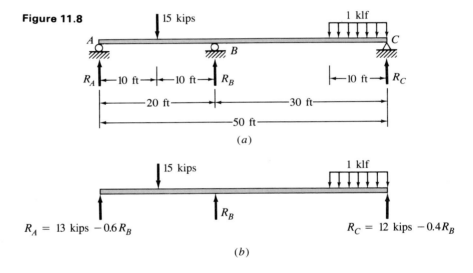

Figure 11.8

B, and C. By consistent deformations, a redundant reaction is removed to provide a determinate and stable structure, and the redundant is determined using the principle of superposition. The method of least work does not require the removal of a redundant and the formation of a primary structure. Instead, all support reactions are expressed in terms of both the real loads and the unknown redundant force treated as an applied load on the beam. Then, the internal strain energy is expressed in terms of the known loads and the unknown redundant force. For example, in Fig. 11.8b, the redundant reaction is selected at support B, where the displacement is known to be zero. Therefore, the deflection at B, from Castigliano's second theorem, is

$$y_B = \frac{\partial W_i}{\partial R_B} = 0$$

Castigliano's Theorem of Least Work

This expression can be interpreted to suggest:

1 A structure will do the *least* work necessary to maintain equilibrium while supporting its loads. That is, a truss, beam, or frame will deflect the least possible amount and do no more work than is required to maintain static equilibrium.

2 The magnitude of the redundant force is that value which minimizes the strain energy. Thus, the derivatives of the internal strain energy of deformation is a minimum when taken with respect to redundant supports (or redundant members).

Where two or more redundants exist in a structure, such as R_B and R_C shown in Fig. 11.9, the remaining reactions can be written in terms of the real loads and the redundant forces at B and C. Then, partial derivatives of the internal strain energy, W_i, with respect to R_B and R_C, are made and set equal to zero. This will provide two equations in terms of the two unknown reactions which can then be solved.

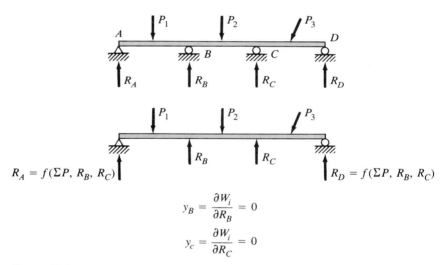

$$R_A = f(\Sigma P, R_B, R_C)$$

$$R_D = f(\Sigma P, R_B, R_C)$$

$$y_B = \frac{\partial W_i}{\partial R_B} = 0$$

$$y_C = \frac{\partial W_i}{\partial R_C} = 0$$

Figure 11.9

The method of least work was proven by Castigliano to apply to trusses with redundant members. It can also be used for stress analysis of statically indeterminate beams, frames, and structures which contain both axially loaded and flexural members. Examples 11.9, 11.10, and 11.11 provide a few cases of the application of the method of least work to structures which are statically indeterminate to the first degree. These examples involve either a redundant reaction component or a redundant member. Example 11.11 shows that the method of least work is also suitable to evaluate structures which are composed of both axial and flexural members.

EXAMPLE 11.9

Determine the reaction at support A by the method of least work. E and I are constant.

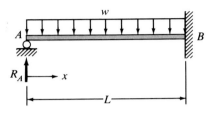

Solution

Consider the redundant support force at A where $\delta_A = 0$. Therefore, by Castigliano's second theorem,

$$\delta_A = 0 = \frac{\partial W_i}{\partial R_A} = \int_0^L M\left(\frac{\partial M}{\partial R_A}\right) \cdot \frac{dx}{EI}$$

where

$$M = R_A x - \frac{wx^2}{2}$$

and

$$\frac{\partial M}{\partial R_A} = x$$

Therefore,

$$\delta_A = \int_0^L \left(R_A x - \frac{wx^2}{2}\right)(x) \frac{dx}{EI} = 0$$

$$= \frac{1}{EI}\left[\frac{R_A x^3}{3} - \frac{wx^4}{8}\right]_0^L = 0$$

$$= \frac{1}{EI}\left[\frac{R_A L^3}{3} - \frac{wL^4}{8}\right] = 0$$

or $R_A = +\frac{3}{8}wL$ $\therefore$ $R_A = \frac{3}{8}wL\uparrow$ Answer

EXAMPLE 11.10

Determine the force in redundant CE of the truss. E is constant. Cross-sectional areas are shown encircled in units of in².

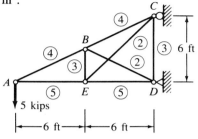
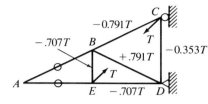

Solution

The member forces (S) with the redundant CE removed can be found by Example 10.11. The redundant force in CE can be signified by T; the effect of T on the other truss members is shown above.

Member	$\dfrac{L}{A} \sim \dfrac{1}{in^2}$	S (kips)	$\partial S/\partial T$	$S \cdot \dfrac{L}{A} \cdot \dfrac{\partial S}{\partial T}$
AB	20.1	$11.18 + 0$	0	0
BC	20.1	$11.18 - 0.791T$	-0.791	$-177.8 + 12.6T$
AE	14.4	$-10.0 + 0$	0	0
ED	14.4	$-10.0 - 0.707T$	-0.707	$+101.8 + 7.2T$
BE	12.0	$0 - 0.707T$	-0.707	$0 + 6.0T$
CD	24.0	$-5 - 0.353T$	-0.353	$+42.4 + 3.0T$
BD	40.2	$0 + 0.791T$	$+0.791$	$0 + 25.1T$
CE	50.9	$+T$	$+1.0$	$+50.9T$
Σ				$-33.6 + 104.8T$

By least work, it is required that

$$\dfrac{\partial W_i}{\partial T} = \Sigma S \cdot \dfrac{\partial S}{\partial T} \cdot \dfrac{L}{AE} = 0$$

Thus,

$$\dfrac{-33.6 + 104.8T}{E} = 0 \quad \text{or} \quad T = \dfrac{+33.6}{104.8} = +0.321 \text{ kips (tension)}$$

EXAMPLE 11.11

By the method of least work, find the tensile force in the steel support cable.

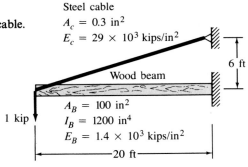

Solution

Designate the force in the cable by T. Also define the beam length by L and the cable length by l.

By least work,

$$\frac{\partial W_i}{\partial T} = 0:$$

$$\frac{\partial W_i}{\partial T} = \int_0^L M\left(\frac{\partial M}{\partial T}\right)\frac{dx}{E_B I_B} + \int_0^L S\left(\frac{\partial S}{\partial T}\right)\frac{dx}{A_B E_B}$$
$$+ \int_0^l S\left(\frac{\partial S}{\partial T}\right)\frac{dx}{A_c E_c} = 0$$

Thus,

$$\int_0^{20\,\text{ft}} (0.287Tx - 1x)(0.287x)\frac{dx}{E_B I_B}$$
$$+ \int_0^{20\,\text{ft}} (0.958T)(0.958)\frac{dx}{A_B E_B} + \int_0^{20.88\,\text{ft}} (T)(1)\frac{dx}{A_c E_c} = 0$$

$$\left(\frac{220T - 765}{E_B I_B}\right)\left(\frac{1728\ \text{in}^3}{\text{ft}^3}\right) + \left(\frac{18.36T}{A_B E_B}\right)\left(\frac{12\ \text{in}}{\text{ft}}\right) + \left(\frac{20.88T}{A_c E_c}\right)\left(\frac{12\ \text{in}}{\text{ft}}\right) = 0$$

$$0.226T - 0.787 + 0.002T + 0.029T = 0$$

From which we find

$$T = +3.0 \text{ kips (tension)}$$

11.7. INFLUENCE LINES FOR INDETERMINATE STRUCTURES

The main purpose of influence lines for both statically determinate and indeterminate structures is to determine where to position moving loads to cause the maximum effect of a design function. For example, an influence line can be used to forecast where the design live load should be placed on a continuous beam of a building floor system to cause maximum positive bending moment; another influence line can be made for a bridge truss member to determine where to place the live loads that will result in maximum member force. The same methods used in Chaps. 6 and 7 for determinate structures can also be used to construct influence lines for indeterminate structures. Influence lines of indeterminate structures can also be determined by other methods, such as the use of moment distribution and influence tables [11.5]. Although these traditional methods involve tedious hand computations, they provide analytic verification of the influence-line construction and serve to enhance a student's understanding of a structure's behavior.

In the paragraphs that follow, two approaches are presented for influence line construction; the first approach develops *quantitative* influence lines using the prior methods of structural analysis conveyed in this text; the second approach provides *qualitative* influence-line construction by use of the Müller-Breslau principle presented in Sec. 6.5. The objective of both methods remains the same, that is, to provide a visual aid which will establish where to position design live loads for maximum moment, shear, force, and so on.

Cantilevered lift frame structure, MASSPORT, Logan Airport, Boston. (*Courtesy of Fay, Spofford & Thorndike, Engineers, Lexington, Massachusetts.*)

11.7.1. Quantitative Influence Lines—Continuous Beams

In the preceding sections, the student observed that the solution of an indeterminate structure focuses on a procedure to evaluate the redundant forces and/or members. Thus, the construction of influence lines for either bending moment or shear in an indeterminate beam, or for axial force in the members of an indeterminate truss, usually begins by finding the influence lines of the redundants.

Influence line for support reaction Consider the continuous beam of Figure 11.10a. By the method of consistent deformations, the support reaction at B, R_B, is removed to provide a stable determinate beam; then, the deflection at B due to the real loading, δ_B, and the deflection at B due to a unit load placed at B, δ_{bb}, can be determined by one of the deflection methods of Chap. 10. Then, the deflection equation at B can be expressed as

$$\delta_B + \delta_{bb} \cdot R_B = 0$$

from which

$$R_B = -\delta_B/\delta_{bb} \qquad \text{(see Figs. 11.10b and 11.10c)}$$

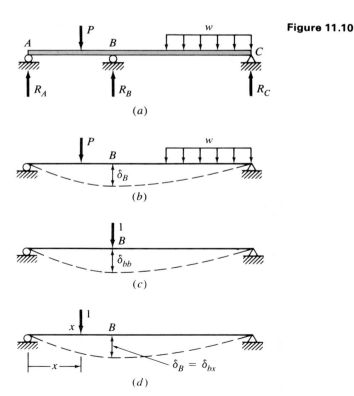

Figure 11.10

335 Influence Lines for Indeterminate Structures

If the influence line of R_B is desired, the real load on the beam is a moving-unit concentrated load (see Fig. 11.10d). Thus, $\delta_B = \delta_{bx}$ where δ_{bx} denotes the deflection at B due to a unit load placed at x. Then, the deflection equation can be expressed as

$$\delta_{bx} + \delta_{bb} \cdot R_B = 0 \quad \text{or} \quad R_B = -\delta_{bx}/\delta_{bb}$$

From Maxwell's reciprocal theorem, $\delta_{bx} = \delta_{xb}$. Therefore, the expression for R_B can be rewritten as

$$R_B = -\delta_{xb}/\delta_{bb}$$

This equation reveals that the influence line for R_B is found by placing the unit load on the determinate beam at point B, followed by computation of the deflection at several x-beam locations including point B. In addition, it is observed that influence lines for indeterminate beams are nonlinear functions since they depend on deflection methods for their solution.

Influence lines for beam shear and bending moment Example 11.12 uses the conjugate-beam method to determine the influence line for the reaction at B. After the influence line for R_B is constructed, influence lines for the remaining reactions at R_A and R_C are determined by statics. For each position of the unit load on the beam, the reactions are known; therefore, the values of shear and bending moment at interior point D on the continuous beam can easily be found. Subsequently, the influence lines of shear and moment at D are easily drawn (see tabular computations for V_D and M_D in Example 11.12).

EXAMPLE 11.12

Determine the ordinate values at 5-ft intervals and draw the influence lines for all support reactions and for shear and bending moment at point D on the continuous beam. E and I are constant.

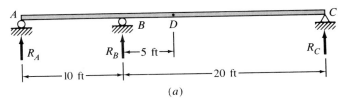

(a)

336 Statically Indeterminate Structures

Solution

Remove the redundant at B and solve for R_B using the conjugate-beam method.

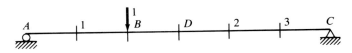

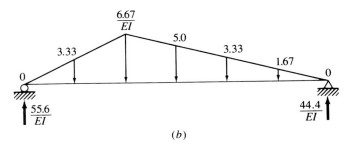

(b)

$$\delta_1 = M_{CB_1} = \frac{55.6}{EI}(5) - \frac{1}{2}\left(\frac{3.33}{EI}\right)(5) \cdot \frac{5}{3} = \frac{264}{EI}$$

$$\delta_B = \delta_{bb} = M_{CB_B} = \frac{55.6}{EI}(10) - \frac{1}{2}\left(\frac{6.67}{EI}\right)(10) \cdot \frac{10}{3} = \frac{445}{EI}$$

$$\delta_D = M_{CB_D} = \frac{44.4}{EI}(15) - \frac{1}{2}\left(\frac{5}{EI}\right)(15) \cdot 5 = \frac{479}{EI}$$

$$\delta_2 = M_{CB_2} = \frac{44.4}{EI}(10) - \frac{1}{2}\left(\frac{3.33}{EI}\right)(10) \cdot \frac{10}{3} = \frac{389}{EI}$$

$$\delta_3 = M_{CB_3} = \frac{44.4}{EI}(5) - \frac{1}{2}\left(\frac{1.67}{EI}\right)(5) \cdot \frac{5}{3} = \frac{215}{EI}$$

Influence line for R_B is found from

$$R_B = -\frac{\delta_{xb}}{\delta_{bb}}$$ Influence line for $R_B(\uparrow)$

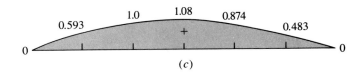

(c)

With the influence line of R_B known, the influence line for R_A is found by $\Sigma M_C = 0$.

Influence line for $R_A(\uparrow)$

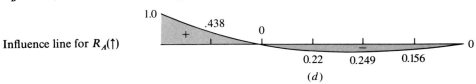

(d)

337 Influence Lines for Indeterminate Structures

Since $R_A + R_B + R_C - 1 = 0$, the influence line for R_C is found ($\Sigma F_y = 0$).

Influence line for $R_C(\uparrow)$

(e)

V_D—Influence line for shear at point D:

Unit Load at:	Shear at D
1	$V_D = R_A + R_B - 1$ $= 0.593 + 0.438 - 1 = +0.031$
B	$V_D = R_A + R_B - 1 = 0$
Left of D	$V_D = R_A + R_B - 1 = -0.14$
Right of D	$V_D = R_A + R_B = +0.86$
2	$V_D = R_A + R_B = +0.625$
3	$V_D = R_A + R_B = +0.327$

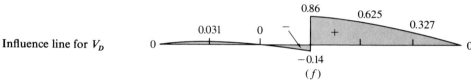

Influence line for V_D

(f)

M_D—Influence line for bending moment at point D:

Unit Load at:	Bending Moment at D
1	$M_D = 15R_A + 5R_B - 1(10)$ $= 15(0.438) + 5(0.593) - 10 = -0.54$
B	$M_D = 15R_A + 5R_B - 1(5) = 0$
D	$M_D = 15R_A + 5R_B = +2.1$
2	$M_D = 15R_A + 5R_B = +0.635$
3	$M_D = 15R_A + 5R_B = +0.075$

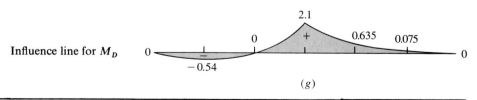

Influence line for M_D

(g)

Now, consider drawing the influence lines for the four-span continuous beam of Fig. 11.11 which has three redundant reactions. Following the removal of R_B, R_C, and R_D as the redundants, deflections (δ_{xb}) can be computed at selected locations along the beam for a unit load placed at B.

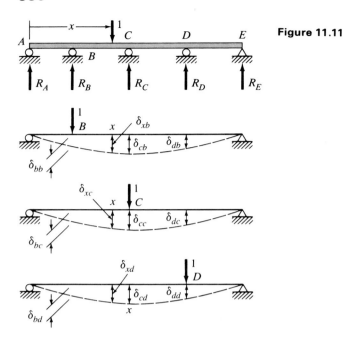

Figure 11.11

Deflections (δ_{xc}) can also be computed at the same locations on the beam for a unit load placed at C; the deflection procedure is repeated again for a unit load placed D (δ_{xd}). By Maxwell's reciprocal law, it is evident that $\delta_{xb} = \delta_{bx}$, $\delta_{xc} = \delta_{cx}$, and $\delta_{xd} = \delta_{dx}$, which suggests that deflections are easily found at several x locations due to a unit load placed either at B, C, or D. After the deflection ordinate values are determined, deflection equations can be expressed as

$$\delta_{xb} + \delta_{bb} \cdot R_B + \delta_{bc} \cdot R_C + \delta_{bd} \cdot R_D = 0$$
$$\delta_{xc} + \delta_{cb} \cdot R_B + \delta_{cc} \cdot R_C + \delta_{cd} \cdot R_D = 0$$
$$\delta_{xd} + \delta_{db} \cdot R_B + \delta_{dc} \cdot R_C + \delta_{dd} \cdot R_D = 0$$

In the equations above, it is also evident from Maxwell's law that $\delta_{bc} = \delta_{cb}$, $\delta_{bd} = \delta_{db}$, and $\delta_{cd} = \delta_{dc}$. Simultaneous solution of the equations are repeated for each set of variables δ_{xb}, δ_{xc}, and δ_{xd} to determine the influence ordinate values of R_B, R_C, and R_D. After the reaction influence lines are drawn, influence lines for shear, bending moment, and so forth, can be found by statics.

11.7.2. Quantitative Influence Lines—Indeterminate Trusses

Influence lines for indeterminate trusses are found by the same procedures of Sec. 11.5 and the concepts presented for continuous beams, above. For example, in Fig. 11.12, the redundant reaction at B is removed and deflection values at joint B are computed for the placement of a unit load at each joint on the loaded chord of the truss. The redundant force for each unit load position is found from the familiar expression

$$R_B = -\delta_{xb}/\delta_{bb} = -\frac{(\Sigma u_x u_b L/AE)}{[\Sigma(u_b)^2 L/AE]}$$

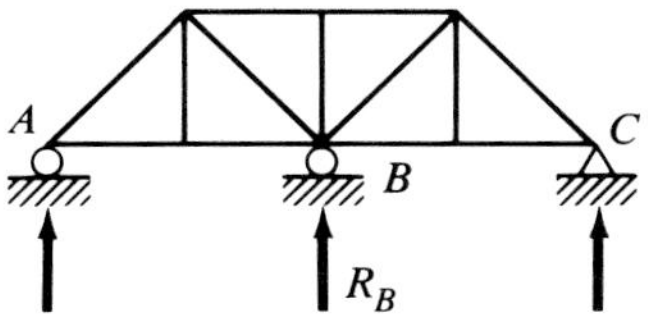

Figure 11.12

After the influence line for the redundant reaction is drawn, the remaining reaction influence lines can be determined by statics. Thereafter, the influence lines for axial force in the truss members can be found by statics. This procedure can also be used for trusses with a redundant member and for trusses with many redundants. Additional attention to influence line development for indeterminate trusses can be found in Refs. [11.5], [11.6], and [11.7].

When many bar-force influence lines of an indeterminate truss are required, the author recommends the use of a computer program such as IMAGES. Typically, the truss geometry and its restraints are initially defined and saved on the computer. Then, several load cases are presented, where each load case considers the unit load at one joint on the loaded chord of the truss. Computer solutions of truss member forces (and support reactions) are obtained for each load case. Therefore, the influence line ordinates for each truss member force can be drawn simultaneously from the computer output data.

11.7.3. Qualitative Influence Lines—Müeller-Breslau

The importance of the influence line is fully realized when an analyst must determine where to position design live loads on a highly indeterminate structure that will maximize a design function. The design function may be shear, bending moment, axial force, slope change, deflection, and so forth. The structure in question may be a multistory building, a bridge girder resting on many supports, or a highly redundant truss. In any case, the analyst must consider where the live loads are to be placed, regardless of whether the analysis is performed by classical hand-calculation methods or by use of a computer code. The studies of influence lines, presented above and in Chaps. 6 and 7, serve to develop a student's ability to construct many quantitative influence lines. Now, the important principle of Müeller-Breslau, first presented in Sec. 6.5, can be recalled for the purpose of drawing qualitative influence lines for indeterminate structures.

The Müeller-Breslau principle can be stated as follows:

If a function at a point on a structure, such as reaction, or shear, or moment, is allowed to act without restraint, the deflected shape of the structure, to some scale, represents the influence line of the function.

The shape of the qualitative influence line can be drawn quickly and accurately when compared to the work required to compute the ordinates of the quantitative influence line. This is significant since the major reason for drawing influence lines is to locate critical live-load positions for a particular function.

Recall the quantitative influence lines constructed for the two-span continuous beam in Example 11.12. By use of the Müeller-Breslau principle,

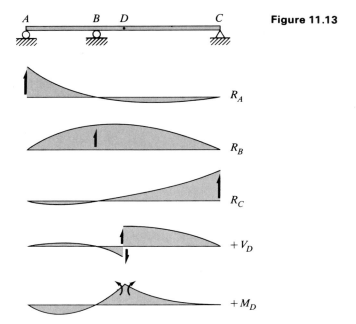

Figure 11.13

qualitative influence lines are drawn in Fig. 11.13 for the support reactions and for shear and bending moment at interior point D. The influence line for R_A is drawn by removing the restraint at A and permitting the beam to displace freely in the direction of R_A; influence lines for R_B and R_C are drawn in the same way. The influence line for positive shear at point D can be drawn by assuming the beam to be cut at D with positive shear force applied at the cut ends. In a similar manner, the influence line for positive moment at D is drawn by applying positive moments to the assumed "cut" ends of the beam at D. The qualitative influence-line shapes of Fig. 11.13 are seen to compare very well with the respective shapes of the quantitative influence lines constructed in Example 11.12.

11.7.4. Critical Live-Load Placement

Figures 11.14a and 11.14b show sketches of influence lines for the reactions at A and B, respectively, of a four-span continuous beam. The maximum reaction at A will occur for distributed live-load placement in spans 1 and 3. The maximum reaction at B will occur when live loading is placed on spans 1, 2, and 4. From Fig. 11.14c, the positive shear influence line at point i has a maximum value when uniform live loading is placed over the positive regions of the influence line. Maximum negative shear at point i is found when the uniform live loading is placed over the negative regions of the influence line. Figures 11.14d and 11.14e show sketches of influence lines for positive bending moment at point j and for negative bending moment at C, respectively; maximum values of these functions are achieved when the live load is placed over the positive influence line regions along the beam. In general, live loads placed at least two or more spans away from a span of

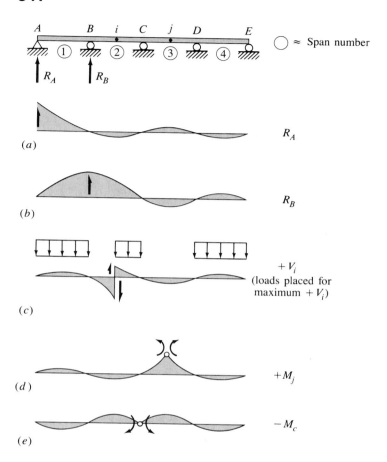

Figure 11.14

interest will have little or negligible effect on a shear or moment function within the span.

Figure 11.15a shows the influence line for positive shear in a first floor outer span of a three-story rigid frame; the uniform live-load pattern for maximum shear is shown in Fig. 11.15b. Figure 11.15c shows the influence line for positive bending moment in the second story middle span; the live-load pattern for maximum moment is shown in Fig. 11.15d.

11.7.5. Moment Envelope Curve

A design engineer often uses influence lines to construct a moment envelope curve for continuous beams in a building. The envelope curve defines the extreme boundary values of bending moment along the beam due to critical placements of design live loading. For example, Fig. 11.16a shows a three-span continuous beam along with qualitative influence lines for maximum positive bending in the center span, and maximum negative bending at interior support b. Figure 11.16b shows bending-moment curves for three load conditions: load case 1 provides the maximum positive moment in the center span; load case 2 results in maximum negative moment at an interior

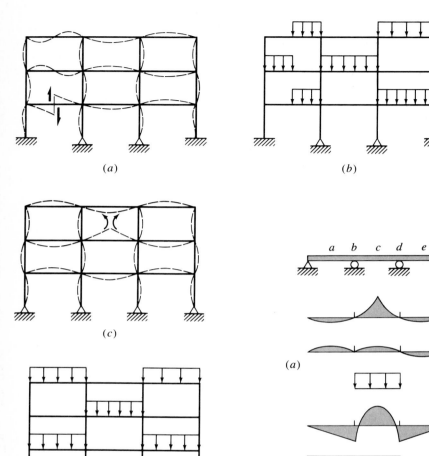

Figure 11.15

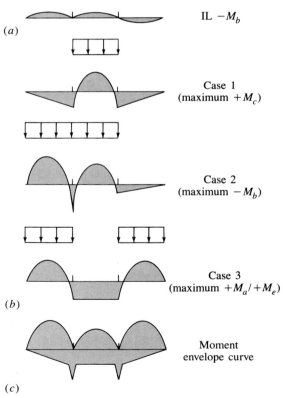

Figure 11.16

support (shown here for support *b*). The mirror image of the curve for case 2 yields the maximum negative moment at *d*. Load case 3 shows maximum positive moment in each end span. The superposition of these moment diagrams result in the envelope curve shown shaded in Fig. 11.16c for live load. *However, the envelope curve is not complete until the moment curve due to design dead load is superimposed with the live-load curve.* Design envelope curves can also be developed for other design functions of interest such as beam shear.

11.8. QUALITATIVE INFLUENCE LINES BY COMPUTER

Initially, the use of a computer to obtain a qualitative influence line may seem impractical. However, accurate construction of some qualitative influence lines may be difficult to accomplish. McCormac [11.6] gives a warning that difficulty arises in attempts to draw influence lines near quarter points of beam sections of continuous spans.

The IMAGES-2D computer program can be used to draw qualitative influence lines for indeterminate structures in the following manner:

Figure 11.17

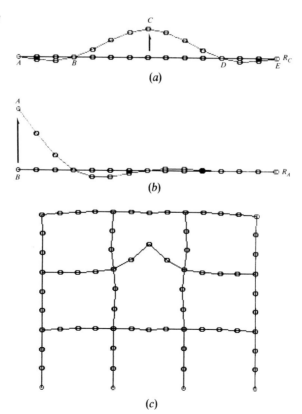

344 Statically Indeterminate Structures

After the input to the Geometry Definition Menu has been completed, enter the Static Menu and proceed to select Create/Edit Loads. This action provides a submenu where *Define Enforced Displacements* is listed. A nominal value of enforced displacement can be input at any *free* or *released* node point. Figures 11.17a and 11.17b show IMAGES computer plots of the beam of Fig. 11.14 subjected to enforced displacements at nodes C and A, respectively. Figure 11.17c gives the enforced displacement IMAGES plot for the building frame shown in Fig. 11.15c; the enforced displacement is shown to compare favorably with the influence line of Fig. 11.15c.

Problems

11.1 to 11.5 Using the methods of consistent distortion and moment area, compute all reactions and draw shear and moment diagrams.

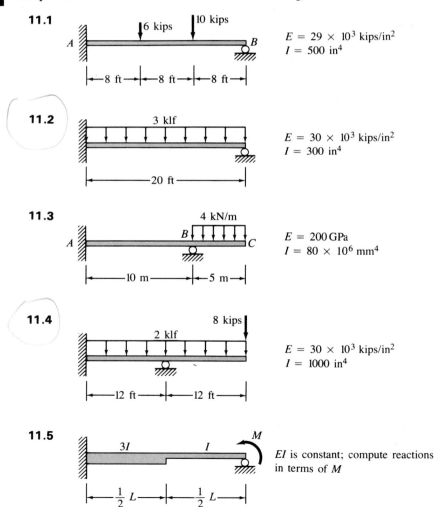

11.1 $E = 29 \times 10^3$ kips/in^2; $I = 500$ in^4

11.2 $E = 30 \times 10^3$ kips/in^2; $I = 300$ in^4

11.3 $E = 200$ GPa; $I = 80 \times 10^6$ mm^4

11.4 $E = 30 \times 10^3$ kips/in^2; $I = 1000$ in^4

11.5 EI is constant; compute reactions in terms of M

345 Problems

11.6 to 11.10 Using the methods of consistent distortion and conjugate beam, compute all reactions and draw shear and moment diagrams. E is constant. I is constant except where noted.

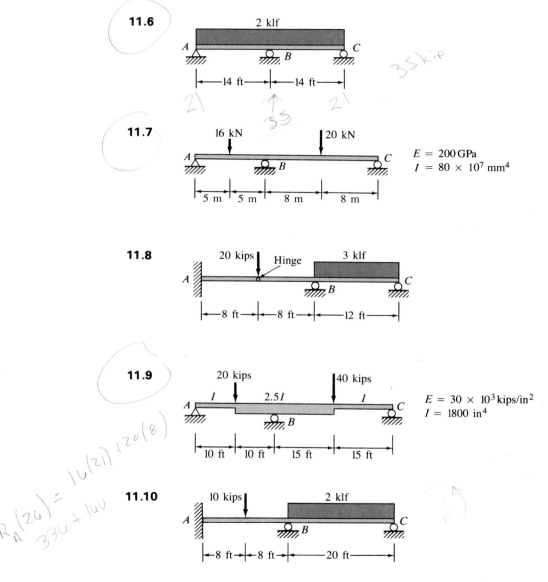

11.11 Rework Prob. 11.1 if support B settles 1.50 in

11.12 Rework Prob. 11.9 if support B settles 3.00 in

11.13 Rework Prob. 11.7 if support A settles 10 mm, B settles 30 mm, and C settles 20 mm.

11.14 to 11.18 Determine all support reactions using the methods of consistent distortions and virtual work. *EI* is constant for all problems.

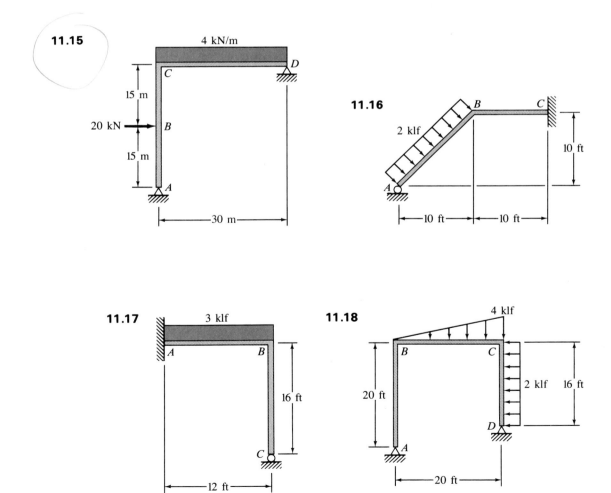

347 Problems

11.19 to 11.29 Determine all reaction components using the methods of consistent distortions and virtual work. E is constant for all problems. All areas are equal unless shown in circles (in in^2).

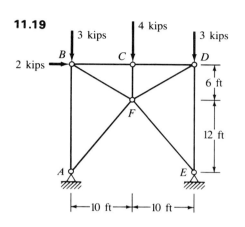

11.19

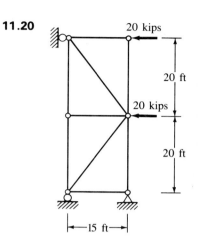

11.20

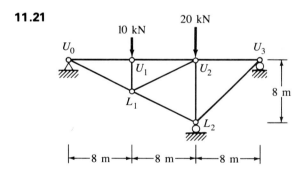

11.21

11.22 Areas in in^2, shown in circles

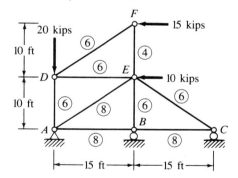

11.23 Areas in in^2, shown in circles

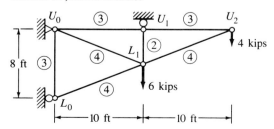

11.24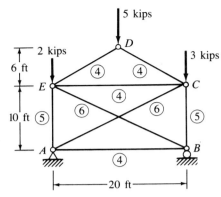

11.25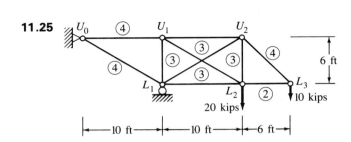

11.24 to 11.29 Compute member forces and reactions.
11.26 to 11.29 All member areas are equal.

11.26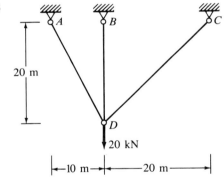

11.27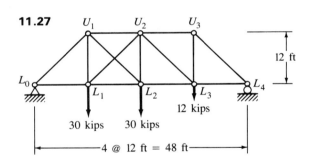

11.28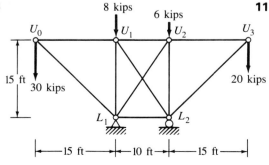

11.29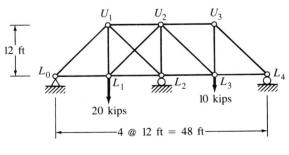

11.30 to 11.32 For each problem, determine the tensile force in the steel cable by use of Castigliano's theorem of least work. Only consider bending and axial energy. $E = 30 \times 10^6$ psi for all members. Cable area = 1.0 in². $I = 2000$ in⁴ and $A = 16$ in² for all bending-member elements.

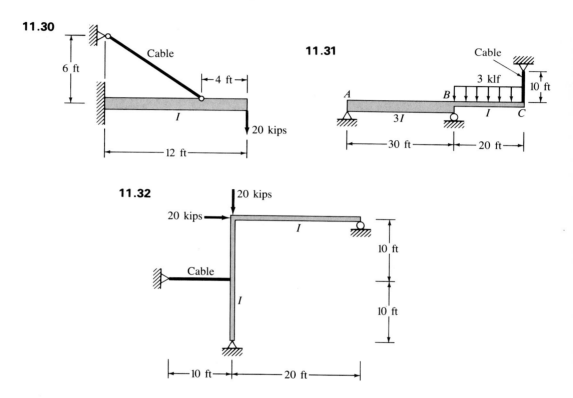

11.33 to 11.36 Quantitative influence lines of continuous beams.

11.33 Draw quantitative influence lines at midsupport intervals for all support reactions of Prob. 11.7.

11.34 Draw quantitative influence lines at midsupport intervals for all support reactions of Prob. 11.8.

11.35 Draw quantitative influence lines at 5-meter intervals for shear immediately to left of B and moment at support A of Prob. 11.3.

11.36 Draw quantitative influence lines for reactions at A and B, shear at C and moment at C, for the beam shown below (provide values at A, B, C, D, and E locations).

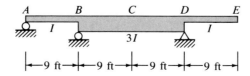

11.37 to 11.40 Draw qualitative influence lines for the specified parameters using the Müeller-Breslau principle.

11.37 Shear and moment under each load on the beam of Prob. 11.1.

11.38 Shear and moment under each load on the beam of Prob. 11.7.

11.39 Shear at the interior hinge and shear just to the right of support B of the beam of Prob. 11.8.

11.40 For the building frame below, negative shear and positive moment at A, negative shear and negative moment at B, and negative moment immediately to the right of the joint at C.

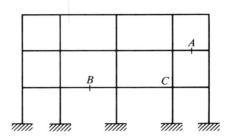

References

11.1 James Clerk Maxwell, "On the Calculation of the Equilibrium and Stiffness of Frames," *Philosophical Magazine*, vol. 4, no. 27, 1864.

11.2 Otto Mohr, "Beitrag zur Theorie des Fachwerks," *Zeitschrift des Architekten und Ingenieur-Vereins zu Hannover*, vol. 20, 1874.

11.3 L. F. Ménabréa, "Nouveau principe sur la distribution des tensions dans les systémes élastiques," *Comptes Rendus des Séances de l'Académie des Sciences*, Paris, vol. 46, 1858.

11.4 Carlo Alberto Castigliano, "Théorie de l'equilibre des systémes élastiques et ses applications," Turin, 1879; translated into English by E. S. Andrews, *Elastic Stresses in Structures*, Scott, Greenwood, and Son, London, 1919.

11.5 C. H. Norris, J. B. Wilbur, and S. Utku, *Elementary Structural Analysis*, 3d ed., McGraw-Hill, New York, 1976, pp. 361–364.

11.6 Jack McCormac and Rudolf E. Elling, *Structural Analysis*, Harper & Row, New York, 1988, pp. 346–350.

11.7 J. C. Smith, *Structural Analysis*, Harper & Row, New York, 1988, pp. 482–489.

CHAPTER TWELVE

Slope Deflection

12.1. INTRODUCTION

Slope deflection is a displacement method of analysis for statically indeterminate beam and rigid frame structures. This method was originally presented in 1880 by Heinrich Manderla [12.1] to consider the secondary effects of bending in trusses. In 1892, Otto Mohr [12.2] presented a slope and deflection procedure to study secondary stresses in a truss. In 1914, a publication by Alex Bendixen [12.3] revealed his extended studies of this method to frame structures and named it the slope-deflection method. In 1915, George Maney of the University of Minnesota independently developed the slope-deflection method and published his findings which included applications to frame and beam structures [12.4].

This classical method was considered the most popular exact method of analysis in the United States until 1930, when the Hardy Cross method of moment distribution (Chap. 13) was introduced. Slope deflection continues to be an important analytical procedure for the following reasons:

1. A knowledge of the slope-deflection method can improve a student's understanding of structural behavior.
2. It is an easy hand-analysis method for small indeterminate structures when few deformations are to be found.
3. An understanding of the slope-deflection method provides an excellent introduction to learning the Hardy Cross method of moment distribution.

Tied-arch bridge at the end of a long floating bridge, Seattle. (*Courtesy of International Structural Slides, Berkeley.*)

4 It serves to introduce a student to the stiffness method of matrix structural analysis (Chap. 14) which forms the base upon which many structural-analysis computer programs are written.

5 The slope-deflection method can easily be programmed for a computer analysis of indeterminate flexural structures—especially where large numbers of unknown deformations are to be determined.

In the slope deflection method, fundamental equations are developed which express the end moments of a flexural member in terms of the rotations and displacements at each member end. Slope deflection is known as a *displacement method* since it involves the solution of moment-equilibrium equations written at member joints in terms of the known and unknown rotations (slopes) and displacements at the member ends. After the solution of unknown joint rotations and displacements is completed, bending moments are computed from the slope-deflection equations. Then, the remaining internal forces of the structure are found by principles of statics.

12.2. SLOPE-DEFLECTION EQUATIONS

Slope-deflection equations are developed below in terms of an interior span *ij* of a continuous beam subjected to an arbitrary load (see Fig. 12.1a). Moreover, the equations are derived in such a manner that deformations due to axial and shear forces are assumed to be negligible. Therefore, the slope-deflection equations are expressed in terms of the bending moments and the transverse loads which induce bending on the beam.

Additional assumptions

1 The beam is initially straight in the unloaded condition.
2 Each beam span between supports is prismatic (constant *EI*). However, *EI* can vary between beam spans.
3 All beam elements consist of linear elastic material properties and obey Hooke's law.

The assumption of elastic behavior infers that the principle of superposition can be used to combine the effects of individual load components on a structure.

Sign convention All clockwise directed rotations (slopes) and bending moments are positive (+).

In Fig. 12.1a, the elastic curve of the loaded beam is shown to deflect, rotate, and experience a support settlement at *j* and *k*; support *i* remains stationary. Let us consider beam section *ij*. As the loaded beam deforms to a new equilibrium position, unknown internal moments develop at ends *i* and *j* which can be related to the end rotations and displacements as the sum of the following effects:

1 Imagine that *i* and *j* are unyielding fixed-end supports (e.g., zero-end displacements and rotations as shown in Fig. 12.1b). Then, bending moments would occur at each fixed end of beam section *ij* due to transverse loads. These moments at *i* and *j*, called *fixed-end moments*, can be readily computed from tabulated formulas for specified types of loading (see Table 12.1). Fixed-end moment equations can be derived using the methods of Chaps. 10 and 11. Observe that the fixed-end moments at $i(\text{FEM}_{ij})$ and at $j(\text{FEM}_{ji})$ contain a double subscript notation, where the first subscript signifies the location of the FEM, and the double subscript defines the member.

2 Additional moments also develop at *i* and *j* due to the relative settlement (without rotation) between *j* and *i*, as shown in Fig. 12.1c. In effect, this relative settlement causes a rotation of the beam segment, referred to as the average chord rotation, ψ_{ij}, which is expresed as

$$\psi_{ij} \approx \tan \psi_{ij} = \frac{\Delta_{ji}}{L} \qquad (12.1)$$

354 Slope Deflection

Figure 12.1

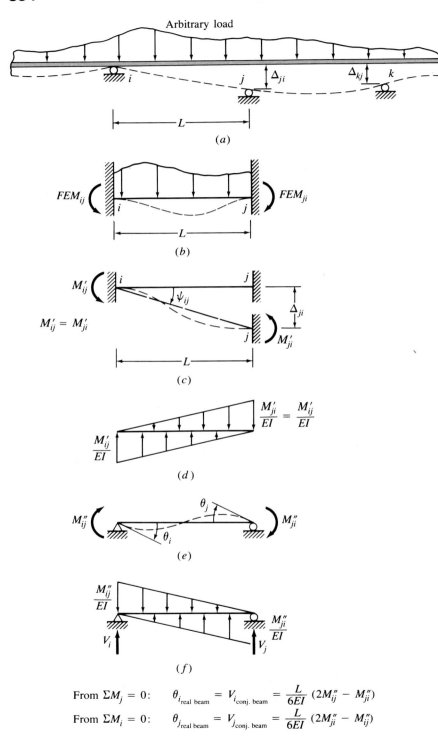

From $\Sigma M_j = 0$: $\theta_{i_{\text{real beam}}} = V_{i_{\text{conj. beam}}} = \dfrac{L}{6EI}(2M''_{ij} - M''_{ji})$

From $\Sigma M_i = 0$: $\theta_{j_{\text{real beam}}} = V_{j_{\text{conj. beam}}} = \dfrac{L}{6EI}(2M''_{ji} - M''_{ij})$

Table 12.1 Fixed-End Moments

$\text{FEM}_{AB} = \dfrac{PL}{8}$ — point load P at midspan — $\text{FEM}_{BA} = \dfrac{PL}{8}$

$\text{FEM}_{AB} = \dfrac{Pb^2 a}{L^2}$ — point load P at distance a from A, b from B — $\text{FEM}_{BA} = \dfrac{Pa^2 b}{L^2}$

$\text{FEM}_{AB} = \dfrac{wL^2}{12}$ — uniform load w over full span — $\text{FEM}_{BA} = \dfrac{wL^2}{12}$

$\text{FEM}_{AB} = \dfrac{11\,wL^2}{192}$ — uniform load w over left half — $\text{FEM}_{BA} = \dfrac{5wL^2}{192}$

$\text{FEM}_{AB} = \dfrac{wL^2}{20}$ — triangular load, max w at A — $\text{FEM}_{BA} = \dfrac{wL^2}{30}$

$\text{FEM}_{AB} = \dfrac{5wL^2}{96}$ — triangular load, max w at midspan — $\text{FEM}_{BA} = \dfrac{5wL^2}{96}$

$\text{FEM}_{AB} = \dfrac{6EI\Delta}{L^2}$ — support settlement Δ — $\text{FEM}_{BA} = \dfrac{6EI\Delta}{L^2}$

$\text{FEM}_{AB} = M(1-k)(1-3k)$ — applied moment M at kL from A — $\text{FEM}_{BA} = Mk(2-3k)$

$\text{FEM}_{AB} = \dfrac{wL^2}{60} \cdot k^2(10 - 10k + 3k^2)$ — triangular load over length kL from A — $\text{FEM}_{BA} = \dfrac{wL^2}{60} \cdot k^3(5 - 3k)$

Both end moments are equal and can be determined by the conjugate beam method where the fixed-fixed beam is replaced by a free-free conjugate beam (see Fig. 12.1d). Then, it is evident that the moment of the M/EI loads on the conjugate beam, taken about j, is equal to Δ_{ji} shown in Fig. 12.1c. Rearranging terms leads to

$$M'_{ij} = M'_{ji} = -6EI \frac{\Delta_{ji}}{L^2} \qquad (12.2)$$

3 In reality, the ends of segment ij are not fixed supports and will rotate when load is applied to the beam. The method of conjugate beam can also be used to formulate the bending moments associated with these end rotations. Figure 12.1e shows beam segment ij with end moments M''_{ij} and M''_{ji} applied positively (note: pin supports shown at i and j provide the freedom to rotate without displacement due to the applied end moments). Figure 12.1f illustrates the associated M/EI loads on the conjugate beam with simple end reactions V_i and V_j. From conjugate-beam theory, it is known that V_i, the reaction at i on the conjugate beam, is equal to θ_i of the real beam, and V_j, the reaction at j on the conjugate beam, is equal to θ_j on the real beam. The conjugate-beam reactions are found by use of statics. Then, simultaneous solution of the θ_i and θ_j equations (see Fig. 12.1f) for M''_{ij} and M''_{ji} reveal the following relationships:

$$M''_{ij} = 2EI \cdot \frac{(2\theta_i + \theta_j)}{L}$$

and

$$M''_{ji} = 2EI \cdot \frac{(2\theta_j + \theta_i)}{L} \qquad (12.3)$$

By superposition, the final moments at i and j are given as

$$M_{ij} = M'_{ij} + M''_{ij} + \text{FEM}_{ij}$$

and

$$M_{ji} = M'_{ji} + M''_{ji} + \text{FEM}_{ji} \qquad (12.4)$$

The equations can be written in terms of the end slopes and relative end displacement as

$$M_{ij} = \frac{2EI}{L} \left[2\theta_i + \theta_j - \left(\frac{3\Delta_{ji}}{L}\right) \right] + \text{FEM}_{ij}$$

and

$$M_{ji} = \frac{2EI}{L} \left[2\theta_j + \theta_i - \left(\frac{3\Delta_{ji}}{L}\right) \right] + \text{FEM}_{ji} \qquad (12.5)$$

By letting $K = I/L$ (called relative member stiffness), and noting that $\psi_{ji} = \Delta_{ji}/L$, the average chord rotation due to settlement, Eqs. (12.5) above can be redefined to provide the general slope-deflection equations.

General slope-deflection equations

and
$$M_{ij} = 2EK \cdot (2\theta_i + \theta_j - 3\psi_{ij}) + \text{FEM}_{ij}$$
$$M_{ji} = 2EK \cdot (2\theta_j + \theta_i - 3\psi_{ij}) + \text{FEM}_{ji} \quad (12.6)$$

The title, slope deflection, is appropriate since it relates unknown moments at member ends in terms of their respective end rotations (slopes) and displacements.

12.3. BEAM APPLICATIONS

In Example 12.1, it is noted that the continuous beam does not experience any support settlement ($\psi_{AB} = \psi_{BC} = 0$). Also note that the rotations at fixed ends A and C are zero and that the relative member stiffness, K, of both beam segments are equal since they have equal length and I values. Fixed-end moments associated with the applied loads are computed from formulas of Table 12.1. Then, the bending moment at each member end is expressed by direct substitution into Eq. (12.6). The values of these support moments are easily obtained following the solution of $EK\theta_B$ obtained from the moment equilibrium equation written at B.

EXAMPLE 12.1

Compute all support moments by use of the general slope-deflection equations [Eqs. (12.6)]. E and I are constant. Supports are unyielding.

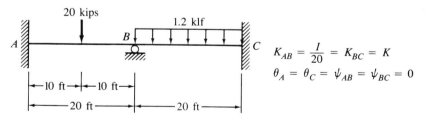

See Table 12.1:

$$\text{FEM}_{AB} = -\frac{20(10)(10)^2}{(20)^2} = -50 \text{ kip} \cdot \text{ft}$$

$$\text{FEM}_{BA} = +\frac{20(10)(10)^2}{(20)^2} = +50 \text{ kip} \cdot \text{ft}$$

$$\text{FEM}_{BC} = -\frac{1.2(20)^2}{12} = -40 \text{ kip} \cdot \text{ft}$$

$$\text{FEM}_{CB} = +\frac{1.2(20)^2}{12} = +40 \text{ kip} \cdot \text{ft}$$

Substitution into Eqs. (12.6) yields

$$M_{AB} = 2EK[2\theta_A + \theta_B - 3\psi_{AB}] + FEM_{AB} = 2EK[0 + \theta_B - 0] - 50$$

$$\therefore \quad M_{AB} = 2EK\theta_B - 50$$
$$M_{BA} = 4EK\theta_B + 50$$
$$M_{BC} = 4EK\theta_B - 40$$
$$M_{CB} = 2EK\theta_B + 40$$

$$\zeta + \Sigma M_B = 0 = M_{BA} + M_{BC} = 8EK\theta_B + 10 = 0$$

$$\therefore \quad EK\theta_B = -1.25$$

Upon substitution, final moments are

$$M_{AB} = 2(-1.25) - 50 = -52.5 \text{ kip·ft}$$
$$M_{BA} = 4(-1.25) + 50 = +45 \text{ kip·ft}$$
$$M_{BC} = 4(-1.25) - 40 = -45 \text{ kip·ft}$$
$$M_{CB} = 2(-1.25) + 40 = +37.5 \text{ kip·ft}$$

EXAMPLE 12.2

Find reactions and draw V and M diagrams for the beam loaded as shown below. EI is constant.

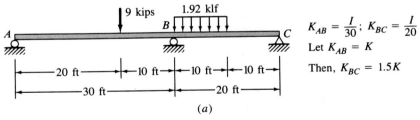

(a)

$K_{AB} = \dfrac{I}{30}$; $K_{BC} = \dfrac{I}{20}$

Let $K_{AB} = K$

Then, $K_{BC} = 1.5K$

Fixed-End Moments (Refer to Table 12.1):

$$FEM_{AB} = -\frac{Pab^2}{L^2} = -\frac{9(20)(10)^2}{(30)^2} = -20 \text{ kip·ft}$$

$$FEM_{BA} = +\frac{Pa^2b}{L^2} = +\frac{9(20)^2(10)}{(30)^2} = +40 \text{ kip·ft}$$

$$FEM_{BC} = -\frac{11wL^2}{192} = -\frac{11(1.92)(20)^2}{192} = -44 \text{ kip·ft}$$

$$FEM_{CB} = +\frac{5wL^2}{192} = +\frac{5(1.92)(20)^2}{192} = +20 \text{ kip·ft}$$

Substituting into general slope-deflection equations [Eqs. (12.6)]:

$$M_{AB} = 2EK[2\theta_A + \theta_B - 0] - 20 = 4EK\theta_A + 2EK\theta_B - 20 \quad (1)$$

$$M_{BA} = 2EK\theta_A + 4EK\theta_B + 40 \quad (2)$$

$$M_{BC} = 2E(1.5K)[2\theta_B + \theta_C - 0] - 44 = 6EK\theta_B + 3EK\theta_C - 44 \quad (3)$$

$$M_{CB} = 3EK\theta_B + 6EK\theta_C + 20 \quad (4)$$

Since $M_{AB} = M_{CB} = 0$, Eqs. (1) and (4) can be rewritten as

$$M'_{AB} = 0 = 4EK\theta_A + 2EK\theta_B - 20 \quad (1a)$$

and $\quad M_{CB} = 0 = 3EK\theta_B + 6EK\theta_C + 20 \quad (4a)$

Equilibrium at joint B:

$$\Sigma M_B = 0 = M_{BA} + M_{BC} = 2EK\theta_A + 10EK\theta_B + 3EK\theta_C - 4 = 0 \quad (5)$$

By solving Eqs. (1a), (4a) and (5) simultaneously, we arrive at

$$EK\theta_A = +4.733, \quad EK\theta_B = +0.533, \quad \text{and} \quad EK\theta_C = -3.600$$

Substituting $EK\theta_A$, $EK\theta_B$, and $EK\theta_C$ into Eqs. (1) to (4):

$M_{AB} = 4(4.733) + 2(0.533) - 20 = 0$ ✓
$M_{BA} = 2(4.733) + 4(0.533) + 40 = +52$ ft·kips
$M_{BC} = 6(0.533) + 3(-3.60) - 44 = -52$ ft·kips
$M_{CB} = 3(0.533) + 6(-3.60) + 20 = 0$ ✓

Reactions (by superposition)

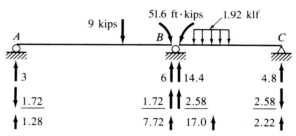

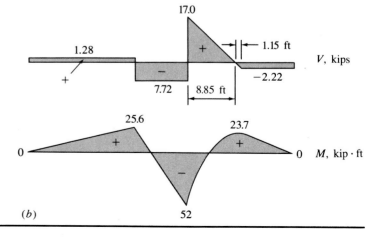

(b)

In Example 12.2, the relative member stiffness of BC is expressed in relation to the stiffness of member AB. It is noted that all joint rotations are unknown. Yet, the moments at simple end supports A and C are known to be zero. Consequently, the three unknowns in Eqs. (1) through (4) in Example 12.2 are found from three equations, namely, (1a) and (4a), obtained by setting Eqs. (1) and (4) equal to zero, respectively, and the moment-equilibrium equation [Eq. (5)] written at support B.

Example 12.3 demonstrates two ways to determine the support moments for the beam of Example 12.1 if a settlement of 1 in occurs at support B. The first approach involves direct substitution into the general slope-deflection equations [Eqs. (12.6)]. Observe that EK is numerically expressed in units of kips·ft which is consistent with the units for the FEMs. The alternate approach uses Table 12.1 formulas to compute FEMs due to the settlement at B and for the applied loads. Since moments due to settlement are determined from Table 12.1, the values of ψ_{AB} and ψ_{BC} are set equal to zero in the alternate method.

EXAMPLE 12.3

If the beam of Example 12.1 has a 1-in settlement at support B, compute all support moments. $E = 29{,}000$ ksi and $I = 1000$ in^4.

Solution

$$\psi_{AB} = +\frac{\Delta}{L_{AB}} = +\frac{1}{20 \times 12} = +4.167 \times 10^{-3}$$

$$\psi_{BC} = -\frac{\Delta}{L_{BC}} = -\frac{1}{20 \times 12} = -4.167 \times 10^{-3}$$

$$M_{AB} = 2EK(\theta_B - 3\psi_{AB}) + \text{FEM}_{AB}$$

$$M_{AB} = 2EK\theta_B - 6EK\psi_{AB} + \text{FEM}_{AB}$$

$$= 2EK\theta_B - 6EK(+4.167 \times 10^{-3}) - 50$$

$$M_{AB} = 2EK\theta_B - 0.025EK - 50$$

$$M_{BA} = 4EK\theta_B - 0.025EK + 50$$

$$M_{BC} = 4EK\theta_B + 0.025EK - 40$$

$$M_{CB} = 2EK\theta_B + 0.025EK + 40$$

$\curvearrowleft + \Sigma M_B = 0 = M_{BA} + M_{BC} = 8EK\theta_B + 10$

$\therefore\ EK\theta_B = -1.25$

$$EK = \frac{29{,}000(1000)}{20 \times 12 \times 12}$$

$$= 10{,}069 \text{ kip·ft (consistent units with FEMs)}$$

Thus,

$M_{AB} = 2(-1.25) - 0.025(10,069) - 50 = -304$ kip·ft

$M_{BA} = 4(-1.25) - 0.025(10,069) + 50 = -207$ kip·ft

$M_{BC} = 4(-1.25) + 0.025(10,069) - 40 = +207$ kip·ft

$M_{CB} = 2(-1.25) + 0.025(10,069) + 40 = 289$ kip·ft

Alternate Approach

In the general slope-deflection equation [Eq. (12.6)], set $\psi_{AB} = \psi_{BC} = 0$ and use Table 12.1 to compute additional FEMs due to settlement at B. Thus,

$$\text{FEM}_{AB} = -\frac{Pab^2}{L^2} - \frac{6EI\Delta}{L^2}$$

$$= -50 \text{ kip·ft} - \frac{6(29,000)(1000)(1)}{(20 \times 12)^2 \times 12}$$

$\text{FEM}_{AB} = -50 - 251.7 = -301.7$ kip·ft

$\text{FEM}_{BA} = +50 - 251.7 = -201.7$ kip·ft

$\text{FEM}_{BC} = -40 + 251.7 = +211.7$ kip·ft

$\text{FEM}_{CB} = +40 + 251.7 = +291.7$ kip·ft

Then from use of Eq. (12.6),

$M_{AB} = 2EK\theta_B - 301.7$

$\left. \begin{array}{l} M_{BA} = 4EK\theta_B - 201.7 \\ M_{BC} = 4EK\theta_B + 211.7 \end{array} \right\}$ $\Sigma M_B = 0 = M_{BA} + M_{BC} = 8EK\theta_B + 10$

$\therefore \quad EK\theta_B = -1.25$

$M_{CB} = 2EK\theta_B + 291.7$

Then,

$M_{AB} = 2(-1.25) - 301.7 \approx -304$ kip·ft

$M_{BA} = 4(-1.25) - 201.7 \approx -207$ kip·ft

$M_{BC} = 4(-1.25) + 211.7 \approx +207$ kip·ft

$M_{CB} = 2(-1.25) + 291.7 \approx +289$ kip·ft

Additional comments on beam settlement

1 When relative support settlement occurs between the ends of a beam span, it should be observed that the average chord rotation at both ends of the span is identical, both in magnitude and direction; that is, $\psi_{ij} = \psi_{ji}$.

2 Figure 12.2 exhibits three possible settlement conditions that can occur to a continuous beam under gravity load. The settlement displacements are exaggerated as shown and focus on average chord rotation of beam segments.

  a Figure 12.2a shows a horizontal beam that undergoes the same amount of settlement at all supports. For this special condition, the beam remains horizontal and undeflected by the settlement. Thus, all relative values of ψ are equal to zero in the slope deflection equations.

  b In Figure 12.2b, support settlements are shown to vary in linear proportion to the span length measured from the left end. Using support A as a reference point, a dashed line is drawn parallel to the unloaded

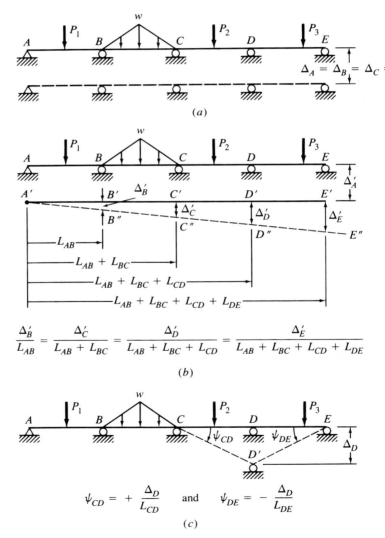

Figure 12.2

beam axis and expressions which relate the displacements at the remaining support points are given. The new beam axis $A'E''$ is assumed to remain straight while it rotates slightly from AE. If the rotation of AE to $A'E''$ is negligible, line $A'E''$ can be viewed as the beam axis from which no relative displacement occurs. Then, the average chord rotations (ψ's) are zero for all spans. Also, if the rotation of the beam due to support settlement is negligible, the FEMs and V and M diagrams can be found with respect to the original beam axis and original load vectors.

c Figure 12.2c shows a continuous beam where all supports are unyielding except at support D. Thus, $\psi_{AB} = \psi_{BC} = 0$, and ψ_{CD} and ψ_{DE} are nonzero and unequal in direction. If $L_{CD} = L_{DE}$, $\psi_{CD} = -\psi_{DE}$.

12.4. SIMPLE END-SUPPORT MODIFICATION

Although the general slope deflection equations can be used to describe the end moments of any continuous beam span, a special condition arises for a span which has a simple end support. Consider a beam-span ij where end i is a simple end support with $M_{ij} = 0$. The general slope deflection equations can be written as

$$M_{ij} = 0 = 2EK(2\theta_i + \theta_j - 3\psi_{ij}) + \text{FEM}_{ij}$$

and

$$M_{ji} = 2EK(2\theta_j + \theta_i - 3\psi_{ij}) + \text{FEM}_{ji}$$

If the second equation is multiplied by 2 and then subtracted from the first equation to eliminate the unknown θ_i, the resulting equation is

$$M_{ji} = 3EK(\theta j - \psi_{ij}) + \text{FEM}_{ji} - \tfrac{1}{2}\text{FEM}_{ij} \qquad (12.7)$$

Since M_{ij} is known to be zero, the modified equation [Eq. (12.7)] offers a simplified approach which eliminates the need to find θ_i in order to evaluate M_{ji}. Equation (12.7) can be used to express the moment at the end of any beam span when the opposite end is a simple end support. In Example 12.4, the support moments for the beam of Example 12.2 are computed by use of the modified slope-deflection equation. Example 12.5 illustrates the combined use of both the general and the modified equations to compute the support moments of a three-span continuous beam.

EXAMPLE 12.4

Reevaluate the support moments for the beam of Example 12.2 by use of the modified slope-deflection equation [Eq. (12.7)].

364 Slope Deflection

$M_{AB} = 0$ (by inspection)

$M_{BA} = 3EK(\theta_B - \psi_{AB}) + FEM_{BA} - \frac{1}{2}FEM_{AB}$

$M_{BA} = 3EK\theta_B + 40 - \frac{1}{2}(-20) = 3EK\theta_B + 50$

$M_{BC} = 3EK(\theta_B - \psi_{BC}) + FEM_{BC} - \frac{1}{2}FEM_{CB}$

$M_{BC} = 3EK\theta_B - 44 - \frac{1}{2}(20) = 3EK\theta_B - 54$

$M_{CB} = 0$ (by inspection)

$\Sigma M_B = 0 = M_{BA} + M_{BC} = 6EK\theta_B - 4 \quad \therefore \quad EK\theta_B = +\frac{2}{3}$

Then,

$M_{BA} = 3(+0.6667) + 50 = 52 \text{ kip} \cdot \text{ft}$

$M_{BC} = 3(+0.6667) - 54 = -52 \text{ kip} \cdot \text{ft}$

EXAMPLE 12.5

The three-span beam has unyielding supports. EI is constant over entire span length. Determine all support moments by the slope-deflection method. Relative beam stiffnesses shown.

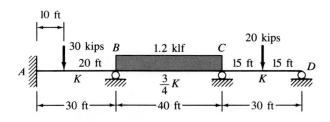

$FEM_{AB} = -\dfrac{30(10)(20)^2}{(30)^2} = -133 \text{ kip} \cdot \text{ft}$

$FEM_{BA} = \dfrac{30(20)(10)^2}{(30)^2} = +66.7 \text{ kip} \cdot \text{ft}$

$FEM_{BC} = -\dfrac{1.2(40)^2}{12} = -160 \text{ kip} \cdot \text{ft}; \quad FEM_{CB} = +160 \text{ kip} \cdot \text{ft}$

$FEM_{CD} = -\dfrac{20(15)(15)^2}{(30)^2} = -75 \text{ kip} \cdot \text{ft}; \quad FEM_{DC} = +75 \text{ kip} \cdot \text{ft}$

Note

$\psi_{AB} = \psi_{BC} = \psi_{CD} = \theta_A = M_{DC} = 0$

Substitution into slope-deflection equations yield

$M_{AB} = 2EK\theta_B - 133$

$M_{BA} = 4EK\theta_B + 66.7$

$M_{BC} = 4E(\frac{3}{4}K)\theta_B + 2E(\frac{3}{4}K)\theta_C - 160$

$M_{BC} = 3EK\theta_B + 1.5EK\theta_C - 160$

$M_{CB} = 3EK\theta_C + 1.5EK\theta_B + 160$

$M_{CD} = 3EK\theta_C - 75 - \frac{1}{2}(75) = 3EK\theta_C - 112.5$

$\Sigma M_B = 0 = M_{BA} + M_{BC} = 7EK\theta_B + 1.5EK\theta_C - 93.3 = 0$ (1)

$\Sigma M_C = 0 = M_{CB} + M_{CD} = 1.5EK\theta_B + 6EK\theta_C + 47.5 = 0$ (2)

Simultaneous solution of (1) and (2) yields

$EK\theta_B = -11.9$ and $EK\theta_C = +15.9$

Final Moments

$M_{AB} = -101$ kip·ft; $M_{BC} = -130$ kip·ft; $M_{DC} = 0$

$M_{BA} = +130$ kip·ft; $M_{CB} = +148$ kip·ft

12.5. FRAMES WITHOUT SIDESWAY

The general and modified slope deflection equations can also be used for the analysis of statically indeterminate frames. Consider the plane frames shown in Fig. 12.3. Figures 12.3a and 12.3b show frames with restraints that prevent them from leaning to one side (no sidesway). Figure 12.3c shows a frame that appears to be free to move, but theoretically will not sway since the loading, geometry (I's and L's), material, and type of supports are symmetric. The same procedure that is used above for indeterminate beams can also be used for frames without sidesway at the free joints.

Figure 12.3d shows a frame of unsymmetrical geometry acted upon by unsymmetrical loads which cause it to lean or sway to one side. A method of analysis for frames with sidesway is presented in Sec. 12.7.

Examples 12.6 and 12.7 demonstrate the analysis of frames without sidesway by the direct use of the slope deflection equations. Carefully note in Example 12.6 that the signs associated with the FEMs of member AB are consistent with the convention established in Sec. 12.2. In Example 12.7, AB

366 Slope Deflection

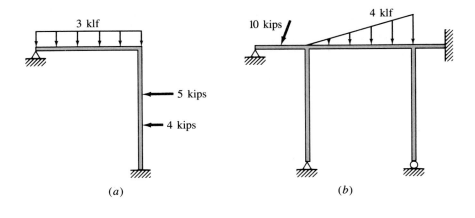

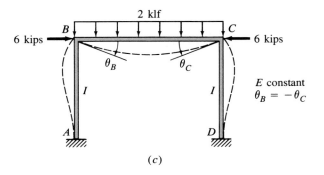

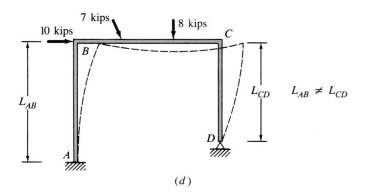

Figure 12.3

and CD are cantilever members with known moments at both ends ($M_{AB} = M_{DC} = 0$, $M_{BA} = +40$ kip·ft, and $M_{CD} = -40$ kip·ft). Therefore, slope-deflection equations are only written for the remaining members in terms of the unknown rotations at B and C.

EXAMPLE 12.6

Compute all joint moments for the frame which is restrained against sidesway. E and I are constant.

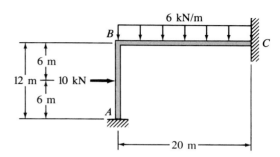

$$K_{AB} = \frac{I}{12}; \quad K_{BC} = \frac{I}{20} = K$$

$$\therefore \quad K_{AB} = \tfrac{5}{3} K_{BC} = \tfrac{5}{3} K$$

$$FEM_{AB} = -\frac{10(6)(6)^2}{(12)^2} = -15 \text{ kN} \cdot \text{m}$$

$$FEM_{BA} = +15 \text{ kN} \cdot \text{m}$$

$$FEM_{BC} = -\frac{6(20)^2}{12} = -200 \text{ kN} \cdot \text{m}$$

$$FEM_{CB} = +200 \text{ kN} \cdot \text{m}$$

$$\theta_A = \theta_C = \psi_{AB} = \psi_{BC} = 0$$

Application of Eqs. (12.6):

$$M_{AB} = 2E\left(\frac{5K}{3}\right)[2(0) + \theta_B - 3(0)] - 15$$

$$M_{AB} = 3.33 EK\theta_B - 15 \tag{1}$$

$$M_{BA} = 2E\left(\frac{5K}{3}\right)[(0) + 2\theta_B - 3(0)] + 15$$

$$M_{BA} = 6.67 EK\theta_B + 15 \tag{2}$$

$$M_{BC} = 2E(K)[2\theta_B + (0) - 3(0)] - 200$$

$$M_{BC} = 4EK\theta_B - 200 \tag{3}$$

$$M_{CB} = 2EK[2(0) + \theta_B - 3(0)] + 200$$

$$M_{CB} = 2EK\theta_B + 200 \tag{4}$$

$$\zeta + \Sigma M_B = 0 = M_{BA} + M_{BC} = 10.67 EK\theta_B - 185 \quad \therefore \quad EK\theta_B = +17.34$$

Substituting into Eqs. (1) to (4), we get

$$M_{AB} = 3.33(17.34) - 15 = 42.7 \text{ kN} \cdot \text{m}$$

$$M_{BA} = 6.67(17.34) + 15 = 130.6 \text{ kN} \cdot \text{m}$$

$$M_{BC} = 4(17.34) - 200 = -130.6 \text{ kN} \cdot \text{m}$$

$$M_{CB} = 2(17.34) + 200 = 234.7 \text{ kN} \cdot \text{m}$$

EXAMPLE 12.7

Determine the joint moments of the frame shown below which has no sidesway due to symmetry of load, geometry, support, and material. E and I are constant.

$K_{BE} = K_{CF} = \dfrac{I}{15}$

$K_{BC} = K = \dfrac{I}{20}$

$\therefore \quad K_{BE} = K_{CF} = \tfrac{4}{3}K$

$FEM_{BC} = -\dfrac{3(20)^2}{12} = -100 \text{ kip} \cdot \text{ft}$

$FEM_{CB} = +100 \text{ kip} \cdot \text{ft}$

All ψ's $= 0$

$M_{AB} = 0; \quad M_{BA} = +40 \text{ kip} \cdot \text{ft}$

$M_{CD} = -40 \text{ kip} \cdot \text{ft}; \quad M_{DC} = 0$

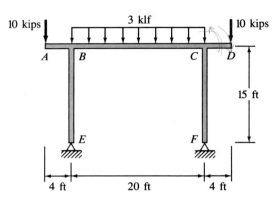

Using Eqs. (12.6):

$M_{BC} = 2EK[2\theta_B + \theta_C - 3(0)] - 100$

$M_{BC} = 4EK\theta_B + 2EK\theta_C - 100$ \hfill (1)

$M_{CB} = 2EK[2\theta_C + \theta_B - 3(0)] + 100$

$M_{CB} = 4EK\theta_C + 2EK\theta_B + 100$ \hfill (2)

Using Eq. 12.7:

$M_{BE} = 3E(\tfrac{4}{3}K)[\theta_B - (0)] + (0) - \tfrac{1}{2}(0)$

$M_{BE} = 4EK\theta_B$ \hfill (3)

Also, $M_{CF} = 4EK\theta_C$ \hfill (4)

$\curvearrowleft + \Sigma M_B = 0 = M_{BA} + M_{BC} + M_{BE} = 8EK\theta_B + 2EK\theta_C - 60$ \hfill (5)

$\curvearrowleft + \Sigma M_C = 0 = M_{CB} + M_{CD} + M_{CF} = 2EK\theta_B + 8EK\theta_C + 60$ \hfill (6)

Solving Eqs. (5) and (6) simultaneously yields $EK\theta_B = +10$ and $EK\theta_C = -10$. Final moments are found after substitution as

$M_{BC} = 4(10) + 2(-10) - 100 = -80$ kip·ft
$M_{CB} = 4(-10) + 2(10) + 100 = +80$ kip·ft
$M_{BE} = 4(+10) \qquad\qquad = +40$ kip·ft
$M_{CF} = 4(-10) \qquad\qquad = -40$ kip·ft
$M_{EB} = M_{FC} = 0$

12.6. AWARENESS OF SYMMETRY AND ANTISYMMETRY

The analysis of a beam or frame by the slope-deflection method is a simple hand procedure, especially when there are few unknown deformations to resolve. When symmetric or antisymmetric loads are applied to a symmetric structure, the number of independent unknown deformations are reduced and the analysis is easier to perform. For the frame of Fig. 12.3c, only one unknown exists since $\theta_B = -\theta_C$. Symmetry is also apparent in Example 12.7 where $EK\theta_B = -EK\theta_C = 10$; the mirror-image effect associated with symmetry is also evident by observing the joint-moment values. Ideally, a symmetric beam or frame under symmetric loads will deform so that the rotations and joint moments left of the centerline are equal in magnitude and opposite in direction to the respective rotations and moments right of the centerline. Therefore, when symmetry is evident, only half the number of slope deflection equations and joint moment equilibrium equations are required for problem resolution.

When antisymmetric loading is applied to a symmetric beam or frame structure, the number of independent unknown joint deformations are less than the total number of free joints. Figure 12.4 shows a beam and a frame structure acted upon by antisymmetric loads; for the beam, it is seen that $\theta_A = \theta_D$ and $\theta_B = \theta_C$; for the frame, $\theta_B = \theta_C$. Therefore, structural analysis by the slope deflection method is simplified by the awareness of symmetry and antisymmetry. (Additional comments, discussions, and references are cited in a text by Norris, Wilber, and Utku [12.5].)

12.7. FRAMES WITH SIDESWAY

A frame will go sidesway, or lean to one side, if it is not restrained to prevent the sidesway, and if one or more of the following conditions exist:

370 Slope Deflection

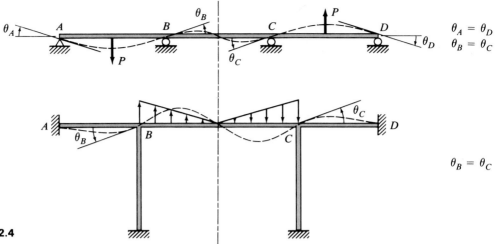

Figure 12.4

1. Applied loads are unsymmetric.
2. Geometry (I's and L's) is unsymmetric.
3. Supports are unsymmetric.
4. Material properties are unsymmetric.

The frame shown in Fig. 12.5 undergoes sidesway deformation due to non-symmetric loads and supports. The end moments of the frame members are expressed in terms of θ_B, θ_C, and the horizontal sidesway Δ. Since the axial deformation of member BC is assumed to be negligible, the unknown sidesway of each joint is identical ($\Delta_B = \Delta_C$). Therefore, slope deflection equations written at joints B and C contain three independent unknowns. The three unknowns are determined from the following three equations of statics:

Figure 12.5

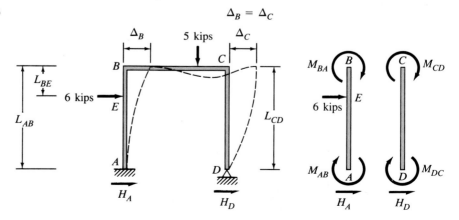

371 Frames with Sidesway

$$\Sigma M_B = 0 = M_{BA} + M_{BC}$$
$$\Sigma M_C = 0 = M_{CB} + M_{CD}$$
and
$$\Sigma F_x = 0 = H_A + H_D + 6$$
where
$$H_A = \frac{M_{AB} + M_{BA} - 6L_{BE}}{L_{AB}}$$
and
$$H_D = \frac{M_{DC} + M_{CD}}{L_{CD}}$$

(Note: H_A and H_D are directed to satisfy moment equilibrium of each vertical member about joints B and C, respectively.) Thus, the horizontal equilibrium equation can be rewritten as

$$\Sigma F_x = 0 = \frac{(M_{AB} + M_{BA} - 6L_{BE})}{L_{AB}} + \frac{(M_{DC} + M_{CD})}{L_{CD}} + 6$$

EXAMPLE 12.8

The frame shown below will sway due to unsymmetrical loading. E is constant. Values of inertia (I) are given as shown. Determine all joint moments by slope deflection.

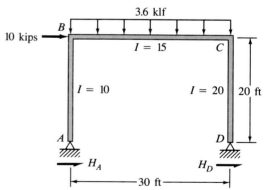

$\text{FEM}_{BC} = -\frac{3.6(30)^2}{12} = -270 \text{ kip} \cdot \text{ft}$

$\text{FEM}_{CB} = +270 \text{ kip} \cdot \text{ft}$

$K_{AB} = K_{BC} = \frac{1}{2} \sim K$

$K_{CD} = 1 \sim 2 \text{ K}$

$M_{AB} = M_{DC} = \psi_{BC} = 0$

$\psi_{AB} = \psi_{CD} = \psi$

By use of Eqs. (12.6) and (12.7):

$\left. \begin{array}{l} M_{BA} = 3EK(\theta_B - \psi) \\ M_{BC} = 2EK(2\theta_B + \theta_C) - 270 \end{array} \right\} \begin{array}{l} \Sigma M_B = 0 = M_{BA} + M_{BC} \\ = 7EK\theta_B + 2EK\theta_C - 3EK\psi - 270 = 0 \quad (1) \end{array}$

$\left. \begin{array}{l} M_{CB} = 2EK(2\theta_C + \theta_B) + 270 \\ M_{CD} = 3E(2 \text{ K})(\theta_C - \psi) \end{array} \right\} \begin{array}{l} \Sigma M_C = 0 = M_{CB} + M_{CD} \\ = 2EK\theta_B + 10EK\theta_C - 6EK\psi + 270 = 0 \quad (2) \end{array}$

$\Sigma F_x = 0 = H_A + H_D + 10 \quad \text{where} \quad H_A = \frac{M_{AB} + M_{BA}}{20} = \frac{M_{BA}}{20}$

and $H_D = \dfrac{M_{CD} + M_{DC}}{20} = \dfrac{M_{CD}}{20}$

$\therefore\ F_x = 0 = \dfrac{M_{BA}}{20} + \dfrac{M_{CD}}{20} + 10$

which can be reduced to

$$3EK\theta_B + 6EK\theta_C - 9EK\psi + 200 = 0 \qquad (3)$$

Simultaneous solution of Eqs. (1), (2), and (3) yield $EK\theta_B = +56.11$, $EK\theta_C = -22.78$, and $EK\psi = +25.74$.

Final Moments

$M_{AB} = 0$

$M_{BA} = +91.1$ kip·ft

$M_{BC} = -91.1$ kip·ft

$M_{CB} = +291$ kip·ft

$M_{CD} = -291$ kip·ft

$M_{DC} = 0$

The slope-deflection method can also be applied to frames that have multiple storys and bays. In Fig. 12.6a, a one-story frame is shown to consist of two bays with arbitrary column lengths. Observe that the sidesway deflections at the free joints are identical since the axial deformation of each horizontal member is assumed to be negligible. Thus, slope-deflection equations can be written in terms of four unknowns, θ_B, θ_C, θ_D, plus Δ. These four unknowns can be solved from four available equations of statics, namely, three moment equilibrium equations written at joints B, C, and D, plus a horizontal equilibrium equation which states that the applied horizontal loading must be resisted by the shear forces in the columns. The addition of another bay to the frame will introduce a new joint and therefore add one more unknown θ; however, it also provides another joint-moment equilibrium equation. Therefore, equal number of equations and unknowns will exist for the expanded frame. Consequently, one-story frames of multiple bays can be resolved by slope deflection.

Figure 12.6b shows a two story, one bay frame with lateral loads applied at both levels. In this case, the extra story introduces three additional unknowns, namely, θ_C, θ_D, and Δ_2. However, three more equations can be written: $\Sigma M_C = 0$, $\Sigma M_D = 0$, and another horizontal equilibrium equation to reflect that the applied loads which act above level BE are resisted by the shear forces in the second story columns. Observe that the first-story column shears must resist both P_1 and P_2.

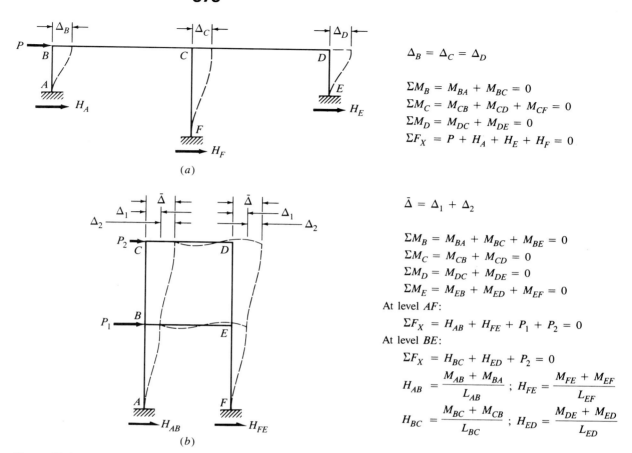

Figure 12.6

The author recommends the use of a computer to perform structural analysis for frames with many bays and storys, and to limit the use of slope deflection for the analysis of small frames which have few unknowns.

12.8. FRAMES WITH SUPPORT SETTLEMENT

By neglecting axial deformations, the horizontal members of frames with vertical columns and unyielding supports are assumed to experience no chord rotation ($\psi = 0$). However, if support settlement occurs, as shown in Fig. 12.7, slope-deflection equations can be written to reflect notable chord rotation of the horizontal members. If the frames shown in Fig. 12.7 experience both sidesway and support settlement, nonzero values of ψ will exist for *all* frame members.

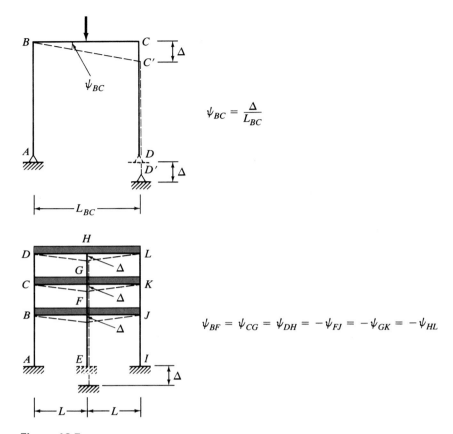

Figure 12.7

12.9. FRAMES WITH SLOPING LEG MEMBERS

When a frame has one or more sloping leg members and is subject to sidesway, the horizontal members may undergo notable chord rotation. The slant leg frame of Example 12.9 undergoes sidesway with joints B and C displaced to B' and C', respectively. If axial deformations are neglected, both ends of member BC will experience the same lateral displacement Δ (Δ represents both the total deformation of joint B to B' and the horizontal component of the deformation of joint C to C'). Since these deformations are assumed to be very small, the true arc distance from points B to B' and C to C' are replaced by chords drawn normal to AB and CD. In Example 12.9, relative ψ values of the frame members are easily found after a trigonometric study of the motion at joint C is complete. It is important to note that the shear component of the reaction at D is required in the slope-deflection expressions for the frame.

EXAMPLE 12.9

Slant leg frame by slope deflection.

$EI = $ constant

$\theta_A = \theta_D = 0$

Relative stiffnesses

$K_{AB} = \dfrac{I}{15} = \dfrac{20I}{300} \sim K$

$K_{BC} = \dfrac{I}{20} = \dfrac{15I}{300} \sim \dfrac{3}{4}K$

$K_{CD} = \dfrac{I}{25} = \dfrac{12I}{300} \sim \dfrac{3}{5}K$

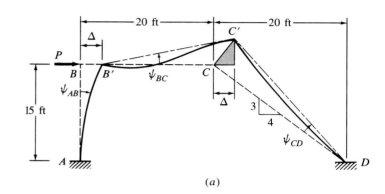

(a)

To determine ψ_{BC} and ψ_{CD} let us study joint C:

Then $\psi_{AB} = +\dfrac{\Delta}{15}$

$\psi_{BC} = -\dfrac{4\Delta/3}{20} = -\dfrac{\Delta}{15}$

$\psi_{CD} = +\dfrac{5\Delta/3}{25} = +\dfrac{\Delta}{15}$

Use relative ψ's:

$\psi_{AB} = +\psi$

$\psi_{BC} = -\psi$

$\psi_{CD} = +\psi$

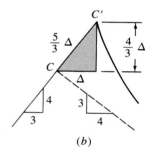

(b)

Slope deflection equations:

$M_{AB} = 2EK(\theta_B - 3\psi) = 2EK\theta_B - 6EK\psi$

$M_{BA} = 2EK(2\theta_B - 3\psi) = 4EK\theta_B - 6EK\psi$

$M_{BC} = 2E(\tfrac{3}{4}K)[2\theta_B + \theta_C + 3\psi] = 3EK\theta_B + 1.5EK\theta_C + 4.5EK\psi$

$M_{CB} = 2E(\tfrac{3}{4}K)[\theta_B + 2\theta_C + 3\psi] = 1.5EK\theta_B + 3EK\theta_C + 4.5EK\psi$

$M_{CD} = 2E(\tfrac{3}{5}K)[2\theta_C - 3\psi] = 2.4EK\theta_C - 3.6EK\psi$

$M_{DC} = 2E(\tfrac{3}{5}K)[\theta_C - 3\psi] = 1.2EK\theta_C - 3.6EK\psi$

$\Sigma M_B = 0 = M_{BA} + M_{BC} = 7EK\theta_B + 1.5EK\theta_C - 1.5EK\psi = 0$ (1)

$\Sigma M_C = 0 = M_{CB} + M_{CD} = 1.5EK\theta_B + 5.4EK\theta_C + 0.9EK\psi = 0$ (2)

The third equation can be complex since sloping member shear contains both horizontal and vertical reaction components. The following approach is recommended.

Resolve forces at D into axial and normal components S_D and N_D as shown. (Note: Length $OD = 50$ ft.)

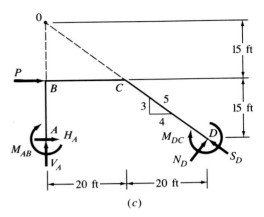

(c)

$$\Sigma M_O = 0 = -15P - 30H_A - 50N_D + M_{AB} + M_{DC} = 0$$

$$= -15P - 30\left(\frac{M_{AB} + M_{BA}}{15}\right) - 50\left(\frac{M_{CD} + M_{DC}}{25}\right) + M_{AB} + M_{DC} = 0$$

which reduces to

$$-15P - M_{AB} - 2M_{BA} - 2M_{CD} - M_{DC} = 0 \qquad (3)$$

Substituting the slope-deflection equations into Eq. (3) above, yields

$$-15P - 2EK\theta_B + 6EK\psi - 8EK\theta_B + 12EK\psi$$
$$-4.8EK\theta_C + 7.2EK\psi - 1.2EK\theta_C - 3.6EK\psi = 0$$

or $\quad -10EK\theta_B - 6EK\theta_C + 28.8EK\psi - 15P = 0 \qquad (3)$

For a given load P these equations can be solved to find $EK\theta_B$, $EK\theta_C$, and $EK\psi$. Then, moments at joints are computed. Finally, forces H_A and N_D are computed from $\Sigma M_B = 0$ and $\Sigma M_C = 0$, respectively. S_D is found from the vertical equilibrium equation.

Problems

12.1 to 12.12 Compute all support moments by the method of slope deflection. Also, draw shear and moment diagrams. E is constant.

12.1

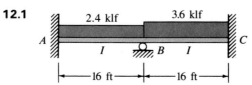

12.2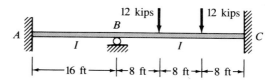

12.3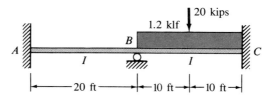

12.4 I is constant and equal to 1000 in^4.
$E = 30 \times 10^3$ ksi

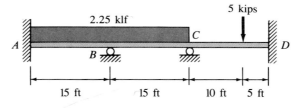

12.5 $I = 8 \times 10^8$ mm^4
$E = 200$ GPa

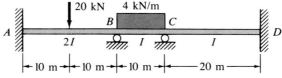

12.6

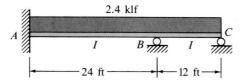

12.7 I constant

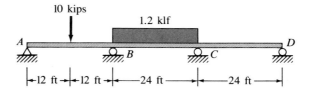

12.8 *I* constant

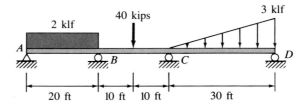

12.9

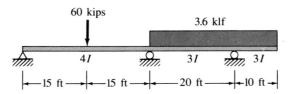

12.10 $I = 400$ in^4
$E = 30 \times 10^3$ ksi and support B settles 2.4 in.

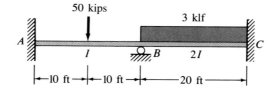

12.11 The beam of Prob. 12.4 where support B settles 3.0 in

12.12 The beam of Prob. 12.5 where support B settles 40 mm and support C settles 20 mm.

12.13 to 12.20 Compute all end moments. E is constant.

12.13 *EI* constant

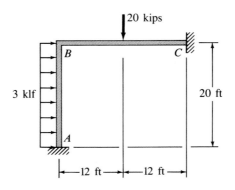

379 Problems

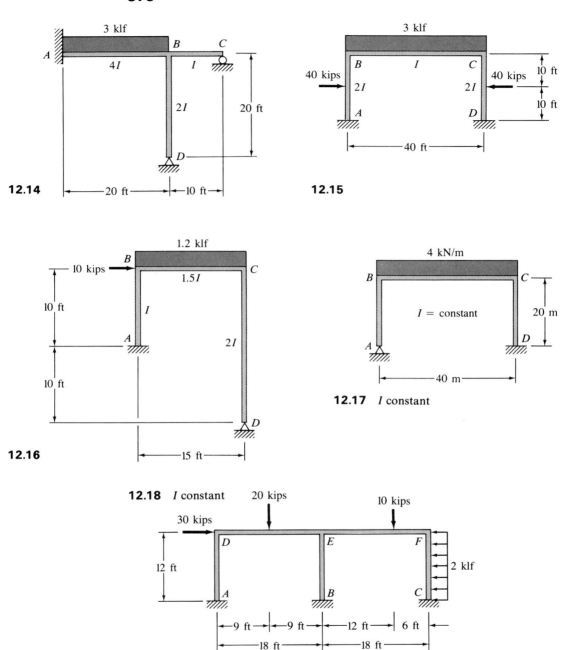

12.14

12.15

12.16

12.17 *I* constant

12.18 *I* constant

12.19 Resolve the frame of Prob. 12.14 if support *D* settles 2.4 in. $I = 250$ in^4. $E = 30 \times 10^3$ ksi.

12.20 Rework the frame of Prob. 12.16 if support *D* settles 3.0 in. $I = 800$ in^4. $E = 30 \times 10^3$ ksi.

References

12.1 Heinrich Manderla, "Die Berechnung der Sekundärspannungen, welche im einfachen Fachwerke in Folge starrer Knotenverbindungen auftreten," *Forster's Bauzeitung*, Munich, **45** (1880), p. 34.

12.2 Otto Mohr, "Die Berechnung der Fachwerke mit starren Knotenverbindungen," *Der Civilingenieur*, vol. 38, 1892, pp. 577–594.

12.3 Axel Bendixen, *Die Methode der Alpha-Gleichungen zur Berechnung von Rahmenkonstruktionen*, Berlin, 1914.

12.4 George A. Maney, *Studies in Engineering*, no. 1, University of Minnesota, March 1915.

12.5 C. H. Norris, J. B. Wilbur, and S. Utku, *Elementary Structural Analysis*, 3d ed., McGraw-Hill, New York, 1976, pp. 350–351.

CHAPTER THIRTEEN
Moment Distribution

13.1. INTRODUCTION

The moment-distribution method was first introduced in 1924 by Professor Hardy Cross to his students at the University of Illinois and received rapid national acceptance following publication in 1930 [13.1] and 1932 [13.2]. It was the dominant method of analysis for statically indeterminate continuous beams and plane frames from the early 1930s until the emergence of matrix computer methods in the late 1950s. Moment distribution became very popular since, unlike the preceding methods, the final end moments are obtained by a quick and simple iteration procedure that avoids simultaneous equation solution.

Let us preface the study of moment distribution with a review of the behavior of an elastic structure under static loading. Consider a structure that is at rest, unloaded, and composed of straight members in a geometrically stable configuration. Also, assume that the structural deformations are primarily due to bending forces. Now, as static loads are applied, the structure will gradually deform until it reaches a new state of equilibrium. That is, *a structure will continue to deflect until internal moments are fully developed to resist the external moments due to the applied loads*. Therefore, the bending moments that develop are directly related to the final rotations and displacements of a structure in its deformed equilibrium position.

Both slope deflection (Chap. 12) and moment distribution are displacement methods which describe the end moments of a member as the algebraic sum of three distinct effects:

Arch Bridge, Dunkeld, Scotland. (*Courtesy of International Structural Slides, Berkeley.*)

1 Fixed-end moments (FEMs) due to transverse loads on a member if both member ends are assumed to have fixed supports (end rotation prevented). The FEMs are easily computed for the load cases shown in Table 12.1.

2 End moments due to actual rotations at the member ends.

3 End moments due to relative translation between the ends of the member.

The moment distribution method is a successive correction procedure that begins with the assumption that all joints are fixed supports. This assumption permits an analyst to compute fixed-end moments due to applied loads on the members. The method also includes additional joint moments for structural members that experience relative joint translation (without rotation) between its member ends. However, all the joints are truly not fixed supports and may be free to rotate. Therefore, the restraint is released at a joint that is free to rotate. This, in effect, will cause its attached members to deform to a new equilibrium position and develop additional moments at the member ends. Then, the rotated joint is reclamped and another joint that can rotate is released, and so forth. As the process is repeated from joint to joint, and back again, the end-moment corrections become smaller and smaller. Thus, the internal joint moments are successively corrected until moment equilibrium exists at all structural joints. It is worth noting that convergence to an exact solution can be achieved by moment distribution if enough iteration cycles are performed (two to three iteration cycles have been known to produce an approximate solution that is within 1 to 3 percent of the exact solution).

At present, moment distribution (also known as the Hardy Cross method) continues to be used for preliminary analysis and design, and also serves to check the accuracy of computer results. The method is suitable to investigate small simple structures or substructures, such as a continuous beam in a multistory building frame. In fact, a moment distribution analysis can be as efficient and, in some cases, more economical than a computer analysis for small-structure applications.

13.2. BASIC ASSUMPTIONS AND FUNDAMENTAL CONCEPTS

In this section, structures with members that have relative joint translation are omitted (studies involving joint translation are presented in Secs. 13.5 and 13.7). The sign convention and basic assumptions used for moment distribution are as follows:

Sign convention Clockwise internal moments and rotations of a member are considered positive; therefore, counterclockwise moments and rotations are negative.

Basic assumptions

1. All members of a structure are prismatic (constant EI).
2. All members are primarily subjected to bending.
3. The structure behaves elastically (obeys Hooke's law).
4. Axial deformations are negligible.
5. Rigid body rotations occur at member-to-member joints.

A physical interpretation of the moment-distribution method is presented as follows:

Figure 13.1a shows a plane cruciform frame made of vertical and horizontal members which are rigidly joined together at E. The frame has a pin support at A and fixed supports at B, C, and D. Joint E is free to rotate but will not translate (see assumption 5 above). Assume that joints A and E are clamped to prevent rotation. Then, fixed-end moments (FEMs) can easily be computed at C and E due to the applied load P, and the frame will deform as shown in Fig. 13.1b. However, joints A and E are truly able to rotate. Let us first consider joint E. Imagine that you remove (unlock) the clamp at E which, in effect, will cause the fixed-end moment at E to be released and act opposite in direction. The released *unbalanced moment* will rotate the joint until equilibrium is restored by internal moments of the members at joint E. Concurrently, moments are developed at the far ends of the members joined at E. The internal moments at E are known as *resisting or balancing moments* and are denoted as M_{EA}, M_{EB}, M_{EC}, and M_{ED} in Fig. 13.1c. The moment that develops at the far end of each member joined at E is called a *carryover moment*. These far end supports are assumed to prevent rotation.

384 Moment Distribution

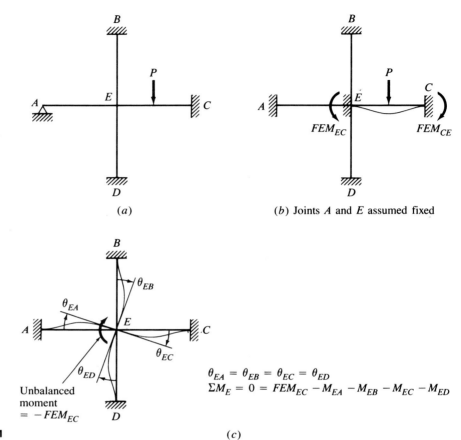

Figure 13.1

Now, let us consider member AE. The rotation at E will theoretically result in a carryover moment at the assumed fixed end at A. However, joint A is a pin support that can rotate (also, $M_{AE} = 0$). Thus, if joint A is unlocked, the released moment is unbalanced and causes joint A to rotate until $M_{AE} = 0$. This action, in turn, causes a new unbalanced moment to occur back at E. As stated earlier, the activities of locking or unlocking joints, computing the unbalanced moments, the resisting moments, and the carryover moments, and so on, are recurring events in the moment distribution procedure which continue until moment equilibrium is evident at all joints. It now remains for us to determine the following:

1 How the resisting moment is distributed among the members at a joint that is unclamped and free to rotate

2 What is the carryover moment at one end of a member when a resisting moment is applied at the other end

385 Basic Assumptions and Fundamental Concepts

Figure 13.2

Carryover moment Member ij of Fig. 13.2a represents a *prismatic beam member* (constant EI) with unyielding supports (that is, no relative joint translation). When a moment (M_{ij}) is applied at the pin supported beam end, the member rotates (θ_i) and develops a moment (M_{ji}) at the fixed support j as shown in Fig. 13.2b. By use of the slope deflection equations (Chap. 12), the following relationships are obtained:

$$M_{ij} = 2EK \cdot [2\theta_i + (0) - 3(0)] + 0 = 4EK\theta_i \tag{13.1}$$

$$M_{ji} = 2EK \cdot [2(0) + \theta_i - 3(0)] + 0 = 2EK\theta_i \tag{13.2}$$

Thus,
$$M_{ji} = +\frac{1}{2} M_{ij} \tag{13.3}$$

Equation (13.3) states that an applied moment at the pin-end support of a prismatic member results in a moment at the fixed-end support equal in direction and one-half the magnitude of the applied moment. Consequently, *a prismatic member is said to have a carryover factor of plus one-half* where the plus sign indicates the same direction for both moments.

In Eq. (13.1) above, the term $4EK$ (or $4EI/L$) is called the *absolute stiffness* and is symbolized by K^A. Absolute stiffness is defined as the end moment that will produce a unit rotation at the pin end of a pin-fixed member. Also, the symbol K represents the *relative member stiffness* which is often defined with subscripts to identify each member stiffness.

Distribution factors Let us recall the frame of Fig. 13.1 where it is shown that an unbalanced moment occurs when joint E is theoretically unlocked. Moment equilibrium at E requires that

$$\Sigma M_E = 0 = M_{EA} + M_{EB} + M_{EC} + M_{ED} + \Sigma \text{FEM}_E \tag{13.4}$$

where $\Sigma(\text{FEM})_E$ represents the algebraic sum of the FEMs at joint E due to transverse loads on the joining members at E. Since the members at joint E are assumed to undergo rigid rotation, it is evident from Fig. 13.1c that

$$\theta_{EA} = \theta_{EB} = \theta_{EC} = \theta_{ED} \tag{13.5}$$

The general equation [Eq. (13.1)] can be rewritten in the form

$$\theta_{ij} = \frac{M_{ij}}{4EK} \tag{13.6}$$

Substitution of Eq. (13.6) into Eq. (13.5) yields

$$\frac{M_{EA}}{4EK_{EA}} = \frac{M_{EB}}{4EK_{EB}} = \frac{M_{EC}}{4EK_{EC}} = \frac{M_{ED}}{4EK_{ED}}$$

In standard designs, E is the same for all members; thus, the equation above can be simplified as

$$\frac{M_{EA}}{K_{EA}} = \frac{M_{EB}}{K_{EB}} = \frac{M_{EC}}{K_{EC}} = \frac{M_{ED}}{K_{ED}} \qquad (13.7)$$

These relationships show that the member moments at joint E are in direct proportion to each other by their relative stiffness values. Substitution of Eq. (13.7) into Eq. (13.4) can be done in terms of one member; successive substitutions for each member will provide the following results:

$$M_{EA} = [K_{EA}/\Sigma K_E] \cdot \Sigma \text{FEM}_E = \text{DF}_{EA} \cdot \Sigma \text{FEM}_E$$
$$M_{EB} = [K_{EB}/\Sigma K_E] \cdot \Sigma \text{FEM}_E = \text{DF}_{EB} \cdot \Sigma \text{FEM}_E$$
$$M_{EC} = [K_{EC}/\Sigma K_E] \cdot \Sigma \text{FEM}_E = \text{DF}_{EC} \cdot \Sigma \text{FEM}_E$$
$$M_{ED} = [K_{ED}/\Sigma K_E] \cdot \Sigma \text{FEM}_E = \text{DF}_{ED} \cdot \Sigma \text{FEM}_E \qquad (13.8)$$

where $\Sigma K_E = K_{EA} + K_{EB} + K_{EC} + K_{ED}$

and
$$\text{DF}_{EA} = K_{EA}/\Sigma K_E$$
$$\text{DF}_{EB} = K_{EB}/\Sigma K_E$$
$$\text{DF}_{EC} = K_{EC}/\Sigma K_E$$
$$\text{DF}_{ED} = K_{ED}/\Sigma K_E \qquad (13.9)$$

Equations (13.8) above, provide the expressions that relate the resisting (balancing) moments at the unlocked joint E to the corresponding unbalanced moment (ΣFEM_E). The resisting moments of Eqs. (13.8) are also known as *distributed moments*. The stiffness ratios given by Eqs. (13.9) are referred to as *distribution factors* where the sum of the DFs at a joint is always equal to 1.0.

The moment distribution procedure can be summarized as follows:

1 Lock all joints. Compute FEMs due to applied loads for each member. FEMs due to individual load types can be determined and combined by superposition (see Table 12.1).

2 Unlock (release) one joint which is free to rotate.

3 Determine the unbalanced moment at the released joint.

4 Compute the distributed (balanced) moment for each member at the released joint.

5 Then, carryover one-half of the balanced moment to the joint support at the opposite member end. Lock the released joint in its new rotated position.

6 Repeat steps 2 to 5 at another supported joint that is free to rotate.

7 Steps 2 to 6 are repeated until the magnitudes of the joint moments are within the accuracy desired.

Additional comments

1 Steps 2 to 5 above represent an iteration cycle performed for one joint. *A recommended alternate approach is to perform the operations of computing distributed moments and carryover moments at all joints simultaneously rather than by random locking and unlocking of different joints, one-by-one.* Example problems in this text use this approach.

2 For an approximate analysis, the author suggests a preset tolerance equal to 5 percent of the largest ΣFEM value. This level of accuracy is usually achieved in two or three cycles of moment distribution.

3 Fixed-end moments (FEMs) for load cases other than those shown in Table 12.1 may be available from other sources. The methods of moment area, consistent deformations, and so on, can also be used to verify or establish FEMs for various loads.

4 The carryover moment, derived from the slope-deflection equations in Eqs. (13.1) and (13.2) above, can also be obtained by use of the methods described in Chaps. 9, 10, and 11.

Alternate derivation of carryover factor and absolute stiffness

Consider couple M applied at pin-support A as shown below. Employ the moment-area theorems and the principle of superposition.

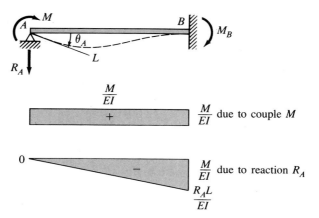

By the second moment-area theorem and use of the principle of superposition:

$$\delta_{AtB} = 0 = \left(\frac{M}{EI} \cdot L\right)\left(\frac{L}{2}\right) - \left(\frac{1}{2} \cdot \frac{R_A L}{EI} \cdot L\right)\frac{2L}{3}$$

$$\therefore \quad R_A = \frac{3M}{2L}$$

$$\zeta + \Sigma M_B = 0 = M + M_B - R_A L = M + M_B - \frac{3}{2}\frac{M}{L}(L)$$

$$\therefore \quad M_B = +\frac{1}{2}M$$

Thus the carryover factor $= +\frac{1}{2}$

By the first moment-area theorem and the principle of superposition:

$$\theta_A - \theta_B = \theta_A - (0) = \theta_A = A_{M/EI}\Big|_A^B = \frac{M}{EI} \cdot L - \frac{1}{2} \cdot \frac{R_A L}{EI} \cdot L$$

$$= \frac{ML}{EI} - \frac{1}{2}\left(\frac{3M}{2L}\right)L^2$$

thus $\quad \theta_A = \frac{ML}{EI} - \frac{3ML}{4EI} = \frac{ML}{4EI} = \frac{M}{4EK} \quad$ (see Eq. 13.1)

where $\quad K = \dfrac{M}{4E\theta_A}$

When $\theta_A = 1$, $M = 4EK(1) = 4EK$, also called *absolute stiffness* and symbolized by K^A.

13.3. APPLICATIONS: BEAMS WITHOUT SUPPORT SETTLEMENT

The introductory examples presented in this section also contain step-by-step comments—some general, and others particular to each example.
  Additional comments for Example 13.1 are listed as follows:

1 Distribution factors (DFs):

  a Distribution factors are only determined at joint *B*. DFs are not computed at fixed supports *A* and *C* since fixed supports remain "locked" and do not rotate.

  b It is customary to construct a block around the DFs for easy referral.

2 A good "bookkeeping" practice is to draw a horizontal line beneath the distributed moments at B to show that the joint has been balanced and restrained again.

3 A customary visual aid is to draw an arrow from a distributed moment to its associated carryover moment.

4 In this example, notice that the iteration process ends with the carryover (CO) operation since all joints are thereafter in equilibrium. In subsequent examples, it is usually found that equilibrium is obtained when the iteration cycle ends with distribution of the balanced moments.

5 It is also customary to draw a double horizontal line after the last iteration step before final moment summation.

6 A free-body diagram is shown below the final moment summation to reflect the importance of the sign ($+$ or $-$) to establish the correct direction of each final joint moment.

EXAMPLE 13.1

Determine all joint moments for the structure shown. EI is constant.

Comments

1. $DF_{BA} = \dfrac{K_{BA}}{K_{BA} + K_{BC}} = \dfrac{3}{5} = 0.6;\quad DF_{BC} = \dfrac{K_{BC}}{K_{BA} + K_{BC}} = \dfrac{2}{5} = 0.4$

2. $\Sigma DF = 1.0$ at each joint.

3. $FEM_{CB} = \dfrac{wL^2}{12} = -FEM_{BC}$ (see Table 12.1).

4. Distributed moments (Dist.) are of opposite sign to unbalanced joint moment at B. Note: No distribution at C.

5. Carryover moment (CO) is $\frac{1}{2}$ in magnitude and of same sign as the distributed moment.

6. Fixed supports do not rotate, and restrain all moments transmitted. Fixed supports are never "unlocked."

7. Final moments are the algebraic sum of FEM, Dist., and CO at each joint.

8. Observe moment equilibrium at joint B.

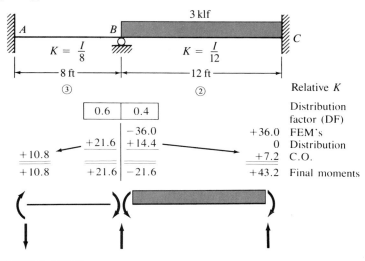

EXAMPLE 13.2

Determine all joint moments for the structure below. *I* values shown below.

Comments

1. DF at *D* shows that only member *DC* can provide balanced moment at *D*.
2. Since $M_D = 0$ (roller-end support), distributed moment at *D* must negate any unbalanced moment at *D*.
3. Note: Distributed moment at fixed-end *A* is zero since fixed supports do not rotate and are never "unlocked."
4. Observe that final moments at joints *B* and *C* are in equilibrium and boundary condition at joint *D* is satisfied ($M_D = 0$).

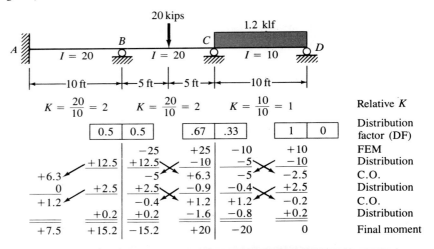

EXAMPLE 13.3

Determine all joint moments for the continuous beam shown on the next page. *I* is constant.

Comments

1. By statics, $M_{CD} = 4$ kips $\cdot$ 5 ft $= 20$ kip $\cdot$ ft. If joint *C* is locked, $FEM_{CD} = -20$ kip $\cdot$ ft.
2. Only the normal component of the 25 kips load on member *AB* contributes to the FEMs.

391 Applications: Beams without Support Settlement

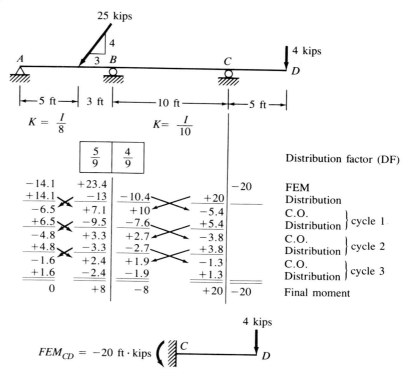

3. $M_A = 0$ (pin-end support). All distributions at A are shown to negate unbalanced moments at A.
4. Since $\Sigma M_C = 0 = M_{CD} + M_{CB}$, $M_{CB} = -M_{CD} = +20$ kip·ft. All subsequent distributions after the first one are shown to negate unbalanced moments at C.
5. Note: The last two iteration cycles are not required since they do not contribute to the final results.

Examples 13.2 and 13.3 The moment at the simple end support is known to be zero. Consequently, all FEMs and carryover moments at the simple end supports are seen to be canceled by distributed moments of the same magnitude and opposite sign. A display of DFs at a simple end support is optional.

Example 13.3 considers a continuous beam with an overhanging member. By statics, it is evident that $M_{CD} = 20$ kip·ft. By assuming joint C is "locked," $FEM_{CD} = -20$ kip·ft (see sketch of CD presented with the solution). Thus, moment equilibrium is maintained at C by setting the resisting moment M_{CB} equal to $+20$ kip·ft. Subsequent distributions of M_{CB} are shown to cancel the CO moments in order to maintain the known moment value of $+20$ kip·ft. Since the overhanging member provides no restraint on the rotation of joint C, it, in effect, has a stiffness factor of zero; thus, no DFs are shown at joint C.

13.4. STIFFNESS MODIFICATIONS

13.4.1. Simple End Modification

The moment is zero at a simple end support of a beam (pin or roller)—a fact that is distinguished by the freedom of the joint to rotate without resistance. In the previous moment-distribution examples, it is obvious that each carryover moment operation continues to disrupt the equilibrium at a simple end support. The remedy is obvious: After each carryover moment to a simple end, the joint is unlocked which releases the carryover moment. Then, a distributed moment is introduced to restore the moment back to zero. This extra work can be avoided by use of a modified member stiffness. The modification permits the simple end support to remain unlocked after FEMs are balanced.

Recall that the absolute stiffness (K^A) is equal to $4EK$ for a prismatic beam with the far end fixed (refer to Fig. 13.2b and Eq. (13.1)). Figure 13.3

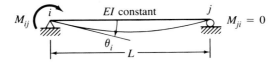

Figure 13.3

shows the prismatic beam of Fig. 13.2b with the fixed support at j replaced by a simple end support ($M_{ji} = 0$). The relationship between θ_i and the applied moment M_{ij}, as expressed by slope-deflection Eq. (12.7), is

$$M_{ij} = 3EK[\theta_i - (0)] + (0) - \tfrac{1}{2}(0) = 3EK\theta_i \qquad (13.10)$$

Therefore, the absolute stiffness (K^A) is equal to $3EK$ for a prismatic beam with a pin support at j. By comparison, the beam with a pin support at j will produce the same unit rotation at i using three-quarters of the applied moment M_{ij} needed by the beam with a fixed support at j. Therefore, the stiffness of a member with the far end pin supported can be modified by three-quarters for simplified moment distribution analysis; the modified relative stiffness is expressed as

$$K^M = \tfrac{3}{4} \cdot K = \frac{3I}{4L} \qquad (13.11)$$

Obviously, joint distribution factors will change when K is replaced by a modified stiffness (K^M) for one or more joint members. Also, when a modified stiffness is used, the pin end support j remains unlocked after FEMs are computed; therefore, $M_{ji} = 0$, and the carryover factor is zero. In Example 13.4, each outer span of the continuous beam has a simple end support. Therefore, distribution factors at joints B and C are based on modified stiffnesses as follows:

$$\mathrm{DF}_{BA} = \frac{K^M_{AB}}{\Sigma K_B} = \frac{11.25}{(11.25 + 12)} = 0.484$$

EXAMPLE 13.4

By use of modified member stiffness, compute all joint moments in the continuous beam. I is constant. Determine reactions and draw V and M diagrams.

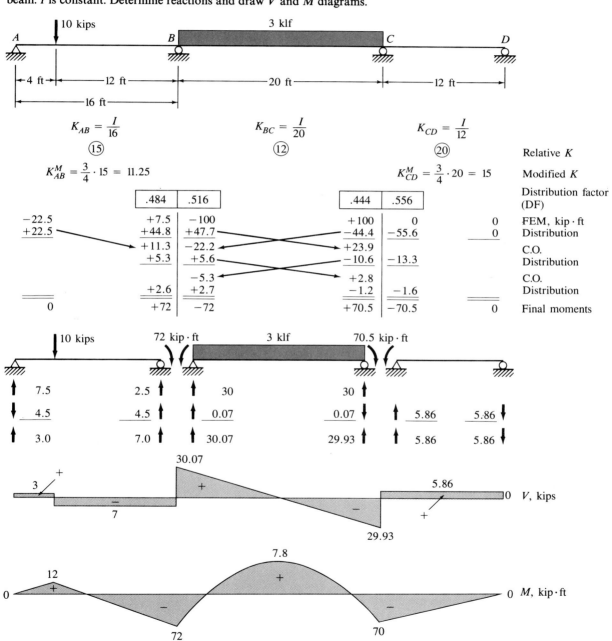

394 Moment Distribution

$$\mathrm{DF}_{BC} = \frac{K_{BC}}{\Sigma K_B} = \frac{12}{(11.25 + 12)} = 0.516$$

$$\mathrm{DF}_{CB} = \frac{K_{BC}}{\Sigma K_C} = \frac{12}{(12 + 15)} = 0.444$$

and

$$\mathrm{DF}_{CD} = \frac{K_{CD}^M}{\Sigma K_C} = \frac{15}{(12 + 15)} = 0.556$$

Recall that the moment-distribution procedure starts by locking all joints and computing FEMs. Then, joints are unlocked and balanced moments are distributed. Next, carryover moments to the interior joints B and C are shown. However, the use of K_{AB}^M and K_{CD}^M allows the simple joints A and D to remain unlocked; therefore, it is observed that moment-distribution cycles continue without carryover moments to the simple end supports.

Shear and moment diagrams After the final moments are found by a moment-distribution analysis, reactions plus V and M diagrams are determined from statics. In Example 13.4, free-body diagrams are shown which consider each of the three spans as simply supported. First, reactions are found due to loads on the spans. Then, the final joint moments are placed and directed on each span end in accordance with the moment-distribution sign convention. Next, reactions due to the end moments are found. Finally, the reactions are combined by superposition. The V and M diagrams are drawn using the same sign convention presented in Sec. 5.2.

Figure 13.4

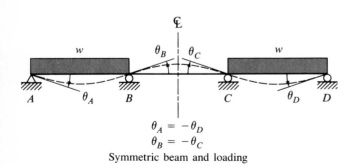

13.4.2. Symmetry and Antisymmetry Modifications

Symmetric loading A symmetric beam or frame under symmetric loads is noted to have symmetric joint deformations and bending moments about its centerline. For example, a study of the symmetric beam shown in Fig. 13.4a reveals that $\theta_A = -\theta_D$ and $\theta_B = -\theta_C$. An isolated view of the center span also shows that $M_{BC} = -M_{CB}$ (see Fig. 13.4b). Therefore, a stiffness modification of the center span can be made which will only require analysis for one-half of the symmetric structure. To this end, the modified absolute stiffness of the center span can be established from the general slope deflection equation by setting $\theta_B = 1$ and $\theta_C = -1$. That is, the modified absolute stiffness $K^M_{BC} = M_{BC}$ when $\theta_B = 1$. Therefore,

$$K^M_{BC} = M_{BC} = 2EK \cdot [2(1) + (-1) - 3(0)] = 2EK \quad (13.12)$$

By comparing Eq. (13.12) with the absolute stiffness $K^A_{BC} = 4EK$ from Eq. (13.1), we obtain the general relationship for the *member with symmetry about both ends* as

$$K^M = \tfrac{1}{2} \cdot K \quad (13.13)$$

Example 13.5 demonstrates a simplified moment-distribution analysis of a symmetric beam under symmetric loading by use of the modified stiffness of Eq. (13.13) for the center span.

EXAMPLE 13.5

Determine final joint moments by moment distribution with use of modified stiffness. I is given below.

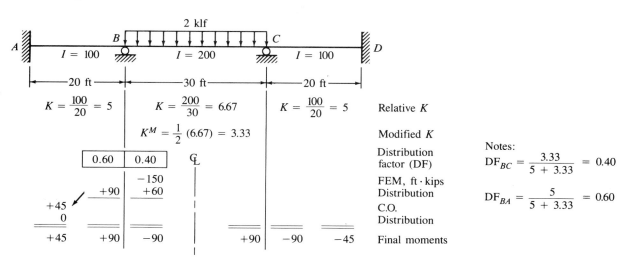

EXAMPLE 13.6

Determine final joint moments by moment distribution. I is constant. Use modified stiffnesses for simplification.

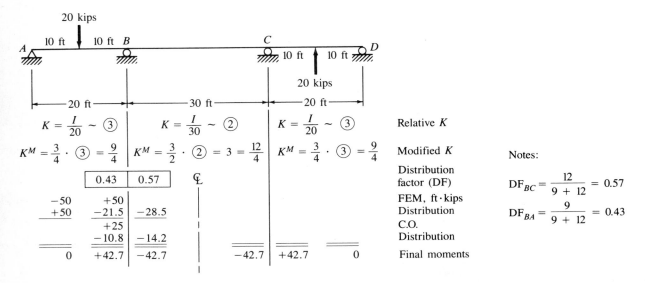

Antisymmetric loading Figure 13.5a illustrates a symmetric continuous beam under antisymmetric loading. A study of the center span in Fig. 13.5b reveals that $M_{BC} = M_{CB}$ and $\theta_B = \theta_C$. The modified stiffness is defined as the end moment required to produce a unit-end rotation. Therefore, by use of the general slope-deflection equation with $\theta_B = \theta_C = 1$, we obtain

$$M_{BC} = 2EK \cdot [2(1) + (1) - 3(0)] = 6EK \tag{13.14}$$

By comparison with the absolute stiffness ($K_{BC}^A = 4EK$) from Eq. (13.1), we obtain the general relationship for the *member with one end antisymmetric about the other end* as

$$K^M = \frac{3}{2} K \tag{13.15}$$

Example 13.6 demonstrates a simplified moment-distribution analysis of a symmetric beam under antisymmetric loading by use of the modified stiffness of Eq. (13.15) for the center span. Observe that both simple end and antisymmetric stiffness modifications are employed in the analysis.

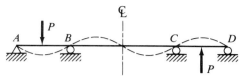

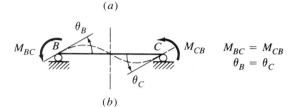

Symmetric beam under antisymmetric loading

$M_{BC} = M_{CB}$
$\theta_B = \theta_C$

Figure 13.5

13.5. BEAMS WITH SUPPORT SETTLEMENT

Thus far, applications of the moment-distribution method have been limited to continuous beams with unyielding supports. However, most civil engineering structures rest on soil or rock foundations which can deform and possibly result in differential settlement of supports. In this event, added bending may develop in the structural members due to the relative joint translation associated with the support settlement.

Figure 13.6 shows a prismatic beam with both ends fixed against rotation. Assume that support i remains stationary while support j undergoes a settlement of Δ. This action results in equal fixed-end bending moment resistance at both joints. The joint moments, as found from the general slope deflection equations, are

$$M_{AB} = M_{BA} = 2EK \cdot [2(0) + (0) - 3\Delta/L] + 0 = -6EK\Delta/L$$

or $M_{AB} = M_{BA} = -6EI\Delta/L^2$ (13.16)

(*Note:* These moment values conform to the sign convention established for moment distribution above.)

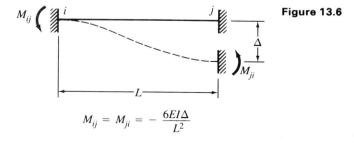

Figure 13.6

$$M_{ij} = M_{ji} = -\frac{6EI\Delta}{L^2}$$

398 Moment Distribution

In moment-distribution analysis, FEMs due to member loads are generally computed separately from FEMs due to known relative support settlement values. Subsequently, these FEMs are combined by superposition before the joints are unlocked for the first distribution. Example 13.7 presents a moment-distribution analysis of a continuous beam under load which, after a time, experiences support settlements.

EXAMPLE 13.7

Determine the joint moments for the continuous beam if supports B and C settle 30 mm and 50 mm, respectively. $EI = 8000$ kN·m². Relative Is are shown below.

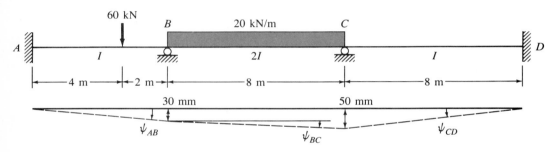

$K = \frac{I}{6}$ ~ ④			$K = \frac{I}{4}$ ~ ⑥			$K = \frac{I}{8}$ ~ ③		Relative K
	0.4	0.6		0.67	0.33			Distribution factor (DF)
−26.7	+53.3	−106.7		+106.7				FEM$_{loads}$
−40		−40	−15		−15	+37.5	+37.5	FEM$_{settlement}$
−66.7	+13.3	−121.7		+91.7	+37.5		+37.5	ΣFEM
0	+43.4	+65		−86.6	−42.6		0	Distribution
+21.7		−43.3		+32.5			−21.3	C.O.
0	+17.3	+26		−21.8	−10.7		0	Distribution
+8.6		−10.9		+13			−5.3	C.O.
	+4.4	+6.5		−8.7	−4.3		0	Distribution
+2.2		−4.4		+3.3			−2.2	C.O.
0	+1.8	+2.6		−2.2	−1.1		0	Distribution
−34.2	+80.2	−80.2		+21.2	−21.2		+8.7	Final moments

$$DF_{BA} = \frac{4}{4+6} = 0.4$$

$$DF_{BC} = \frac{6}{4+6} = 0.6$$

$$DF_{CB} = \frac{6}{6+3} = 0.67$$

$$DF_{CD} = \frac{3}{6+3} = 0.33$$

FEMs due to load:

$$FEM_{AB} = -\frac{60(4)(2)^2}{6^2} = -26.7$$

$$FEM_{BA} = +\frac{60(2)(4)^2}{6^2} = +53.3$$

$$FEM_{BC} = -\frac{20(8)^2}{12} = -106.7$$

$$FEM_{CB} = +\frac{20(8)^2}{12} = +106.7$$

FEMs due to settlement:

$$FEM_{AB} = FEM_{BA} = -\frac{6EI\Delta}{L^2} = -\frac{6(8000)(0.03)}{(6)^2} = -40$$

$$FEM_{BC} = FEM_{CD} = -\frac{6EI\Delta}{L^2} = -\frac{6(8000)(0.02)}{(8)^2} = -15$$

$$FEM_{CD} = FEM_{DC} = +\frac{6EI\Delta}{L^2} = +\frac{6(8000)(0.05)}{(8)^2} = +37.5$$

$$\psi_{AB} = \frac{\Delta_B}{L_{AB}} = +\frac{0.03}{6}$$

$$\psi_{BC} = +\frac{(\Delta_C - \Delta_B)}{L_{BC}} = +\frac{0.02}{8}$$

$$\psi_{CD} = -\frac{\Delta_C}{L_{CD}} = -\frac{0.05}{8}$$

13.6. FRAMES WITHOUT SIDESWAY

In Sec. 12.5, it was learned that a frame will not lean to one side (sway) if it has symmetric geometry and load, or if it is restrained against sidesway. For these conditions, the same moment-distribution method presented in Secs. 13.3 and 13.4 above can be applied to frames. Example 13.8 shows a moment-distribution analysis of a symmetric simple frame under symmetric loading. Example 13.9 provides a moment-distribution analysis for a frame with sidesway prevented. The computations of moment-distribution frame analyses in this text are usually displayed in an upright orientation. However, alternate formats for recording the data can be found in other books on this subject.

EXAMPLE 13.8

Compute the joint moments of the symmetric frame.

$\text{FEM}_{AB} = +\dfrac{4(10)(10)^2}{20^2} = -10 \text{ kip} \cdot \text{ft}$

$\text{FEM}_{BA} = -\dfrac{4(10)(10)^2}{20^2} = +10 \text{ kip} \cdot \text{ft}$

$\text{FEM}_{BC} = -\dfrac{2.4(25)^2}{12} = -125 \text{ kip} \cdot \text{ft}$

$\text{FEM}_{CB} = +\dfrac{2.4(25)^2}{12} = +125 \text{ kip} \cdot \text{ft}$

$\text{FEM}_{CD} = -\dfrac{4(10)(10)^2}{20^2} = -10 \text{ kip} \cdot \text{ft}$

$\text{FEM}_{DC} = +\dfrac{4(10)(10)^2}{20^2} = +10 \text{ kip} \cdot \text{ft}$

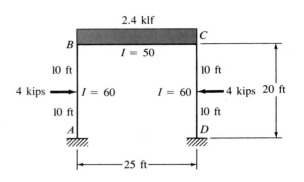

Solution A

Without stiffness modification

$K_{AB} = \dfrac{I}{L} = \dfrac{60}{20} = 3 = K_{CD}$

$K_{BC} = \dfrac{I}{L} = \dfrac{50}{25} = 2$

$\text{DF}_{BA} = \dfrac{K_{AB}}{\Sigma K_B} = \dfrac{3}{3+2} = 0.6; \quad \text{DF}_{BC} = \dfrac{K_{BC}}{\Sigma K_B} = \dfrac{2}{3+2} = 0.4$

Similarly, $\text{DF}_{CB} = \dfrac{2}{5} = 0.6$ and $\text{DF}_{CD} = \dfrac{3}{6} = 0.4$

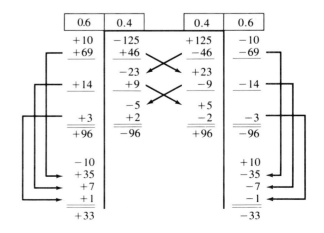

401 Frames without Sidesway

Solution B

Using modified stiffness due to symmetry

$$K_{AB} = \frac{60}{20} = 3$$

$$K_{BC}^M = \frac{1}{2} K_{BC} = \frac{1}{2} \cdot \frac{50}{25} = 1$$

$$DF_{BA} = \frac{3}{3+1} = 0.75$$

$$DF_{BC} = \frac{1}{3+1} = 0.25$$

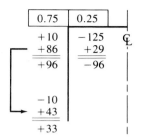

Note
Additional iteration cycles in solution *A* above are required to provide identical moment values given in solution *B*.

EXAMPLE 13.9

Compute all end moments for the frame shown with sidesway prevented. *I* is constant.

$$K_{BA} = K_{BC} = \frac{I}{12} \sim \text{③}$$

$$K_{BD} = \frac{I}{9} \sim \text{④}$$

$$K_{BA}^M = \frac{3}{4} \cdot \text{③} = 2.25$$

$$DF_{BA} = \frac{K_{BA}}{\Sigma K_B} = \frac{2.25}{2.25 + 3 + 4} = 0.243$$

$$DF_{BC} = \frac{K_{BC}}{\Sigma K_B} = \frac{3}{2.25 + 3 + 4} = 0.324$$

$$DF_{BD} = \frac{K_{BD}}{\Sigma K_B} = \frac{4}{2.25 + 3 + 4} = 0.433$$

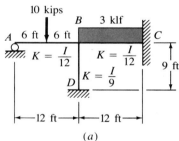

(a)

Solution

$$FEM_{AB} = -FEM_{BA} = -\frac{10(6)(6)^2}{(12)^2} = -15 \text{ kip} \cdot \text{ft}$$

$$FEM_{BC} = -FEM_{CB} = -\frac{3(12)^2}{12} = -36 \text{ kip} \cdot \text{ft}$$

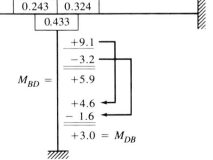

(b)

Note
$\Sigma M_B = 0 = M_{BA} + M_{BC} + M_{BD} = 25.8 - 31.6 + 5.9 \approx 0$ ✓

13.7. FRAMES WITH SIDESWAY

Thus far, the moment-distribution method has been presented for statically indeterminate beams and frames without sidesway. When beam structures experience known support settlements, moment distribution is performed in the usual manner. In these cases, member FEMs due to relative joint translation are computed and summed with the FEMs due to applied loads; then, moment distribution and carryover iteration cycles are continued in normal fashion (see Example 13.7). However, this approach is not suitable for frames with sidesway because the sidesway FEM values ($6EI\Delta/L^2$) are unknowns which depend on the unknown Δ displacements of the free joints. It is possible to first perform displacement analyses, then compute the FEMs due to the sidesway, and finally proceed with moment distribution in the usual manner. However, this method would be tedious and impractical. The following example presents an orderly three-step approach for the analysis of single-story frames with sidesway by the moment-distribution method.

Step one: distribution of FEMs due to applied loads—sidesway prevented Consider the symmetric frame of Fig. 13.7a which sways due to the off-centered vertical load. Initially, FEMs due to the applied load are computed, and final joint moments are found by moment distribution in the usual manner. Then, horizontal support components at the base of each column are computed. This analysis results in a horizontal force imbalance of 0.82 kip to the right. The sum of the horizontal forces on the structure does not equal zero since the moment effects due to sidesway are not included. Therefore, a sidesway correction must be performed for the frame.

Step two: distribution of assumed FEMs due to sidesway Let us suppose that a support is placed at joint C to prevent sidesway of the frame (see Fig. 13.7b). Then, the same results shown in Fig. 13.7a are found and the imaginary support at C will furnish the horizontal restraint required for equilibrium. Now, if we remove the imaginary support at C, the horizontal force at C is released, and the frame will sway to the right (see Fig. 13.7c). The joints are assumed to be locked against rotations during the frame sidesway. By neglecting axial deformations, the free joint of each column undergoes the same lateral motion, Δ. Then, the sidesway FEMs that develop at each column end can be expressed from Eq. (13.16) as

$$M_L = -6E\Delta(I/L^2)_L$$

and
$$M_R = -6E\Delta(I/L^2)_R \qquad (13.17)$$

where M_L and M_R represent the moments in the left and right column,

Figure 13.7 Moment distribution; Frame with sidesway

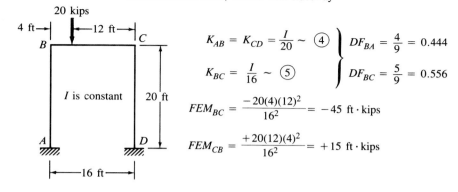

$$K_{AB} = K_{CD} = \frac{I}{20} \sim \text{\textcircled{4}}$$

$$K_{BC} = \frac{I}{16} \sim \text{\textcircled{5}}$$

$$DF_{BA} = \frac{4}{9} = 0.444$$

$$DF_{BC} = \frac{5}{9} = 0.556$$

$$FEM_{BC} = \frac{-20(4)(12)^2}{16^2} = -45 \text{ ft} \cdot \text{kips}$$

$$FEM_{CB} = \frac{+20(12)(4)^2}{16^2} = +15 \text{ ft} \cdot \text{kips}$$

Step 1 Distribution of FEMs due to load

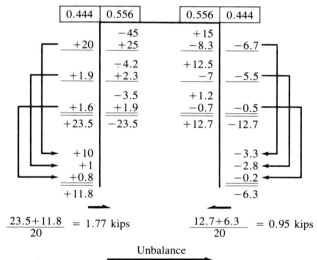

(a)

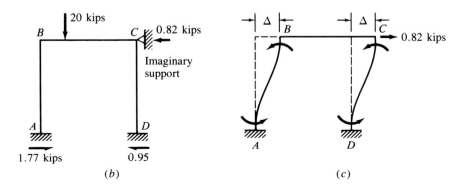

(b) (c)

Step 2 Distribution of assumed sidesway FEMs

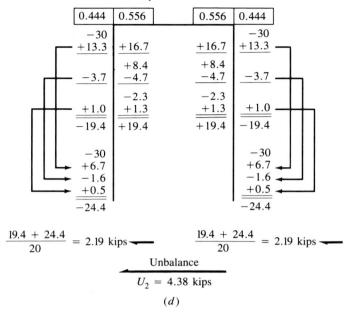

(d)

Step 3 Final moments by superposition

$H_A = 1.77 - 0.187(2.19) = 1.36$ kips → $H_D = 1.36$ kips ← $= 0.95 + 0.187(2.19)$

(e)

respectively. Thus, the proportional relationship between the fixed-end column sidesway moments is

$$M_L/M_R = (I/L^2)_L/(I/L^2)_R \qquad (13.18)$$

Since all columns sway in the same direction, *the sign of the sidesway FEMs is identical at all column ends.* From Eq. (13.18), $M_L/M_R = (I/400)/(I/400) = 1/1$. Therefore, an assumed sidesway FEM value of -30 kip·ft is selected for M_R and M_L at both column ends (see Fig. 13.7d). Other values of sides-

405 Frames with Sidesway

way column moments can be used as long as they are proportional in accordance with Eq. (13.18) above. This analysis results in an unbalanced horizontal force of 4.38 kips to the left.

Step three: superposition The correct final joint moments of the frame under sidesway loading is obtained by the principle of superposition. Since $U_1/U_2 = 0.82/4.38 = 0.187$, multiply the final moments shown in Fig. 13.7d by 0.187 and add them to the respective final moment values from Fig. 13.7a. These results are verified by evidence of horizontal equilibrium. The final moments and horizontal forces are shown in Fig. 13.7e. Note: the final horizontal support forces are shown as the algebraic sum of the forces of Fig. 13.7d multiplied by 0.187, and the corresponding forces of Fig. 13.7a.

Example 13.10 demonstrates the sidesway correction method for a single-story frame with both fixed and pin column supports. When sidesway occurs to the right, the moment at the fixed end of a pin-fixed column is expressed as $M = -3EI/L^2$ in Example 13.10. Since both columns have the same I and L, the ratio of the sidesway FEMs of the left column (fixed-support end) to the right column (pin-end support) is expressed by $M_L/M_R = (-6EI/L^2)_L/(-3EI/L^2)_R = 2/1$, where M_L and M_R denote the left and right column FEM, respectively. The complete frame analysis is presented in Example 13.10.

EXAMPLE 13.10

Find all joint moments using the sidesway correction method. I is constant.

$K_{AB} = \dfrac{I}{20} = \dfrac{12I}{240} \sim 12$

$K_{BC} = \dfrac{I}{30} = \dfrac{8I}{240} \sim 8$

$K_{CD}^M = \dfrac{3}{4} \cdot \dfrac{I}{20} = \dfrac{3I}{80} = \dfrac{9I}{240} \sim 9$

$DF_{BA} = \dfrac{12}{20} = 0.6$

$DF_{BC} = \dfrac{8}{20} = 0.4$

$DF_{CB} = \dfrac{8}{17} = 0.47$

$DF_{CD} = \dfrac{9}{17} = 0.53$

$FEM_{AB} = -FEM_{BA} = -\dfrac{20(10)(10)^2}{20^2} = -50 \text{ kip} \cdot \text{ft}$

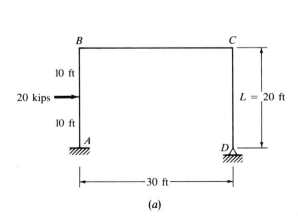

(a)

406 Moment Distribution

Step 1: Distribution of FEMs due to loads

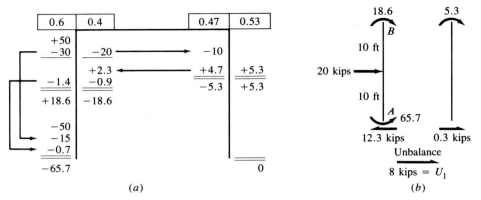

(a)

(b)

$$M_{AB} = M_{BA} = -\frac{6EI\Delta}{L^2}$$

$$M_{CD} = -\frac{3EI\Delta}{L^2} \quad \text{(from modified slope deflection equation)}$$

use $M_{AB} = M_{BA} = -60$ kip · ft

$M_{CD} = -30$ kip · ft

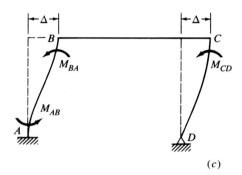

(c)

Step 2: Distribution of assumed sidesway FEMs

0.6	0.4		0.47	0.53
−60				−30
+36	+24		+14.1	+15.9
	+7		+12	
−4.2	−2.8		−5.6	−6.4
	−2.8		−1.4	
+1.7	+1.1		+0.7	+0.7
−26.5	+26.5		+19.8	−19.8
−60				
+18				
−2.1				
+0.8				
−43.3				0

3.5 kips   1.0 kips

Unbalance
$U_2 = 4.5$ kips

Factor $\dfrac{U_1}{U_2} = \dfrac{8}{4.5} = 1.78$

(d)

407 Frames with Sidesway

Step 3: Final moments by superposition

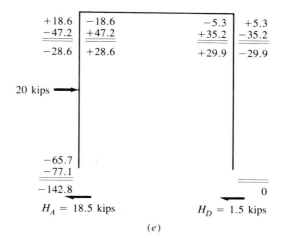

(e)

EXAMPLE 13.11

Compute all joint moments for the frame shown.

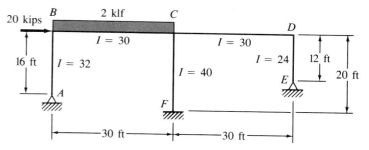

Thus far, the moment-distribution format has been easy to use for one-story, one-bay frames. However, this format becomes awkward to deal with as frames with more than one bay are studied. Therefore, a tabular format is presented in Example 13.11 as an alternate way to record the computations of a moment-distribution analysis. Example 13.11 considers a single-story, two-bay frame under sidesway loading. Since the frame sways to the right, all the sidesway FEMs are negative. Obviously, $M_{AB} = M_{ED} = 0$ (pin supports). The relationship of the sidesway FEMs in the columns are

$$
\begin{array}{ccccc}
M_{BA} & : & M_{CF} & : & M_{DE} \\
-6(32)/16^2 & : & -3(40)/20^2 & : & -6(24)/12^2 \\
-30 & : & -12 & : & -40
\end{array}
$$

Observe that the final results are verified by a horizontal equilibrium check.

408 Moment Distribution

Step 1: Distribution of FEM due to load

	Joint B		Joint C			Joint D		Joint F
Member	BA	BC	CB	CF	CD	DC	DE	FC
DF	0.60*	0.40*	0.25	0.50	0.25	0.40*	0.60*	—
FEM	0	−150	+150	0	0	0	0	0
Dist.	+90	+60	−37.5	−75	−37.5			
CO		−18.8	+30			−18.8		−37.5
Dist.	+11.3	+7.5	−7.5	−15	−7.5	+7.5	+11.3	
CO		−3.7	+3.7		+3.7	−3.7		−7.5
Dist.	+2.2	+1.5	−1.8	−3.7	−1.7	+1.5	+2.2	
Σ①	+103.5	−103.5	+136.9	−93.7	−43.2	−13.5	+13.5	−45

$$H_A = \frac{103.5 + 0}{16} = 6.5 \text{ kips} \rightarrow ; \quad H_F = \frac{93.7 + 45}{20} = 6.9 \text{ kips} \leftarrow ; \quad H_E = \frac{13.5 + 0}{12} = 1.1 \text{ kips} \rightarrow$$

$$\Sigma H = 20 + 6.5 - 6.9 + 1.1 = 20.7 \text{ kips} \rightarrow \ = U_1 \text{ (unbalanced)}$$

* DF based on modified stiffness.

Step 2: Distribution of assumed FEMs due to sidesway

FEM	−30	—	—	−12	—	—	−40	−12
Dist.	+18	+12	+3	+6	+3	+16	+24	0
CO		+1.5	+6		+8	+1.5		+3
Dist.	−0.9	−0.6	−3.5	−7	−3.5	−0.6	−0.9	0
CO		−1.8	−0.3		−0.3	−1.8		−3.5
Dist.	+1.1	+0.7	+0.1	+0.3	+0.2	+0.7	+1.1	0
Σ②	−11.8	+11.8	+5.3	−12.7	+7.4	+15.8	−15.8	−12.5

$$H_A = \frac{11.8 + 0}{16} = 0.7 \text{ kips} \leftarrow ; \quad H_F = \frac{12.7 + 12.5}{20} = 1.3 \text{ kips} \leftarrow ; \quad H_E = \frac{15.8 + 0}{12} = 1.3 \text{ kips} \leftarrow$$

$$\Sigma H = 0.7 + 1.3 + 1.3 = 3.3 \text{ kips} \leftarrow \ = U_2 \text{ (unbalanced)}$$

Step 3: Final moments by superposition

$$\text{Sidesway correction factor} = U_1/U_2 = 20.7/3.3 = 6.27$$

Σ①	+103.5	−103.5	+136.9	−93.7	−43.2	−13.5	+13.5	−45
Σ② × 6.27	−74	+74	+33.2	−79.7	+46.4	+99.1	−99.1	−78.4
Final moments	+29.5	−29.5	+170.1	−173.4	+3.2	+85.6	−85.6	−123.4

$$H_A = 2.1 \text{ kips} \rightarrow ; \quad H_F = 15.1 \text{ kips} \leftarrow ; \quad H_E = 7.0 \text{ kips} \leftarrow ; \quad \Sigma H = 20 + 2.1 - 15.1 - 7.0 = 0 \ \checkmark$$

Frames with sloping leg members To this point, sidesway analysis has been limited to frames constructed from horizontal beams and vertical columns. Since axial deformations are neglected, the sidesway causes the columns to bend while the beams remain horizontal. Hence, sidesway does not produce FEMs in the beams. It was previously shown that the sidesway FEMs in the columns are proportional to their respective $kEI\Delta/L^2$ values ($k = 6$ for a fixed-supported column and $k = 3$ for a pin-supported column).

When a frame with one or more sloping legs is subjected to sidesway,

EXAMPLE 13.12

Compute final joint moments in the frame by the sidesway correction method. $I = 900$ for all members.

Step 1: Distribution of FEMs due to load.

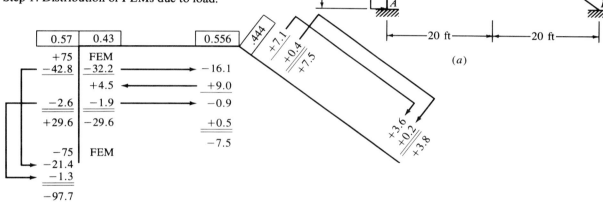

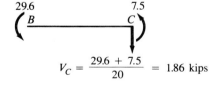

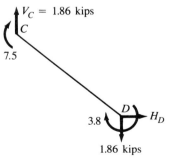

$$H_D = \frac{1.86(20) + 3.8 + 7.5}{15} = 3.23 \text{ kips}$$

$$\Sigma H = 4(15) - 34.5 + 3.23 = 28.7 \text{ kips} = U_1$$

(b)

410 Moment Distribution

Step 2: Distribution of assumed FEMs due to sidesway.

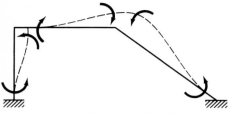
Sidesway FEM vector directions
Schematic A

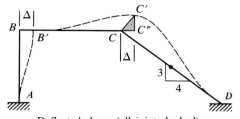
Deflected shape (all joints locked)
Schematic B

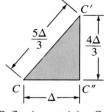
Deflections at joint C
Schematic C

(c)

Relative sidesway FEMs:

$$M_{AB} = -\frac{6(900)}{15^2} \cdot \Delta = -24\Delta \sim -24$$

$$M_{BC} = +\frac{6(900)}{20^2} \cdot \frac{4\Delta}{3} = +18\Delta \sim +18$$

$$M_{CD} = -\frac{6(900)}{20^2} \cdot \frac{5\Delta}{3} = -22.5\Delta \sim -22.5$$

	0.57	0.43		0.556	.444	
	−24	+18		+18	−22.5	
	+3.4	+2.6		+2.5	+2.0	
		+1.2		+1.3	−0.6	
	−0.7	−0.5		−0.7	−21.1	
	−21.3	+21.3		+21.1		
	−24				−22.5	
	+1.7				+1.0	
	+0.3				−0.3	
	−22.6				−21.8	

$H_A = 2.93$ kips $H_D = 5.69$ kips

$\Sigma H = 2.93 + 5.69 = 8.62$ kips $= U_2$

$\therefore$ Sidesway correction factor $= \dfrac{U_1}{U_2} = \dfrac{28.7}{8.62} = 3.33$

(d)

Step 3: Final moments by superposition.

B
+29.6   −29.6
−71.0   +71.0
$M_{BA} = -41.4$ $+41.4 = M_{CB}$

C
−7.5   +7.5
+70.3   −70.3
$+62.8 = M_{CB}$ $-62.8 = M_{CD}$

−97.7
−75.3
$M_{AB} = -173$ A

+3.8
−72.7
$-68.9 = M_{DC}$ D

$H_A = 44.3$ kips $H_D = 15.7$ kips

$\Sigma H = 60 - 44.3 - 15.7 = 60 - 60 = 0$ ✓

(e)

each frame member undergoes notable chord rotation. Therefore, sidesway FEMs develop in all members of a frame with sloping legs. The three-step procedure used above for vertical column frames can also be used to analyze frames with sloping legs.

Consider the rigid frame of Example 13.12. Schematic A shows the vector directions of the sidesway FEMs in each member. Schematic B shows that distance $BB' = CC''$ when joints B and C move to B' and C', respectively. Let us assume that the distances BB' and CC' are small with respect to the member lengths. Also, assume that the arc motions of AB to AB' and DC to DC' are replaced by perpendicular chords to AB and CD, respectively. Then, proportional relationships between the sidesway FEMs of the members are determined by use of trigonometry (see schematic C of step 2). The relative proportions of the sidesway FEMs are given in step 2.

The moment distribution of the applied loading (step 1) is carried out in the usual manner after which horizontal support forces are computed. The horizontal force at support D can be computed after the vertical force at D is known. This is accomplished using statics as follows: First, place the final joint moments from step 1 on a free-body diagram of beam BC and develop the moment equilibrium equation about B to determine V_C. Then, from a free-body diagram of member CD, it is evident that $V_D = V_C$. Finally, using the free-body diagram of member CD with V_D known, H_D is found by $\Sigma M_C = 0$. The unbalanced horizontal force for the step 1 analysis is $U_1 = 28.7$ kips to the right.

A moment distribution of the FEMs due to sidesway (step 2) results in an unbalanced horizontal force of $U_2 = 8.62$ kips to the left. The sidesway correction factor of $U_1/U_2 = +3.33$. The final joint moments in step 3 can be expressed as

$$\text{Final moments} = M_{\text{step 1}} + \frac{U_1}{U_2} M_{\text{step 2}}$$

The analysis is concluded with a verification of horizontal equilibrium.

Comments on multistory frame analysis For several decades, the methods of moment distribution and slope deflection have been commonly used for the analysis of small multistory buildings. Obviously, these hand analyses are cumbersome, time-consuming, and prone to error for frames with many bays and stories. Each added story above the first floor provides another degree of freedom of sidesway joint translation. In a slope deflection analysis, additional bays and stories in a building frame simply result in more ($n \times n$) equations to be solved simultaneously. However, a sidesway analysis of a multistory building by moment distribution can be an awkward, impractical, and agonizing task.

In 1932, C. T. Morris introduced a successive sidesway correction approach that simplified the task of analyzing multistory frames [13.3]. The interested reader can find the Morris method presented in several texts on this subject.

Matrix structural computer analysis provides an efficient and economical alternative for the analysis of multistory frame buildings. An analyst can write a program or use one of the many available mainframe or minicomputer codes. The IMAGES-2D program represents one choice for multistory plane-frame structural analyses. Laursen [13.4] provides listings of two microcomputer BASIC and FORTRAN matrix programs for the solution of simultaneous equations and moment-distribution analysis. Lefter and Bergin [13.5] also present a BASIC program for the solution of simultaneous equations along with software to perform plane-frame analysis. Consequently, this author strongly recommends the use of a computer for the analysis of multistory frames with sidesway.

13.8. NONPRISMATIC BEAMS

Thus far, the method of moment distribution has been limited to prismatic structures. Yet, bridge and building structures often contain nonprismatic members which are identified by a variable depth along their span lengths (see Fig. 13.8). A variable member depth is usually selected to lower the stress at points of high bending moment and to maintain deflections within acceptable limits. Now, we will study the application of moment distribution for the analysis of nonprismatic indeterminate beam structures.

The basic concepts of moment distribution remain the same for both prismatic and nonprismatic members. However, the fixed-end moments,

Figure 13.8 A three-span continuous concrete highway bridge in Wilmore, Kentucky, with an increasing variable depth shown over the two interior supports. (*Courtesy of International Structural Slides, Berkeley.*)

413 Nonprismatic Beams

distribution, and carryover factors of prismatic members are not valid for nonprismatic members. Therefore, the FEMs, distribution and carryover factors must first be established for each nonprismatic member before a moment distribution analysis of the structure can be done.

Consider the uniformly loaded nonprismatic beam shown in Example 13.13 where it is our mission to determine the carryover factors, absolute stiffnesses, and fixed-end moments. These parameters are found by the method of consistent deformations and the moment-area theorems. However, the parameters could also be determined using any suitable deflection method found in Chaps. 9 and 10. First, let us reestablish the following definitions:

A *carryover factor* is the ratio of the moment that develops at the fixed end of a member to the applied moment at the simply supported end. Let C_{AB} = carryover factor of member AB at end A, and C_{BA} = carryover factor of member AB at end B.

EXAMPLE 13.13

Determine carryover factors, absolute stiffnesses, and FEMs at each end of the nonprismatic beam shown in (a) below, for a uniform load (w). $L = 20$ ft.

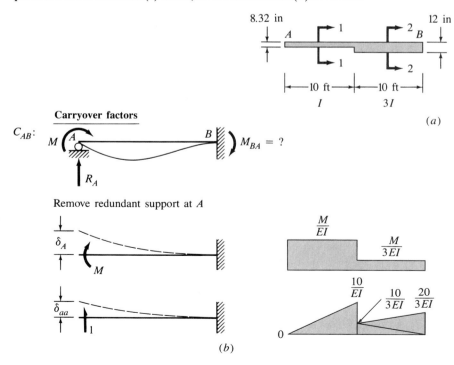

Using the second moment-area theorem: ΣM of M/EI areas about A yields

$$\delta_A = \frac{M}{EI}(10)5 + \frac{M}{3EI}(10)5 = 100M/EI$$

and $\quad \delta_{aa} = \frac{1}{2}(10)\left(\frac{10}{EI}\right)6.67 + \frac{1}{2}(10)\left(\frac{20}{3EI}\right)16.67 + \frac{1}{2}(10)\left(\frac{10}{3EI}\right)13.33 = \frac{111}{EI}$

$$R_A = -\frac{\delta_A}{\delta_{aa}} = -\frac{100M/EI}{1111/EI} = -0.09M$$

$\circlearrowleft + \Sigma M_B = 0 = M + 20R_A + M_{BA} = M + 20(-0.09M) + M_{BA}$

or $\quad M_{BA} = +0.80M \quad \therefore \quad C_{AB} = +0.80$

C_{BA}:

Remove redundant support at B

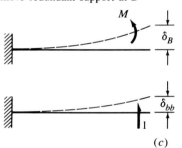

(c)

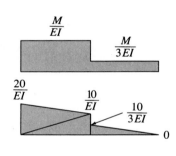

Using the second moment-area theorem: ΣM of M/EI areas about B yields

$$\delta_B = \frac{M}{EI}(10)15 + \frac{M}{3EI}(10)5 = 166.7\frac{M}{EI}$$

and $\quad \delta_{bb} = \frac{1}{2}(10)\left(\frac{10}{3EI}\right)6.67 + \frac{1}{2}(10)\left(\frac{10}{EI}\right)13.33 + \frac{1}{2}(10)\left(\frac{20}{EI}\right)16.67 = \frac{2444}{EI}$

$$R_B = -\frac{\delta_B}{\delta_{bb}} = -\frac{166.7M/EI}{2444/EI} = -0.0682M$$

$\circlearrowleft + \Sigma M_A = 0 = M_{AB} + M + 20R_B = M_{AB} + M + 20(-0.0682M)$

or $\quad M_{AB} = +0.364M \quad \therefore \quad C_{BA} = +0.364$

Absolute stiffness

At end A:

$\theta_A = 1^{RAD}$, $\theta_B = 0$

(d)

Refer to the M/EI diagrams of (b) above. Using the first moment-area theorem (ΣM of the M/EI areas between A and $B = \theta_A - \theta_B$) plus the principle of superposition, we find

$$\theta_A - \theta_B = \theta_A - (0) = \frac{M}{EI}(10) + \frac{M}{3EI}(10) + R_A\left[\frac{1}{2}\left(\frac{10}{EI}\right)10 + \frac{1}{2}\left(\frac{10}{3EI}\right)10 + \frac{1}{2}\left(\frac{20}{3EI}\right)10\right]$$

$$\theta_A = \frac{13.33M}{EI} + (-0.09M)\left[\frac{50}{EI} + \frac{16.7}{EI} + \frac{33.3}{EI}\right]$$

$$= \frac{13.33M}{EI} - \frac{9M}{EI} = \frac{4.33M}{EI}$$

$$\text{or} \quad M = \frac{EI\theta_A}{4.33} = \frac{EI\theta_A}{4.33} \cdot \frac{L}{L} = \frac{EI(1^{\text{radian}})20}{4.33L} = 4.62\frac{EI}{L}$$

where $k_{AB} = 4.62$

At end B:

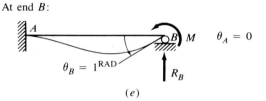

(e)

Refer to M/EI diagrams of Fig. c. Using the first moment-area theorem (ΣM of the M/EI areas between A and $B = \theta_B - \theta_A$) plus the principle of superposition, we find

$$\theta_B - \theta_A = \theta_B - (0) = \frac{13.33M}{EI} + R_B\left[\frac{1}{2}\left(\frac{10}{EI}\right)10 + \frac{1}{2}\left(\frac{10}{3EI}\right)10 + \frac{1}{2}\left(\frac{20}{3EI}\right)10\right]$$

$$\theta_B = \frac{13.33M}{EI} + (-0.0682M)\left[\frac{16.7}{EI} + \frac{50}{EI} + \frac{100}{EI}\right]$$

$$= \frac{13.33M}{EI} - \frac{11.37M}{EI} = \frac{1.96M}{EI}$$

$$\text{or} \quad M = \frac{EI\theta_B}{1.96} = \frac{EI\theta_B}{1.96} \cdot \frac{L}{L} = \frac{EI(1^{\text{radian}}) \cdot 20}{1.96L} = 10.2\frac{EI}{L}$$

where $k_{BA} = 10.2$

Fixed-End Moments Due to Uniform Load (w)

Using superposition, the fixed-fixed beam under uniform load can be expressed as

416 Moment Distribution

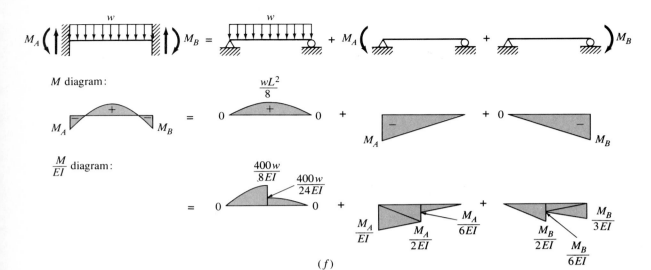

(f)

The relative slopes between the tangents drawn to the elastic curve at A and B is zero. Thus, using the M/EI diagrams above and the first moment-area principle,

$$\theta_A - \theta_B = 0 = \frac{2}{3}\left(\frac{400w}{8EI}\right)10 + \frac{2}{3}\left(\frac{400w}{24EI}\right)10 - \frac{1}{2}\left(\frac{M_A}{EI}\right)10 - \frac{1}{2}\left(\frac{M_A}{2EI}\right)10 - \frac{1}{2}\left(\frac{M_A}{6EI}\right)10$$

$$- \frac{1}{2}\left(\frac{M_B}{2EI}\right)10 - \frac{1}{2}\left(\frac{M_B}{6EI}\right)10 - \frac{1}{2}\left(\frac{M_B}{3EI}\right)10$$

which reduces to

$$\frac{444w}{EI} - \frac{8.33 M_A}{EI} - \frac{5 M_B}{EI} = 0 \qquad (1)$$

A tangent drawn from either A or B to the elastic curve at the opposite end results in zero relative displacement. Thus, the moment of the M/EI diagram between A and B about either end is zero, that is, from moment summation about end A:

$$\delta_{AtB} = 0 = \left[\frac{2}{3}\left(\frac{400w}{8EI}\right)10\right]6.25 + \left[\frac{2}{3}\left(\frac{400w}{24EI}\right)10\right]13.75 - \left[\frac{1}{2}\left(\frac{M_A}{EI}\right)10\right]3.33$$

$$- \left[\frac{1}{2}\left(\frac{M_A}{2EI}\right)10\right]6.67 - \left[\frac{1}{2}\left(\frac{M_A}{6EI}\right)10\right]13.33 - \left[\frac{1}{2}\left(\frac{M_B}{2EI}\right)10\right]6.67$$

$$- \left[\frac{1}{2}\left(\frac{M_B}{6EI}\right)10\right]13.33 - \left[\frac{1}{2}\left(\frac{M_B}{3EI}\right)10\right]16.7$$

which reduces to

$$\frac{3611w}{EI} - \frac{44.5 M_A}{EI} - \frac{55.6 M_B}{EI} = 0 \qquad (2)$$

Solving Eqs. (1) and (2) simultaneously, we arrive at

$M_A = 27.7w$ and $M_B = 42.7w$

which are the FEMs at A and B, respectively. We know from Table 12.1 that the FEM for a prismatic uniformly loaded beam is $= \pm wL^2/12 = \pm 0.0833wL^2$. Thus, we can express the FEMs at ends A and B in similar form as follows:

$$\text{FEM}_{AB} = M_A = 27.7w = 27.7w \cdot \frac{L^2}{L^2} = \frac{27.7wL^2}{(20)^2} = 0.069wL^2$$

and $$\text{FEM}_{BA} = M_B = 42.7w = 42.7w \cdot \frac{L^2}{L^2} = \frac{42.7wL^2}{(20)^2} = 0.107wL^2$$

Absolute stiffness is defined as the moment applied at the simple end of a pin-fixed member that will produce a unit rotation at the simply supported end. Let k_{AB} = stiffness factor of member AB at end A where absolute stiffness = $k_{AB} \cdot EI_C/L$ and I_C is the moment of inertia at minimum depth. Let k_{BA} equal the stiffness factor of member AB at end B.

For any prismatic member, we have found that the carryover factor is $+\frac{1}{2}$ and the absolute stiffness is equal to $4EI/L$. In Example 13.13, the carryover factor and the absolute stiffness are found to have different values at each end due to the unsymmetric nature of the nonprismatic beam. Recall that the FEM at both ends of a prismatic beam under uniform loading w has a known absolute value of $wL^2/12$ (refer to Table 12.1). Observe that the FEM at one end is different from the FEM at the other end of the nonprismatic beam. Distribution factors are determined in the usual manner. It should be noted that this effort is required for each nonprismatic member in a structure before moment distribution can begin.

It is evident from Example 13.13 that a significant amount of preparation work is required to perform a moment-distribution analysis of a nonprismatic structure. In 1958, the Portland Cement Association issued the publication, *Handbook of Frame Constants* which contains a series of tables that list carryover factors, stiffness factors, and FEM coefficients for commonly used nonprismatic shapes. The PCA tables are established for constant width beams that have either straight, parabolic, or prismatic haunches as shown in representative Tables 13.1, 13.2, and 13.3. The PCA handbook table nomenclature is generally understood from the figures shown above the tables. The terms r_A and r_B equal ratios for rectangular cross sections, where $r_A = (h_A - h_C)/h_C$ and $r_B = (h_B - h_C)/h_C$. M_{AB} and M_{BA} are the fixed-end moments at ends A and B, respectively, for uniform load w or concentrated force P.

TABLE 13.1
Straight Haunches—Constant Width*

Right Haunch		Carryover Factors		Stiffness Factors		Unif. Load FEM Coef. × wL^2		Concentrated load FEM—Coef. × PL										Haunch Load at			
																		Left		Right	
																		FEM Coef. × $W_A L^2$		FEM Coef. × $W_B L^2$	
									b												
							0.1		0.3		0.5		0.7		0.9						
a_B	r_B	C_{AB}	C_{BA}	k_{AB}	k_{BA}	M_{AB}	M_{BA}	M_{AB}	M_{BA}	M_{AB}	M_{BA}	M_{AB}	M_{BA}	M_{AB}	M_{BA}	M_{AB}	M_{BA}	M_{AB}	M_{BA}	M_{AB}	M_{BA}
		$a_A = 0.5$					a_B = variable				$r_A = 2.0$						r_B = variable					
0.1	0.4	0.390	1.230	22.23	7.05	0.1657	0.0555	0.0954	0.0017	0.2491	0.0186	0.3046	0.0628	0.1737	0.1199	0.0190	0.0836	0.0357	0.0022	0.0001	0.0016
	0.6	0.403	1.226	22.85	7.51	0.1633	0.0584	0.0953	0.0019	0.2478	0.0197	0.3003	0.0665	0.1661	0.1265	0.0154	0.0867	0.0355	0.0023	0.0001	0.0016
	1.0	0.420	1.219	23.66	8.15	0.1587	0.0625	0.0951	0.0020	0.2458	0.0215	0.2941	0.0720	0.1554	0.1358	0.0107	0.0907	0.0353	0.0025	0.0001	0.0016
	1.5	0.433	1.215	24.39	8.69	0.1551	0.0656	0.0950	0.0021	0.2445	0.0226	0.2894	0.0760	0.1468	0.1432	0.0071	0.0938	0.0351	0.0026	0.0000	0.0016
	2.0	0.441	1.212	24.83	9.04	0.1527	0.0677	0.0949	0.0022	0.2436	0.0234	0.2863	0.0788	0.1414	0.1480	0.0050	0.0955	0.0350	0.0027	0.0000	0.0016
0.2	0.4	0.424	1.194	23.57	8.37	0.1614	0.0606	0.0952	0.0020	0.2463	0.0213	0.2957	0.0716	0.1606	0.1325	0.0187	0.0836	0.0353	0.0025	0.0013	0.0056
	0.6	0.450	1.179	24.85	9.49	0.1555	0.0660	0.0949	0.0023	0.2436	0.0228	0.2870	0.0796	0.1466	0.1454	0.0148	0.0869	0.0350	0.0028	0.0010	0.0058
	1.0	0.488	1.157	26.88	11.34	0.1464	0.0745	0.0945	0.0026	0.2393	0.0279	0.2732	0.0926	0.1248	0.1657	0.0096	0.0913	0.0345	0.0032	0.0007	0.0060
	1.5	0.519	1.141	28.73	13.06	0.1385	0.0820	0.0941	0.0030	0.2355	0.0316	0.2609	0.1044	0.1060	0.1836	0.0060	0.0945	0.0341	0.0037	0.0005	0.0052
	2.0	0.539	1.130	30.02	14.31	0.1332	0.0872	0.0939	0.0033	0.2328	0.0342	0.2524	0.1128	0.0931	0.1961	0.0040	0.0964	0.0337	0.0040	0.0003	0.0064
0.3	0.4	0.449	1.147	24.37	9.54	0.1594	0.0625	0.0950	0.0022	0.2447	0.0232	0.2908	0.0769	0.1561	0.1362	0.0193	0.0825	0.0352	0.0027	0.0042	0.0110
	0.6	0.488	1.117	26.18	11.43	0.1521	0.0694	0.0946	0.0026	0.2409	0.0270	0.2791	0.0886	0.1393	0.1520	0.0156	0.0854	0.0347	0.0031	0.0036	0.0115
	1.0	0.548	1.076	29.43	14.99	0.1401	0.0814	0.0940	0.0031	0.2342	0.0339	0.2587	0.1095	0.1116	0.1790	0.0106	0.0898	0.0339	0.0039	0.0026	0.0124
	1.5	0.601	1.043	32.75	18.88	0.1285	0.0933	0.0934	0.0039	0.2274	0.0412	0.2381	0.1317	0.0858	0.2053	0.0070	0.0930	0.0331	0.0048	0.0018	0.0132
	2.0	0.638	1.022	35.37	22.09	0.1199	0.1024	0.0929	0.0046	0.2220	0.0471	0.2221	0.1495	0.0671	0.2249	0.0049	0.0951	0.0325	0.0054	0.0013	0.0137
0.4	0.4	0.462	1.099	24.84	10.44	0.1584	0.0624	0.0949	0.0023	0.2437	0.0242	0.2882	0.0789	0.1542	0.1336	0.0200	0.0810	0.0350	0.0028	0.0096	0.0167
	0.6	0.510	1.053	26.98	13.06	0.1507	0.0697	0.0945	0.0028	0.2394	0.0288	0.2748	0.0927	0.1373	0.1484	0.0165	0.0837	0.0345	0.0033	0.0084	0.0178
	1.0	0.591	0.989	31.05	18.56	0.1376	0.0831	0.0937	0.0037	0.2311	0.0380	0.2500	0.1196	0.1099	0.1743	0.0118	0.0877	0.0336	0.0044	0.0065	0.0195
	1.5	0.670	0.939	35.71	25.51	0.1241	0.0978	0.0928	0.0048	0.2218	0.0491	0.2229	0.1511	0.0843	0.2005	0.0082	0.0911	0.0325	0.0057	0.0048	0.0211
	2.0	0.730	0.905	39.81	32.10	0.1134	0.1103	0.0919	0.0058	0.2136	0.0593	0.1997	0.1793	0.0655	0.2208	0.0060	0.0933	0.0316	0.0068	0.0036	0.0224
0.5	0.4	0.462	1.058	25.34	11.05	0.1566	0.0617	0.0949	0.0023	0.2426	0.0245	0.2839	0.0790	0.1511	0.1306	0.0203	0.0801	0.0348	0.0029	0.0173	0.0222
	0.6	0.513	0.997	27.69	14.23	0.1485	0.0686	0.0943	0.0028	0.2377	0.0295	0.2689	0.0929	0.1341	0.1438	0.0168	0.0825	0.0344	0.0034	0.0153	0.0237
	1.0	0.605	0.910	32.23	21.45	0.1349	0.0817	0.0934	0.0039	0.2286	0.0398	0.2421	0.1208	0.1076	0.1666	0.0122	0.0861	0.0333	0.0046	0.0123	0.0263
	1.5	0.705	0.838	37.58	31.64	0.1213	0.0967	0.0924	0.0053	0.2180	0.0533	0.2128	0.1551	0.0838	0.1899	0.0088	0.0892	0.0321	0.0061	0.0096	0.0289
	2.0	0.789	0.789	42.61	42.61	0.1103	0.1103	0.0914	0.0067	0.2083	0.0667	0.1875	0.1875	0.0667	0.2083	0.0067	0.0914	0.0310	0.0075	0.0075	0.0310

419 Nonprismatic Beams

		$a_A = 0$			a_B = variable								$r_A = 0$					r_B = variable			
0.1	0.4	0.556	0.495	4.14	4.64	0.0780	0.0946	0.0804	0.0103	0.1426	0.1164	0.0724	0.1432	0.0534	0.1672	0.0052	0.0889	0.0000	0.0000	0.0000	0.0016
	0.6	0.573	0.495	4.19	4.85	0.0763	0.0981	0.0802	0.0108	0.1412	0.1137	0.0754	0.1490	0.0505	0.1735	0.0042	0.0911	0.0000	0.0000	0.0000	0.0016
	1.0	0.596	0.493	4.25	5.14	0.0741	0.1029	0.0799	0.0114	0.1393	0.1100	0.0795	0.1568	0.0464	0.1822	0.0028	0.0940	0.0000	0.0000	0.0000	0.0016
	1.5	0.613	0.492	4.30	5.36	0.0724	0.1056	0.0797	0.0118	0.1378	0.1072	0.0826	0.1629	0.0434	0.1887	0.0018	0.0960	0.0000	0.0000	0.0000	0.0016
	2.0	0.624	0.491	4.33	5.50	0.0714	0.1088	0.0795	0.0121	0.1369	0.1055	0.0846	0.1667	0.0415	0.1928	0.0013	0.0972	0.0000	0.0000	0.0000	0.0016
0.2	0.4	0.605	0.486	4.26	5.31	0.0747	0.1025	0.0798	0.0116	0.1391	0.1102	0.0806	0.1581	0.0476	0.1809	0.0050	0.0893	0.0000	0.0000	0.0003	0.0059
	0.6	0.642	0.481	4.35	5.81	0.0717	0.1093	0.0794	0.0125	0.1363	0.1049	0.0871	0.1701	0.0423	0.1929	0.0033	0.0917	0.0000	0.0000	0.0003	0.0061
	1.0	0.694	0.475	4.49	6.57	0.0673	0.1192	0.0788	0.0140	0.1321	0.0971	0.0968	0.1881	0.0346	0.2105	0.0023	0.0948	0.0000	0.0000	0.0002	0.0063
	1.5	0.736	0.470	4.61	7.22	0.0638	0.1274	0.0783	0.0152	0.1287	0.0908	0.1049	0.2030	0.0285	0.2247	0.0014	0.0969	0.0000	0.0000	0.0001	0.0064
	2.0	0.764	0.467	4.68	7.66	0.0616	0.1327	0.0780	0.0164	0.1264	0.0866	0.1104	0.2129	0.0246	0.2339	0.0009	0.0980	0.0000	0.0000	0.0001	0.0065
0.3	0.4	0.648	0.470	4.34	5.98	0.0730	0.1059	0.0795	0.0127	0.1368	0.1064	0.0870	0.1684	0.0453	0.1865	0.0051	0.0887	0.0000	0.0000	0.0012	0.0123
	0.6	0.704	0.461	4.48	6.84	0.0690	0.1162	0.0789	0.0141	0.1327	0.0990	0.0970	0.1862	0.0387	0.2017	0.0039	0.0911	0.0000	0.0000	0.0009	0.0128
	1.0	0.791	0.449	4.71	8.29	0.0630	0.1311	0.0779	0.0166	0.1262	0.0874	0.1134	0.2150	0.0289	0.2252	0.0024	0.0943	0.0000	0.0000	0.0006	0.0135
	1.5	0.866	0.439	4.91	9.68	0.0577	0.1442	0.0770	0.0190	0.1203	0.0771	0.1286	0.2412	0.0208	0.2452	0.0015	0.0965	0.0000	0.0000	0.0004	0.0140
	2.0	0.918	0.433	5.06	10.72	0.0542	0.1534	0.0764	0.0210	0.1161	0.0700	0.1397	0.2599	0.0155	0.2585	0.0010	0.0977	0.0000	0.0000	0.0003	0.0143
0.4	0.4	0.679	0.453	4.39	6.59	0.0722	0.1084	0.0793	0.0133	0.1355	0.1044	0.0911	0.1734	0.0449	0.1849	0.0053	0.0877	0.0000	0.0000	0.0027	0.0199
	0.6	0.754	0.438	4.57	7.86	0.0678	0.1192	0.0785	0.0153	0.1305	0.0958	0.1041	0.1950	0.0383	0.2001	0.0042	0.0900	0.0000	0.0000	0.0023	0.0209
	1.0	0.879	0.418	4.87	10.24	0.0607	0.1376	0.0772	0.0190	0.1219	0.0815	0.1273	0.2327	0.0283	0.2241	0.0028	0.0931	0.0000	0.0000	0.0016	0.0224
	1.5	0.996	0.403	5.18	12.82	0.0541	0.1554	0.0758	0.0228	0.1134	0.0679	0.1513	0.2704	0.0199	0.2453	0.0018	0.0954	0.0000	0.0000	0.0011	0.0237
	2.0	1.082	0.392	5.42	14.94	0.0494	0.1688	0.0748	0.0258	0.1070	0.0578	0.1702	0.2993	0.0144	0.2597	0.0012	0.0968	0.0000	0.0000	0.0009	0.0244
0.5	0.4	0.697	0.434	4.43	7.12	0.0718	0.1079	0.0791	0.0137	0.1346	0.1032	0.0930	0.1733	0.0448	0.1812	0.0055	0.0869	0.0000	0.0000	0.0051	0.0280
	0.6	0.788	0.413	4.52	8.81	0.0672	0.1191	0.0783	0.0161	0.1291	0.0941	0.1079	0.1958	0.0384	0.1950	0.0044	0.0890	0.0000	0.0000	0.0044	0.0295
	1.0	0.948	0.385	4.99	12.28	0.0597	0.1390	0.0767	0.0208	0.1192	0.0788	0.1364	0.2371	0.0288	0.2175	0.0030	0.0919	0.0000	0.0000	0.0033	0.0321
	1.5	1.114	0.363	5.39	16.52	0.0524	0.1599	0.0749	0.0263	0.1087	0.0636	0.1688	0.2812	0.0207	0.2382	0.0020	0.0943	0.0000	0.0000	0.0024	0.0344
	2.0	1.245	0.349	5.73	20.42	0.0458	0.1770	0.0735	0.0311	0.1001	0.0519	0.1969	0.3174	0.0153	0.2529	0.0014	0.0958	0.0000	0.0000	0.0017	0.0365
0.75	0.4	0.691	0.393	4.56	8.02	0.0692	0.1041	0.0785	0.0139	0.1297	0.0972	0.0915	0.1648	0.0430	0.1737	0.0054	0.0856	0.0000	0.0000	0.0143	0.0462
	0.6	0.788	0.359	4.80	10.65	0.0641	0.1134	0.0774	0.0165	0.1227	0.0871	0.1058	0.1824	0.0369	0.1837	0.0045	0.0871	0.0000	0.0000	0.0126	0.0491
	1.0	0.982	0.311	5.25	16.60	0.0562	0.1302	0.0755	0.0222	0.1108	0.0717	0.1343	0.2137	0.0282	0.2002	0.0032	0.0894	0.0000	0.0000	0.0101	0.0540
	1.5	1.225	0.272	5.76	25.92	0.0491	0.1490	0.0733	0.0238	0.0986	0.0580	0.1694	0.2471	0.0212	0.2160	0.0022	0.0915	0.0000	0.0000	0.0081	0.0589
	2.0	1.461	0.247	6.24	36.95	0.0437	0.1659	0.0713	0.0380	0.0886	0.0481	0.2037	0.2756	0.0167	0.2281	0.0017	0.0928	0.0000	0.0000	0.0066	0.0629
1.00	0.4	0.642	0.388	5.17	8.57	0.0675	0.1011	0.0766	0.0139	0.1243	0.0953	0.0885	0.1583	0.0434	0.1689	0.0055	0.0850	0.0000	0.0000	0.0256	0.0586
	0.6	0.709	0.350	5.74	11.63	0.0618	0.1086	0.0744	0.0168	0.1154	0.0850	0.1001	0.1717	0.0375	0.1766	0.0048	0.0858	0.0000	0.0000	0.0230	0.0621
	1.0	0.834	0.294	6.86	19.46	0.0529	0.1216	0.0706	0.0224	0.1005	0.0691	0.1221	0.1951	0.0289	0.1893	0.0035	0.0877	0.0000	0.0000	0.0190	0.0680
	1.5	0.981	0.247	8.23	32.69	0.0450	0.1352	0.0664	0.0236	0.0860	0.0555	0.1475	0.2184	0.0221	0.2010	0.0026	0.0893	0.0000	0.0000	0.0156	0.0738
	2.0	1.119	0.214	9.57	50.13	0.0392	0.1466	0.0527	0.0370	0.0752	0.0460	0.1682	0.2371	0.0176	0.2097	0.0020	0.0905	0.0000	0.0000	0.0131	0.0786
		$a_A = 0$			$a_B = 0$									$r_A = 0$					$r_B = 0$			
0.00	0.0	0.500	0.500	4.00	4.00	0.0833	0.0833	0.0810	0.0090	0.1470	0.1250	0.0630	0.1250	0.0630	0.1470	0.0090	0.0810	0.0000	0.0000	0.0000	0.0000

*Note: All carryover factors and fixed-end moment coefficients are negative and all stiffness factors are positive.

TABLE 13.2
Parabolic Haunches—Constant Width*

Concentrated load FEM—Coef. × PL

Right Haunch				Carryover Factors		Stiffness Factors		Unif. Load FEM Coef. × wL²		0.1		0.3		0.5		0.7		0.9		Haunch Load at Left FEM Coef. × $W_A L^2$		Right FEM Coef. × $W_B L^2$	
a_B	r_B	C_{AB}	C_{BA}	k_{AB}	k_{BA}	M_{AB}	M_{BA}	M_{AB}	M_{BA}	M_{AB}	M_{BA}	M_{AB}	M_{BA}	M_{AB}	M_{BA}	M_{AB}	M_{BA}	M_{AB}	M_{BA}	M_{AB}	M_{BA}
		$a_A = 0.3$																r_B = variable					
0.1	0.4	0.506	0.702	7.03	5.07	0.1147	0.0746	0.0941	0.0029	0.2069	0.0392	0.1824	0.1079	0.0884	0.1463	0.0096	0.0856	0.0069	0.0003	0.0000	0.0008
	0.6	0.518	0.701	7.11	5.25	0.1131	0.0769	0.0940	0.0030	0.2060	0.0406	0.1799	0.1115	0.0852	0.1510	0.0082	0.0876	0.0069	0.0003	0.0000	0.0008
	1.0	0.534	0.699	7.23	5.52	0.1108	0.0803	0.0939	0.0031	0.2047	0.0425	0.1764	0.1168	0.0806	0.1578	0.0063	0.0905	0.0069	0.0003	0.0000	0.0008
	1.5	0.547	0.698	7.33	5.75	0.1090	0.0831	0.0938	0.0033	0.2036	0.0441	0.1735	0.1212	0.0770	0.1634	0.0048	0.0927	0.0068	0.0003	0.0000	0.0008
	2.0	0.556	0.697	7.40	5.91	0.1078	0.0849	0.0938	0.0033	0.2029	0.0453	0.1715	0.1242	0.0745	0.1672	0.0039	0.0941	0.0068	0.0003	0.0000	0.0008
0.2	0.4	0.539	0.693	7.25	5.64	0.1113	0.0799	0.0939	0.0032	0.2046	0.0429	0.1765	0.1174	0.0815	0.1573	0.0084	0.0873	0.0068	0.0003	0.0002	0.0030
	0.6	0.563	0.689	7.42	6.06	0.1084	0.0844	0.0938	0.0034	0.2028	0.0458	0.1716	0.1249	0.0756	0.1664	0.0067	0.0898	0.0068	0.0003	0.0002	0.0031
	1.0	0.598	0.683	7.68	6.73	0.1042	0.0911	0.0936	0.0037	0.2000	0.0502	0.1643	0.1365	0.0669	0.1801	0.0044	0.0932	0.0068	0.0004	0.0001	0.0031
	1.5	0.628	0.678	7.92	7.33	0.1004	0.0970	0.0934	0.0040	0.1975	0.0542	0.1579	0.1467	0.0594	0.1920	0.0029	0.0956	0.0068	0.0004	0.0001	0.0032
	2.0	0.649	0.674	8.08	7.77	0.0980	0.1010	0.0932	0.0042	0.1957	0.0571	0.1534	0.1541	0.0542	0.2003	0.0019	0.0970	0.0068	0.0004	0.0001	0.0032
0.3	0.4	0.568	0.681	7.42	6.19	0.1093	0.0834	0.0938	0.0034	0.2029	0.0460	0.1722	0.1248	0.0774	0.1640	0.0084	0.0872	0.0068	0.0003	0.0008	0.0063
	0.6	0.604	0.672	7.68	6.90	0.1054	0.0896	0.0936	0.0038	0.2002	0.0505	0.1653	0.1360	0.0697	0.1764	0.0066	0.0898	0.0068	0.0004	0.0006	0.0065
	1.0	0.660	0.660	8.10	8.10	0.0994	0.0994	0.0932	0.0043	0.1958	0.0578	0.1543	0.1543	0.0578	0.1958	0.0043	0.0932	0.0068	0.0004	0.0004	0.0068
	1.5	0.710	0.649	8.49	9.28	0.0940	0.1085	0.0929	0.0049	0.1917	0.0648	0.1442	0.1716	0.0473	0.2133	0.0027	0.0957	0.0068	0.0005	0.0003	0.0070
	2.0	0.745	0.642	8.79	10.22	0.0903	0.1149	0.0927	0.0053	0.1886	0.0702	0.1366	0.1846	0.0399	0.2259	0.0018	0.0971	0.0067	0.0005	0.0002	0.0071
0.4	0.4	0.592	0.666	7.55	6.71	0.1081	0.0854	0.0937	0.0036	0.2016	0.0484	0.1693	0.1299	0.0754	0.1666	0.0085	0.0867	0.0068	0.0004	0.0018	0.0103
	0.6	0.640	0.652	7.88	7.73	0.1035	0.0928	0.0934	0.0041	0.1982	0.0543	0.1609	0.1441	0.0669	0.1802	0.0068	0.0893	0.0068	0.0004	0.0015	0.0108
	1.0	0.717	0.632	8.45	9.58	0.0963	0.1050	0.0929	0.0049	0.1924	0.0646	0.1468	0.1685	0.0533	0.2027	0.0045	0.0927	0.0068	0.0005	0.0010	0.0116
	1.5	0.789	0.615	9.03	11.58	0.0896	0.1168	0.0925	0.0057	0.1866	0.0751	0.1332	0.1931	0.0413	0.2231	0.0029	0.0952	0.0067	0.0006	0.0008	0.0119
	2.0	0.843	0.603	9.50	13.28	0.0846	0.1258	0.0921	0.0064	0.1820	0.0838	0.1226	0.2127	0.0329	0.2379	0.0020	0.0967	0.0067	0.0007	0.0006	0.0123
0.5	0.4	0.609	0.648	7.62	7.18	0.1077	0.0867	0.0936	0.0038	0.2015	0.0501	0.1674	0.1329	0.0748	0.1667	0.0086	0.0862	0.0068	0.0004	0.0032	0.0149
	0.6	0.669	0.630	8.00	8.50	0.1024	0.0944	0.0932	0.0044	0.1967	0.0570	0.1578	0.1489	0.0655	0.1817	0.0069	0.0888	0.0068	0.0004	0.0028	0.0156
	1.0	0.766	0.603	8.74	11.12	0.0943	0.1085	0.0926	0.0055	0.1899	0.0703	0.1414	0.1779	0.0515	0.2036	0.0046	0.0922	0.0067	0.0005	0.0022	0.0167
	1.5	0.861	0.578	9.50	14.15	0.0863	0.1219	0.0920	0.0066	0.1821	0.0843	0.1251	0.2084	0.0392	0.2245	0.0030	0.0947	0.0067	0.0007	0.0016	0.0177
	2.0	0.938	0.560	10.15	16.90	0.0803	0.1329	0.0915	0.0076	0.1760	0.0968	0.1122	0.2339	0.0306	0.2399	0.0021	0.0963	0.0066	0.0008	0.0013	0.0184
		$a_A = 0.4$																r_B = variable					
0.1	0.4	0.485	0.764	8.25	5.23	0.1213	0.0718	0.0937	0.0030	0.2136	0.0358	0.1982	0.1016	0.0979	0.1426	0.0107	0.0851	0.0118	0.0007	0.0000	0.0008
	0.6	0.496	0.763	8.40	5.43	0.1196	0.0742	0.0936	0.0031	0.2132	0.0371	0.1956	0.1052	0.0945	0.1474	0.0092	0.0873	0.0118	0.0007	0.0000	0.0008
	1.0	0.512	0.761	8.51	5.72	0.1172	0.0775	0.0935	0.0032	0.2118	0.0389	0.1919	0.1103	0.0895	0.1542	0.0070	0.0902	0.0118	0.0007	0.0000	0.0008
	1.5	0.525	0.760	8.63	5.97	0.1152	0.0803	0.0934	0.0034	0.2107	0.0405	0.1888	0.1146	0.0855	0.1598	0.0054	0.0924	0.0118	0.0008	0.0000	0.0008

421 Nonprismatic Beams

		$a_A = 0.5$																			
0.3	0.6	0.540	8.74	0.750	6.29	0.1147	0.0815	0.0933	0.0035	0.2099	0.1869	0.0420	0.1182	0.0842	0.1628	0.0075	0.0895	0.0118	0.0008	0.0002	0.0031
	1.0	0.574	9.07	0.743	7.00	0.1100	0.0882	0.0931	0.0039	0.2070	0.1791	0.0462	0.1295	0.0745	0.1766	0.0050	0.0930	0.0117	0.0008	0.0001	0.0031
	1.5	0.602	9.36	0.737	7.65	0.1060	0.0941	0.0928	0.0042	0.2045	0.1723	0.0500	0.1395	0.0662	0.1887	0.0032	0.0954	0.0117	0.0009	0.0001	0.0032
	2.0	0.622	9.58	0.733	8.13	0.1033	0.0981	0.0927	0.0044	0.2026	0.1674	0.0527	0.1468	0.0605	0.1972	0.0022	0.0969	0.0116	0.0010	0.0001	0.0032
0.4	0.4	0.544	8.73	0.740	6.42	0.1156	0.0805	0.0933	0.0035	0.2100	0.1874	0.0422	0.1180	0.0861	0.1603	0.0094	0.0867	0.0118	0.0008	0.0009	0.0062
	0.6	0.579	9.06	0.731	7.17	0.1116	0.0866	0.0931	0.0039	0.2072	0.1803	0.0463	0.1289	0.0777	0.1727	0.0074	0.0894	0.0117	0.0009	0.0007	0.0065
	1.0	0.632	9.58	0.717	8.45	0.1050	0.0963	0.0927	0.0045	0.2027	0.1685	0.0533	0.1468	0.0646	0.1924	0.0049	0.0929	0.0116	0.0010	0.0005	0.0068
	1.5	0.680	10.09	0.705	9.73	0.0991	0.1054	0.0923	0.0051	0.1985	0.1577	0.0600	0.1639	0.0530	0.2101	0.0031	0.0955	0.0115	0.0011	0.0003	0.0070
	2.0	0.714	10.47	0.696	10.74	0.0951	0.1118	0.0920	0.0055	0.1953	0.1496	0.0652	0.1769	0.0448	0.2231	0.0021	0.0969	0.0115	0.0012	0.0002	0.0071
0.4	0.4	0.567	8.89	0.723	6.96	0.1143	0.0824	0.0932	0.0037	0.2087	0.1843	0.0444	0.1228	0.0839	0.1628	0.0095	0.0863	0.0117	0.0008	0.0020	0.0102
	0.6	0.612	9.30	0.708	8.04	0.1095	0.0896	0.0929	0.0042	0.2049	0.1756	0.0500	0.1367	0.0749	0.1767	0.0076	0.0888	0.0117	0.0009	0.0017	0.0107
	1.0	0.686	10.02	0.686	10.02	0.1017	0.1017	0.0924	0.0051	0.1992	0.1607	0.0597	0.1607	0.0597	0.1992	0.0051	0.0924	0.0116	0.0011	0.0011	0.0116
	1.5	0.755	10.75	0.667	12.17	0.0944	0.1136	0.0918	0.0060	0.1932	0.1460	0.0698	0.1850	0.0465	0.2200	0.0033	0.0950	0.0114	0.0013	0.0009	0.0119
	2.0	0.807	11.34	0.653	14.01	0.0891	0.1226	0.0914	0.0067	0.1884	0.1346	0.0782	0.2045	0.0371	0.2352	0.0022	0.0965	0.0114	0.0015	0.0007	0.0122
0.5	0.4	0.586	8.95	0.703	7.43	0.1137	0.0834	0.0930	0.0039	0.2080	0.1824	0.0461	0.1255	0.0833	0.1627	0.0098	0.0858	0.0116	0.0009	0.0036	0.0148
	0.6	0.639	9.43	0.681	8.85	0.1083	0.0908	0.0926	0.0045	0.2039	0.1724	0.0526	0.1410	0.0732	0.1778	0.0077	0.0882	0.0115	0.0012	0.0031	0.0155
	1.0	0.731	10.37	0.653	11.62	0.0994	0.1050	0.0920	0.0056	0.1965	0.1551	0.0652	0.1698	0.0578	0.1999	0.0052	0.0918	0.0114	0.0013	0.0024	0.0165
	1.5	0.823	11.28	0.624	14.95	0.0911	0.1185	0.0913	0.0068	0.1888	0.1374	0.0785	0.1996	0.0441	0.2213	0.0034	0.0944	0.0112	0.0015	0.0018	0.0176
	2.0	0.895	12.10	0.606	17.90	0.0847	0.1295	0.0907	0.0079	0.1823	0.1233	0.0906	0.2251	0.0346	0.2371	0.0024	0.0960	0.0111	0.0018	0.0014	0.0183

		a_B = variable														$r_A = 1.0$				r_B = variable		
0.1	0.4	0.458	9.56	0.818	5.36	0.1253	0.0698	0.0932	0.0031	0.2152	0.2085	0.0340	0.0971	0.1060	0.1398	0.0120	0.0846	0.0171	0.0017	0.0000	0.0008
	0.6	0.471	9.67	0.817	5.57	0.1236	0.0725	0.0932	0.0032	0.2140	0.2059	0.0357	0.1008	0.1024	0.1447	0.0102	0.0870	0.0170	0.0018	0.0000	0.0008
	1.0	0.489	9.84	0.816	5.88	0.1215	0.0757	0.0931	0.0033	0.2129	0.2023	0.0378	0.1057	0.0975	0.1512	0.0077	0.0899	0.0170	0.0018	0.0000	0.0008
	1.5	0.499	10.00	0.814	6.11	0.1192	0.0787	0.0930	0.0035	0.2118	0.1990	0.0392	0.1099	0.0928	0.1571	0.0059	0.0922	0.0169	0.0019	0.0000	0.0008
	2.0	0.506	10.10	0.812	6.30	0.1175	0.0805	0.0928	0.0037	0.2115	0.1968	0.0396	0.1127	0.0900	0.1610	0.0047	0.0937	0.0168	0.0019	0.0000	0.0008
0.2	0.4	0.488	9.85	0.807	5.97	0.1214	0.0753	0.0929	0.0034	0.2131	0.2021	0.0371	0.1061	0.0979	0.1506	0.0105	0.0863	0.0171	0.0017	0.0003	0.0030
	0.6	0.515	10.10	0.803	6.45	0.1183	0.0795	0.0928	0.0036	0.2110	0.1969	0.0404	0.1136	0.0917	0.1600	0.0083	0.0892	0.0170	0.0018	0.0002	0.0030
	1.0	0.547	10.51	0.796	7.22	0.1138	0.0865	0.0926	0.0040	0.2079	0.1890	0.0448	0.1245	0.0809	0.1740	0.0056	0.0928	0.0168	0.0020	0.0001	0.0031
	1.5	0.571	10.90	0.786	7.90	0.1093	0.0922	0.0923	0.0043	0.2055	0.1818	0.0485	0.1344	0.0719	0.1862	0.0035	0.0951	0.0167	0.0021	0.0001	0.0032
	2.0	0.590	11.17	0.784	8.40	0.1063	0.0961	0.0922	0.0046	0.2041	0.1764	0.0506	0.1417	0.0661	0.1948	0.0025	0.0968	0.0166	0.0022	0.0001	0.0032
0.3	0.4	0.513	10.08	0.793	6.55	0.1191	0.0786	0.0928	0.0037	0.2111	0.1974	0.0400	0.1132	0.0932	0.1576	0.0103	0.0863	0.0170	0.0018	0.0009	0.0062
	0.6	0.551	10.47	0.784	7.35	0.1150	0.0848	0.0926	0.0040	0.2083	0.1901	0.0446	0.1241	0.0842	0.1697	0.0082	0.0890	0.0168	0.0020	0.0008	0.0064
	1.0	0.603	11.12	0.766	8.74	0.1085	0.0943	0.0922	0.0046	0.2036	0.1779	0.0515	0.1414	0.0703	0.1899	0.0055	0.0926	0.0167	0.0022	0.0005	0.0067
	1.5	0.642	11.74	0.751	10.08	0.1021	0.1036	0.0917	0.0053	0.1995	0.1665	0.0581	0.1583	0.0576	0.2079	0.0035	0.0953	0.0165	0.0024	0.0003	0.0069
	2.0	0.677	12.25	0.744	11.15	0.0977	0.1100	0.0914	0.0057	0.1965	0.1578	0.0628	0.1714	0.0485	0.2208	0.0024	0.0969	0.0163	0.0026	0.0003	0.0071
0.4	0.4	0.537	10.30	0.774	7.17	0.1178	0.0802	0.0927	0.0039	0.2097	0.1946	0.0422	0.1178	0.0906	0.1600	0.0105	0.0859	0.0169	0.0018	0.0020	0.0102
	0.6	0.583	10.80	0.759	8.28	0.1128	0.0874	0.0924	0.0043	0.2061	0.1853	0.0482	0.1317	0.0809	0.1738	0.0084	0.0883	0.0167	0.0020	0.0019	0.0107
	1.0	0.653	11.62	0.731	10.37	0.1050	0.0994	0.0918	0.0052	0.1999	0.1698	0.0578	0.1551	0.0652	0.1965	0.0056	0.0920	0.0165	0.0022	0.0013	0.0114
	1.5	0.717	12.50	0.710	12.60	0.0971	0.1117	0.0911	0.0062	0.1941	0.1543	0.0677	0.1792	0.0506	0.2178	0.0037	0.0948	0.0163	0.0024	0.0010	0.0118
	2.0	0.764	13.25	0.698	14.55	0.0915	0.1206	0.0907	0.0070	0.1894	0.1420	0.0756	0.1987	0.0403	0.2330	0.0024	0.0963	0.0161	0.0027	0.0008	0.0122
0.5	0.4	0.554	10.42	0.753	7.66	0.1170	0.0811	0.0926	0.0040	0.2087	0.1924	0.0442	0.1205	0.0898	0.1595	0.0107	0.0853	0.0169	0.0020	0.0022	0.0145
	0.6	0.606	10.96	0.730	9.12	0.1115	0.0889	0.0922	0.0046	0.2045	0.1820	0.0506	0.1360	0.0791	0.1738	0.0086	0.0878	0.0167	0.0022	0.0036	0.0152
	1.0	0.694	12.03	0.694	12.03	0.1025	0.1025	0.0915	0.0057	0.1970	0.1639	0.0626	0.1639	0.0626	0.1970	0.0057	0.0915	0.0164	0.0028	0.0028	0.0164
	1.5	0.781	13.12	0.664	15.47	0.0937	0.1163	0.0908	0.0070	0.1891	0.1456	0.0759	0.1939	0.0479	0.2187	0.0039	0.0940	0.0160	0.0034	0.0021	0.0174
	2.0	0.850	14.09	0.642	18.64	0.0870	0.1275	0.0901	0.0082	0.1825	0.1307	0.0877	0.2193	0.0376	0.2348	0.0027	0.0957	0.0157	0.0039	0.0016	0.0181

* Note: All carryover factors and fixed-end moment coefficients are negative and all stiffness factors are positive.

TABLE 13.3a
Prismatic Haunch at One End*

Right Haunch		Carryover Factors		Stiffness Factors		Unif. Load FEM Coef. × wL^2		Concentrated Load FEM—Coef. × PL													Moment, M at $b = (1 - a_B)$ FEM Coef. × M		Haunch Load, Both Haunches FEM Coef. × $W_B L^2$	
								b																
								0.1		0.3		0.5		0.7		0.9		$1 - a_B$						
a_B	r_B	C_{AB}	C_{BA}	k_{AB}	k_{BA}	M_{AB}	M_{BA}	M_{AB}	M_{BA}	M_{AB}	M_{BA}	M_{AB}	M_{BA}	M_{AB}	M_{BA}	M_{AB}	M_{BA}	M_{AB}	M_{BA}	M_{AB}	M_{BA}	M_{AB}	M_{BA}
												$r_A = 0$												
												$a_A = 0$												
0.1	0.4	0.593	0.491	4.24	5.12	0.0749	0.1016	0.0799	0.0113	0.1397	0.0788	0.1110	0.1553	0.0478	0.1798	0.0042	0.0911	0.0042	0.0911	0.0793	0.8275	0.0001	0.0047
	0.6	0.615	0.490	4.30	5.40	0.0727	0.1062	0.0797	0.0119	0.1378	0.0828	0.1074	0.1630	0.0439	0.1881	0.0029	0.0937	0.0029	0.0937	0.0561	0.8780	0.0001	0.0048
	1.0	0.639	0.488	4.37	5.72	0.0703	0.1114	0.0794	0.0125	0.1358	0.0873	0.1035	0.1716	0.0396	0.1974	0.0016	0.0966	0.0016	0.0966	0.0304	0.9339	0.0001	0.0049
	1.5	0.652	0.487	4.40	5.89	0.0690	0.1143	0.0792	0.0129	0.1346	0.0898	0.1012	0.1764	0.0373	0.2026	0.0008	0.0982	0.0008	0.0982	0.0161	0.9651	0.0000	0.0049
	2.0	0.658	0.487	4.42	5.97	0.0684	0.1156	0.0791	0.0131	0.1341	0.0910	0.1002	0.1786	0.0361	0.2050	0.0005	0.0990	0.0005	0.0990	0.0094	0.9795	0.0000	0.0050
0.2	0.4	0.677	0.469	4.42	6.37	0.0706	0.1126	0.0791	0.0134	0.1345	0.0925	0.1020	0.1788	0.0409	0.1975	0.0050	0.0890	0.0182	0.1581	0.1640	0.6037	0.0013	0.0171
	0.6	0.730	0.463	4.56	7.18	0.0664	0.1225	0.0785	0.0149	0.1302	0.1025	0.0942	0.1972	0.0335	0.2148	0.0037	0.0917	0.0137	0.1684	0.1241	0.7006	0.0010	0.0178
	1.0	0.793	0.458	4.74	8.22	0.0610	0.1353	0.0777	0.0168	0.1248	0.1154	0.0843	0.2207	0.0242	0.2368	0.0022	0.0951	0.0080	0.1815	0.0728	0.8245	0.0006	0.0187
	1.5	0.831	0.455	4.86	8.88	0.0576	0.1434	0.0772	0.0180	0.1214	0.1235	0.0781	0.2355	0.0182	0.2507	0.0012	0.0973	0.0044	0.1897	0.0403	0.9029	0.0003	0.0193
	2.0	0.849	0.453	4.91	9.20	0.0559	0.1473	0.0769	0.0186	0.1197	0.1276	0.0750	0.2429	0.0153	0.2576	0.0007	0.0984	0.0026	0.1939	0.0242	0.9418	0.0002	0.0196
0.3	0.4	0.741	0.439	4.52	7.63	0.0698	0.1155	0.0787	0.0149	0.1319	0.1013	0.0987	0.1899	0.0420	0.1929	0.0056	0.0868	0.0420	0.1929	0.2371	0.3457	0.0045	0.0338
	0.6	0.831	0.427	4.75	9.24	0.0642	0.1296	0.0777	0.0175	0.1255	0.1182	0.0877	0.2185	0.0338	0.2130	0.0045	0.0893	0.0338	0.2130	0.1935	0.4682	0.0036	0.0359
	1.0	0.954	0.415	5.09	11.69	0.0559	0.1511	0.0762	0.0215	0.1158	0.1440	0.0711	0.2621	0.0217	0.2436	0.0028	0.0930	0.0217	0.2436	0.1261	0.6548	0.0023	0.0391
	1.5	1.036	0.409	5.34	13.53	0.0497	0.1673	0.0751	0.0245	0.1085	0.1633	0.0587	0.2948	0.0128	0.2665	0.0017	0.0959	0.0128	0.2665	0.0750	0.7952	0.0014	0.0415
	2.0	1.078	0.407	5.48	14.54	0.0464	0.1762	0.0745	0.0262	0.1045	0.1740	0.0520	0.3129	0.0080	0.2792	0.0010	0.0974	0.0080	0.2792	0.0467	0.8725	0.0008	0.0448

423 Nonprismatic Beams

0.4	0.4	0.774	0.405	4.55	8.70	0.0703	0.1117	0.0786	0.0156	0.1315	0.1035	0.0992	0.1855	0.0445	0.1773	0.0059	0.0849	0.0713	0.1938	0.2780	0.0876	0.0106	0.0509
	0.6	0.901	0.386	4.83	11.28	0.0646	0.1269	0.0774	0.0192	0.1240	0.1254	0.0875	0.2182	0.0377	0.1932	0.0049	0.0869	0.0611	0.2204	0.2456	0.2035	0.0089	0.0547
	1.0	1.102	0.367	5.33	16.03	0.0549	0.1548	0.0752	0.0257	0.1105	0.1658	0.0671	0.2780	0.0267	0.2222	0.0034	0.0904	0.0438	0.2689	0.1817	0.4177	0.0063	0.0616
	1.5	1.260	0.357	5.79	20.46	0.0462	0.1807	0.0732	0.0319	0.0982	0.2035	0.0485	0.3339	0.0173	0.2491	0.0022	0.0938	0.0284	0.3142	0.1198	0.6183	0.0037	0.0679
	2.0	1.349	0.352	6.09	23.32	0.0407	0.1975	0.0719	0.0358	0.0903	0.2278	0.0367	0.3699	0.0113	0.2664	0.0014	0.0959	0.0187	0.3434	0.0793	0.7479	0.0027	0.0720
0.5	0.4	0.768	0.371	4.56	9.45	0.0700	0.1048	0.0786	0.0154	0.1312	0.0993	0.0983	0.1679	0.0442	0.1663	0.0059	0.0836	0.0983	0.1679	0.2710	+0.1319	0.0189	0.0656
	0.6	0.919	0.343	4.84	12.94	0.0651	0.1176	0.0774	0.0193	0.1240	0.1218	0.0884	0.1935	0.0386	0.1769	0.0051	0.0849	0.0884	0.1935	0.2593	+0.0493	0.0167	0.0702
	1.0	1.200	0.316	5.42	20.61	0.0561	0.1451	0.0749	0.0280	0.1096	0.1709	0.0706	0.2486	0.0299	0.1993	0.0038	0.0877	0.0706	0.2486	0.2203	0.1356	0.0131	0.0802
	1.5	1.470	0.301	6.10	29.74	0.0466	0.1777	0.0720	0.0384	0.0934	0.2290	0.0516	0.3137	0.0215	0.2255	0.0027	0.0909	0.0516	0.3137	0.1663	0.3579	0.0094	0.0918
	2.0	1.647	0.295	6.63	37.04	0.0393	0.2036	0.0698	0.0466	0.0807	0.2755	0.0370	0.3655	0.0153	0.2463	0.0019	0.0934	0.0370	0.3655	0.1209	0.5361	0.0067	0.1011
0.6	0.4	0.726	0.341	4.62	9.84	0.0675	0.0986	0.0782	0.0146	0.1280	0.0916	0.0923	0.1519	0.0419	0.1603	0.0056	0.0829	0.1154	0.1276	0.2103	+0.2862	0.0283	0.0769
	0.6	0.872	0.305	4.88	13.97	0.0630	0.1072	0.0771	0.0183	0.1214	0.1096	0.0835	0.1664	0.0368	0.1666	0.0048	0.0837	0.1068	0.1463	0.2221	+0.2453	0.0254	0.0813
	1.0	1.196	0.267	5.43	24.35	0.0560	0.1277	0.0748	0.0274	0.1092	0.1537	0.0705	0.1999	0.0299	0.1804	0.0038	0.0854	0.0926	0.1910	0.2190	0.1321	0.0212	0.0913
	1.5	1.588	0.247	6.18	39.79	0.0482	0.1572	0.0718	0.0408	0.0939	0.2183	0.0572	0.2478	0.0237	0.1997	0.0030	0.0878	0.0762	0.2559	0.1926	0.0433	0.0171	0.1055
	2.0	1.905	0.237	6.92	55.51	0.0412	0.1870	0.0688	0.0544	0.0792	0.2839	0.0455	0.2960	0.0186	0.2189	0.0023	0.0901	0.0611	0.3215	0.1589	0.2243	0.0136	0.1197
0.7	0.4	0.657	0.321	4.86	9.96	0.0631	0.0954	0.0770	0.0138	0.1175	0.0846	0.0844	0.1461	0.0392	0.1582	0.0053	0.0827	0.1175	0.0846	0.0959	+0.3666	0.0372	0.0854
	0.6	0.770	0.275	5.14	14.39	0.0580	0.1006	0.0758	0.0167	0.1097	0.0955	0.0745	0.1543	0.0335	0.1621	0.0045	0.0832	0.1097	0.0955	0.1322	+0.3615	0.0330	0.0890
	1.0	1.056	0.224	5.62	26.45	0.0516	0.1122	0.0738	0.0243	0.0992	0.1213	0.0626	0.1710	0.0269	0.1694	0.0035	0.0841	0.0992	0.1213	0.1655	+0.3228	0.0280	0.0965
	1.5	1.491	0.196	6.24	47.48	0.0463	0.1304	0.0714	0.0371	0.0890	0.1633	0.0537	0.1959	0.0223	0.1796	0.0028	0.0854	0.0890	0.1633	0.1731	+0.2367	0.0241	0.1076
	2.0	1.944	0.183	6.95	73.85	0.0417	0.1523	0.0687	0.0530	0.0793	0.2149	0.0468	0.2255	0.0191	0.1915	0.0024	0.0869	0.0793	0.2149	0.1646	+0.1219	0.0210	0.1210
0.8	0.4	0.583	0.319	5.46	9.97	0.0585	0.0951	0.0741	0.0137	0.1040	0.0837	0.0793	0.1456	0.0380	0.1580	0.0053	0.0826	0.1023	0.0461	+0.0804	+0.3734	0.0452	0.0917
	0.6	0.645	0.263	5.89	14.44	0.0516	0.0990	0.0721	0.0160	0.0921	0.0907	0.0667	0.1520	0.0311	0.1614	0.0043	0.0831	0.0950	0.0517	+0.0150	+0.3956	0.0388	0.0951
	1.0	0.818	0.196	6.47	27.06	0.0435	0.1053	0.0696	0.0211	0.0781	0.1025	0.0521	0.1615	0.0232	0.1660	0.0031	0.0838	0.0863	0.0628	0.0588	+0.4118	0.0314	0.1004
	1.5	1.128	0.155	6.98	50.85	0.0385	0.1130	0.0676	0.0296	0.0692	0.1175	0.0432	0.1715	0.0184	0.1705	0.0024	0.0844	0.0802	0.0793	0.0990	+0.4009	0.0268	0.1064
	2.0	1.533	0.135	7.47	84.60	0.0355	0.1222	0.0658	0.0412	0.0638	0.1357	0.0384	0.1824	0.0159	0.1750	0.0020	0.0849	0.0759	0.1009	0.1150	+0.3684	0.0242	0.1133
0.9	0.4	0.524	0.356	6.87	10.10	0.0604	0.0948	0.0674	0.0157	0.1031	0.0835	0.0844	0.1439	0.0418	0.1568	0.0059	0.0824	0.0674	0.0157	+0.3652	+0.2913	0.0550	0.0942
	0.6	0.542	0.295	7.95	14.58	0.0497	0.0991	0.0623	0.0184	0.0866	0.0913	0.0691	0.1510	0.0339	0.1605	0.0048	0.0830	0.0623	0.0184	+0.2658	+0.3364	0.0460	0.0985
	1.0	0.594	0.206	9.44	27.16	0.0372	0.1052	0.0553	0.0226	0.0642	0.1023	0.0484	0.1609	0.0231	0.1656	0.0032	0.0837	0.0553	0.0226	+0.1311	+0.3969	0.0337	0.1044
	1.5	0.695	0.142	10.48	51.25	0.0289	0.1098	0.0506	0.0266	0.0492	0.1105	0.0346	0.1680	0.0159	0.1692	0.0021	0.0842	0.0505	0.0266	+0.0410	+0.4351	0.0255	0.1089
	2.0	0.842	0.107	11.07	86.80	0.0245	0.1147	0.0481	0.0306	0.0414	0.1159	0.0274	0.1723	0.0121	0.1714	0.0016	0.0845	0.0481	0.0306	0.0049	+0.4515	0.0213	0.1117

* Note: All carryover factors are negative and all stiffness factors are positive. All fixed-end moment coefficients are negative except where plus sign is shown.

TABLE 13.3b
Prismatic Haunch at Both Ends*

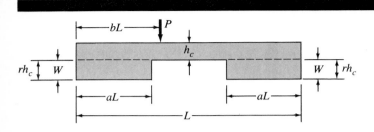

a	r	Carryover Factors $C_{AB}=C_{BA}$	Stiffness Factors $k_{AB}=k_{BA}$	Unif. Load FEM Coef. × wL^2 $M_{AB}=M_{BA}$	Concentrated Load FEM—Coef. × PL, b										Haunch Load, Both Haunches FEM Coef. × WL^2 $M_{AB}=M_{BA}$
					0.1		0.3		0.5		0.7		0.9		
					M_{AB}	M_{BA}	M_{AB}	M_{BA}	M_{AB}	M_{BA}	M_{AB}	M_{BA}	M_{AB}	M_{BA}	
0.1	0.4	0.583	5.49	0.0921	0.0905	0.0053	0.1727	0.0606	0.1396	0.1396	0.0606	0.1727	0.0053	0.0905	0.0049
	0.6	0.603	5.93	0.0940	0.0932	0.0040	0.1796	0.0589	0.1428	0.1428	0.0589	0.1796	0.0040	0.0932	0.0049
	1.0	0.624	6.45	0.0961	0.0962	0.0023	0.1873	0.0566	0.1462	0.1462	0.0566	0.1873	0.0023	0.0962	0.0050
	1.5	0.636	6.75	0.0972	0.0980	0.0013	0.1918	0.0551	0.1480	0.1480	0.0551	0.1918	0.0013	0.0980	0.0050
	2.0	0.641	6.90	0.0976	0.0988	0.0008	0.1939	0.0543	0.1489	0.1489	0.0543	0.1939	0.0008	0.0988	0.0050
0.2	0.4	0.634	7.32	0.0970	0.0874	0.0079	0.1852	0.0623	0.1506	0.1506	0.0623	0.1852	0.0079	0.0874	0.0187
	0.6	0.674	8.80	0.1007	0.0899	0.0066	0.1993	0.0584	0.1575	0.1575	0.0584	0.1993	0.0066	0.0899	0.0191
	1.0	0.723	11.09	0.1049	0.0935	0.0046	0.2193	0.0499	0.1654	0.1654	0.0499	0.2193	0.0046	0.0935	0.0195
	1.5	0.752	12.87	0.1073	0.0961	0.0029	0.2338	0.0420	0.1699	0.1699	0.0420	0.2338	0.0029	0.0961	0.0197
	2.0	0.765	13.87	0.1084	0.0976	0.0018	0.2410	0.0372	0.1720	0.1720	0.0372	0.2410	0.0018	0.0976	0.0198
0.3	0.4	0.642	9.02	0.0977	0.0845	0.0097	0.1763	0.0707	0.1558	0.1558	0.0707	0.1763	0.0097	0.0845	0.0397
	0.6	0.697	12.09	0.1027	0.0861	0.0095	0.1898	0.0700	0.1665	0.1665	0.0700	0.1898	0.0095	0.0861	0.0410
	1.0	0.775	18.68	0.1091	0.0890	0.0084	0.2136	0.0627	0.1803	0.1803	0.0627	0.2136	0.0084	0.0890	0.0426
	1.5	0.828	26.49	0.1132	0.0920	0.0065	0.2376	0.0492	0.1891	0.1891	0.0492	0.2376	0.0065	0.0920	0.0437
	2.0	0.855	32.77	0.1153	0.0943	0.0048	0.2555	0.0366	0.1934	0.1934	0.0366	0.2555	0.0048	0.0943	0.0442
0.4	0.4	0.599	10.15	0.0937	0.0825	0.0101	0.1601	0.0732	0.1509	0.1509	0.0732	0.1601	0.0101	0.0825	0.0642
	0.6	0.652	14.52	0.0986	0.0833	0.0106	0.1668	0.0776	0.1632	0.1632	0.0776	0.1668	0.0106	0.0833	0.0668
	1.0	0.744	26.06	0.1067	0.0847	0.0112	0.1790	0.0835	0.1833	0.1833	0.0835	0.1790	0.0112	0.0847	0.0711
	1.5	0.827	45.95	0.1131	0.0862	0.0113	0.1919	0.0852	0.1995	0.1995	0.0852	0.1919	0.0113	0.0862	0.0746
	2.0	0.878	71.41	0.1169	0.0876	0.0108	0.2033	0.0822	0.2089	0.2089	0.0822	0.2033	0.0108	0.0876	0.0766
0.5	0.0	0.500	4.00	0.0833	0.0810	0.0090	0.1470	0.0630	0.1250	0.1250	0.0630	0.1470	0.0090	0.0810	0.0833

* Note: All carryover factors and fixed-end moment coefficients are negative and all stiffness factors are positive.

Example 13.14 demonstrates the relative ease in using the PCA tables to establish the parameter values for the nonprismatic beam shown in Example 13.13.

Example 13.15 illustrates the use of the PCA tables for a moment-distribution analysis of a nonprismatic continuous bridge girder under uniform loading. Also, it is seen that the stiffness modifications of Sec. 13.4 can be used for nonprismatic as well as prismatic structures.

EXAMPLE 13.14

By use of PCA data from Table 13.3, determine carryover factors, stiffness factors, and FEMs at both ends of the nonprismatic beam of Fig. *a* in Example 13.13. Load is uniformly distributed.

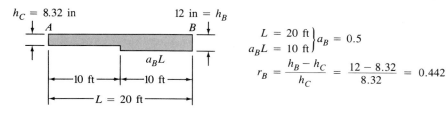

Entering Table 13.3a with $a_B = 0.50$, $r_A = 0$, $a_A = 0$, and $r_B = 0.442$, we find

$C_{AB} = 0.80$ FEMs for uniform load:

$C_{BA} = 0.365$ $M_{AB} = 0.067wL^2$

$k_{AB} = 4.62$ $M_{BA} = 0.107wL^2$

$k_{BA} = 10.18$

These values were obtained by linear interpolation between $r_B = 0.4$ and $r_B = 0.6$. The comparison with data obtained in Example 13.13 is excellent.

EXAMPLE 13.15

Compute all joint moments by moment distribution using the PCA tables for the continuous bridge girder under uniform loading.

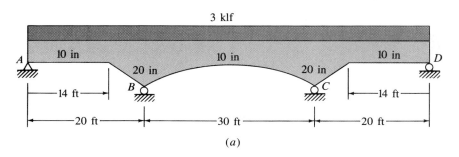

(a)

Solution

Member AB (refer to Table 13.1):

$$a_B = \frac{a_B L}{L} = \frac{6}{20} = 0.3$$

$$a_A = 0$$

$$r_B = \frac{20 - 10}{10} = 1.0$$

$C_{AB} = 0.791; \quad M_{AB} = 0.0630wL^2 = 75.6 \text{ kip·ft}$

$C_{BA} = 0.449; \quad M_{BA} = 0.1311wL^2 = 157 \text{ kip·ft}$

$k_{BA} = 8.29 \quad \therefore \quad K_{BA} = k_{BA}\dfrac{EI}{L} = \dfrac{8.29EI}{20} = 0.4145EI$

$$K_{BA}^M = \frac{3}{4}(0.4145EI) = 0.311EI$$

Member BC (refer to Table 13.2):

$$a_A = \frac{a_A L}{L} = \frac{15}{30} = 0.5$$

$$a_B = \frac{a_B L}{L} = \frac{15}{30} = 0.5$$

$$r_A = r_B = \frac{20 - 10}{10} = 1.0$$

$C_{BC} = 0.694; \quad M_{BC} = M_{CB} = 0.1025wL^2 = 277 \text{ kip·ft}$

$k_{BC} = 12.03 \quad \therefore \quad K_{BC} = k_{BC}\dfrac{EI}{L} = \dfrac{12.03EI}{30} = 0.401EI$

Due to symmetry, $K_{BC}^M = \dfrac{1}{2}(0.401EI) = 0.20EI$

Distribution factors (DF):

$$DF_{BA} = \frac{K_{BA}^M}{K_{BA}^M + K_{BC}^M} = \frac{0.311EI}{(0.311 + 0.20)EI} = 0.609$$

$$DF_{BC} = \frac{K_{BC}^M}{K_{BA}^M + K_{BC}^M} = \frac{0.20EI}{(0.311 + 0.20)EI} = 0.391$$

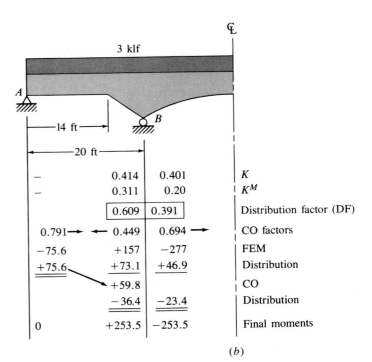

(b)

Computer check by IMAGES Figure 13.9 shows a computer model used on the IMAGES-2D program to verify the moment-distribution results from Example 13.15. This model considers a rectangular cross section of uniform 4 in width and variable depth. A constant value of moment of inertia was determined for each member element at its midspan location by use of the following PCA equations:

$$I_x = I_c \cdot [1 + r(1 - x/aL)^2]^3 \quad \text{for parabolic haunches}$$

and

$$I_x = I_c \cdot [1 + r(1 - x/aL)]^3 \quad \text{for straight haunches}$$

where x = distance from variable point to its related end (A or B).

The IMAGES-2D program produced final moments at supports B and C equal to 257.9 kip·ft compared to 253.5 kip·ft from the moment-distribution analysis of Example 13.15.

Figure 13.9

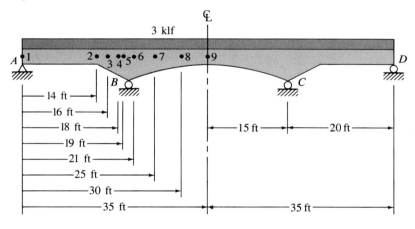

IMAGES model of nonprismatic girder of example 13-15

$I_1 = 333 \text{ in}^4$
$I_2 = 2667 \text{ in}^4$
$I_3 = 530 \text{ in}^4$
$I_4 = 1125 \text{ in}^4$
$I_5 = 1788 \text{ in}^4$
$I_6 = 1470 \text{ in}^4$
$I_7 = 651 \text{ in}^4$
$I_8 = 343 \text{ in}^4$

$$I_1 = I_9 = \frac{bd^3}{12} = \frac{4(10)^3}{12} = 333 \text{ in}^4$$

$M_B = M_C = 257.9$ kip·ft (By IMAGES)
$ = 253.5$ kip·ft (By Example 13-15)

Problems

13.1 to 13.19 Compute all joint moments by the method of moment distribution and draw shear and moment diagrams. E is constant.

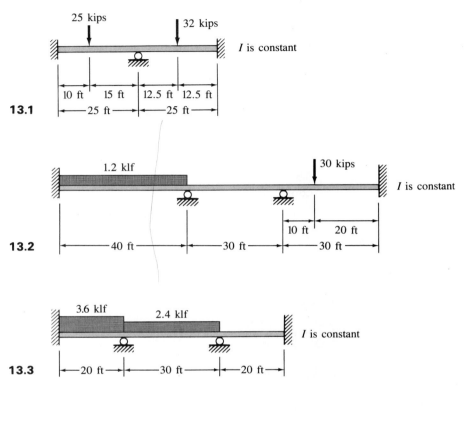

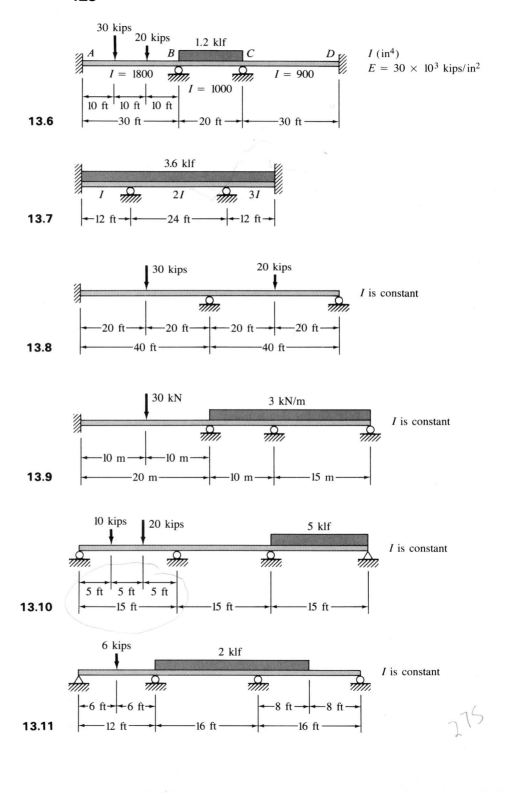

13.12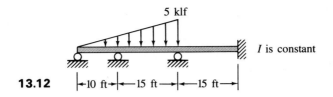

13.13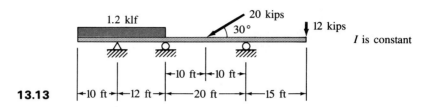

13.14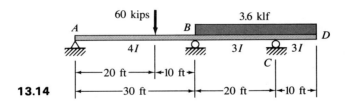

13.15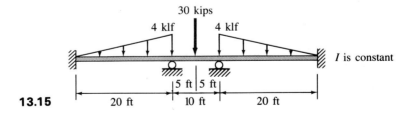

13.16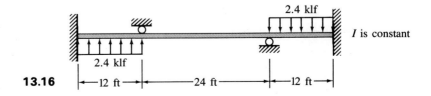

13.17 Rework Prob. 13.6 if support B settles 2.5 in.

13.18 Rework Prob. 13.14 if support B settles 1.5 in and support C settles 3.0 in. $I = 1000$ in^4, $E = 29 \times 10^3$ ksi.

431 Problems

13.19 Draw shear and moment diagrams if (a) no settlement occurs, and (b) if support C settles 30 mm. $E = 200$ GPa, $I = 80 \times 10^6$ mm^4.

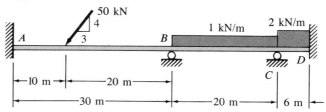

13.20 to 13.36 Determine final joint moments by the moment distribution method. E and I constant unless otherwise noted.

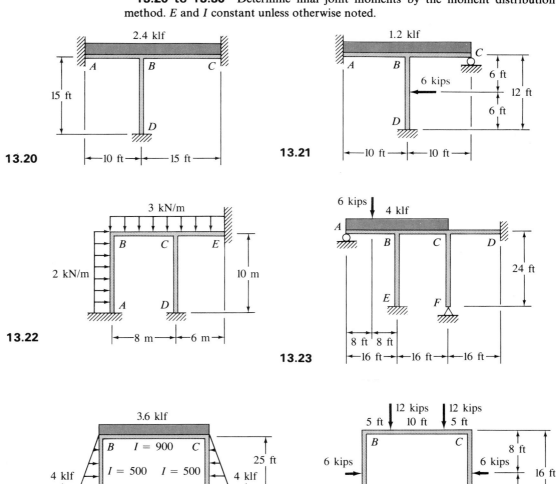

432 Moment Distribution

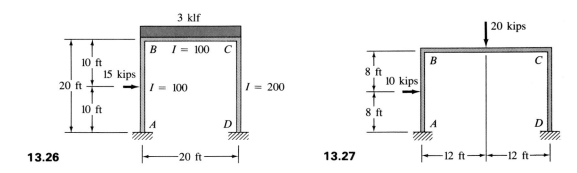

13.26

13.27

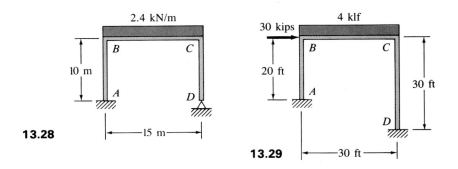

13.28

13.29

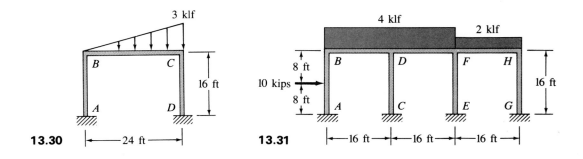

13.30

13.31

13.32 Rework Prob. 13.29 if right support is replaced by a hinge support.

13.33 Rework Prob. 13.31 if $I = 400$ in^4 for all columns and $I = 600$ in^4 for all beams.

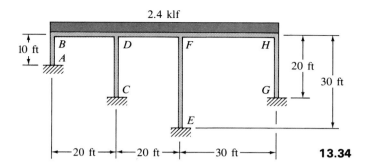

13.34

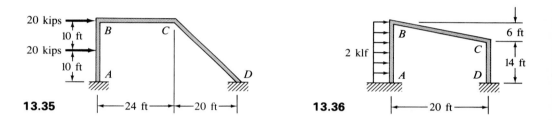

13.35   **13.36**

13.37 to 13.38 Determine final moments and draw shear and moment diagrams using the moment-distribution method and the PCA tables to determine FEMs, distribution and carryover factors, and so on. (Hint: Assume uniform beam width of 1 ft over entire length.)

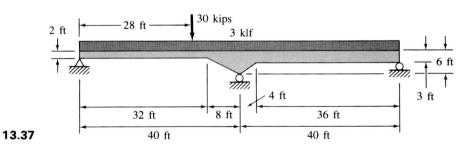

13.37

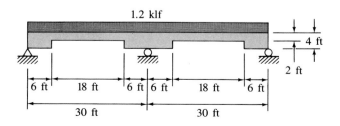

13.38

References

13.1 Hardy Cross, "Analysis of Continuous Frames by Distributing Fixed-End Moments," *Proceedings ASCE*, vol. 56, May 1930, pp. 919–928.

13.2 Hardy Cross, "Analysis of Continuous Frames by Distributing Fixed-End Moments," *Transactions ASCE*, vol. 96, 1932, pp. 1–156.

13.3 C. T. Morris, "Morris on Analysis of Continuous Frames," *Transactions ASCE*, vol. 96, 1932, pp. 66–69.

13.4 Harold I. Laursen, *Structural Analysis*, 3d ed., McGraw-Hill, New York, 1988, Appendix B.

13.5 James Lefter and Thomas J. Bergin, *A Microcomputer-Based Primer on Structural Behavior*, Prentice-Hall, Englewood Cliffs, N.J., 1986. pp. 357–358 and pp. 195–212.

CHAPTER FOURTEEN

Matrix Structural Analysis

14.1. INTRODUCTION

Over the last few decades, it has become common practice for an engineer to use a high-speed computer for the solution of complex structural systems. Until the 1950s, the classical methods presented in the preceding chapters were the "bread and butter" tools used by an engineer for the analysis of building or bridge structures. Hand analysis of complex structures was often performed by either making conservative assumptions that simplified the effort, or by separate analysis of subdivisions of the structure assembly (called *substructure analysis*). These classical hand methods are still used to solve small simple structures and are very important in helping a student develop a sound understanding of structural behavior. During the 1950s and 1960s, computer analysis was usually performed by large engineering firms that could bear the hardware and development costs of computer analysis. By the 1970s, most engineers replaced their slide rule with a pocket calculator and many small design offices provided computer services by time-share rental on available mainframe computers. However, the emergence in the 1980s of inexpensive high-speed microcomputers and the availability of quality software has encouraged the expanded use of computer analysis by both small and large design offices.

Matrix algebra is the mathematical language used to perform the operations of a computer structural analysis. As before, the basic principles of structural engineering are used to write a system of linear elastic equations for a structure. In matrix structural analysis, these equations are assembled into an efficient format for solution by matrix algebra operations on a high-

Colorado Bridge over Colorado River, on Highway 95 in Utah. (*Courtesy of U.S. Department of the Interior, Bureau of Reclamation; photo by Mel Davis.*)

speed digital computer. Sec. 14.2 presents the basic laws and operations of matrix algebra used in matrix structural analysis. The reader with an education in linear or matrix algebra may elect to review Sec. 14.2 or skip to the introduction of matrix structural analysis in Sec. 14.3.

Some of the many factors which affect the accuracy of a computer structural analysis include the ability to do the following:

1. Correctly model the geometry of the real structure
2. Define and model the restraints of the real structure
3. Identify the material properties of the real structure
4. Select a mathematical algorithm to solve the governing equations with a minimum degree of computational error

These same factors are valid for hand methods as well as for computer methods of structural analysis. Although the computer may relieve the structural engineer of the tedium associated with routine analytical operations, the structural engineer remains responsible for the design of the structure. The examples in this chapter are simple and involve few unknowns. Although these examples could easily be solved by a classical method, they serve to illustrate the primary operations of matrix structural analysis with a minimum amount of number crunching. Moreover, a student should be aware that engineering structures usually contain many unknown member and support forces. Therefore, structural analysis often involves an enormous amount of computations whereby the only practical solution available is by computer.

It is the belief of this author that a better understanding of structural behavior is obtained by an initial study of structures by classical methods followed by studies of matrix methods. Maguire and Gallagher stated that "although adopting the matrix approach does influence the way we think about structural behavior, its main virtue is that it provides the best format presently known for analyzing structures by computer" [14.1].

14.2. MATRIX ALGEBRA

14.2.1. Definitions

A matrix is a rectangular array of elements arranged in rows and columns. A general form of presenting a matrix **A** having m rows and n columns is as follows:

$$\mathbf{A} = \begin{bmatrix} a_{11} & a_{12} & \cdots & a_{1n} \\ a_{21} & a_{22} & \cdots & a_{2n} \\ \cdots & \cdots & \cdots & \cdots \\ a_{m1} & a_{m2} & \cdots & a_{mn} \end{bmatrix} \qquad (14.1)$$

In general, matrices are denoted by boldface letters. A matrix element is usually a single scalar quantity that is designated by lightface letters and located within the matrix by subscript notation. The first subscript is its row number (consecutively numbered from top to bottom) and the second subscript is its column number (consecutively numbered from left to right). Thus, a_{21} represents the element in row 2 and column 1. The size of a matrix is designated as its *order*. That is, the order of matrix **A** is $m \times n$ where m is the number of rows and n is the number of columns.

In the set of simultaneous equations

$$\begin{aligned} X_1 - 2X_2 + 3X_3 &= 6 \\ -X_1 + X_2 - 2X_3 &= 4 \\ 3X_1 + 2X_2 + X_3 &= 0 \end{aligned} \qquad (14.2)$$

coefficients of the unknowns can be represented as matrix **A** in the form

$$\mathbf{A} = \begin{bmatrix} 1 & -2 & 3 \\ -1 & 1 & -2 \\ 3 & 2 & 1 \end{bmatrix} \quad (14.3)$$

The order of **A** is 3×3 where $a_{13} = 3$, $a_{32} = 2$, and so forth. In general, an element denoted as a_{ij} is located in the ith row and jth column. Those elements with repeated subscripts $(i = j)$ are located along the main diagonal of the matrix.

14.2.2. Types of Matrices

Square matrix When the number of rows and columns are equal, the matrix is said to be square. That is, $m = n$. Matrix **A** of Eq. (14.3) above is square since $m = n = 3$.

Column and row matrices When a matrix has one column ($n = 1$), it is called a column matrix of order $m \times 1$. The constant values on the right side of Eqs. (14.2) can be expressed in column matrix form as

$$\mathbf{B} = \begin{bmatrix} b_1 \\ b_2 \\ b_3 \end{bmatrix} = \begin{bmatrix} 6 \\ 4 \\ 0 \end{bmatrix} \quad (14.4)$$

where it is seen that only one subscript is required to locate the position of any element in the column. Column matrices are often denoted by { } braces. Similarly, when a matrix has one row ($m = 1$), it is called a row matrix. The order of a row matrix is $1 \times n$.

Symmetric matrix When the elements of a square matrix are symmetric about the main diagonal, it is called a *symmetric matrix*. Therefore, $a_{ij} = a_{ji}$ in a symmetric matrix. An example of a 3×3 symmetric matrix is given by

$$\mathbf{C} = \begin{bmatrix} 1 & 4 & 3 \\ 4 & 2 & -6 \\ 3 & -6 & 1 \end{bmatrix} \quad (14.5)$$

Diagonal matrix A diagonal matrix is defined as a square matrix with all elements equal to zero *except* along the main diagonal.

Identity matrix A square matrix that has all main diagonal elements equal to unity and all other elements equal to zero, is called an *identity* (or *unit*) matrix. An identity matrix is noted by the symbol **I** and is shown in a 4×4 array as

$$\mathbf{I} = \begin{bmatrix} 1 & 0 & 0 & 0 \\ 0 & 1 & 0 & 0 \\ 0 & 0 & 1 & 0 \\ 0 & 0 & 0 & 1 \end{bmatrix} \qquad (14.6)$$

Null matrix A null matrix is noted by **O** and has all elements equal to zero.

14.2.3. Matrix Operations

A. Equality of matrices Two matrices are equal if the corresponding elements of the two matrices are equal and both matrices are of equal orders.

B. Addition and subtraction of matrices Matrices can be added or subtracted only if they have the same order. The addition of matrix **A** and matrix **B** results in matrix **C**. That is, $\mathbf{A} + \mathbf{B} = \mathbf{C}$. Similarly, the difference between matrices **A** and **B** is another matrix **D** which can be written as $\mathbf{A} - \mathbf{B} = \mathbf{D}$. An example of matrix addition and subtraction are given as follows:

$$\mathbf{A} = \begin{bmatrix} 4 & 5 \\ 6 & 3 \end{bmatrix} \qquad \mathbf{B} = \begin{bmatrix} 1 & -1 \\ 3 & 2 \end{bmatrix}$$

$$\mathbf{A} + \mathbf{B} = \mathbf{C} = \begin{bmatrix} 5 & 4 \\ 9 & 5 \end{bmatrix} \qquad \mathbf{A} - \mathbf{B} = \mathbf{D} = \begin{bmatrix} 3 & 6 \\ 3 & 1 \end{bmatrix}$$

The commutative and associative laws are valid for the addition and subtraction of matrices. That is,

$$\mathbf{A} + \mathbf{B} = \mathbf{B} + \mathbf{A} \quad \text{(commutative law)}$$

and

$$\mathbf{A} + (\mathbf{B} + \mathbf{C}) = (\mathbf{A} + \mathbf{B}) + \mathbf{C} \quad \text{(associative law)} \qquad (14.7)$$

C. Scalar multiplication of a matrix When a matrix is multiplied by a scalar, each element of the matrix is multiplied by the same scalar. Therefore,

$$\beta \mathbf{A} = \beta \begin{bmatrix} a_{11} & a_{12} & \cdots & a_{1n} \\ a_{21} & a_{22} & \cdots & a_{2n} \\ \cdots & \cdots & \cdots & \cdots \\ a_{m1} & a_{m2} & \cdots & a_{mn} \end{bmatrix} = \begin{bmatrix} \beta a_{11} & \beta a_{12} & \cdots & \beta a_{1n} \\ \beta a_{21} & \beta a_{22} & \cdots & \beta a_{2n} \\ \cdots & \cdots & \cdots & \cdots \\ \beta a_{m1} & \beta a_{m2} & \cdots & \beta a_{mn} \end{bmatrix}$$

where β is constant.

D. Multiplication of matrices The operation of matrix multiplication obeys both the distributive and associative laws but not the commutative law as follows:

By the distributive law: $\mathbf{A(B + C) = AB + AC}$

By the associative law: $\mathbf{A(BC) = (AB)C = ABC}$

However, the product of two matrices is not commutative except when both $\mathbf{A}$ and $\mathbf{B}$ are diagonal matrices. That is, the product $\mathbf{AB}$ is not equal to the product $\mathbf{BA}$.

The product $\mathbf{AB}$ of two matrices $\mathbf{A}$ and $\mathbf{B}$ can only exist if the number of columns of $\mathbf{A}$ is equal to the number of rows of $\mathbf{B}$. When two matrices satisfy this rule, they are said to be *conformable* for multiplication. The product $\mathbf{AB}$ forms a new matrix $\mathbf{C}$. However, if an attempt is made to multiply $\mathbf{A}$ and $\mathbf{B}$ in reverse order, that is, $\mathbf{BA}$, the two matrices may not be conformable. Consider the following matrices:

$$\mathbf{A} = \begin{bmatrix} 4 & 5 \\ 6 & 8 \end{bmatrix} \quad \mathbf{B} = \begin{bmatrix} 1 & -1 & 2 \\ 3 & 2 & 4 \end{bmatrix} \tag{14.8}$$

Since the number of columns of $\mathbf{A}$ equals the number of rows of $\mathbf{B}$, the two matrices are conformable and the product $\mathbf{AB}$ can be determined. If we attempt to find the product in reverse matrix order, namely $\mathbf{BA}$, we notice that the number of columns of $\mathbf{B}$ is three and the number of rows of $\mathbf{A}$ is two; therefore, the matrices are not conformable and the product $\mathbf{BA}$ does not exist. A formal definition of the product of two conformable matrices $(\mathbf{C = AB})$ may be given as

$$c_{ij} = \sum_{k=1}^{n} a_{ik} b_{kj} \tag{14.9}$$

where n is the number of columns in $\mathbf{A}$ (which is equal to the number of rows in $\mathbf{B}$) and c_{ij} represents any element in $\mathbf{C}$. Each element c_{ij} is found by summing the products of elements in the ith row of $\mathbf{A}$ with elements in the jth column of $\mathbf{B}$. For example, the product of the matrices given in Eq. (14.8) is

$$\mathbf{C = AB} = \begin{bmatrix} 4 & 5 \\ 6 & 8 \end{bmatrix} \begin{bmatrix} 1 & -1 & 2 \\ 3 & 2 & 4 \end{bmatrix} = \begin{bmatrix} c_{11} & c_{12} & c_{13} \\ c_{21} & c_{22} & c_{23} \end{bmatrix} = \begin{bmatrix} 19 & 6 & 28 \\ 30 & 10 & 44 \end{bmatrix}$$

$$2 \times 2 \qquad 2 \times 3 \qquad\qquad 2 \times 3$$

where $c_{11} = 4(1) + 5(3) = 19$

$c_{12} = 4(-1) + 5(2) = 6$

$c_{13} = 4(2) + 5(4) = 28$

$c_{21} = 6(1) + 8(3) = 30$

$c_{22} = 6(-1) + 8(2) = 10$

$c_{23} = 6(2) + 8(4) = 44$

From the rules of matrix multiplication presented above, it can be shown that the product of the identity matrix **I** and any square matrix **A** can be expressed as

$$\mathbf{IA} = \mathbf{AI} = \mathbf{A} \tag{14.10}$$

For example,

$$\begin{bmatrix} 1 & 0 & 0 \\ 0 & 1 & 0 \\ 0 & 0 & 1 \end{bmatrix} \begin{bmatrix} 1 & 4 & 3 \\ 4 & 2 & -6 \\ 3 & -6 & 1 \end{bmatrix} = \begin{bmatrix} 1 & 4 & 3 \\ 4 & 2 & -6 \\ 3 & -6 & 1 \end{bmatrix}$$
$$\quad\mathbf{I}\quad\quad\quad\quad\mathbf{A}\quad\quad\quad\quad\mathbf{A}$$

E. Transposed matrix When the rows and columns of a matrix are interchanged (that is, first row becomes first column, first column becomes first row, etc.) the resulting matrix is called the transposed matrix. The transpose of matrix **A** is denoted by $\mathbf{A}^T$. The transpose of two matrices are shown as follows:

$$\mathbf{A} = \begin{bmatrix} 5 & 4 \\ 9 & 6 \end{bmatrix} \quad \mathbf{A}^T = \begin{bmatrix} 5 & 9 \\ 4 & 6 \end{bmatrix}$$

$$\mathbf{B} = \begin{bmatrix} 1 & 4 & 5 \\ 4 & 2 & 6 \\ 3 & -6 & 1 \end{bmatrix} \quad \mathbf{B}^T = \begin{bmatrix} 1 & 4 & 3 \\ 4 & 2 & -6 \\ 5 & 6 & 1 \end{bmatrix}$$

An important property of transposing a matrix is to make it conformable for multiplication. Consider matrices **C**, **D**, and $\mathbf{D}^T$ shown below.

$$\mathbf{C} = \begin{bmatrix} c_{11} & c_{12} & c_{13} \\ c_{21} & c_{22} & c_{23} \\ c_{31} & c_{32} & c_{33} \end{bmatrix} \quad \mathbf{D} = [d_1 \ d_2 \ d_3] \quad \mathbf{D}^T = \begin{bmatrix} d_1 \\ d_2 \\ d_3 \end{bmatrix}$$

C and **D** are not conformable for multiplication since **C** has three columns and **D** has one row. However, the product $\mathbf{CD}^T$ exists and can be determined since the number of columns in **C** equals the number of rows in $\mathbf{D}^T$.

F. Determinant of a square matrix A determinant is a square array of numbers which has a unique value. In particular, a determinant is an important scalar used in matrix inverse operations. Examples of determinants of square matrices presented in both symbolic and numerical form are

$$|\mathbf{A}| = \begin{vmatrix} a_{11} & a_{12} & a_{13} \\ a_{21} & a_{22} & a_{23} \\ a_{31} & a_{32} & a_{33} \end{vmatrix} \quad |\mathbf{B}| = \begin{vmatrix} 3 & 5 \\ 1 & 4 \end{vmatrix} \tag{14.11}$$

Observe that the array of numbers in a determinate are enclosed by vertical lines whereas the elements of a matrix are enclosed in [ ] brackets. It is worth repeating that a determinant such as $|\mathbf{A}|$ has a single numerical value whereas the matrix $\mathbf{A}$ cannot be reduced to a single value. The value of the 2×2 determinant $|\mathbf{B}|$ of Eq. (14.11) is easily found as the product of the numbers on the main diagonal minus the product of the numbers on the secondary diagonal. That is,

$$|\mathbf{B}| = \begin{vmatrix} 3 & 5 \\ 1 & 4 \end{vmatrix} = 3(4) - 1(5) = 7$$

Unfortunately, this procedure is only valid for determinants of order 2.

Laplace expansion Laplace expansion represents a practical way of computing the determinant of a square matrix of higher order. A Laplace expansion involves the continued replacement of a high-order determinant by a set of lower order determinants until the expanded form is able to be solved by the diagonal procedure shown above. The expanded terms are expressed by minors and cofactors of each element of a determinant which can be described by example in the following way:

$$|\mathbf{A}| = \begin{vmatrix} a_{11} & a_{12} & a_{13} \\ a_{21} & a_{22} & a_{23} \\ a_{31} & a_{32} & a_{33} \end{vmatrix} = \begin{vmatrix} 1 & 2 & 1 \\ 2 & 2 & 3 \\ 3 & -5 & 1 \end{vmatrix} \qquad (14.12)$$

Minor of an element The minor of an element of a determinant is the determinant that remains after the row and column in which the element occurs are deleted. If we define the minor of element a_{ij} as m_{ij}, we find that

$$m_{11} = \begin{vmatrix} 2 & 3 \\ -5 & 1 \end{vmatrix} \qquad m_{32} = \begin{vmatrix} 1 & 1 \\ 2 & 3 \end{vmatrix} \qquad m_{22} = \begin{vmatrix} 1 & 1 \\ 3 & 1 \end{vmatrix}$$

and so forth.

The minors m_{11}, m_{32}, and m_{22} can easily be evaluated by the diagonal method above as $+17$, $+1$, and -2, respectively.

Cofactor of an element A cofactor is the minor of an element of a determinant that is preceded by a proper sign. The sign of the minor of an element a_{ij} is determined by $(-1)^{i+j}$ which visual provides the following sign pattern for a 3×3 determinant array

$$\begin{vmatrix} + & - & + \\ - & + & - \\ + & - & + \end{vmatrix} \qquad \begin{array}{l} i\text{---row number} \\ j\text{---column number} \end{array}$$

Matrix Algebra

Let the cofactor of element a_{ij} of a determinant be denoted by C_{ij}. Then, we can define the cofactor of element a_{ij} as

$$C_{ij} = (-1)^{i+j} m_{ij}$$

The cofactors of elements a_{11}, a_{32}, and a_{22} are $+17$, -1, and -2, respectively. The remaining cofactors can easily be found. A complete presentation of the cofactors for the array shown in Eq. (14.12) is given as

$$[\mathbf{C}] = \begin{vmatrix} 1 & 2 & 1 \\ 2 & 2 & 3 \\ 3 & -5 & 1 \end{vmatrix} = \begin{bmatrix} +\begin{vmatrix} 2 & 3 \\ -5 & 1 \end{vmatrix} & -\begin{vmatrix} 2 & 3 \\ 3 & 1 \end{vmatrix} & +\begin{vmatrix} 2 & 2 \\ 3 & -5 \end{vmatrix} \\ -\begin{vmatrix} 2 & 1 \\ -5 & 1 \end{vmatrix} & +\begin{vmatrix} 1 & 1 \\ 3 & 1 \end{vmatrix} & -\begin{vmatrix} 1 & 2 \\ 3 & -5 \end{vmatrix} \\ +\begin{vmatrix} 2 & 1 \\ 2 & 3 \end{vmatrix} & -\begin{vmatrix} 1 & 1 \\ 2 & 3 \end{vmatrix} & +\begin{vmatrix} 1 & 2 \\ 2 & 2 \end{vmatrix} \end{bmatrix}$$

Finally, *the value of a determinant by Laplace expansion is equal to the sum of the products of the elements in any row or column with their respective cofactors.* For example, the determinant of Eq. (14.12) can be found by using the elements and cofactors of row 1 as

$$1\begin{vmatrix} 2 & 3 \\ -5 & 1 \end{vmatrix} - 2\begin{vmatrix} 2 & 3 \\ 3 & 1 \end{vmatrix} + 1\begin{vmatrix} 2 & 2 \\ 3 & -5 \end{vmatrix} = 1(17) - 2(-7) + 1(-16) = 15$$

or by use of the elements and cofactors of column 2 as

$$-2\begin{vmatrix} 2 & 3 \\ 3 & 1 \end{vmatrix} + 2\begin{vmatrix} 1 & 1 \\ 3 & 1 \end{vmatrix} + 5\begin{vmatrix} 1 & 1 \\ 2 & 3 \end{vmatrix} = -2(-7) + 2(-2) + 5(1) = 15$$

The Laplace expansion as applied to the 3 × 3 determinant above is fairly simple. However, the task becomes tedious as the determinant becomes large and undergoes numerous expansions until the minors are reduced to order 2. A Laplace expansion computer program can easily be written to evaluate an $n \times n$ determinant $|\mathbf{A}|$ with reference to any row (i) as

$$|\mathbf{A}| = \sum_{j=1}^{n} a_{ij} C_{ij} \qquad (14.13)$$

or with reference to any column (j) by

$$|\mathbf{A}| = \sum_{i=1}^{n} a_{ij} C_{ij} \qquad (14.14)$$

Important properties of determinants The value of a determinant

1. Is unchanged if all rows and columns are interchanged
2. Changes sign if two rows or two columns are interchanged
3. Is zero if all elements in any row or column are zero
4. Is zero if two rows or two columns are proportional
5. Is multiplied by a factor β if each element of a row or column of the determinant is multiplied by the factor β

G. Adjoint matrix If the elements of a square matrix **A** are replaced by their respective cofactors and then transposed, the resulting matrix is called the adjoint matrix of **A**. The adjoint of the square matrix of Eq. (14.12) is found as follows:

$$\mathbf{A} = \begin{bmatrix} 1 & 2 & 1 \\ 2 & 2 & 3 \\ 3 & -5 & 1 \end{bmatrix} \quad \text{Adjoint of } \mathbf{A} = \begin{bmatrix} C_{11} & C_{12} & C_{13} \\ C_{21} & C_{22} & C_{23} \\ C_{31} & C_{32} & C_{33} \end{bmatrix}^T$$

$$\text{Adjoint of } \mathbf{A} = \begin{bmatrix} C_{11} & C_{21} & C_{31} \\ C_{12} & C_{22} & C_{32} \\ C_{13} & C_{23} & C_{33} \end{bmatrix} = \begin{bmatrix} +17 & -7 & +4 \\ +7 & -2 & -1 \\ -16 & +11 & -2 \end{bmatrix}$$

H. Inverse of a matrix The operation of division does not exist in matrix algebra. That is, matrix **A** cannot be divided by matrix **B** in the usual form **A/B**. Instead, division is performed by an inverse operation, similar to ordinary algebra, which involves the use of a determinant and an adjoint matrix. In elementary algebra, the value of x may be found for the equation $5x = 9$ as $x = 9/5$, or by the expression $x = (5)^{-1} \cdot 9$ which resembles the matrix inverse form presented below.

Let the inverse of a square matrix **A** be denoted by $\mathbf{A}^{-1}$ and the adjoint of **A** by adj**A**. In matrix algebra, it is defined that

$$\mathbf{A}^{-1}\mathbf{A} = \mathbf{I} = \mathbf{A}\mathbf{A}^{-1} \quad \text{and} \quad (\text{adj}\mathbf{A})\mathbf{A} = |\mathbf{A}|\mathbf{I} \tag{14.15}$$

Multiplying both terms in the second definition of Eq. (14.15) by $\mathbf{A}^{-1}$ yields

$$(\text{adj}\mathbf{A})\mathbf{A}\mathbf{A}^{-1} = |\mathbf{A}|\mathbf{I}\mathbf{A}^{-1}$$

Recalling that $\mathbf{A} \cdot \mathbf{A}^{-1} = \mathbf{I}$, the above expression can be rewritten as

$$(\text{adj}\mathbf{A})\mathbf{I} = |\mathbf{A}|\mathbf{I}\mathbf{A}^{-1}$$

The removal of **I** and multiplication by the reciprocal of the scalar $|\mathbf{A}|$ to both sides of this equation results in the following inverse relationship:

$$\mathbf{A}^{-1} = \frac{\operatorname{adj}\mathbf{A}}{|\mathbf{A}|} \qquad (14.16)$$

Consider the set of simultaneous equations

$$\begin{aligned} x_1 + 2x_2 + x_3 &= 1 \\ 2x_1 + 2x_2 + 3x_3 &= 1 \\ 3x_1 - 5x_2 + x_3 &= 2 \end{aligned} \qquad (14.17)$$

which can be written in matrix form $\mathbf{AB} = \mathbf{C}$ as

$$\overset{\mathbf{A}}{\begin{bmatrix} 1 & 2 & 1 \\ 2 & 2 & 3 \\ 3 & -5 & 1 \end{bmatrix}} \overset{\mathbf{B}}{\begin{bmatrix} x_1 \\ x_2 \\ x_3 \end{bmatrix}} = \overset{\mathbf{C}}{\begin{bmatrix} 1 \\ 1 \\ 2 \end{bmatrix}}$$

$$\mathbf{B} = \begin{bmatrix} x_1 \\ x_2 \\ x_3 \end{bmatrix} = \mathbf{A}^{-1}\mathbf{C} = \begin{bmatrix} 1 & 2 & 1 \\ 2 & 2 & 3 \\ 3 & -5 & 1 \end{bmatrix}^{-1} \begin{bmatrix} 1 \\ 1 \\ 2 \end{bmatrix}$$

The adjoint and determinant value of matrix $\mathbf{A}$ were determined earlier as

$$\operatorname{adj}\mathbf{A} = \begin{bmatrix} +17 & -7 & +4 \\ +7 & -2 & -1 \\ -16 & 11 & -2 \end{bmatrix} \quad \text{and} \quad |\mathbf{A}| = 15$$

Therefore, the inverse of $\mathbf{A}$ is expressed as

$$\mathbf{A}^{-1} = \frac{\operatorname{adj}\mathbf{A}}{|\mathbf{A}|} = \frac{1}{15} \begin{bmatrix} 17 & -7 & 4 \\ 7 & -2 & -1 \\ -16 & 11 & -2 \end{bmatrix}$$

and the solution for x_1, x_2, and x_3 is

$$\begin{bmatrix} x_1 \\ x_2 \\ x_3 \end{bmatrix} = \mathbf{A}^{-1}\mathbf{C} = \frac{1}{15} \begin{bmatrix} 17 & -7 & 4 \\ 7 & -2 & -1 \\ -16 & 11 & -2 \end{bmatrix} \begin{bmatrix} 1 \\ 1 \\ 2 \end{bmatrix} = \begin{bmatrix} 1.2 \\ 0.2 \\ -0.6 \end{bmatrix}$$

It is important to note that if the determinant of $\mathbf{A}$ is zero, the division operation above cannot be done. A square matrix whose determinant is equal to zero is said to be *singular*. A square matrix is referred to as *nonsingular* if the value of its determinant is nonzero. The importance and significance of singular matrices in structural analysis will be discussed later.

I. Partitioning of matrices The computations associated with matrix operations are vastly simplified when a high-order matrix can be subdivided into a number of lower order submatrices. This procedure is called matrix partitioning. For example, the 3×3 matrix **A** shown below is divided into lower order submatrices where partition lines are noted by dashed lines.

$$\mathbf{A} = \begin{bmatrix} a_{11} & a_{12} & a_{13} \\ a_{21} & a_{22} & a_{23} \\ a_{31} & a_{32} & a_{33} \end{bmatrix} = \begin{bmatrix} \mathbf{A}_{11} & \mathbf{A}_{12} \\ \mathbf{A}_{21} & \mathbf{A}_{22} \end{bmatrix} \qquad (14.18)$$

where the submatrices are

$$\mathbf{A}_{11} = [a_{11}] \qquad \mathbf{A}_{12} = [a_{12} \quad a_{13}]$$

$$\mathbf{A}_{21} = \begin{bmatrix} a_{21} \\ a_{31} \end{bmatrix} \quad \text{and} \quad \mathbf{A}_{22} = \begin{bmatrix} a_{22} & a_{23} \\ a_{32} & a_{33} \end{bmatrix} \qquad (14.19)$$

The rules of matrix algebra apply to the submatrices as long as they are conformable submatrices. The matrices of Eq. (14.20) below are partitioned into comformable submatrices by placing horizontal partition lines at the same location in all matrices of Eq. (14.20). Also, observe that the position of the vertical partition line in the square matrix satisfies the conformance requirement for matrix multiplication.

The set of equations described in Eq. (14.17) is presented in matrix form as

$$\begin{bmatrix} 1 & 2 & 1 \\ 2 & 2 & 3 \\ 3 & -5 & 1 \end{bmatrix} \begin{bmatrix} x_1 \\ x_2 \\ x_3 \end{bmatrix} = \begin{bmatrix} 1 \\ 1 \\ 2 \end{bmatrix} \qquad (14.20)$$

which can be replaced by two matrices as follows:

$$1 x_1 + [2 \quad 1] \begin{bmatrix} x_2 \\ x_3 \end{bmatrix} = 1 \qquad (14.21)$$

and

$$\begin{bmatrix} 2 \\ 3 \end{bmatrix} x_1 + \begin{bmatrix} 2 & 3 \\ -5 & 1 \end{bmatrix} \begin{bmatrix} x_2 \\ x_3 \end{bmatrix} = \begin{bmatrix} 1 \\ 2 \end{bmatrix} \qquad (14.22)$$

Equation (14.21) can be rewritten for x_1 in terms of x_2 and x_3 as

$$x_1 = 1 - [2 \quad 1] \begin{bmatrix} x_2 \\ x_3 \end{bmatrix} \qquad (14.23)$$

Substitution of Eq. (14.23) into Eq. (14.22) results in an equation from which the values of x_2 and x_3 are directly solved:

$$\begin{bmatrix} 2 \\ 3 \end{bmatrix} \left[1 - \begin{bmatrix} 2 & 1 \end{bmatrix} \begin{bmatrix} x_2 \\ x_3 \end{bmatrix} \right] + \begin{bmatrix} 2 & 3 \\ -5 & 1 \end{bmatrix} \begin{bmatrix} x_2 \\ x_3 \end{bmatrix} = \begin{bmatrix} 1 \\ 2 \end{bmatrix}$$

which can be reduced to

$$-\begin{bmatrix} 4 & 2 \\ 6 & 3 \end{bmatrix} \begin{bmatrix} x_2 \\ x_3 \end{bmatrix} + \begin{bmatrix} 2 & 3 \\ -5 & 1 \end{bmatrix} \begin{bmatrix} x_2 \\ x_3 \end{bmatrix} = \begin{bmatrix} 1 \\ 2 \end{bmatrix} - \begin{bmatrix} 2 \\ 3 \end{bmatrix}$$

Collecting terms, we have

$$\begin{bmatrix} -2 & 1 \\ -11 & -2 \end{bmatrix} \begin{bmatrix} x_2 \\ x_3 \end{bmatrix} = \begin{bmatrix} -1 \\ -1 \end{bmatrix}$$

Then, the solution of x_2 and x_3 is found as

$$\begin{bmatrix} x_2 \\ x_3 \end{bmatrix} = \begin{bmatrix} -2 & 1 \\ -11 & -2 \end{bmatrix}^{-1} \begin{bmatrix} -1 \\ -1 \end{bmatrix} = \begin{bmatrix} 0.2 \\ -0.6 \end{bmatrix}$$

The value of x_1 can now be found by substitution of the values of x_2 and x_3 into Eq. (14.23) as follows:

$$x_1 = 1 - \begin{bmatrix} 2 & 1 \end{bmatrix} \begin{bmatrix} 0.2 \\ -0.6 \end{bmatrix} = 1 - (-0.2) = 1.2$$

14.2.4. Solution of Simultaneous Linear Equations

The Gauss elimination method represents one of the most widely used techniques for the solution of simultaneous linear algebraic equations. Although the method has been used for hand solution, it is particularly well-suited for computer solution of large systems of equations. The Gauss method can easily be illustrated by example using the equations of Eq. (14.17) above.

$$\begin{aligned} x_1 + 2x_2 + x_3 &= 1 \\ 2x_1 + 2x_2 + 3x_3 &= 1 \\ 3x_1 - 5x_2 + x_3 &= 2 \end{aligned} \quad (14.17)$$

First, divide each equation by the coefficient of its respective x_1 term so that the value of all leading coefficients is one. This yields

$$\begin{aligned} 1.000x_1 + 2.000x_2 + 1.000x_3 &= 1.000 \\ 1.000x_1 + 1.000x_2 + 1.500x_3 &= 0.500 \\ 1.000x_1 - 1.667x_2 + 0.333x_3 &= 0.667 \end{aligned} \quad (14.17a)$$

Now, subtract the first equation from the other equations. This action eliminates the x_1 term from the remaining two equations.

$$1.000x_1 + 2.000x_2 + 1.000x_3 = 1.000$$
$$-1.000x_2 + 0.500x_3 = -0.500$$
$$-3.667x_2 - 0.667x_3 = -0.333$$

Starting with the second equation, divide each equation by the coefficient of its respective x_2 term so that the value of all leading coefficients is $+1.0$. This yields

$$1.000x_1 + 2.000x_2 + 1.000x_3 = 1.000$$
$$1.000x_2 - 0.500x_3 = 0.500$$
$$1.000x_2 + 0.182x_3 = 0.091$$

Then, subtract the second equation from the third equation. This action eliminates the x_2 term from the third equation and the set appears as

$$1.000x_1 + 2.000x_2 + 1.000x_3 = 1.000$$
$$1.000x_2 - 0.500x_3 = 0.500 \tag{14.24}$$
$$0.682x_3 = -0.409$$

Next, division of the third equation of Eq. (14.24) by the leading coefficient 0.682 results in the solution of x_3. That is,

$$x_3 \approx -0.600$$

Back substitution of x_3 into the second equation of Eq. (14.24) is shown as

$$1.000x_2 - 0.500(-0.6) = 0.500$$

from which
$$x_2 = 0.200$$

Finally, x_1 is solved by substituting the values of x_2 and x_3 into the first equation of Eq. (14.24):

$$1.000x_1 + 2.000(0.2) + 1.000(-0.6) = 1.000$$

Thus,
$$x_1 = 1.2$$

In summary, the Gauss method performs a successive reduction of the number of unknowns in each equation of a set of $n \times n$ equations, starting with the second equation. The process is continued until the number of unknowns in the nth equation is reduced to one and the nth unknown is solved. Then, starting with equation $(n - 1)$, the remaining unknowns are found by successive back substitution. The Gauss elimination method is regarded by many as the standard method for computer use. The Gauss-Jordan and Cholesky methods represent variations to the basic Gauss algorithm which provide added reductions in the computational work. These methods are recommended for additional study.

14.3. MATRIX STRUCTURAL ANALYSIS

Analyses of statically indeterminate structures are generally performed by force and displacement methods. The basic difference between the two approaches is that the force method initially determines unknown forces and then joint displacements, whereas the displacement method initially determines joint displacements and then computes the forces (i.e., reactions, shears, moments, axial forces, etc.).

In this chapter, our attention is limited to linear elastic behavior which implies the following conditions of analysis:

1 Equilibrium and displacement compatibility equations are written at joints (called nodes) with respect to the original undeformed geometry of the structure.

2 Translational force vector components at a node remain unchanged as the node moves during structural deformation.

3 The principle of superposition can be employed.

14.3.1. Basic Concepts of the Force (Flexibility) Method

In a force method of structural analysis, selected redundant forces are first removed to form a *primary* structure which is statically determinate and geometrically stable. Then, the force method procedure continues as follows:

1 Displacements due to applied loads are computed at each redundant location in the direction of each redundant.

2 A unit load is placed on the primary structure at a redundant location in the direction of the redundant. Then, displacements due to the unit load are computed at each redundant location in the direction of each redundant, respectively. These unit-load displacements are called *flexibility coefficients*. This step is repeated to determine flexibility coefficients associated with a unit load placed at each of the other redundant locations.

3 Deflection compatibility equations are written at each redundant location in terms of the unknown redundant forces. For this method, the term *compatibility* implies that displacements at a redundant location due to both applied and redundant forces will result in geometric continuity of the structure, and with its support system. For example, at a rigid

redundant support, structural continuity exists when the sum of displacements due to all redundant forces and applied loads equals zero. Thus, the redundants found by simultaneous solution of these equations, are the forces required to maintain consistent deformations and forces between the released members and/or supports with the primary structure. This approach is called the *method of consistent deformations* (Chap. 11) and is often referred to as the *compatibility* or *flexibility method*. An illustrated example follows.

The continuous beam of Fig. 14.1 contains four supports that are identified by a joint (node) number of 1 to 4, from left to right. If the roller reactions at nodes 1, 2, and 3 are the selected redundants, the deformed primary structure under load appears as shown in Fig. 14.1a. Observe that the free displacements at redundant locations due to the applied loads are symbolized by single subscript notation (δ_1, δ_2, and δ_3). Figures 14.1b to 14.1d show free nodal displacements due to placement of a unit load at a redundant location in the direction of the unknown redundant. These displacements (flexibility coefficients) are designated by double subscript notation; i.e., f_{13} denotes the displacement at 1 due to a unit load at 3; f_{21} denotes the displacement at 2 due to a unit load at 1, and so forth.

Deflection equations can be written at each support node as

$$f_{11} Y_1 + f_{12} Y_2 + f_{13} Y_3 = \Sigma\delta_1 + \delta_1$$
$$f_{21} Y_1 + f_{22} Y_2 + f_{23} Y_3 = \Sigma\delta_2 + \delta_2 \quad (14.25)$$
$$f_{31} Y_1 + f_{32} Y_2 + f_{33} Y_3 = \Sigma\delta_3 + \delta_3$$

where $\Sigma\delta_1$, $\Sigma\delta_2$, and $\Sigma\delta_3$ are the support displacements at 1, 2, and 3, respectively (if nodes 1, 2, and 3 are unyielding supports, $\Sigma\delta_1 = \Sigma\delta_2 = \Sigma\delta_3 = 0$). These equations can be written in matrix form as

$$\begin{bmatrix} \delta_1 \\ \delta_2 \\ \delta_3 \end{bmatrix} + \begin{bmatrix} f_{11} + f_{12} + f_{13} \\ f_{21} + f_{22} + f_{23} \\ f_{31} + f_{32} + f_{33} \end{bmatrix} \begin{bmatrix} Y_1 \\ Y_2 \\ Y_3 \end{bmatrix} = \begin{bmatrix} \Sigma\delta_1 \\ \Sigma\delta_2 \\ \Sigma\delta_3 \end{bmatrix} \quad (14.26)$$

$$[\delta] \qquad [f] \qquad [Y] \qquad [\Sigma\delta]$$

or in a compact symbolic form as

$$[\delta] + [f][Y] = [\Sigma\delta] \quad (14.27)$$

where $[f]$ is a square matrix of flexibility coefficients.

The Maxwell-Betti reciprocal theorem (Sec. 10.8) ensures that $f_{ij} = f_{ji}$ for linear elastic structures which undergo small displacements; therefore, the flexibility matrix $[f]$ is symmetric. It will be subsequently shown that the inverse of a symmetric matrix is also a symmetric matrix. This matrix property can greatly simplify the computations, assemblage, and storage of the

Matrix Structural Analysis

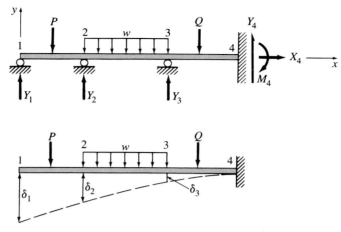

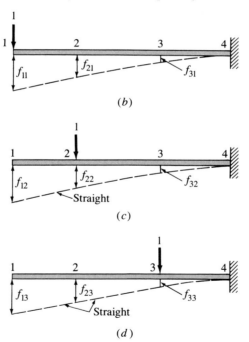

Figure 14.1

flexibility matrix and force-displacement relations in computer structural analysis procedures. If all elements in matrices $[\delta]$, $[\Sigma\delta]$ and $[f]$ are known, $[Y]$ forces can be found by use of the matrix inverse operation or by direct solution using a Gauss elimination routine.

The flexibility method is less amenable to computer usage when compared with the displacement (stiffness) method discussed below. In a flexibility method of analysis, we must initially establish the degree of statical indeterminacy of the structure. For a complex structure with many members and constraints, it can be very difficult to determine how many redundants exist or which ones should be removed to produce a statically determinate primary structure. In the stiffness method, it is not necessary to select redundants or to know whether the structure is statically determinate or indeterminate. Also, if a structure is unstable, the solution cannot be obtained. In fact, most matrix stiffness computer programs will warn the user that an instability exists. Consequently, the stiffness method is generally preferred and has become the dominant choice for computer structural analysis.

At this point in the text, the author has elected to curtail discussions on the flexibility method. References [14.2] to [14.4] at the end of this chapter are recommended for added study of the flexibility method. The remainder of this chapter is devoted to studies of the displacement (stiffness) method of structural analysis.

14.3.2. Global and Local Coordinate Systems

Before we begin our study of the displacement method, it is appropriate to establish a coordinate axes system which will provide a frame of reference to identify the node force and node displacement vectors of a structure. The joint deformations of a structure are usually described with reference to *global* and *local* coordinate systems. The total structure is referenced to global coordinates which are generally defined as an xyz right-hand system; for planar structures, xy defines the plane of the member and z is normal to the plane. Each member of the structure is also identified by an $x'y'z'$ local coordinate system, where x' is located along the member axis, and y' and z' are directed along its principal cross-section axes. Both coordinate systems are illustrated for a plane-truss structure in Fig. 14.2 where the joints are identified by consecutive numbering from 1 to 5. The origin of the local coordinates is usually set at a member end (observe the local $x'y'$ axes of member 2–3 in Fig. 14.2 with origin at node 2). In the global coordinate system, positive joint deformations are defined as u, v, and w, along the positive x, y, and z axes, respectively. Similarly, positive joint displacements u', v', and w' occur in positive directions along the respective x', y', and z' axes of the local coordinate system (see Fig. 14.2).

Figure 14.2

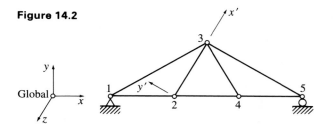

Axis	Displacement
x	u
y	v
z	w
x'	u'
y'	v'
z'	w'

14.3.3. Degrees of Freedom

In the process of modeling a structure for matrix analysis, all member nodes are initially treated as unrestrained and free to displace and rotate as the structure deforms under load. Each deformation component is called a degree of freedom. For a spacial structure, each node may have six degrees of freedom (three translation components and three rotation components). For a planar beam structure, three degrees of freedom exist at a member node, namely, translation in the x and y (or x' and y') directions and rotation about the z (or z') axis. An ideal plane truss contains two force members which have two translation degrees of freedom at each pin end. Obviously, some of the nodal degrees of freedom on an assembled structural model represent support points which must be restrained to prevent rigid-body motion and ensure geometric stability.

14.3.4. Basic Concepts of the Displacement (Stiffness) Method

Overview In a displacement method of structural analysis, a set of force equilibrium equations is written at each joint in terms of unknown displacements. The equations are cast after both member force-displacement relationships at each joint, and deflection compatibility between external nodes with corresponding member ends are satisfied. The term deflection compatibility indicates that the integral structure and each of its members fit together at their respective node points both before and after loads are applied. Obviously, some of the nodal displacements must be restrained to prevent rigid body motion and insure geometric stability of the structure.

After known displacements are prescribed (usually zero at supports), the unknown joint displacements are determined by simultaneous solution of the equation set. Then, joint member forces are obtained from the member force-displacement relationships. Subsequently, support reactions are found by the laws of statics. Slope deflection (Chap. 12) and moment distribution (Chap. 13) are displacement methods.

A. Spring analogy The objective of this section is to develop the fundamentals of the displacement method of structural analysis by use of elementary examples. The reader will come to realize that the basic concepts developed in these early examples are employed, but not limited to matrix analysis of truss-, beam-, and frame-type structures. Therefore, let us begin our study by considering an ideal pin-connected truss model as a system of linear elastic springs (see Fig. 14.3). This analogy is valid since ideal truss members only support axial force. From basic mechanics, the force-deformation relationship of an axially loaded member is $\delta = FL/AE$ or $F = AE\delta/L$. Furthermore, we know that the force F required to produce a unit deformation is called a spring constant, namely, $k = AE(1)/L = AE/L$. In matrix structural analysis, a member force which produces a corresponding unit displacement is called a *stiffness influence coefficient* or an *element stiffness coefficient*.

Now, let us attempt to write the force-displacement relationship for an arbitrary truss member ij. If node-end j is held fixed and an axial force is applied at node-end i, the force is

454 Matrix Structural Analysis

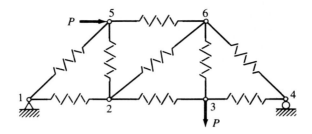

Figure 14.3   Planar truss spring model

$$F_i = k_{ij}\delta_i$$

where F_i = the internal force of member ij at node i
           k_{ij} = the element stiffness coefficient
           δ_i = the displacement at node i (assuming node j is fixed)

    However, the truss model of Fig. 14.3 reveals that both ends of four truss members will displace when the truss deflects due to the P loads. Therefore, a truss-member axial force is dependant on the relative displacement of *both* member ends and its element stiffness value. Our next objective is to formulate a force-displacement relationship based on the relative motion at both truss member ends.

B. Stiffness matrix for a horizontal truss member Figure 14.4*a* is an isolated sketch of horizontal member 1–2 of the truss shown in Fig. 14.3.

Figure 14.4

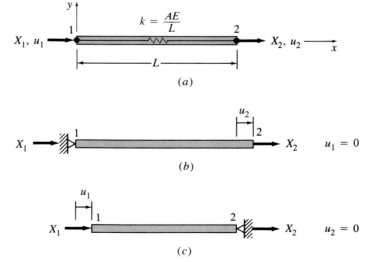

Once again, this truss member is illustrated as a linear elastic spring that is able to transmit force and deform in its axial direction. The ends of the member are the node locations where forces are applied and displacements are measured. Observe that both the node forces and node displacements are shown in the positive x axis direction (in this case, the global and local axes are coincident). In addition, AE is constant along the length L of the member.

Consider a member along the global x axis. The member force-displacement relationship is the sum of the two following conditions:

Condition 1—Place a restraint at node 1 (see Fig. 14.4b)

The member node forces, X_i and X_j represent the respective components of F_i and F_j in the global x direction. Likewise, u_i and u_j represent the respective components of nodal displacements δ_i and δ_j in the global x direction.

If node 1 is prevented from moving, and force X_2 is applied, member 1–2 is in tension and

$$u_2 = \frac{X_2 L}{AE} \quad \text{or} \quad X_2 = \frac{AEu_2}{L} \tag{14.28}$$

Horizontal force equilibrium requires that

$$X_1 = -X_2 = -\frac{AEu_2}{L} \tag{14.29}$$

(*Note:* signs of force conform to positive axis direction shown in Fig. 14.4a.)

Condition 2—Place a restraint at node 2 (see Fig. 14.4c)

If node 2 is prevented from moving, and force X_1 is applied, member 1–2 is in compression. Using the sign convention of Fig. 14.4a above, we find that

$$X_1 = \frac{AEu_1}{L} \quad \text{and} \quad X_2 = -X_1 = -\frac{AEu_1}{L} \tag{14.30}$$

By the principle of superposition, the force-displacement relationship of member 1–2 is summed in matrix notation as

$$\begin{bmatrix} X_1 \\ X_2 \end{bmatrix} = \begin{bmatrix} \dfrac{AE}{L} & -\dfrac{AE}{L} \\ -\dfrac{AE}{L} & \dfrac{AE}{L} \end{bmatrix} \begin{bmatrix} u_1 \\ u_2 \end{bmatrix} \tag{14.31}$$

$$\quad \textbf{X} \qquad\qquad \textbf{K} \qquad\quad \textbf{u}$$

456 Matrix Structural Analysis

which in compact matrix form is

$$[X] = [K][u] \qquad (14.32)$$

where $[X]$ = the column matrix of nodal member forces
$[K]$ = the element stiffness matrix
$[u]$ = the column matrix of nodal displacements

Observe that the element stiffness matrix $[K]$ is a square symmetric matrix. If we replace each element stiffness coefficient value (AE/L) in the **K** matrix by k_{12}, Eq. (14.31) can be rewritten as

$$\begin{bmatrix} X_1 \\ X_2^1 \end{bmatrix} = \begin{bmatrix} \dfrac{AE}{L} & -\dfrac{AE}{L} \\ -\dfrac{AE}{L} & \dfrac{AE}{L} \end{bmatrix} \begin{bmatrix} u_1 \\ u_2 \end{bmatrix} = \underbrace{\begin{bmatrix} k_{12} & -k_{12} \\ -k_{12} & k_{12} \end{bmatrix}}_{\mathbf{K}_{12}} \begin{bmatrix} u_1 \\ u_2 \end{bmatrix} \qquad (14.33)$$

where k_{12} represents the axial element stiffness coefficient of member 1–2. X_2^1 in Eq. (14.33) above is shown in Fig. 14.5.

Now consider two horizontal truss members pin connected at node 2 as shown in Fig. 14.5 where k_{12} and k_{23} represent the respective spring constants of members 1–2 and 2–3, and X_1, X_2, and X_3 represent internal member forces. As before, the members are perceived as two linear springs in series which experience nodal forces and nodal displacements at nodes 1, 2, and 3. The element stiffness matrix of truss-member 2–3 is found by the same approach used for member 1–2. Thus, Eq. (14.32) can be written for member 2–3 as

$$\begin{bmatrix} X_2^2 \\ X_3 \end{bmatrix} = \begin{bmatrix} \dfrac{AE}{L} & -\dfrac{AE}{L} \\ -\dfrac{AE}{L} & \dfrac{AE}{L} \end{bmatrix} \begin{bmatrix} u_2 \\ u_3 \end{bmatrix} = \underbrace{\begin{bmatrix} k_{23} & -k_{23} \\ -k_{23} & k_{23} \end{bmatrix}}_{\mathbf{K}_{23}} \begin{bmatrix} u_2 \\ u_3 \end{bmatrix} \qquad (14.34)$$

Figure 14.5

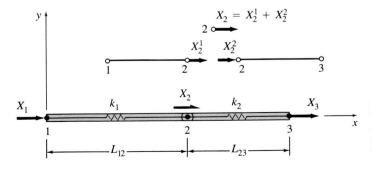

The total stiffness matrix for the two member assembly of Fig. 14.5 contains elements of u_1, u_2, and u_3; thus, the matrix is of order 3. The total matrix is easily found by superposition after the elements of both the $\mathbf{K}_{12}$ and $\mathbf{K}_{23}$ matrices are expanded into a 3×3 array as follows:

$$\begin{array}{ccc} u_1 & u_2 & u_3 \end{array} \quad \begin{array}{ccc} u_1 & u_2 & u_3 \end{array} \quad \begin{array}{ccc} u_1 & u_2 & u_3 \end{array}$$

$$\begin{bmatrix} k_{12} & -k_{12} & 0 \\ -k_{12} & k_{12} & 0 \\ 0 & 0 & 0 \end{bmatrix} + \begin{bmatrix} 0 & 0 & 0 \\ 0 & k_{23} & -k_{23} \\ 0 & -k_{23} & k_{23} \end{bmatrix} = \begin{bmatrix} k_{12} & -k_{12} & 0 \\ -k_{12} & (k_{12}+k_{23}) & -k_{23} \\ 0 & -k_{23} & k_{23} \end{bmatrix} \quad (14.35)$$

$$\mathbf{K}_{12} \qquad\qquad \mathbf{K}_{23} \qquad\qquad\qquad \mathbf{K}$$

Observe that only the order of $\mathbf{K}_{12}$ and $\mathbf{K}_{23}$ is changed by the addition of a row and column of zero elements shown above. The order of the stiffness matrix in Eq. (14.35) will increase to 4×4 if, for example, the lower chord member 3–4 is joined to the existing two lower chord members at node 3.

It is interesting to note that the stiffness matrices of Eqs. (14.33), (14.34), and (14.35) are symmetric. This fact is evident by the Maxwell-Betti reciprocal law where $k_{ij} = k_{ji}$. Also, all elements along the main diagonal are positive. This indicates that a positive directed force produces a corresponding positive displacement which is physically evident.

C. Stiffness matrices for truss members in local coordinates

It is obvious that our truss structure model of Fig. 14.3 contains a mix of horizontal, vertical, and inclined members. In general, a global coordinate system offers a convenient and efficient way to define the total structure, the support restraints, and the applied load vectors. However, there are many occasions where it is convenient to define a structure in both the local and global coordinate systems. As an example, consider a truss with members fastened together by rigid end connections. When loads are applied, the structure will develop shear and bending, as well as axial member force. Thus, it is convenient to reference node coordinates, applied loads, and support restraints to global axes, whereas member forces are clearly defined with reference to the local axes. Nevertheless, each element stiffness matrix is generally written in terms of the local axes and subsequently transformed with reference to a set of global axes for assembly of the overall stiffness matrix of the structure. The approach that follows is developed in the same manner as presented by Martin [14.2].

Figure 14.6 shows a pin-connected planar truss member in both global and local axis orientation. In local coordinates, we notice two nodal forces, X'_1 and X'_2; in global coordinates, X'_1 is replaced by its components X_1 and Y_1, and X'_2 is replaced by components X_2 and Y_2. Local nodal displacement components are specified as u' and v' for the x' and y' axis, respectively. Equations (14.31) can be redefined in terms of local coordinates as

Figure 14.6

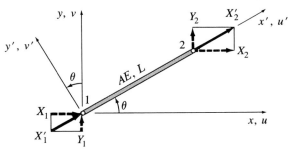

Local (x', y') and global (x, y) coordinate systems for plane truss member

$$\begin{bmatrix} X'_1 \\ X'_2 \end{bmatrix} = \begin{bmatrix} \dfrac{AE}{L} & -\dfrac{AE}{L} \\ -\dfrac{AE}{L} & \dfrac{AE}{L} \end{bmatrix} \begin{bmatrix} u'_1 \\ u'_2 \end{bmatrix} = \dfrac{AE}{L} \begin{bmatrix} 1 & -1 \\ -1 & 1 \end{bmatrix} \begin{bmatrix} u'_1 \\ u'_2 \end{bmatrix} \quad (14.36)$$

Observe that $Y'_1 = Y'_2 = 0$, since a pin-connected truss member can only support axial force components and undergo axial displacements. Also, there are four force and four displacement components (degrees of freedom) in terms of global coordinates.

In order to permit proper transformation from local to global coordinates, the 2×2 matrix of Eq. (14.36) must be expanded to a 4×4 matrix which include Y'_1, Y'_2, and their associated displacements, v'_1 and v'_2. In reality, Eqs. (14.36) remain the same with values of zero in the row and column positions where Y'_1, Y'_2, v'_1, and v'_2 reside. This expanded form of Eq. (14.36) is rewritten as

$$\begin{bmatrix} X'_1 \\ Y'_1 \\ X'_2 \\ Y'_2 \end{bmatrix} = \dfrac{AE}{L} \begin{bmatrix} 1 & 0 & -1 & 0 \\ 0 & 0 & 0 & 0 \\ -1 & 0 & 1 & 0 \\ 0 & 0 & 0 & 0 \end{bmatrix} \begin{bmatrix} u'_1 \\ v'_1 \\ u'_2 \\ v'_2 \end{bmatrix} \quad (14.37)$$

$$\mathbf{X'} \qquad\qquad\qquad \mathbf{K'} \qquad\qquad \mathbf{u'}$$

which in compact symbolic matrix form is

$$[\mathbf{X'}] = [\mathbf{K'}][\mathbf{u'}] \quad (14.38)$$

D. Transformation of truss stiffness matrix From Fig. 14.6, it is seen that

$$\begin{aligned} X'_1 &= X_1 \cos\theta + Y_1 \sin\theta \\ Y'_1 &= -X_1 \sin\theta + Y_1 \cos\theta \\ X'_2 &= X_2 \cos\theta + Y_2 \sin\theta \\ Y'_2 &= -X_2 \sin\theta + Y_2 \cos\theta \end{aligned} \quad (14.39)$$

If we let

$$s = \sin\theta \quad \text{and} \quad c = \cos\theta \quad (14.40)$$

the force relationships of Eq. (14.39) may be written as

$$\begin{bmatrix} X'_1 \\ Y'_1 \\ X'_2 \\ Y'_2 \end{bmatrix} = \begin{bmatrix} c & s & 0 & 0 \\ -s & c & 0 & 0 \\ 0 & 0 & c & s \\ 0 & 0 & -s & c \end{bmatrix} \begin{bmatrix} X_1 \\ Y_1 \\ X_2 \\ Y_2 \end{bmatrix} \quad (14.41)$$
$$\quad \mathbf{X'} \qquad\qquad \mathbf{T} \qquad\quad \mathbf{X}$$

or
$$[\mathbf{X'}] = [\mathbf{T}][\mathbf{X}] \quad (14.42)$$

where **T** is called the transformation matrix.

The partition lines in Eq. (14.41) show that the **T** matrix is symmetric. This property greatly simplifies the inverse operations involving **T** since the inverse of **T** is also its transpose. Thus, to solve Eq. (14.42) in terms of **X**, we get

$$[\mathbf{X}] = [\mathbf{T}]^{-1}[\mathbf{X'}] = [\mathbf{T}]^{T}[\mathbf{X'}] \quad (14.43)$$

Similarly, the displacement vectors will transform from local to global coordinates in the same way as the forces. That is,

$$[\mathbf{u'}] = [\mathbf{T}][\mathbf{u}] \quad (14.44)$$

Now, we can use the transformation matrix **T** to establish the stiffness matrix in global coordinates. Substituting Eqs. (14.42) and (14.44) into Eq. (14.38) gives

$$\mathbf{TX} = \mathbf{K'u'} = \mathbf{K'Tu} \quad (14.45)$$

Solving for **X**:

$$\mathbf{X} = [\mathbf{T}]^{-1}\mathbf{K'Tu} = \mathbf{T}^{T}\mathbf{K'Tu} = \mathbf{Ku} \quad (14.46)$$

where the element stiffness matrix in global coordinates is defined as

$$\mathbf{K} = \mathbf{T}^{T}\mathbf{K'T} \quad (14.47)$$

Therefore, after the stiffness matrix **K'** is found in terms of convenient local coordinates, it is easy to transform **K'** to an arbitrary global reference system.

We can find **K** by substituting **K′** from Eq. (14.37) and **T** from Eq. (14.41) into Eq. (14.47). The result is

$$\mathbf{K}_{12} = \frac{AE}{L} \begin{array}{c} \\ \end{array} \begin{matrix} u_1 & v_1 & u_2 & v_2 \end{matrix} \\ \begin{bmatrix} c^2 & sc & -c^2 & -sc \\ sc & s^2 & -sc & -s^2 \\ -c^2 & -sc & c^2 & sc \\ -sc & -s^2 & sc & s^2 \end{bmatrix} \tag{14.48}$$

where $s = \sin\theta$ and $c = \cos\theta$.

Eqs. (14.48) can be rewritten in a general form as

$$\mathbf{K}_{ij} = \frac{AE}{L} \begin{matrix} u_i & v_i & u_j & v_j \end{matrix} \\ \begin{bmatrix} c^2 & sc & -c^2 & -sc \\ sc & s^2 & -sc & -s^2 \\ -c^2 & -sc & c^2 & sc \\ -sc & -s^2 & sc & s^2 \end{bmatrix} \tag{14.49}$$

where $s = \sin\theta$ and $c = \cos\theta$ and where K_{ij} is the element stiffness matrix of member ij.

Consequently, the stiffness matrix of a truss member may be determined in terms of global coordinates in two ways: (1) by the direct use of Eq. (14.49), which is often referred to as the *direct stiffness method*, and (2) by first deriving **K′** and **T**, and then performing the transformation of Eq. (14.47). Although the first option appears easy for truss members, it becomes difficult for more complex structural units (beams, plates, shells, etc.). Many computer codes have been written to provide computer algorithms to form the **K′** and **T** matrices from geometric property input data and perform the transformation operation of Eq. (14.47). Yet, other computer codes exist which directly form the stiffness matrix **K** from Eq. (14.49). Hence, the name direct stiffness method is apt.

E. Formulation of the joint-equilibrium equations After the element stiffness matrices (force-displacement relationships) of a structural system are formed, we can proceed to write force-equilibrium equations at each node. These force equations are written to satisfy the conditions of static equilibrium and nodal displacement compatibility at each joint. A careful study of the matrix construction of these equations is mandatory since it provides the reader with a clear understanding of the fundamental approach to assemble a structure-stiffness matrix. An example follows:

Development of a force-equilibrium equation The planar truss of Fig. 14.3 is redrawn in Fig. 14.7 with degrees of freedom specified at all member nodes and each member assigned a letter name. Recall that each

Figure 14.7

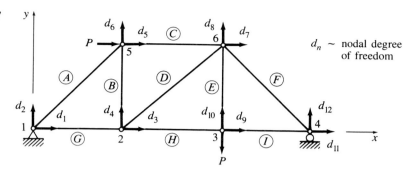

nodal displacement component is a degree of freedom. The objective of the following study is to formulate the force-equilibrium equation at node 5 in the global x direction.

Figure 14.8 shows the three members which are joined at node 5. For sake of clarity and generalization, only global x internal-force components are shown for members A, B, and C.

Observe that all member node forces are given in the positive global x direction. An external force P, which happens to be directed in the positive global x direction, also exists at node 5. Horizontal equilibrium at node 5 requires that

$$P = X_5^A + X_5^B + X_5^C \qquad (14.50)$$

The force-displacement relationship for a truss element can be expressed in terms of its nodal degrees of freedom as

$$X = k_{5n} \cdot d_n \qquad (14.51)$$

where k_{5n} = an element stiffness at node 5,
n = a nodal degree of freedom
d_n = a degree of freedom displacement (see Fig. 14.7).

Substitution of Eq. (14.51) for each member at node 5 into Eq. (14.50)

Figure 14.8

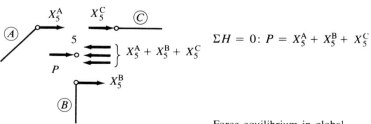

Force equilibrium in global x direction at node 5

yields

$$P = k_{51}^A \cdot d_1^A + k_{52}^A \cdot d_2^A + k_{55}^A \cdot d_5^A + k_{56}^A \cdot d_6^A$$
$$+ k_{53}^B \cdot d_3^B + k_{54}^B \cdot d_4^B + k_{55}^B \cdot d_5^B + k_{56}^B \cdot d_6^B$$
$$+ k_{55}^C \cdot d_5^C + k_{56}^C \cdot d_6^C + k_{57}^C \cdot d_7^C + k_{58}^C \cdot d_8^C \quad (14.52)$$

Since the member displacements at node 5 must be compatible, we can state that $d_5^A = d_5^B = d_5^C = d_5$ and $d_6^A = d_6^B = d_6^C = d_6$. By combining terms,

$$P = (k_{55}^A + k_{55}^B + k_{55}^C)d_5 + (k_{56}^A + k_{56}^B + k_{56}^C)d_6$$
$$+ k_{51}^A \cdot d_1^A + \cdots + k_{58}^C \cdot d_8^C$$

The element stiffness coefficients with the same subscripts have a common degree of freedom. Thus, they can be added to form one coefficient of the equation. In essence, this is the basis for assembly of the structure stiffness matrix. As an example, let

$$K_{55} = k_{55}^A + k_{55}^B + k_{55}^C \quad \text{and} \quad K_{56} = k_{56}^A + k_{56}^B + k_{56}^C$$

Then,

$$P = K_{51} \cdot d_1 + K_{52} \cdot d_2 + K_{53} \cdot d_3 + \cdots + K_{58} \cdot d_8 \quad (14.53)$$

The capital terms, $K_{51}, K_{52}, K_{53}, \ldots, K_{58}$, are the global stiffness coefficients for the horizontal force-equilibrium equation at node 5. Similar equations can be written at all truss structure nodes in both the x and y directions.

An assembly of all node force equilibrium equations, cast in matrix form, can be expressed as

$$\begin{bmatrix} P_1 \\ \vdots \\ P_n \end{bmatrix} = [n \times n] \begin{bmatrix} d_1 \\ \vdots \\ d_n \end{bmatrix} \quad (14.54)$$
$$[\mathbf{P}] \quad \quad [\mathbf{K_S}] \quad [\mathbf{d}]$$

where $[\mathbf{P}]$ = the column matrix of external node forces
$[\mathbf{K_S}]$ = the global structure stiffness matrix
$[\mathbf{d}]$ = the column matrix of nodal displacements

F. Assembly of the structure stiffness matrix Example 14.1 illustrates the formation of the element and overall stiffness matrices of a five-member truss structure. Each element matrix is found directly by use of Eq.

(14.49). The assembly of a structure stiffness matrix K_S is similar to sorting letters at a post office; that is, there is a special cell where each element coefficient must be placed. Figure (b) in Example 14.1 illustrates the formation of K_S. Observe that each element value of K_S is expressed in terms of length L_{2-4}.

Example 14.2 illustrates an overlay approach to form the structure stiffness matrix for the plane truss of Example 14.2. Each element matrix is expanded into a 6 × 6 array format. Then, each cell in K_S is viewed as the sum of three overlays.

EXAMPLE 14.1

Determine the overall (structural) stiffness matrix for the truss shown below. The origin of the global axes is given at node 1. Load and reaction forces are shown for effect. AE is constant.

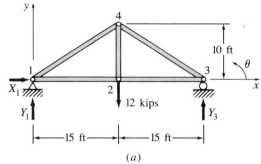

(a)

Solution

Member	$\theta°$	s	c	sc	c^2	s^2
1–2	0	0	1	0	1	0
1–4	33.7	0.555	0.832	0.462	0.692	0.308
2–3	0	0	1	0	1	0
2–4	90	1	0	0	0	1
3–4	146.3	0.555	−0.832	−0.462	0.692	0.308

$s = \sin\theta; c = \cos\theta$

$$K_{ij} = \frac{AE}{L} \begin{bmatrix} c^2 & sc & -c^2 & -sc \\ sc & s^2 & -sc & -s^2 \\ -c^2 & -sc & c^2 & sc \\ -sc & -s^2 & sc & s^2 \end{bmatrix} \begin{matrix} u_i \\ v_i \\ u_j \\ v_j \end{matrix} \quad (14.49)$$

Let $L_{2-4} = L$. Then, $L_{1-2} = L_{2-3} = 1.5L$ and $L_{1-4} = L_{3-4} \cong 1.8L$

$$\mathbf{K}_{1-2} = \frac{AE}{1.5L} \begin{matrix} & u_1 & v_1 & u_2 & v_2 \\ & \begin{bmatrix} 1 & 0 & -1 & 0 \\ 0 & 0 & 0 & 0 \\ -1 & 0 & 1 & 0 \\ 0 & 0 & 0 & 0 \end{bmatrix} \end{matrix}; \quad \mathbf{K}_{1-4} = \frac{AE}{1.8L} \begin{matrix} & u_1 & v_1 & u_4 & v_4 \\ & \begin{bmatrix} 0.692 & 0.462 & -0.692 & -0.462 \\ 0.462 & 0.308 & -0.462 & -0.308 \\ -0.692 & -0.462 & 0.692 & 0.462 \\ -0.462 & -0.308 & 0.462 & 0.308 \end{bmatrix} \end{matrix};$$

$$\mathbf{K}_{2-3} = \frac{AE}{1.5L} \begin{matrix} & u_2 & v_2 & u_3 & v_3 \\ & \begin{bmatrix} 1 & 0 & -1 & 0 \\ 0 & 0 & 0 & 0 \\ -1 & 0 & 1 & 0 \\ 0 & 0 & 0 & 0 \end{bmatrix} \end{matrix}$$

$$\mathbf{K}_{2-4} = \frac{AE}{L} \begin{matrix} & u_2 & v_2 & u_4 & v_4 \\ & \begin{bmatrix} 0 & 0 & 0 & 0 \\ 0 & 1 & 0 & -1 \\ 0 & 0 & 0 & 0 \\ 0 & -1 & 0 & 1 \end{bmatrix} \end{matrix}; \quad \mathbf{K}_{3-4} = \frac{AE}{1.8L} \begin{matrix} & u_3 & v_3 & u_4 & v_4 \\ & \begin{bmatrix} 0.692 & -0.462 & -0.692 & 0.462 \\ -0.462 & 0.308 & 0.462 & -0.308 \\ -0.692 & 0.462 & 0.692 & -0.462 \\ 0.462 & -0.308 & -0.462 & 0.308 \end{bmatrix} \end{matrix}$$

The structure matrix $\mathbf{K}_s$ is easily obtained by expanding each of the element matrices above into a form suitable for overlay. The order of $\mathbf{K}$ is 8×8 which is shown in grid form as follows:

in expanded form,

$\mathbf{K}_{1-2}$ populates the upper left quadrant cells.
$\mathbf{K}_{3-4}$ populates the lower right quadrant cells.
$\mathbf{K}_{2-3}$ populates the 16 central cells.
$\mathbf{K}_{1-4}$ populates the shaded cells.
$\mathbf{K}_{2-4}$ populates the cross-hatched cells.

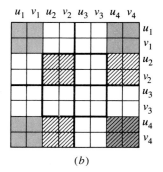

(b)

The assembled stiffness matrix is

$$\mathbf{K}_s = \frac{AE}{L} \begin{matrix} & u_1 & v_1 & u_2 & v_2 & u_3 & v_3 & u_4 & v_4 \\ & \begin{bmatrix} 1.051 & 0.256 & -0.667 & 0 & 0 & 0 & -0.384 & -0.256 \\ 0.256 & 0.171 & 0 & 0 & 0 & 0 & -0.256 & -0.171 \\ -0.667 & 0 & 1.334 & 0 & -0.667 & 0 & 0 & 0 \\ 0 & 0 & 0 & 1.0 & 0 & 0 & 0 & -1.0 \\ 0 & 0 & -0.667 & 0 & 1.051 & -0.256 & -0.384 & 0.256 \\ 0 & 0 & 0 & 0 & -0.256 & 0.171 & 0.256 & -0.171 \\ -0.384 & -0.256 & 0 & 0 & -0.384 & 0.256 & 0.768 & 0 \\ -0.256 & -0.171 & 0 & -1.0 & 0.256 & -0.171 & 0 & 1.342 \end{bmatrix} \end{matrix}$$

EXAMPLE 14.2

Determine the element stiffness matrix of each truss member and the overall stiffness matrix of the truss structure shown below. Let $AE = 8000$ kN for all members.
Solution: Set origin of global axes at node 1.

Member	$\theta°$	c^*	s^*	c^2	s^2	sc
12	0	1	0	1	0	0
13	90	0	1	0	1	0
23	135	$-1/\sqrt{2}$	$1/\sqrt{2}$	1/2	1/2	$-1/2$

* ($s = \sin \theta$ and $c = \cos \theta$)

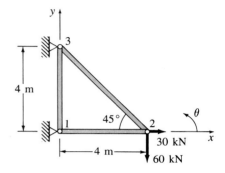

Element stiffness matrices: [Refer to Eqs. (14.49)]

Let $L = L_{12} = L_{13}$

Then, $L_{23} = \sqrt{2}L$

$$K_{1-2} = \frac{AE}{L} \begin{bmatrix} u_1 & v_1 & u_2 & v_2 & u_3 & v_3 \\ 1 & 0 & -1 & 0 & & \\ 0 & 0 & 0 & 0 & & \\ -1 & 0 & 1 & 0 & & \\ 0 & 0 & 0 & 0 & & \\ & & & & & \\ & & & & & \end{bmatrix} ; \quad K_{1-3} = \frac{AE}{L} \begin{bmatrix} u_1 & v_1 & u_2 & v_2 & u_3 & v_3 \\ 0 & 0 & & & 0 & 0 \\ 0 & 1 & & & 0 & -1 \\ & & & & & \\ & & & & & \\ 0 & 0 & & & 0 & 0 \\ 0 & -1 & & & 0 & 1 \end{bmatrix} ;$$

$$K_{2-3} = \frac{AE}{L} \begin{bmatrix} u_1 & v_1 & u_2 & v_2 & u_3 & v_3 \\ & & & & & \\ & & & & & \\ & & 1/2\sqrt{2} & -1/2\sqrt{2} & -1/2\sqrt{2} & 1/2\sqrt{2} \\ & & -1/2\sqrt{2} & 1/2\sqrt{2} & 1/2\sqrt{2} & -1/2\sqrt{2} \\ & & -1/2\sqrt{2} & 1/2\sqrt{2} & 1/2\sqrt{2} & -1/2\sqrt{2} \\ & & 1/2\sqrt{2} & -1/2\sqrt{2} & -1/2\sqrt{2} & 1/2\sqrt{2} \end{bmatrix}$$

The overall (structure) stiffness matrix is assembled by merging the element matrices in an overlay-type fashion. Since $\mathbf{K}_s = \mathbf{K}_{12} + \mathbf{K}_{13} + \mathbf{K}_{23}$ where $\mathbf{K}_s$ represents the structure stiffness matrix, we arrive at the following 6×6 matrix:

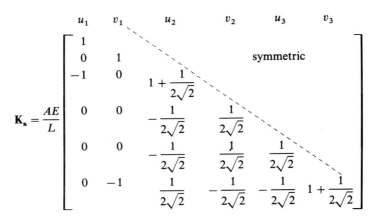

By final substitution of AE and L, we arrive at

$$\mathbf{K}_s = \begin{matrix} & u_1 & v_1 & u_2 & v_2 & u_3 & v_3 \end{matrix}$$

$$\mathbf{K}_s = \begin{bmatrix} 2000 & 0 & -2000 & 0 & 0 & 0 \\ 0 & 2000 & 0 & 0 & 0 & -2000 \\ -2000 & 0 & 2707 & -707 & -707 & 707 \\ 0 & 0 & -707 & 707 & 707 & -707 \\ 0 & 0 & -707 & 707 & 707 & -707 \\ 0 & -2000 & 707 & -707 & -707 & 2707 \end{bmatrix} \quad \text{(kN/m)}$$

G. Applications for truss analysis After the structural stiffness matrix is formed, we can proceed to determine the joint displacements, the support reactions, and the internal member forces. The truss of Example 14.2 and the corresponding stiffness matrix of Example 14.2 are used as follows:

The force-displacement relationships at the truss joints are shown in matrix form as

$$\begin{bmatrix} X_1 \\ Y_1 \\ X_2 \\ Y_2 \\ X_3 \\ Y_3 \end{bmatrix} = \begin{bmatrix} 2000 & 0 & -2000 & 0 & 0 & 0 \\ 0 & 2000 & 0 & 0 & 0 & -2000 \\ -2000 & 0 & 2707 & -707 & -707 & 707 \\ 0 & 0 & -707 & 707 & 707 & -707 \\ 0 & 0 & -707 & 707 & 707 & -707 \\ 0 & -2000 & 707 & -707 & -707 & 2707 \end{bmatrix} \begin{bmatrix} u_1 \\ v_1 \\ u_2 \\ v_2 \\ u_3 \\ v_3 \end{bmatrix} \quad (14.55)$$

From the figure in Example 14.2, we observe that pin supports exist at nodes 1 and 3; thus, $u_1 = u_3 = v_1 = v_3 = 0$. Now, the equations of the matrix shown as Eqs. (14.55), can be rewritten in partitioned form to separate the equations with known displacements from those with unknown displacements. The partitioning leads to

$$\begin{array}{c} \mathbf{X_u} \\ \begin{bmatrix} X_2 \\ Y_2 \\ \hdashline X_1 \\ Y_1 \\ X_3 \\ Y_3 \end{bmatrix} \end{array} = \begin{array}{c} \mathbf{K_{11}} \quad\quad\quad\quad\quad \mathbf{K_{12}} \\ \begin{bmatrix} 2707 & -707 & -2000 & 0 & -707 & 707 \\ -707 & 707 & 0 & 0 & 707 & -707 \\ \hdashline -2000 & 0 & 2000 & 0 & 0 & 0 \\ 0 & 0 & 0 & 2000 & 0 & -2000 \\ -707 & 707 & 0 & 0 & 707 & -707 \\ 707 & -707 & 0 & -2000 & -707 & 2707 \end{bmatrix} \\ \mathbf{K_{21}} \quad\quad\quad\quad\quad \mathbf{K_{22}} \end{array} \begin{array}{c} \boldsymbol{\delta_u} \\ \begin{bmatrix} u_2 \\ v_2 \\ \hdashline u_1 \\ v_1 \\ u_3 \\ v_3 \end{bmatrix} \end{array} \begin{array}{l} \\ \\ = 0 \\ = 0 \\ = 0 \\ = 0 \end{array} \quad (14.56)$$

$\mathbf{X_k}$ below, $\boldsymbol{\delta_k}$ below.

where $\mathbf{X_k}$ = the lower column matrix of unknown reactions

$\mathbf{X_u}$ = the upper column matrix of known applied forces

$\boldsymbol{\delta_u}$ = the upper column matrix of free joint displacements

$\boldsymbol{\delta_k}$ = the lower column matrix of zero displacements

and $\mathbf{K_{11}}$, $\mathbf{K_{12}}$, $\mathbf{K_{21}}$, and $\mathbf{K_{22}}$ = submatrices that result from partitioning of the structure stiffness matrix.

It should be noted that partitioning of the equations of Eq. (14.55) involves both a change in the order of the equations to different rows along with a shift of their elements from one column to another. This is necessary to ensure that the element location in Eq. (14.56) corresponds to the same force and displacement components of Eq. (14.55).

Solution of unknown nodal displacements The upper partitioned equations of Eq. (14.56) in expanded compact form is

$$[\mathbf{X_u}] = [\mathbf{K_{11}}][\boldsymbol{\delta_u}] + [\mathbf{K_{12}}][\boldsymbol{\delta_k}] = [\mathbf{K_{11}}][\boldsymbol{\delta_u}]$$

where $\boldsymbol{\delta_k}$ is a null matrix, that is, equal to zero.

Thus, displacements are found by

$$[\boldsymbol{\delta_u}] = [\mathbf{K_{11}}]^{-1}[\mathbf{X_u}]: \begin{bmatrix} u_2 \\ v_2 \end{bmatrix} = \begin{bmatrix} 2707 & -707 \\ -707 & 707 \end{bmatrix}^{-1} \begin{bmatrix} X_2 \\ Y_2 \end{bmatrix} \quad (14.57)$$

for which the simple task of obtaining the inverse of a 2 × 2 matrix remains. Since the determinant $|\mathbf{K_{11}}| = 1.414 \times 10^{+6}$, the inverse $[\mathbf{K_{11}}]^{-1}$ exists and

$$\begin{bmatrix} u_2 \\ v_2 \end{bmatrix} = 10^{-6} \begin{bmatrix} 500 & 500 \\ 500 & 1914 \end{bmatrix} \begin{bmatrix} 30 \\ -60 \end{bmatrix} = \begin{bmatrix} -0.015 \\ -0.100 \end{bmatrix} \quad \text{(meters)} \quad (14.58)$$

Solution of support reactions Two approaches are presented for computing the support reactions. The first approach considers the lower equations of the partitioned matrix of Eq. (14.56) which in compact form is

$$[X_k] = [K_{21}][\delta_u] + [K_{22}][0] = [K_{21}][\delta_u] = [K_{21}][K_{11}]^{-1}[X_u]$$

The results from substitution into this matrix form yields

$$\begin{bmatrix} X_1 \\ Y_1 \\ X_3 \\ Y_3 \end{bmatrix} = \underset{K_{21}}{\begin{bmatrix} -2000 & 0 \\ 0 & 0 \\ -707 & 707 \\ 707 & -707 \end{bmatrix}}_{4 \times 2} \underset{[K_{11}]^{-1}}{\begin{bmatrix} 500 & 500 \\ 500 & 1914 \end{bmatrix}}_{2 \times 2} \times 10^{-6} \underset{X_u}{\begin{bmatrix} X_2 \\ Y_2 \end{bmatrix}}_{2 \times 1}$$

which reduces to

$$\begin{bmatrix} X_1 \\ Y_1 \\ X_3 \\ Y_3 \end{bmatrix}_{4 \times 1} = \begin{bmatrix} -1 & -1 \\ 0 & 0 \\ 0 & 1 \\ 0 & -1 \end{bmatrix}_{4 \times 2} \begin{bmatrix} X_2 \\ Y_2 \end{bmatrix}_{2 \times 1} = \begin{bmatrix} -1 & -1 \\ 0 & 0 \\ 0 & 1 \\ 0 & -1 \end{bmatrix}_{4 \times 2} \begin{bmatrix} 30 \\ -60 \end{bmatrix}_{2 \times 1} = \begin{bmatrix} 30 \\ 0 \\ -60 \\ 60 \end{bmatrix}_{4 \times 1} \quad \text{(kN)}$$

The order of each matrix in the last two operations is shown to illustrate the conformal arrangement required for matrix multiplication.

The support reactions can also be obtained by direct substitution into Eq. (14.55). After we discard all zero products, the reaction equations reduce to

$$X_1 = -2000 u_2 = -2000(-0.015) = +30 \text{ kN}$$

$$Y_1 = 0 \text{ kN}$$

$$X_3 = -707 u_2 + 707 v_2 = -707(-0.015) + 707(-0.100) = -60 \text{ kN}$$

$$Y_3 = 707 u_2 - 707 v_2 = 707(-0.015) - 707(-0.100) = +60 \text{ kN}$$

Solution of truss member forces A pin-connected truss member behaves like a linear spring. Therefore, the axial truss member force can be found from the familiar force-displacement linear-spring equation, $X = k\delta$. Since both ends of a truss member are likely to displace when a truss structure is subject to load, the axial displacement δ is found from the relative motion of both end nodes. Thus, for a horizontal truss member of $k = AE/L$, the axial force can be found as

$$X_{ij} = \frac{AE}{L} \cdot (u_j - u_i)$$

Similarly, for a vertical truss member, the axial force is

$$Y_{ij} = \frac{AE}{L} \cdot (V_j - V_i)$$

These forces are known as equivalent nodal forces which are imagined as the forces that cause the nodal deformations. By substitution of Eq. (14.49) into Eq. (14.46), the equivalent nodal forces of a truss member in any orientation is expressed as

$$S_{ij} = \frac{AE}{L} \underset{1 \times 2}{[c \quad s]} \underset{2 \times 1}{\begin{bmatrix}(uj - ui)\\(vj - vi)\end{bmatrix}} \quad (14.59)$$

where $c = \cos\theta$ and $s = \sin\theta$, and where S_{ij} is the equivalent nodal force in member ij and is referenced to global coordinates. Therefore, the truss members of Example 14.2 are found using Eq. (14.59) as follows:

Member 1–2:

$$S_{1-2} = 2000 \underset{1 \times 2}{[1 \quad 0]} \underset{2 \times 1}{\begin{bmatrix}(-0.015 - 0)\\(-0.100 - 0)\end{bmatrix}} = 2000(-0.015) = -30 \text{ kN}$$

Member 1–3:

$$S_{1-3} = 2000[0 \quad 1]\begin{bmatrix}0\\0\end{bmatrix} = 0 \text{ kN}$$

Member 2–3:

$$S_{2-3} = 1414[-0.707 \quad 0.707]\begin{bmatrix}(0 - (-0.015))\\(0 - (-0.100))\end{bmatrix} = 1414(0.0601)$$

$$= +85 \text{ kN}$$

EXAMPLE 14.3

Use the **K** matrix developed in Example 14.1 to compute the joint displacements and support reactions for the truss loaded as shown. $AE = 90(10^3)$ kips.

Solution

From the truss schematic, we know that external forces

$X_2 = X_3 = X_4 = Y_4 = 0$ and $Y_2 = -12$ kips

Also, $u_1 = v_1 = v_3 = 0$ (support restraints)

The joint-equilibrium equations can be written for the truss in partitioned form as

$$\begin{bmatrix} X_2 = 0 \\ Y_2 = -12 \\ X_3 = 0 \\ X_4 = 0 \\ Y_4 = 0 \\ \hline X_1 \\ Y_1 \\ Y_3 \end{bmatrix} = \frac{AE}{L} \begin{bmatrix} 1.334 & 0 & -0.667 & 0 & 0 & | & -0.667 & 0 & 0 \\ 0 & 1 & 0 & 0 & -1 & | & 0 & 0 & 0 \\ -0.667 & 0 & 1.051 & -0.384 & 0.256 & | & 0 & 0 & -0.256 \\ 0 & 0 & -0.384 & 0.768 & 0 & | & -0.384 & -0.256 & 0.256 \\ 0 & -1 & 0.256 & 0 & 1.342 & | & -0.256 & -0.171 & -0.171 \\ \hline -0.667 & 0 & 0 & -0.384 & -0.256 & | & 1.051 & 0.256 & 0 \\ 0 & 0 & 0 & -0.256 & -0.171 & | & 0.256 & 0.171 & 0 \\ 0 & 0 & -0.256 & 0.256 & -0.171 & | & 0 & 0 & 0.171 \end{bmatrix} \begin{bmatrix} u_2 \\ v_2 \\ u_3 \\ u_4 \\ v_4 \\ \hline u_1 = 0 \\ v_1 = 0 \\ v_3 = 0 \end{bmatrix}$$

Note
The elements of the **K** matrix are rearranged to comply with the shifting of joint equations involved in the partitioning process.

Displacements

$$\begin{bmatrix} u_2 \\ v_2 \\ u_3 \\ u_4 \\ v_4 \end{bmatrix} = \frac{L}{AE} \begin{bmatrix} 1.334 & 0 & -0.667 & 0 & 0 \\ 0 & 1 & 0 & 0 & -1 \\ -0.667 & 0 & 1.051 & -0.384 & 0.256 \\ 0 & 0 & -0.384 & 0.768 & 0 \\ 0 & -1 & 0.256 & 0 & 1.342 \end{bmatrix}^{-1} \left\{ \begin{bmatrix} 0 \\ -12 \\ 0 \\ 0 \\ 0 \end{bmatrix} + \begin{bmatrix} 0 \\ 0 \\ 0 \\ 0 \\ 0 \end{bmatrix} \right\} = \begin{bmatrix} -1.8 \\ -9 \\ 3.6 \\ 1.8 \\ -7.4 \end{bmatrix} \times 10^{-2} \text{ in}$$

Support Reactions (using the results above)

$$\begin{bmatrix} X_1 \\ Y_1 \\ Y_3 \end{bmatrix} = \frac{AE}{L} \begin{bmatrix} -0.667 & 0 & 0 & -0.384 & -0.256 \\ 0 & 0 & 0 & -0.256 & -0.171 \\ 0 & 0 & -0.256 & 0.256 & -0.171 \end{bmatrix} \begin{bmatrix} u_2 \\ v_2 \\ u_3 \\ u_4 \\ v_4 \end{bmatrix} + \begin{bmatrix} 0 \\ 0 \\ 0 \end{bmatrix} = \begin{bmatrix} 0 \\ 6.0 \\ 6.0 \end{bmatrix} \text{(kips)}$$

Equilibrium Check

$\Sigma H \approx 0$

$\Sigma V = 0 = Y_1 + Y_3 - 12 = 6 + 6 - 12$ ✓

H. Settlement, temperature change, and fabrication errors

Matrix structural analysis employs the basic principles of mechanics and structural analysis to evaluate the effects of support settlement, temperature change, and fabrication errors. Example 14.4 presents a stiffness method analysis to determine the combined effects of applied load and support settlement on a truss. The approach is self-explanatory.

By now, we realize that matrix analysis of trusses is reduced to finding unknown forces and displacements at nodes. In the case of a truss subject to temperature change, member forces due to temperature change are determined and applied to the truss nodes as follows:

Let us assume that (1) the node ends of members that undergo temperature change are restrained, and (2) the temperature change is uniform over each member length. The resulting axial member force is computed as follows:

$$N_T = \sigma A = (E\varepsilon)A = E[\alpha(dT)]A = EA\alpha(dT)$$

where N_T = the restrained axial thermal force in the member
E = the modulus of elasticity
A = the cross-section area
α = the coefficient of linear thermal expansion
dT = the uniform temperature change over the member length

If a restrained member undergoes a temperature drop, the thermal force N_T acts in tension at the member ends; thus, a temperature rise results in a compressive N_T force at member ends. *By statics, these thermal member forces are placed in opposite directions on the joint nodes.* Consequently, a stiffness structural analysis is performed by treating the N_T forces as applied nodal loads. Example 14.5 considers that the truss is resting on unyielding supports and member 2–4 alone subject to a temperature change.

Fabrication errors often lead to mismatch of a truss member at its intended juncture with other members. This situation demands that an initial force be placed on the member to establish the joint connection. This force is computed in a similar fashion to N_T except that the thermal strain of $\alpha(dT)$ is replaced by $\varepsilon = \Delta L/L$ where ΔL is the mismatch length. The references cited at the end of this chapter are recommended for added study of these and other environmental conditions.

14.3.5. Beam Analysis by the Stiffness Method

The basic concepts of the stiffness method presented for trusses in Sec. 14.3.4 are also used for beam analysis. In beam theory, the main interest is with shear and bending-moment forces and their respective relationship to

EXAMPLE 14.4

The truss of the figure below is pin-supported at nodes 1, 2, and 3. A settlement of 2 in occurs at node 2. Determine all support reactions.

$AE = 90 \times 10^3$ kips
(all members)

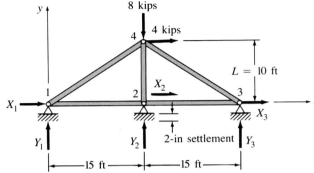

Using the $\mathbf{K}_s$ matrix of Example 14.1 and the method of partitioning,

$$\begin{bmatrix} X_1 \\ Y_1 \\ X_2 \\ Y_2 \\ X_3 \\ Y_3 \\ \hline X_4 = 4 \\ Y_4 = -8 \end{bmatrix} = \frac{AE}{L} \begin{bmatrix} 1.051 & 0.256 & -0.667 & 0 & 0 & 0 & | & -0.384 & -0.256 \\ 0.256 & 0.171 & 0 & 0 & 0 & 0 & | & -0.256 & -0.171 \\ -0.667 & 0 & 1.334 & 0 & -0.667 & 0 & | & 0 & 0 \\ 0 & 0 & 0 & 1.0 & 0 & 0 & | & 0 & -1.0 \\ 0 & 0 & -0.667 & 0 & 1.051 & -0.256 & | & -0.384 & 0.256 \\ 0 & 0 & 0 & 0 & -0.256 & 0.171 & | & 0.256 & -0.171 \\ \hline -0.384 & -0.256 & 0 & 0 & -0.384 & 0.256 & | & 0.768 & 0 \\ -0.256 & -0.171 & 0 & -1.0 & 0.256 & -0.171 & | & 0 & 1.342 \end{bmatrix} \begin{bmatrix} u_1 = 0 \\ v_1 = 0 \\ u_2 = 0 \\ v_2 = -2 \\ u_3 = 0 \\ v_3 = 0 \\ u_4 \\ v_4 \end{bmatrix}$$

Displacements

$$\begin{bmatrix} 4 \\ -8 \end{bmatrix} = \frac{AE}{L} \begin{bmatrix} -0.384 & -0.256 & 0 & 0 & -0.384 & 0.256 \\ -0.256 & -0.171 & 0 & -1.0 & 0.256 & -0.171 \end{bmatrix} \begin{bmatrix} 0 \\ 0 \\ 0 \\ -2 \\ 0 \\ 0 \end{bmatrix} + \frac{AE}{L} \begin{bmatrix} 0.768 & 0 \\ 0 & 1.342 \end{bmatrix} \begin{bmatrix} u_4 \\ v_4 \end{bmatrix}$$

or

$$\begin{bmatrix} 4 \\ -8 \end{bmatrix} = \frac{AE}{L} \begin{bmatrix} 0 \\ 2 \end{bmatrix} + \frac{AE}{L} \begin{bmatrix} 0.768 & 0 \\ 0 & 1.342 \end{bmatrix} \begin{bmatrix} u_4 \\ v_4 \end{bmatrix}$$

Thus,

$$\begin{bmatrix} u_4 \\ v_4 \end{bmatrix} = \frac{L}{AE} \begin{bmatrix} 0.768 & 0 \\ 0 & 1.342 \end{bmatrix}^{-1} \begin{bmatrix} 4 \\ -1508 \end{bmatrix} = \begin{Bmatrix} 6.9 \times 10^{-3} \\ -1.5 \end{Bmatrix} \text{ in}$$

Reactions

$$\begin{bmatrix} X_1 \\ Y_1 \\ X_2 \\ Y_2 \\ X_3 \\ Y_3 \end{bmatrix} = \frac{AE}{L} \begin{bmatrix} 0 \\ 0 \\ 0 \\ -2 \\ 0 \\ 0 \end{bmatrix} + \frac{AE}{L} \begin{bmatrix} -0.384 & -0.256 \\ -0.256 & -0.171 \\ 0 & 0 \\ 0 & -1.0 \\ -0.384 & 0.256 \\ 0.256 & -0.171 \end{bmatrix} \begin{bmatrix} 6.9 \times 10^{-3} \\ -1.5 \end{bmatrix}$$

$$= \frac{AE}{L} \begin{bmatrix} 0.381 \\ 0.255 \\ 0 \\ -0.5 \\ -0.387 \\ 0.256 \end{bmatrix} = \begin{bmatrix} 285 \\ 191 \\ 0 \\ -375 \\ -289 \\ 192 \end{bmatrix}$$

Note
$\Sigma F_y = 0$ and $\Sigma F_x = 0$ are basically satisfied.

EXAMPLE 14.5

Consider the truss in the figure below resting on unyielding supports. $AE = 90 \times 10^3$ kips (all members). If $\alpha = 6 \times 10^{-6}$ in/in/°F and member 2–4 undergoes a temperature change, $dT = +100°F$, determine the displacement at node 4 and support reactions.

Solution

Assume member 2–4 restrained at nodes 2 and 4. Thermal force $N_T = EA\alpha(dT) = 90 \times 10^3 (6 \times 10^{-6})100 = 54$ kips which is compressive on member 2–4 and acts on nodes 2 and 4 as shown below.

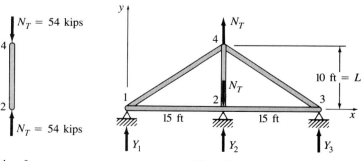

Member force   Node force

Using the force-displacement equations of Example 14.4 and the support conditions above, we have

$$\begin{bmatrix} X_1 \\ Y_1 \\ X_2 \\ Y_2 - 54 \\ X_3 \\ Y_3 \\ X_4 = 0 \\ Y_4 = 54 \end{bmatrix} = \frac{AE}{L} \begin{bmatrix} 1.051 & 0.256 & -0.667 & 0 & 0 & 0 & -0.384 & -0.256 \\ 0.256 & 0.171 & 0 & 0 & 0 & 0 & -0.256 & -0.171 \\ -0.667 & 0 & 1.334 & 0 & -0.667 & 0 & 0 & 0 \\ 0 & 0 & 0 & 1.0 & 0 & 0 & 0 & -1.0 \\ 0 & 0 & -0.667 & 0 & 1.051 & -0.256 & -0.384 & 0.256 \\ 0 & 0 & 0 & 0 & -0.256 & 0.171 & 0.256 & -0.171 \\ -0.384 & -0.256 & 0 & 0 & -0.384 & 0.256 & 0.768 & 0 \\ -0.256 & -0.171 & 0 & -1.0 & 0.256 & -0.171 & 0 & 1.342 \end{bmatrix} \begin{bmatrix} 0 \\ 0 \\ 0 \\ 0 \\ 0 \\ 0 \\ 0^* \\ v_4 \end{bmatrix}$$

The last equation states that

$$54 = \frac{AE}{L}[1.342]v_4 \quad \text{or} \quad v_4 = 5.37 \times 10^{-2} \text{ in}$$

From the second equation:

$$Y_1 = -0.171 v_4 \left(\frac{AE}{L}\right) = -0.171(5.37 \times 10^{-2})\frac{90 \times 10^3}{120} = -6.88 \text{ kips}$$

$$Y_2 - 54 = -1.0 v_4 \left(\frac{AE}{L}\right) = -40.24 \quad \therefore \quad Y_2 = +13.76 \text{ kips}$$

$$Y_3 = -0.171 v_4 \left(\frac{AE}{L}\right) = -6.88 \text{ kips}$$

$$X_1 = -X_3 = 10.31 \text{ kips}; \quad X_2 = 0$$

$$\Sigma F_y = 0 = Y_1 + Y_2 + Y_3 + N_T - N_T = -6.88 + 13.76 - 6.88 = 0 \quad \checkmark$$

* Note: $u_4 = 0$ by symmetry.

---

beam displacements and rotations. In a manner similar to truss structures, beam-element matrices are written in terms of nodal force-displacement relationships. Likewise, nodal force equations are written to satisfy force equilibrium and compatibility of displacements at each node.

Our attention here is limited to horizontal beam structures with coincident global and local axes. Therefore, beam-element stiffness matrices will not have to undergo matrix transformation. Inclined beam elements are studied later in Sec. 14.3.6.

In this section, the following assumptions, idealizations, and conditions are used to formulate the stiffness matrix for a beam element:

1 Attention is limited to planar beams where the effects of shear and bending-moment forces are of primary concern.

2 A beam element is straight and of uniform cross section over its entire length (prismatic).

3 A beam element is made from homogeneous, isotropic, and linear elastic material.

4 All internal and external forces act at element nodes.

5 Beam-element force-displacement relationships are based on conventional small deflection, linear elastic-beam theory.

6 Shear is constant between nodes; shear deformation is ignored.

7 Bending moment varies linearly between element nodes.

8 Axial deformations are negligible (yet, included below).

A. Formation of beam-element stiffness matrix A beam-element matrix consists of the stiffness influence coefficients which relate nodal forces to their corresponding displacements. We may recall that an influence coefficient is a force which produces a unit displacement in its force direction while all other degrees of freedom are held fixed. Figure 14.9 presents the element stiffness influence coefficients for each nodal degree of freedom which can be derived using the classical methods found in Chaps. 10 to 13. Accordingly, Fig. 14.10 establishes the positive sign convention for nodal force, displacement, and rotation associated with the six degrees of

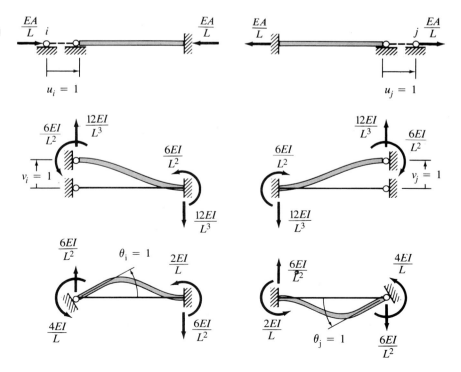

Figure 14.9 Positive sign convention for beam nodal forces and displacement components.

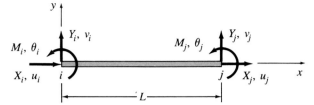

Figure 14.10 Element stiffness influence coefficients.

freedom of a planar beam. Therefore, general force equilibrium equations can be written in terms of arbitrary nodal displacements by superposition of the force data contained in Fig. 14.9.

At node i:

$$\Sigma F_x = 0: \quad X_i = \frac{EAu_i}{L} - \frac{EAu_j}{L} \qquad (14.60)$$

$$\Sigma F_y = 0: \quad V_i = \frac{12EIv_i}{L^3} + \frac{6EI\theta_i}{L^2} - \frac{12EIv_j}{L^3} + \frac{6EI\theta_j}{L^2}$$

$$\Sigma M_z = 0: \quad M_i = \frac{6EIv_i}{L^2} + \frac{4EI\theta_i}{L} - \frac{6EIv_j}{L^2} + \frac{2EI\theta_j}{L}$$

At node j:

$$\Sigma F_x = 0: \quad X_j = -\frac{EAu_i}{L} + \frac{EAu_j}{L} \qquad (14.61)$$

$$\Sigma F_y = 0: \quad V_j = -\frac{12EIv_i}{L^3} - \frac{6EI\theta_i}{L^2} + \frac{12EIv_j}{L^3} - \frac{6EI\theta_j}{L^2}$$

$$\Sigma M_z = 0: \quad M_j = \frac{6EIv_i}{L^2} + \frac{2EI\theta_i}{L} - \frac{6EIv_j}{L^2} + \frac{4EI\theta_i}{L}$$

The element matrix stiffness matrix (in terms of local coordinates) can be written from Eqs. (14.60) and (14.61) as

$$\mathbf{K'_{ij}} = \begin{bmatrix} \frac{EA}{L} & 0 & 0 & -\frac{EA}{L} & 0 & 0 \\ 0 & \frac{12EI}{L^3} & \frac{6EI}{L^2} & 0 & -\frac{12EI}{L^3} & \frac{6EI}{L^2} \\ 0 & \frac{6EI}{L^2} & \frac{4EI}{L} & 0 & -\frac{6EI}{L^2} & \frac{2EI}{L} \\ -\frac{EA}{L} & 0 & 0 & \frac{EA}{L} & 0 & 0 \\ 0 & -\frac{12EI}{L^3} & -\frac{6EI}{L^2} & 0 & \frac{12EI}{L^3} & -\frac{6EI}{L^2} \\ 0 & \frac{6EI}{L^2} & \frac{2EI}{L} & 0 & -\frac{6EI}{L^2} & \frac{4EI}{L} \end{bmatrix} \qquad (14.62)$$

with columns labeled $u_i, v_i, \theta_i, u_j, v_j, \theta_j$.

The stiffness matrix of Eq. (14.62) is both symmetric and singular (no inverse exists). The symmetry is a consequence of the Maxwell-Betti reciprocal law. The matrix is singular since the beam element is not restrained against rigid-body motion. This situation will be remedied after displacement restraints are defined to properly support the structure.

B. Element matrix for a beam without axial force In most beam applications, both axial deformations and axial force are assumed negligible in comparison to the shear and bending force effects. For these cases, the matrix shown in Eq. (14.62) can be reduced to

$$\mathbf{K}'_{ij} = \frac{EI}{L^3} \begin{bmatrix} \overset{v_i}{12} & \overset{\theta_i}{6L} & \overset{v_j}{-12} & \overset{\theta_j}{6L} \\ 6L & 4L^2 & -6L & 2L^2 \\ -12 & -6L & 12 & -6L \\ 6L & 2L^2 & -6L & 4L^2 \end{bmatrix} \qquad (14.63)$$

C. Structure idealization The model selected for matrix structural analysis should characterize the actual behavior of the beam. Clearly, an accurate solution requires proper definition of geometry, support restraints, loads, material properties, and careful attention to use correct signs.

The element stiffness matrix of Eq. (14.62) was derived on the notion that forces are only applied at element nodes. This condition conforms with the requirements of constant shear between nodes and linear variation of bending moment between nodes. Consequently, the analyst can model a beam structure with nodes selected at span locations where concentrated loads and supports are present (see Examples 14.6 and 14.7).

EXAMPLE 14.6

Compute the rotation at node 2 and support reactions for the beam shown in the figure below. $EI = 200$ kip·ft^2. Neglect axial effects and axial forces.

Known forces and displacements:

$X_1 = X_3 = 0$

$u_1 = u_2 = u_3 = v_1 = v_3 = \theta_1 = \theta_3 = 0$

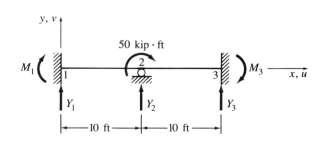

From Eq. (14.63):

$$\mathbf{K_{1-2}} = \frac{EI}{L^3} \begin{array}{c} \\ \end{array} \begin{matrix} v_1 & \theta_1 & v_2 & \theta_2 \end{matrix}$$

$$\mathbf{K_{1-2}} = \frac{EI}{L^3} \begin{bmatrix} 12 & 60 & -12 & 60 \\ 60 & 400 & -60 & 200 \\ -12 & -60 & 12 & -60 \\ 60 & 200 & -60 & 400 \end{bmatrix} \qquad \mathbf{K_{2-3}} = \frac{EI}{L^3} \begin{bmatrix} 12 & 60 & -12 & 60 \\ 60 & 400 & -60 & 200 \\ -12 & -60 & 12 & -60 \\ 60 & 200 & -60 & 400 \end{bmatrix}$$

with columns $v_2, \theta_2, v_3, \theta_3$ for $\mathbf{K_{2-3}}$.

From Eq. (14.54): $[\mathbf{P}] = [\mathbf{k_s}][\mathbf{d}]$

$$\begin{bmatrix} Y_1 \\ M_1 \\ Y_2 \\ M_2 = 50 \\ Y_3 \\ M_3 \end{bmatrix} = \frac{EI}{L^3} \begin{bmatrix} 12 & 60 & -12 & 60 & 0 & 0 \\ 60 & 400 & -60 & 200 & 0 & 0 \\ -12 & -60 & 24 & 0 & -12 & 60 \\ 60 & 200 & 0 & 800 & -60 & 200 \\ 0 & 0 & -12 & -60 & 12 & -60 \\ 0 & 0 & 60 & 200 & -60 & 400 \end{bmatrix} \begin{bmatrix} v_1 \\ \theta_1 \\ v_2 \\ \theta_2 \\ v_3 \\ \theta_3 \end{bmatrix}$$

In partitioned form,

$$\begin{bmatrix} Y_1 \\ M_1 \\ Y_2 \\ Y_3 \\ M_3 \\ \hline 50 \end{bmatrix} = \frac{EI}{L^3} \left[\begin{array}{ccccc|c} 12 & 60 & -12 & 0 & 0 & 60 \\ 60 & 400 & -60 & 0 & 0 & 200 \\ -12 & -60 & 24 & -12 & 60 & 0 \\ 0 & 0 & -12 & 12 & -60 & -60 \\ 0 & 0 & 60 & -60 & 400 & 200 \\ \hline 60 & 200 & 0 & -60 & 200 & 800 \end{array} \right] \begin{bmatrix} 0 \\ 0 \\ 0 \\ 0 \\ 0 \\ \hline \theta_2 \end{bmatrix}$$

Rotation at Node 2:

$$50 = \frac{EI}{L^3}([0] + 800\theta_2) \quad \text{or} \quad \theta_2 = \frac{50L^3}{800EI} = 0.312 \text{ radians} \quad \curvearrowright$$

Support Reactions:

$$Y_1 = \frac{EI}{L^3}([0] + 60\theta_2) = \frac{60EI\theta_2}{L^3} = 3.72 \text{ kips}$$

$$M_1 = \frac{EI}{L^3}([0] + 200\theta_2) = \frac{200EI\theta_2}{L^3} = 12.5 \text{ kip} \cdot \text{ft}$$

$$Y_2 = \frac{EI}{L^3}([0] + [0]) = 0$$

$$Y_3 = \frac{EI}{L^3}([0] - 60\theta_2) = -\frac{60EI\theta_2}{L^3} = -3.72 \text{ kips}$$

$$M_3 = \frac{EI}{L^3}([0] + 200\theta_2) = \frac{200EI\theta_2}{L^3} = 12.5 \text{ kip} \cdot \text{ft}$$

EXAMPLE 14.7

Solve for displacements and reaction forces for the beam shown in the figure below. $E = 200$ GPa, $I_{1-2} = 8 \times 10^{-5}$ m^4, $I_{2-3} = 27 \times 10^{-5}$ m^4. Neglect axial effects. Nodes selected at 1, 2, and 3 as shown.

$$\frac{EI_a}{L_a^3} = \frac{200 \times 10^9 \, (8 \times 10^{-5})}{(2)^3} = 2 \times 10^6 \text{ N/m}$$

$$\frac{EI_b}{L_b^3} = \frac{200 \times 10^9 \, (27 \times 10^{-5})}{(3)^3} = 2 \times 10^6 \text{ N/m}$$

$$\mathbf{K_a} = 2 \times 10^6 \begin{bmatrix} 12 & 6(2) & -12 & 6(2) \\ 6(2) & 4(2)^2 & -6(2) & 2(2)^2 \\ -12 & -6(2) & 12 & -6(2) \\ 6(2) & 2(2)^2 & -6(2) & 4(2)^2 \end{bmatrix} = 2 \times 10^6 \begin{bmatrix} v_1 & \theta_1 & v_2 & \theta_2 \\ 12 & 12 & -12 & 12 \\ 12 & 16 & -12 & 8 \\ -12 & -12 & 12 & -12 \\ 12 & 8 & -12 & 16 \end{bmatrix}$$

$$\mathbf{K_b} = 2 \times 10^6 \begin{bmatrix} 12 & 6(3) & -12 & 6(3) \\ 6(3) & 4(3)^2 & -6(3) & 2(3)^2 \\ -12 & -6(3) & 12 & -6(3) \\ 6(3) & 2(3)^2 & -6(3) & 4(3)^2 \end{bmatrix} = 2 \times 10^6 \begin{bmatrix} v_2 & \theta_2 & v_3 & \theta_3 \\ 12 & 18 & -12 & 18 \\ 18 & 36 & -18 & 18 \\ -12 & -18 & 12 & -18 \\ 18 & 18 & -18 & 36 \end{bmatrix}$$

From Eq. (14.54):

$$\begin{bmatrix} Y_1 \\ M_1 = 0 \\ Y_2 = -20 \times 10^3 \\ M_2 = 0 \\ Y_3 \\ M_3 \end{bmatrix} = 2 \times 10^6 \begin{bmatrix} v_1 & \theta_1 & v_2 & \theta_2 & v_3 & \theta_3 \\ 12 & 12 & -12 & 12 & 0 & 0 \\ 12 & 16 & -12 & 8 & 0 & 0 \\ -12 & -12 & 24 & 6 & -12 & 18 \\ 12 & 8 & 6 & 52 & -18 & 18 \\ 0 & 0 & -12 & -18 & 12 & -18 \\ 0 & 0 & 18 & 18 & -18 & 36 \end{bmatrix} \begin{bmatrix} 0 \\ \theta_1 \\ v_2 \\ \theta_2 \\ 0 \\ 0 \end{bmatrix}$$

In partitioned form:

$$\begin{bmatrix} M_1 = 0 \\ -20 \times 10^3 \\ M_2 = 0 \\ \hdashline Y_1 \\ Y_3 \\ M_3 \end{bmatrix} = 2 \times 10^6 \begin{bmatrix} \theta_1 & v_2 & \theta_2 & v_1 & v_3 & \theta_3 \\ 16 & -12 & 8 & 12 & 0 & 0 \\ -12 & 24 & 6 & -12 & -12 & 18 \\ 8 & 6 & 52 & 12 & -18 & 18 \\ \hdashline 12 & -12 & 12 & 12 & 0 & 0 \\ 0 & -12 & -18 & 0 & 12 & -18 \\ 0 & 18 & 18 & 0 & -18 & 36 \end{bmatrix} \begin{bmatrix} \theta_1 \\ v_2 \\ \theta_2 \\ \hdashline 0 \\ 0 \\ 0 \end{bmatrix}$$

Displacements
Using the inverse matrix operation,

$$\begin{bmatrix} \theta_1 \\ v_2 \\ \theta_2 \end{bmatrix} = \frac{1}{2 \times 10^6} \begin{bmatrix} 16 & -12 & 8 \\ -12 & 24 & 6 \\ 8 & 6 & 52 \end{bmatrix}^{-1} \begin{bmatrix} 0 \\ -20 \times 10^3 \\ 0 \end{bmatrix} = \begin{bmatrix} 7.29 \\ 8.33 \\ -2.08 \end{bmatrix} 10^{-4}$$

Reactions

$$\begin{bmatrix} Y_1 \\ Y_3 \\ M_3 \end{bmatrix} = 2 \times 10^6 \begin{bmatrix} 12 & -12 & 12 \\ 0 & -12 & -18 \\ 0 & 18 & 18 \end{bmatrix} \begin{bmatrix} \theta_1 \\ v_2 \\ \theta_2 \end{bmatrix} + \begin{bmatrix} 0 \\ 0 \\ 0 \end{bmatrix} \cong \begin{matrix} 7500 \text{ N} \\ 12{,}500 \text{ N} \\ -22{,}500 \text{ N} \cdot \text{m} \end{matrix}$$

In matrix analysis, every beam element is assumed to be prismatic. However, if the actual structure has a variable span depth, or has sudden step changes (as found on cover-plated beams), a single beam-element model would misrepresent the true beam. Thus, a variable depth beam could be modeled as a series of shorter length elements with average area properties for each element span. It is also appropriate to locate a node point at each location where a sudden change in geometry occurs (e.g., cover-plated beam), or at a juncture where members of dissimilar materials are joined.

D. Equivalent nodal forces To this point, we have studied beam elements with node point loads. Now, let us study how to model a beam structure with distributed or concentrated load between element nodes. In the case of a simply supported beam under uniform distributed load, we know that shear varies linearly, and bending moment varies parabolically, from end to end. Yet, the conditions stated earlier require that shear must be constant between nodes, and bending moment must vary linearly between nodes. Moreover, these conditions are only satisfied when elements are subject to nodal forces. A practical solution to this problem is to replace the uniform load between element nodes with an equivalent set of fixed-end forces applied at the nodes. The fixed-end forces for a uniformly loaded prismatic beam are shown in Fig. 14.11. Table 12.1 contains additional fixed-end forces for other types of load on prismatic beams.

Equivalent-member node forces are included in a stiffness-method analysis as added forces in the joint-equilibrium equations. That is, let us modify Eq. (14.54) as follows:

$$[P^a] = [K_s][d] + [P^f] \tag{14.64}$$

481 Matrix Structural Analysis

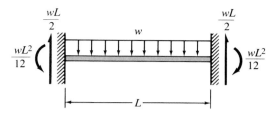

Figure 14.11

where $[\mathbf{P}^a]$ = the column matrix of external node forces
$[\mathbf{K_s}]$ = the global structure stiffness matrix
$[\mathbf{d}]$ = the column matrix of nodal displacements
$[\mathbf{P}^f]$ = the column matrix of fixed-end member forces

Now, if $[\mathbf{P}^e] = [\mathbf{P}^a] - [\mathbf{P}^f]$, Eq. (14.64) can be written in the familiar form of Eq. (14.54) as

$$[\mathbf{P}^e] = [\mathbf{K_s}][\mathbf{d}] \qquad (14.65)$$

Example 14.8 illustrates the stiffness method of analysis for a continuous beam with loads applied between element nodes.

EXAMPLE 14.8

Determine all support reactions of the continuous beam shown in the figure below.
$EI = 30 \times 10^6$ ksi (constant).

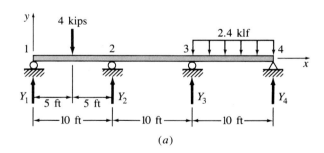

(a)

Solution

$M_1 = M_2 = M_3 = M_4 = 0$

Let $L = 10$ ft

Fixed-End Force Sets:
(Refer to Table 12.1)

Member 1–2

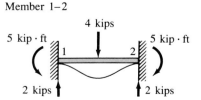

Member 3–4

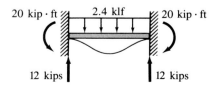

(b)

Since EI/L^3 is identical for each beam segment, all element matrices reduce to the form [Eq. (14.63)]:

$$\mathbf{K}_{ij} = \frac{EI}{L^3} \begin{bmatrix} & v_i & \theta_i & v_j & \theta_j \\ 12 & 60 & -12 & 60 \\ 60 & 400 & -60 & 200 \\ -12 & -60 & 12 & -60 \\ 60 & 200 & -60 & 400 \end{bmatrix}$$

Then, from Eq. (14.64):

$$\begin{bmatrix} Y_1 \\ M_1=0 \\ Y_2 \\ M_2=0 \\ Y_3 \\ M_3=0 \\ Y_4 \\ M_4=0 \end{bmatrix} = \frac{EI}{L^3} \begin{bmatrix} v_1 & \theta_1 & v_2 & \theta_2 & v_3 & \theta_3 & v_4 & \theta_4 \\ 12 & 60 & -12 & 60 & 0 & 0 & 0 & 0 \\ 60 & 400 & -60 & 200 & 0 & 0 & 0 & 0 \\ -12 & -60 & 24 & 0 & -12 & 60 & 0 & 0 \\ 60 & 200 & 0 & 800 & -60 & 200 & 0 & 0 \\ 0 & 0 & -12 & -60 & 24 & 0 & -12 & 60 \\ 0 & 0 & 60 & 200 & 0 & 800 & -60 & 200 \\ 0 & 0 & 0 & 0 & -12 & -60 & 12 & -60 \\ 0 & 0 & 0 & 0 & 60 & 200 & -60 & 400 \end{bmatrix} \begin{bmatrix} v_1=0 \\ \theta_1 \\ v_2=0 \\ \theta_2 \\ v_3=0 \\ \theta_3 \\ v_4=0 \\ \theta_4 \end{bmatrix} + \begin{bmatrix} 2 \\ 5 \\ 2 \\ -5 \\ 12 \\ 20 \\ 12 \\ -20 \end{bmatrix}$$

$$\mathbf{P} \qquad\qquad\qquad\qquad\qquad \mathbf{K}_s \qquad\qquad\qquad\qquad\qquad \mathbf{P}^f$$

which rearranged in partitioned form yields

$$\begin{bmatrix} Y_1 \\ Y_2 \\ Y_3 \\ Y_4 \\ \hline M_1=0 \\ M_2=0 \\ M_3=0 \\ M_4=0 \end{bmatrix} = \frac{EI}{L^3} \left[\begin{array}{cccc|cccc} v_1 & v_2 & v_3 & v_4 & \theta_1 & \theta_2 & \theta_3 & \theta_4 \\ 12 & -12 & 0 & 0 & 60 & 60 & 0 & 0 \\ -12 & 24 & -12 & 0 & -60 & 0 & 60 & 0 \\ 0 & -12 & 24 & -12 & 0 & -60 & 0 & 60 \\ 0 & 0 & -12 & 12 & 0 & 0 & -60 & -60 \\ \hline 60 & -60 & 0 & 0 & 400 & 200 & 0 & 0 \\ 60 & 0 & -60 & 0 & 200 & 800 & 200 & 0 \\ 0 & 60 & 0 & -60 & 0 & 200 & 800 & 200 \\ 0 & 0 & 60 & -60 & 0 & 0 & 200 & 400 \end{array} \right] \begin{bmatrix} 0 \\ 0 \\ 0 \\ 0 \\ \theta_1 \\ \theta_2 \\ \theta_3 \\ \theta_4 \end{bmatrix} + \begin{bmatrix} 2 \\ 2 \\ 12 \\ 12 \\ 5 \\ -5 \\ 20 \\ -20 \end{bmatrix}$$

Displacements

$$\begin{bmatrix} \theta_1 \\ \theta_2 \\ \theta_3 \\ \theta_4 \end{bmatrix} = \begin{bmatrix} 0 \\ 0 \\ 0 \\ 0 \end{bmatrix} + \frac{L^3}{EI} \begin{bmatrix} 400 & 200 & 0 & 0 \\ 200 & 800 & 200 & 0 \\ 0 & 200 & 800 & 200 \\ 0 & 0 & 200 & 400 \end{bmatrix}^{-1} \begin{bmatrix} -5 \\ 5 \\ -20 \\ 20 \end{bmatrix} = \frac{L^3}{EI} \begin{bmatrix} -0.025 \\ +0.025 \\ -0.050 \\ +0.075 \end{bmatrix} = \begin{bmatrix} -1.44 \\ +1.44 \\ -2.88 \\ +4.32 \end{bmatrix} 10^{-3}$$

Reactions

$$\begin{bmatrix} Y_1 \\ Y_2 \\ Y_3 \\ Y_4 \end{bmatrix} = \begin{bmatrix} 0 \\ 0 \\ 0 \\ 0 \end{bmatrix} + \frac{EI}{L^3} \begin{bmatrix} 60 & 60 & 0 & 0 \\ -60 & 0 & 60 & 0 \\ 0 & -60 & 0 & 60 \\ 0 & 0 & -60 & -60 \end{bmatrix} \begin{bmatrix} \theta_1 \\ \theta_2 \\ \theta_3 \\ \theta_4 \end{bmatrix} + \begin{bmatrix} 2 \\ 2 \\ 12 \\ 12 \end{bmatrix} = \begin{bmatrix} 2 \\ 0.5 \\ 15 \\ 10.5 \end{bmatrix}$$

Check Equilibrium

$\Sigma F_y = 0 = Y_1 + Y_2 + Y_3 + Y_4 - 4 - 2.4(10) = 2 + 0.5 + 15 + 10.5 - 4 - 24 = 28 - 28 = 0$ ✓

E. Three-dimensional beam stiffness equations A beam in three-dimensional space can experience torsional, shear, and axial force plus bending about both principle axes. From mechanics, the relation of constant torque and angular deformation for a member of circular cross section is

$$T = \frac{GJ\theta}{L} \tag{14.66}$$

where T = the torsional moment
  G = the shear modulus of elasticity
  J = the torsional constant of the cross section
  θ = the angle of twist
  L = the length of the member

The stiffness matrix for beam torsion can be found using the same approach developed in Sec. 14.3.4 for axial force. The element matrix equations due to torsion for member ij are

$$\begin{bmatrix} T_i \\ T_j \end{bmatrix} = \frac{GJ}{L} \begin{bmatrix} 1 & -1 \\ -1 & 1 \end{bmatrix} \begin{bmatrix} \theta_i \\ \theta_j \end{bmatrix} \tag{14.67}$$

Figure 14.12 illustrates a beam element under the combined actions of shear, bending, torsion, and axial-force loading. The combined 12 × 12 element stiffness matrix of the equation in Fig. 14.12 includes subscript notation to properly identify the axes for bending, torsion, and moment of inertia terms. Applications of three-dimensional beam stiffness analyses are available in numerous referenced documents including Refs. [14.3] and [14.5] cited at the end of this chapter.

14.3.6. Plane-Frame Analysis by the Stiffness Method

In this section, stiffness analysis is limited to plane frames which exhibit linear elastic behavior and obey small deflection theory. In addition, all frame members are assumed to be prismatic and composed of linear elastic material.

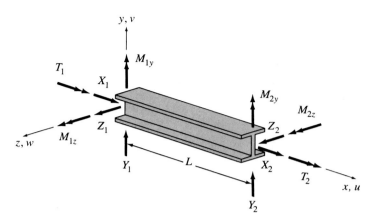

Figure 14.12

Stiffness Relationship for 3-D Beam Element

$$\begin{Bmatrix} X_1 \\ Y_1 \\ Z_1 \\ T_1 \\ M_{1y} \\ M_{1z} \\ X_2 \\ Y_2 \\ Z_2 \\ T_2 \\ M_{2y} \\ M_{2z} \end{Bmatrix} = \begin{bmatrix} EA/L & & & & & & & & & & & \\ 0 & 12EI_z/L^3 & & & & & & & & & & \\ 0 & 0 & 12EI_y/L^3 & & & & \text{Symmetric} & & & & & \\ 0 & 0 & 0 & JG^*/L & & & & & & & & \\ 0 & 0 & -(6EI_y/L^2) & 0 & 4EI_y/L & & & & & & & \\ 0 & 6EI_z/L^2 & 0 & 0 & 0 & 4EI_z/L & & & & & & \\ EA/L & 0 & 0 & 0 & 0 & 0 & EA/L & & & & & \\ 0 & -(12EI_z/L^3) & 0 & 0 & 0 & -(6EI_z/L^2) & 0 & 12EI_z/L^3 & & & & \\ 0 & 0 & -(12EI_y/L^3) & 0 & 6EI_y/L^2 & 0 & 0 & 0 & 12EI_y/L^3 & & & \\ 0 & 0 & 0 & -(JG^*/L) & 0 & 0 & 0 & 0 & 0 & JG^*/L & & \\ 0 & 0 & -(6EI_y/L^2) & 0 & 2EI_y/L & 0 & 0 & 0 & 6EI_y/L^2 & 0 & 4EI_y/L & \\ 0 & 6EI_z/L^2 & 0 & 0 & 0 & 2EI_z/L & 0 & -(6EI_z/L^2) & 0 & 0 & 0 & 4EI_z/L \end{bmatrix} \begin{Bmatrix} u_1 \\ v_1 \\ w_1 \\ \theta_{1x} \\ \theta_{1y} \\ \theta_{1z} \\ u_2 \\ v_2 \\ w_2 \\ \theta_{2x} \\ \theta_{2y} \\ \theta_{2z} \end{Bmatrix}$$

$$(14.68)$$

* Requires modification for noncircular and open sections.

A plane frame consists of horizontal, vertical, and/or inclined beams which, in matrix analysis, are efficiently defined in terms of global axes. To this end, a general frame-element stiffness matrix can be established by transforming the beam element matrix $\mathbf{K}'_{ij}$ of Eq. (14.62) from local to global coordinates. Figures 14.13a and b show a frame member in an arbitrary orientation under the action of local and global node forces, respectively. Each nodal force and corresponding nodal displacement component in these figures are shown as acting in the positive direction. We may recall the relation of local to global node forces as

$$[\mathbf{X}'] = [\mathbf{T}][\mathbf{X}] \qquad (14.42)$$

which upon expansion of $[\mathbf{T}]$ from Eq. (14.41) can be rewritten as

$$\underbrace{\begin{bmatrix} X'_1 \\ Y'_1 \\ M'_1 \\ X'_2 \\ Y'_2 \\ M'_2 \end{bmatrix}}_{\mathbf{X}'} = \underbrace{\begin{bmatrix} c & s & 0 & 0 & 0 & 0 \\ -s & c & 0 & 0 & 0 & 0 \\ 0 & 0 & 1 & 0 & 0 & 0 \\ 0 & 0 & 0 & c & s & 0 \\ 0 & 0 & 0 & -s & c & 0 \\ 0 & 0 & 0 & 0 & 0 & 1 \end{bmatrix}}_{\mathbf{T}} \underbrace{\begin{bmatrix} X_1 \\ Y_1 \\ M_1 \\ X_2 \\ Y_2 \\ M_2 \end{bmatrix}}_{\mathbf{X}} \qquad (14.69)$$

485 Matrix Structural Analysis

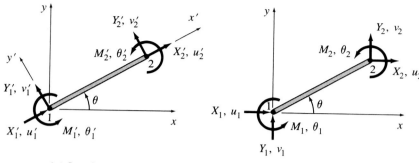

Figure 14.13   (a) Local coordinate member forces   (b) Global coordinate member forces

$$\begin{bmatrix} X_1 \\ Y_1 \\ M_1 \\ X_2 \\ Y_2 \\ M_2 \end{bmatrix} = \begin{bmatrix} [(AE/L)c^2 + (12EI/L^3)s^2] & & & & & \\ [(AE/L) - (12EI/L^3)]sc & [(AE/L)s^2 + (12EI/L^3)c^2] & & \text{Symmetric} & & \\ -(6EI/L^2)s & (6EI/L^2)c & 4EI/L & & & \\ -[(AE/L)c^2 + (12EI/L^3)s^2] & -[(AE/L) - (12EI/L^3)]sc & (6EI/L^2)s & [(AE/L)c^2 + (12EI/L^3)s^2] & & \\ -[(AE/L) - (12EI/L^3)]sc & -[(AE/L)s^2 + (12EI/L^3)c^2] & -(6EI/L^2)c & [(AE/L) - (12EI/L^3)]sc & [(AE/L)s^2 + (12EI/L^3)s^2] & \\ -(6EI/L^2)s & (6EI/L^2)c & 2EI/L & (6EI/L^2)s & -(6EI/L^2)c & 4EI/L \end{bmatrix} \begin{bmatrix} u_1 \\ v_1 \\ \theta_1 \\ u_2 \\ v_2 \\ \theta_2 \end{bmatrix}$$

(14.70)

where $s = \sin\theta$ and $c = \cos\theta$.

From earlier studies, the nodal force-displacement relation in terms of global coordinates was derived as

$$\mathbf{X} = \mathbf{Ku} \qquad (14.46)$$

where **K**, the element stiffness matrix in global coordinates, is defined as

$$\mathbf{K} = \mathbf{T^T K' T} \qquad (14.47)$$

EXAMPLE 14.9

Compute support reactions and displacements for the frame shown. For simplicity, all values of

$$\frac{AE}{L} = \frac{12EI}{L^3} = 1000 \text{ kip} \cdot \text{in.}$$

Forces on member 3–4 due to 1 klf load:

$$M_3^f = -M_4^f = \frac{1(12)^2}{12} = 12 \text{ kip} \cdot \text{ft}$$

$$Y_3^f = Y_4^f = \frac{1(12)}{2} = 6 \text{ kip} \cdot \text{ft}$$

$$u_1 = v_1 = \theta_1 = u_2 = v_2 = \theta_2 = u_4 = v_4 = 0$$

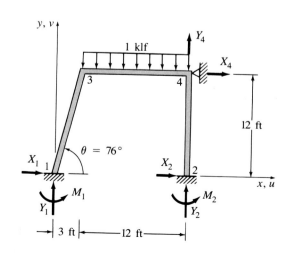

From the equation in Fig. 14.13, element stiffness matrices are

$$\mathbf{K}_{1-3} = \begin{bmatrix} & u_1 & v_1 & \theta_1 & u_3 & v_3 & \theta_3 \\ & 10 & 0 & -60 & -10 & 0 & -60 \\ & 0 & 10 & 15 & 0 & -10 & 15 \\ & -60 & 15 & 510 & 60 & -15 & 255 \\ & -10 & 0 & 60 & 10 & 0 & 60 \\ & 0 & -10 & -15 & 0 & 10 & -15 \\ & -60 & 15 & 255 & 60 & -15 & 510 \end{bmatrix} \times 10^2$$

$$\mathbf{K}_{3-4} = \begin{bmatrix} & u_3 & v_3 & \theta_3 & u_4 & v_4 & \theta_4 \\ & 10 & 0 & 0 & -10 & 0 & 0 \\ & 0 & 10 & 60 & 0 & -10 & 60 \\ & 0 & 60 & 480 & 0 & -60 & 240 \\ & -10 & 0 & 0 & 10 & 0 & 0 \\ & 0 & -10 & -60 & 0 & 10 & -60 \\ & 0 & 60 & 240 & 0 & -60 & 480 \end{bmatrix} \times 10^2$$

$$\mathbf{K}_{2-4} = \begin{bmatrix} & u_2 & v_2 & \theta_2 & u_4 & v_4 & \theta_4 \\ & 10 & 0 & -60 & -10 & 0 & -60 \\ & 0 & 10 & 0 & 0 & -10 & 0 \\ & -60 & 0 & 480 & 60 & 0 & 240 \\ & -10 & 0 & 60 & 10 & 0 & 60 \\ & 0 & -10 & 0 & 0 & 10 & 0 \\ & -60 & 0 & 240 & 60 & 0 & 480 \end{bmatrix} \times 10^2$$

Force displacement equations in assembled partition form:

Applied forces

Fixed-end member forces

$$\begin{bmatrix} X_1 \\ Y_1 \\ M_1 \\ X_2 \\ Y_2 \\ M_2 \\ X_4 \\ Y_4 \\ \hline M_4 = 0 \\ X_3 = 0 \\ Y_3 = 0 \\ M_3 = 0 \end{bmatrix} = 10^2 \begin{bmatrix} 10 & 0 & 60 & & & & & & & -10 & 0 & -60 \\ 0 & 10 & 15 & & & & & & & 0 & -10 & 15 \\ -60 & 15 & 510 & & & & & & & 60 & -15 & 255 \\ & & & 10 & 0 & -60 & -10 & 0 & -60 & & & \\ & & & 0 & 10 & 0 & 0 & -10 & 0 & & & \\ & & & -60 & 0 & 480 & 60 & 0 & 240 & & & \\ & & & -10 & 0 & 60 & 20 & 0 & 60 & -10 & 0 & 0 \\ & & & 0 & -10 & 0 & 0 & 20 & -60 & 0 & -10 & -60 \\ & & & -60 & 0 & 240 & 60 & -60 & 960 & 0 & 60 & 240 \\ -10 & 0 & 60 & & & & -10 & 0 & 0 & 20 & 0 & 60 \\ 0 & -10 & -15 & & & & 0 & -10 & 60 & 0 & 20 & 45 \\ -60 & 15 & 255 & & & & 0 & -60 & 240 & 60 & 45 & 990 \end{bmatrix} \begin{bmatrix} u_1 \\ v_1 \\ \theta_1 \\ u_2 \\ v_2 \\ \theta_2 \\ u_4 \\ v_4 \\ \theta_4 \\ u_3 \\ v_3 \\ \theta_3 \end{bmatrix} + \begin{bmatrix} 0 \\ 0 \\ 0 \\ 0 \\ 0 \\ 0 \\ 0 \\ 6 \\ -12 \\ 0 \\ 6 \\ 12 \end{bmatrix}$$

P $\qquad$ $\mathbf{K}_s$ $\qquad$ $\mathbf{P}^f$

The lower set of partitioned equations reduce to

$$\begin{bmatrix} M_4 = 0 \\ X_3 = 0 \\ Y_3 = 0 \\ M_3 = 0 \end{bmatrix} = \begin{bmatrix} 0 \\ 0 \\ 0 \\ 0 \end{bmatrix} + 10^2 \begin{bmatrix} 960 & 0 & 60 & 240 \\ 0 & 20 & 0 & 60 \\ 60 & 0 & 20 & 45 \\ 240 & 60 & 45 & 990 \end{bmatrix} \begin{bmatrix} \theta_4 \\ u_3 \\ v_3 \\ \theta_3 \end{bmatrix} + \begin{bmatrix} -12 \\ 0 \\ 6 \\ 12 \end{bmatrix}$$

Displacements

$$\begin{bmatrix} \theta_4 \\ u_3 \\ v_3 \\ \theta_3 \end{bmatrix} = 10^{-2} \begin{bmatrix} 960 & 0 & 60 & 240 \\ 0 & 20 & 0 & 60 \\ 60 & 0 & 20 & 45 \\ 240 & 60 & 45 & 990 \end{bmatrix}^{-1} \begin{bmatrix} 12 \\ 0 \\ -6 \\ -12 \end{bmatrix} = \begin{bmatrix} 3.9 \\ 1.1 \\ -40.9 \\ -0.366 \end{bmatrix} \times 10^{-4}$$

Reactions

From the upper set of partitioned equations:

$X_1 = (-10u_3 - 60\theta_3)10^2 = 0.11$ kip

$Y_1 = (-10v_3 + 15\theta_3)10^2 = 4.03$ kips

$M_1 = (60u_3 - 15v_3 + 255\theta_3)10^2 = 5.86$ kip·ft

$X_2 = -60\theta_4 \times 10^{-2} = -2.34$ kips

$Y_2 = 0$

$M_2 = 240\theta_4 \times 10^2 = 9.35$ kip·ft

$X_4 = (60\theta_4 - 10u_3)10^2 = 2.23$ kips

$Y_4 = (-60\theta_4 - 10v_3 - 60\theta_3)10^2 + 6$ kips $= 7.97$ kips

Check Equilibrium

$\Sigma F_x = 0$

$\Sigma F_y = 0$

$\Sigma M_z = 0$

Therefore, the stiffness matrix transformation of Eq. (14.47) is accomplished by using **T** from Eq. (14.69) and **K'** from Eq. (14.62), followed by substitution of **K** into Eq. (14.46). The resulting general force-displacement matrix relationship of a frame member in global coordinates is presented in Fig. 14.13 as Eq. (14.70). A plane frame analysis by the stiffness method is done using the same matrix approaches for beams and trusses. Example 14.9 demonstrates a stiffness method analysis of a plane frame.

14.4. SYMMETRY AND ANTISYMMETRY

Civil engineering structures such as bridges, buildings, and towers are frequently designed as symmetric systems. The choice of a symmetric design usually results in notable time and cost reductions for analysis, design, fabrication, and erection. In structural analyses, we often make use of the principles of symmetry. For example, a simple beam under symmetric load immediately suggests symmetric construction of the V and M diagrams, and equal valued support reactions. In Chaps. 12 and 13, examples are presented to demonstrate that application of the principles of symmetry can result in fewer computations in the solution of classical hand methods of analysis. These same principles can be used for the stiffness analysis of symmetric linear elastic structures. An awareness of structure symmetry often results in an efficient analysis achieved with a half-model (or less) of the structure.

Figures 14.14a to c present three symmetric structures under symmetric loading. In all three cases, the slope change of the structure at the plane of symmetry is zero. The truss model of Fig. 14.14a is shown as one-half of the real truss; observe that nodes b and d on the plane of symmetry are supported by rollers to permit free vertical movement; also, a force of 3 kips (one-half of the 6-kip applied load) is applied on the truss model at node b. Thus, for the one-half symmetric truss model shown in Fig. 14.14a, the following requirements are imposed:

1 Only one-half of the magnitudes of load occurring on the plane of symmetry must be applied.

2 The displacement components normal to the plane of symmetry must be set to zero.

3 In analysis, the magnitude of cross-sectional areas of those truss elements that occur on the plane of symmetry are set equal to one-half of their respective value.

Similarly, the symmetric frame of Fig. 14.14b is represented by a half-frame model with a vertical roller support at point c on the plane of symmetry. The continuous beam of Fig. 14.14c is resting on unyielding supports and has a plane of symmetry passing through the roller at c. Moreover, the beam at point c does not rotate or displace when symmetric loading is applied. Therefore, it is expedient to represent the beam by a model equal to one-half of the real beam with a fixed support placed at point c on the beam (see Fig. 14.14c).

The continuous beam of Fig. 14.14d is shown to support a set of unsymmetric concentrated loads. The elastic curve reveals that point c on the beam will not displace as it rotates. Therefore, a beam model is selected which consists of one-half of the real beam with a pin support placed at point c.

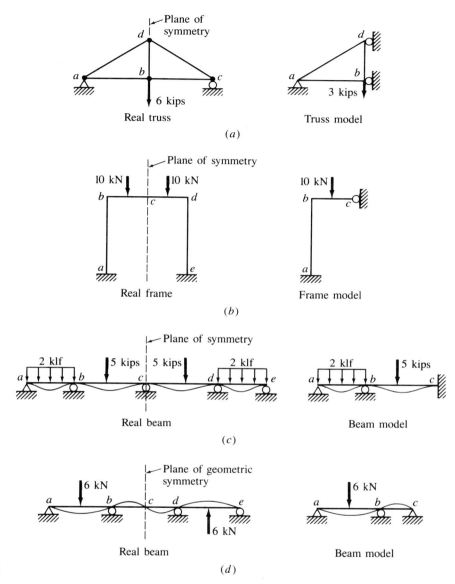

Figure 14.14

All the models shown in Fig. 14.14 contain fewer members and nodes as compared with the corresponding real structures. Moreover, the models yield lower-order structure stiffness matrices, require less computations, and present an efficient and economical means of analyses.

14.5. CLOSING REMARKS

The lessons in this chapter present the concepts of the stiffness method by use of simple structures which contain few members and equally few degrees of freedom. The aim of these lessons is to draw the readers attention to the application of the concepts with minimum effort on number crunching. Therefore, the stiffness method solutions involve simple computations and relative ease of matrix operations. Unfortunately, most engineering structures contain many members and many degrees of freedom so that a practical matrix analysis can only be achieved by use of a computer.

In the execution of a computer program, special programs are commonly used to efficiently store data. Observe that both element and assembled structure-stiffness matrices in all previous examples are symmetric. In these cases, it is only necessary for the computer to store the matrix coefficients that lie along the main diagonal and the terms above or below it (referred to as upper or lower triangular matrices). The *Renumbered Nodes* phase of IMAGES-2D performs efficient data storage; in addition, a routine called *bandwidth minimization* is used for the efficient solution of the simultaneous force equilibrium equations. The term *bandwidth* implies that all nonzero elements of the $\mathbf{K}_s$ matrix are arranged in a banded area surrounding the main diagonal terms (see "Assemble Stiffness Matrix" in App. 1). In large structure applications, a bandwidth minimization routine can result in notable reductions in the number of computations and computer time.

The nodal displacements of a linear elastic structure are computed by the Cholesky decomposition method in the IMAGES-2D computer program. The Cholesky method is an efficient form of the Gauss method which can be applied when $\mathbf{K}_s$ is a symmetric matrix (see Ref. [14.1] and "Theoretical Background" in App. 1).

The challenges and societal demands placed on the engineering community continue to provide excitement and opportunity for professional growth. The popularization of microcomputers will clearly make our tasks more *user friendly*. However, the reader is reminded that computerization does not relieve the engineer of professional responsibility for engineering analysis and design.

Problems

14.1 and 14.2 Establish the overall stiffness matrix $[\mathbf{K}_s]$ in terms of AE/L for the truss shown below. All bars have equal area (A) and modulus (E).

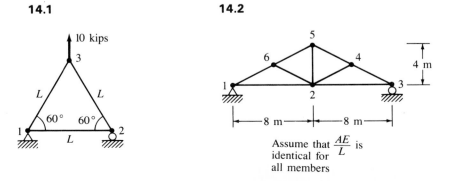

Assume that $\dfrac{AE}{L}$ is identical for all members

491 Problems

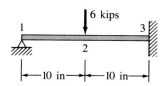

14.3 Find $[\mathbf{K_s}]$. $E = 29 \times 10^3$ ksi, $I = 1200$ in^4.

14.4 Find $[\mathbf{K_s}]$. $E = 20 \times 10^3$ ksi, $I = 1400$ in^4.

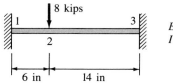

$E = 29 \times 10^3$ ksi
$I = 1400$ in^4

14.5 Determine $[\mathbf{K_s}]$ for frame shown neglecting axial effects. $E = 30 \times 10^3$ ksi, $I = 300$ in^4, both elements. (Hint: See Eq. 14.70 and Fig. 14.13.)

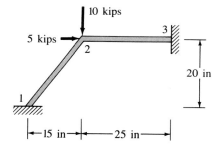

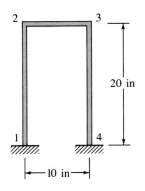

14.6 Determine $[\mathbf{K_s}]$ for frame shown below. Neglect axial effects. Find $[\mathbf{K_s}]$ in terms of EI assumed as constant for both elements. (Hint: See Eq. 14.70 and Fig. 14.13.)

14.7 to 14.17 Find all joint displacements and reactions for each structure defined below.
14.7 The truss of Prob. 14.1.
14.8 The square truss as shown below.
14.9 The beam of Prob. 14.3.
14.10 The beam of Prob. 14.4 if the support at joint 1 is replaced by a hinge support.
14.11 The frame of Prob. 14.5.

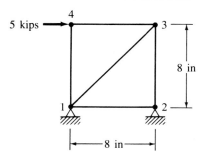

$E = 29 \times 10^3$ ksi
$A = 4$ in^2 for all bars

14.12 The frame shown below.

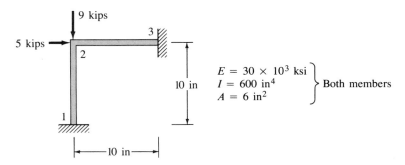

14.13 The frame of Prob. 14.5 if the only loading on the frame is a uniform distributed gravity load of 2 klf between nodes 2 and 3.

14.14 Remove the 5-kip force at node 4 on the square truss of Prob. 14.8. Using the $[K_s]$ matrix determined from Prob. 14.8, find displacements and reactions if diagonal member 1–3 experiences a 61°F temperature increase ($\alpha = 6 \times 10^{-6}$ in/in/°F).

14.15 Redo Prob. 14.8 if support node 2 settles 0.5 in. Do the results change significantly due to this settlement?

14.16 Redo Prob. 14.11 to include axial plus bending effects. Let $A = 5$ in^2 for both frame elements.

14.17 Redo Prob. 14.13 to include both axial and bending effects. Let $A = 5$ in^2 for both frame members.

14.18 Check the results of Prob. 14.8 by the IMAGES-2D program.

14.19 Check the results of Prob. 14.9 by the IMAGES-2D program.

14.20 Redo Prob. 14.12 by use of the IMAGES-2D program.

14.21 Redo Prob. 14.14 by use of the IMAGES-2D program.

References

14.1 William Maguire and Richard H. Gallagher, *Matrix Structural Analysis*, Wiley, New York, 1979, p. 4.

14.2 Harold C. Martin, *Introduction to Matrix Methods of Structural Analysis*, McGraw-Hill, New York, 1966, pp. 233–291.

14.3 W. Weaver and J. M. Gere, *Matrix Analysis of Framed Structures*, 2d ed., Van Nostrand Reinhold, Princeton, N.J., 1980.

14.4 M. Daniel Vanderbilt, *Matrix Structural Analysis*, Quantum Publishers, Inc., New York, 1974, pp. 69–144.

14.5 V. James Meyers, *Matrix Analysis of Structures*, Harper & Row, New York, 1983, pp. 398–434.

APPENDIX ONE

A Tutorial for Use of the IMAGES Program by Celestial Software, Inc.

IMAGES-2D is a copyright software program of Celestial Software, Inc. This appendix is intended for the use and benefit of educators and students. Reproduction of this appendix in any manner without express written permission of Celestial Software, Inc. is prohibited. If problems are encountered with the use of the software, the author should be contacted. That is, support services are *not* provided to educators and students by Celestial Software, Inc., through the purchase of this text.

However, program support services can be purchased by contacting Celestial Software, Inc. at the following address:

IMAGES-2D Book Special
Celestial Software, Inc.
125 University Avenue
Berkeley, CA 94710
(415) 843-0977

Celestial Software, Inc. has a whole family of IMAGES product line which includes IMAGES-2D, IMAGES-3D, IMAGES-THERMAL, IMAGES-AISC, and other utilities programs. For more information please contact Celestial Software Sales Department.

This version of IMAGES-2D is a two-dimensional static structural analysis program written in Microsoft BASIC (a registered trademark of Microsoft Corportation).

IMAGES-2D STATIC VERSION 3.0 HARDWARE REQUIREMENTS:

IBM PC, XT, AT, PS/2 or Compatibles
Two $5\frac{1}{4}$ floppy disk drives or one disk drive with a hard disk
512 kilobytes random access memory
Color graphics adaptor board (CGA, EGA, or VGA)
RGB or composite monitor
Math coprocessor is optional
DOS 2.1 or higher

IMAGES-2D is designed to optimize your time by providing a completely interactive prompting program with instantaneous response. Input prompts are kept brief for the benefit of experienced users, while a HELP command is always available for the new or infrequent user.

All input data are entered interactively through the use of menus, which are subdivided into four major categories: Geometry Definition, Static Analysis,

Dynamic Analysis, and Seismic Analysis. However, Dynamic and Seismic Analysis are not contained on the enclosed program diskettes.

The *Program Menu* listed on the monitor contains:
1. Geometry Definition
2. Static Analysis
3. Dynamic or Seismic Analysis (not included)
4. Restart IMAGES-2D
5. End Session

The program will accept any consistent set of units; however, most defaults are derived from inches, pounds, and seconds.

Geometry plots are displayed on the screen and may be viewed at any time during construction of the model. Users operating under DOS 2.0 or higher can take advantage of the *print screen* option by invoking the *graphics* command from DOS prior to loading IMAGES.

Once you describe the structure's geometry, the program performs a geometry check to help locate any errors that were made during the input process. Bandwidth minimization is also done at the geometry level to assure that there is sufficient memory storage for the problem.

When static analyses are performed, concentrated or distributed weight loadings, as well as axial thermal gradients, may be applied to the model. Loads in both the global and local element coordinate systems and the resulting beam stresses may be calculated. Multiple load cases can be ratioed and summed together.

Throughout this program, all user entries must be followed by a carriage return unless otherwise specified.

A ? appears on the screen and in the manual as a user prompt. Additionally, in the examples, we have used [] in the text as a symbol to indicate specific keys on the keyboard.

PREPARING DATA DISKETTES

Data diskettes must be formatted before they can be used. This process is followed only when a new data diskette is needed.

Use the following procedures to format data diskettes:

2 Floppy System

1. Place a blank diskette in drive B.
2. At the DOS prompt (typically A>), type

   FORMAT B:
   and press [ENTER].

   The screen will display:

   Insert new diskette in drive B
   and strike any key when ready.

3. Press any key since the diskette is already in drive B. When the formatting process is complete, the screen will display:

   FORMAT another (Y/N)?

4. Press the [N] key and then press [ENTER].

Hard Disk System

1. Place a blank diskette in drive A.
2. At the DOS prompt (typically A>), type

   FORMAT A: and press [ENTER].

   The screen will display:

   Insert new diskette in drive A and strike any key when ready.

3. Press any key. When the formatting is complete, the screen will display:

   FORMAT another (Y/N)?

4. Press the [N] key and then [ENTER].

HOW TO RUN IMAGES-2D

To run IMAGES-2D:

2 Floppy System

1. **The first time you use the IMAGES-2D program diskettes,** put disk 1 in drive A, go to the DOS A> prompt, and type

   COPY IMAGES.FLP IMAGES.BAT

   and press [ENTER].

Then (and from now on):

2. Make certain the IMAGES-2D disk 1 is in drive A.
3. Put a formatted data disk in drive B.
4. Type B: and press [ENTER].
5. Type A: IMAGES and press [ENTER].
   The program will be running on drive A and the data will be stored on B.

Hard Disk System

1. **The first time you use the IMAGES-2D program diskettes,** type md\images at the DOS prompt (typically C>) and press [ENTER] to create an IMAGES directory on the hard disks.
2. Type cd\images and press [ENTER] to change to the IMAGES directory. Put the IMAGES-2D disk 1 in drive A, type

   COPY A:*.*

   and press [ENTER] to copy all the files on disk 1 to the hard disk. Replace disk 1 with disk 2 and repeat this copying process.
3. Type COPY IMAGES.HRD IMAGES.BAT and press [ENTER].

   Then (and from now on):
4. Make certain you are in the IMAGES directory of the hard disk.
5. Type IMAGES and press [ENTER].
   The program and data files will be stored on the hard disk. The hard disk version of this program utilizes the DOS ASSIGN command. The directory containing this command must be included in the PATH statement of your AUTOEXEC.BAT file.
6. Type IMAGES and press [ENTER].
   Program and data files will be stored on the hard disk.

AFTER IMAGES is entered, the screen will display:

Enter today's data (or return)

?

- Enter the date in the format of month-day-year (e.g. 2-12-83).
- Next, the copyright notice appears.

- Press any key to continue.
- The program prompts:

Enter name for storing files.

If the name exceeds eight characters, only the first eight will be used (no blank spaces allowed)

Name

?

- Enter the file name.

    The following menu will appear:

Select Printer Type

1. IBM, Epson MX-80 or 100 (default)
2. NEC PC-8023A-C
3. All Others

- Press the menu item number that corresponds to your printer type. For printers other than those listed in menu items 1 or 2, follow the instructions that appear on the screen using your printer manual as an aid in defining the printer parameters.

Note that this version of IMAGES-2D does not support laser printers.

All steps to begin running IMAGES-2D are now complete.

Common Commands

Some commands are common to each section of the program.
These commands are discussed below.

Save Data on Disk

In order to store data on disk for future use:

- Press the appropriate menu item number listed on the screen.

The screen will display a WARNING if you attempt to exit the routine and have not saved the data on disk. If the data had been saved on disk prior to entering the routine and no changes had been made during the routine, you can exit from the routine without saving the data again.

Get Old Data from Disk

This command recalls the previously defined data into the working memory.

To get old data from the disk:

- Press the menu item number on the screen during any of the analyses or geometry definition.

Create/Edit Menu Command Options

The Create/Edit option, the first part of the geometry definition and analyses sections, has available the following commands:

CHANGE, DELETE, END, HELP, LIST, PRINT

CHANGE   changes any previously input item

DELETE   deletes the last item input

END   returns to the previous menu

HELP   displays the required input and the options available

LIST   lists on the screen all currently defined terms

PRINT   lists on the printer all currently defined items

A GENERATE option is available when nodes and elements are being input in the geometry section. This feature speeds the process of entering node and element data.

To use any of these commands, enter the first letter of the command or the entire word. For example, to list on the screen all defined data items, press [L] or type LIST.

Stop Data from Scrolling

In some sections of the program the data will rapidly scroll across the screen.

To temporarily stop the scrolling:

- Press the [PAUSE] key. If your keyboard does not have this key, simultaneously press the [CTRL] and [NUM LOCK] keys.

To resume the scrolling process:

- Press any key on the keyboard.

Note: When scrolling of data occurs, printing will not begin until all data have been displayed on the screen.

Program Execution

IMAGES-2D may be run either by selecting menu items individually or by using the batch execution option. Step-by-step execution of individual menu items gives you full control of the analysis and program flow, while batch execution permits automated execution of all steps of an analysis.

Individual Execution

Any of the menu options may be run by entering the corresponding menu number. The menus are arranged in the logical order of analysis and, in general, should be run in ascending menu order. After a selection is made, the program prompts:

Write output to printer (Y/N)?

and at the bottom of the screen:

Enter E to Exit

- Press [Y] if you want the printed output.
- Press [N] for the screen display only.
- Press [E] to return to the menu from which the routine was called.

Batch Execution

Any of the four parts of IMAGES-2D may be run by using the batch execution feature. Because the individual menus are organized in the order required for execution, any menu selections preceding the ones to be executed in the batch mode must be defined first.

For example, when in the Geometry Menu, the geometry must be defined within the Create/Edit Geometry option before using the [F1] key to run the geometry check and the renumbering portions of the program:

When all of the required input is defined:

- Press the [F1] key followed by a carriage return.

The following prompt appears:

Batch Execution. Do you
wish to continue (Y/N)
?

- Press [Y] to continue in the batch mode.

- Press [N] to return to the previous menu.

Note: Do not select batch mode unless your printer is on line and ready to print. When using the batch execution feature, the program automatically sends output to the printer.

Menu Organization

The following is a list of the geometry and analysis menus. After the data are initially entered through the Create/Edit options, the problem-solving routines may be run by pressing [F1]. The plotting routines and Return to Program Menu items are selected individually.

Geometry Menu
1. Create/Edit Geometry (Must be defined first)
2. Check Geometry
3. Renumber Nodes
4. Plot Geometry
5. Return to Program Menu

Static Menu
1. Create/Edit Loads (Must be defined first)
2. Assemble Stiffness Matrix
3. Solve Displacements
4. Solve Loads, Stresses, and Reactions
5. Sum Static Load Cases
6. Plot Deflected Shapes
7. Return to Program Menu

GEOMETRY DEFINITION

Describing the geometry of the structure you want to analyze is usually the most time consuming step in any finite element analysis. IMAGES-2D is designed to simplify this process through an interactive, menu-driven geometry-generation ability.

Geometry Menu

When item 1, Geometry Definition, is selected from the Program Menu, the following options are displayed:

1. Create/Edit Geometry
2. Check Geometry
3. Renumber Nodes
4. Plot Geometry
5. Return to Program Menu

Creat/Edit Geometry

Define and edit the geometry of a structure.

Menu:
1. Enter Problem Title
2. Define Material Properties
3. Define Node Points
4. Define Elements
5. Define Beam Cross-Sectional Properties
6. Define Restraints
7. Save Geometry on Disk
8. Get Old Geometry File
9. Return to Geometry Menu

To select an item from the menu:

- Enter the number that corresponds to your selection.

Enter Problem Title

Specify a descriptive title for the current analysis problem.

The problem title is used to identify printed output and may be up to sixty characters long. Longer titles will be truncated following the sixtieth character.

When making numerous runs of the same model, this title is a convenient method of differentiating between hard copy output.

- Enter the selected title on the keyboard.

Define Material Properties

Define the modulus of elasticity, density, and coefficient of thermal expansion, and Poisson's ratio for the materials.

The required material properties are:

E = Modulus of Elasticity	(F/L^2)
ρ = Weight Density	(F/L^3)
α = Coefficient of Thermal Expansion	$(L/L/\text{Deg})$
ν = Poisson's ratio	Dimensionless

Material numbers, used to identify the material properties in the element section, are automatically generated by the program.

Caution: IMAGES-2D will accept any consistent set of units. Care must be taken when you use the program default values, which are defined in terms of inches, pounds, and second units.

The following program prompt appears:

Next Material No. 1
?

■ Enter the data in the order of modulus, density, and expansion coefficient, and Poisson's ratio, separating each item with a comma.

■ Pressing a carriage return without making any entries defaults to:

$E = 27.9 \times 10^6$

$\rho = 0$

$\alpha = 0$

$\nu = 0.3$

■ Enter [E] or END to terminate the material property input and return to the previous menu.

Define Node Points

Enter the coordinates of the nodes.

Node point numbers are assigned by the program starting from number one using an increment of one. Nodes not connected to any element will be fully restrained during the geometry check phase, unless they are restrained by the user.

X and Y coordinates are always defined in a right-handed Cartesian coordinate system.

You can automatically generate nodes rather than enter each node individually. (See Example 1.)

The following prompt appears:

Next Node 1
?

■ Enter the X-coordinate and Y-coordinate of the node, separating each item with a comma.

■ Entering a carriage return or blank defaults to $X = 0$ and $Y = 0$.

■ Enter [G] or GENERATE to automatically generate nodes.

■ Enter [E] or END to terminate the nodal input and return to the Create/Edit geometry menu.

Node Generation Example 1

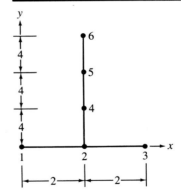

Node Generation Screen Display:

Next Node 1
? 0, 0
1) 0 0
Next Node 2
? G
To Node
? 3

Enter X, Y Coordinates of Node 3
? 4
Starting from Node
? 1
2) 2 0
3) 4 0
Next Node 4
? G
To Node
? 6
Enter X, Y Coordinates of Node 6
? 2, 12
Starting from Node
? 2
4) 2 4
5) 2 8
6) 2 12
Next Node 7
? E

In Example 1, note that all user entries are preceded by the program prompt, ?. The user defined the coordinates of three nodes (i.e., nodes 1, 3, and 6) and used the GENERATE option to supply all intermediate nodes.

Define Elements

Define element type.

Menu:

1. Define Beams
2. Define Plates
3. Define Springs
4. Return to Geometry Menu

■ Enter the number of the desired element type.

Any or all element types may be defined in any order.

BEAM MEMBERS

Define Beam

Define the beam connectivity, cross-sectional property and material property numbers, and any member end releases.

I nodes and J nodes may be any valid node numbers. The program automatically assigns beam numbers.

Member end releases (pin codes) are defined using a three-digit code representing the three DOF, $XY\theta_z$, in local coordinates. A zero digit represents the fixed condition (default), a one represents a released DOF. For example:

100 = Axial (X) release
010 = Shear (Y) release
001 = Moment (θ_z) release.

Combinations such as 101 are valid for representing slotted joints. The release code for node I precedes the release code for node J in the input order.

You can automatically generate beams of the same properties rather than enter each beam individually by using the GENERATE option (see Example 2).

Generation begins from the last beam entered. Cross section and material numbers and pin codes are taken from the last beam entered. Note that at least one beam must be defined before beam generation can occur.

Beams are generated by entering increments for both the I and J ends. The increments may be positive, zero, or negative.

The following program prompt appears:

Next Beam 1
?

■ Enter the I node, J node, cross-section number, material number, pin I, and pin J. Separate all entries with commas.

■ Entering a [0] or no data for the cross section and material numbers will cause the program to default to one.

■ Enter [G] or GENERATE to automatically generate beams.

■ Enter [E] or EXIT or return to the previous menu.

TRUSS MEMBERS

Truss members are defined by use of the *Define Beam* element selection. A truss member is defined as a beam element with a pin code of (001) at the member node end to provide moment release.

However, to insure nodal stability at any joint where two or more truss members are connected, moment release is only provided to $(n-1)$ members at the joint (n = number of members at the joint).

Beam Generation Example 2

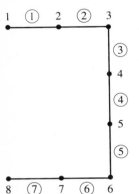

Beams ① and ② have cross-section and material numbers of 1. The cross-section number is 2 for beams ③ through ⑦; however, the material property number remains constant.

Beam Generation Screen Display

```
Next Beam 1
? 1, 2
1) 1 2 1 1 0 0
Next Beam 2
? 2, 3
2) 2 3 1 1 0 0
Next Beam 3
? 3, 4, 2
3) 3 4 2 1 0 0
Next Beam 4
? G
To Beam
? 7
Enter I Increment, J Increment
? 1, 1
4) 4 5 2 1 0 0
5) 5 6 2 1 0 0
6) 6 7 2 1 0 0
7) 7 8 2 1 0 0
Next Beam 8
? E
```

In Example 2, all user entries are preceded by the program prompt, ?. The user defined beams ① and ②, and used the default for material, cross-section and pin code. Beam ③ is defined with cross-sectional property number 2. The remaining beams (④ through ⑦) are generated from beam ③ by incrementing both I and J nodes successively by 1.

PLATE MEMBERS

Define Plates

Define plate connectivity, material number, and plate thickness.

Plates are defined by I node, J node, K node, material number, and thickness (see figure below).

I, J, and K can be any valid nodes forming a triangle with a base to height ratio less than 10. The base being the longest side.

Thickness is the plate thickness in the Z direction and is constant over each plate.

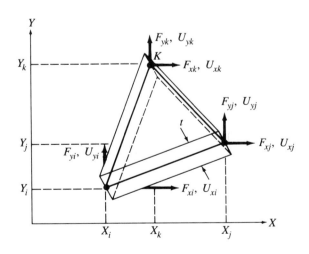

X, Y = global system
I, J, K = node numbers
U_x, U_y = displacement components
F_x, F_y = force components
t = thickness
x, y = coordinates

Plates can be generated by using the GENERATE or [G] command.

To generate plates, the I, J, and K increments are entered. The material and thickness are taken from the preceding element. At least one plate must be defined before generation can occur.

The following program prompt appears when the Define Plates option is selected.

Next Plate 1
?

- Enter the I node, J node, K node, material number, and thickness. Separate all entries with a comma. I, J, and K should represent a counter clockwise progression.

- Entering a [0] or no data for the material number will result in the default value of one. The thickness will default to the last prior value.

- Enter [G] to generate plates.

- Enter [E] or EXIT to return to previous menu.

SPRING MEMBERS

Define Springs

Define the location, stiffness, and orientation of nodal springs.

The spring elements in IMAGES-2D are programmed to provide nodal stiffness modification. An example is the case of modeling some amount of anchor flexibility of a cantilever beam:

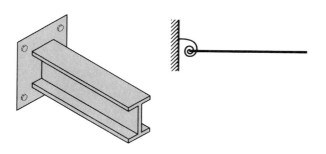

Other cases are where specific joint flexibility is desired which might represent a bolted connection with less than infinite rigidity such as:

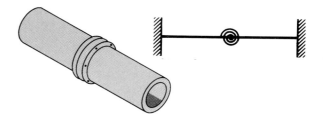

or ground support springs, such as:

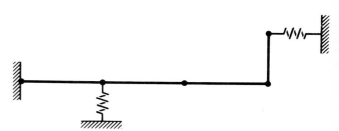

The ground spring elements in IMAGES-2D are defined by designating the node to be stiffness restrained, the direction of the spring, the spring stiffness, and the angle of the spring relative to the input direction.

Valid directions are:

1 = For Gobal X
2 = For Global Y
3 = For Rotational

Spring stiffness is entered as K in units of force per unit length for translational stiffness and moment per unit radian for rotational stiffness.

Angles (in degrees) are measured from the entered direction, for example, a direction 1 spring with an angle of 90 degrees equals a direction 2 spring with 0 angle.

Angle has no meaning for a rotational (direction 3) spring and is disregarded by the program.

For node-to-node springs, enter in the form

I node, J node, Stiffness, DOF

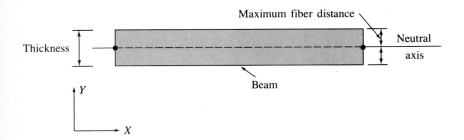

where

    I and J nodes = Node numbers connecting the spring
    Stiffness = Spring stiffness
    DOF = DOF in which the spring acts
        For separated nodes:
          1. Axial spring defined by local axis between nodes I and J
        For coincident nodes:
          1. Spring along global X axis
          2. Spring along global Y axis
          3. Spring about global Z axis

Define Beam Cross-Sectional Properties

Define the cross-sectional properties of the beam elements.

Define the properties in the following order:

    Cross-Sectional Area (L^2)
    Area Moment of Inertia (L^4)
    Maximum Fiber Distance (L)
    Shear Shape Factor (Dimensionless)

The area moment of inertia must be greater than zero for all elements.

The maximum fiber distance is the distance from the neutral axis to the outermost fibers, and is used only for bending stress calculations. (See the drawing above.) For analyses in which bending stresses are of no concern, the default value of one can be taken.

The shear shape factor is for beam bending about the Z axis. The factor is used in conjunction with the cross-sectional area to yield the effective shear area.

If the shear shape factor is zero, then beam shear deformation is neglected. Common shear shape factors are:

    1.18 = solid rectangular cross-sections
    1.12 = solid circular cross-sections
    1.14 = solid ellipse cross-sections
    1.89 = circular tube cross-sections
    2.27 = rectangular tube cross-sections.

Cross-sectional property numbers are automatically incremented, starting with number 1.

The following program prompt will appear:

Next Property 1
?

■ Enter the area, inertia, and maximum fiber distance, and shear shape factor.

■ Enter [E] or EXIT to return to the previous menu. (See figure above.)

Define Restraints

Restrain node points to ground in selected degrees-of-freedom (DOF).

Global DOF are restrained by entering the node number and the DOF to be restrained.
    The DOF are defined as follows:

    1 = X translation
    2 = Y translation
    3 = Z rotation

Restraints are defined only in the global coordinate system. Restraint numbers are automatically

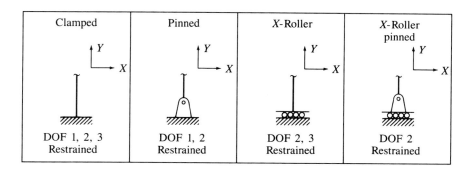

incremented starting with number 1. The drawing below illustrates typical restraints (see figure above).

The following program prompt appears:

Next Restraint 1
?

■ Enter the node to be restrained, and the degree-of-freedom to be restrained.

■ Enter [E] or EXIT to return to the previous menu.

To generate restraints, enter in the form

From node, DOF, To node, Increment

where

From node = Node number at which the DOF are restrained
DOF = Restraint DOF as defined above (Any combination of 1 to 3 may be used. 123 fully restrains the node.)
To node = Node number to which restraints are generated
Increment = Increment between From and To nodes

Check Geometry

Check the geometry for modeling and coding errors.

The program checks all geometry input for potential errors in the model definition. Potential errors include (but are not restricted to):

Unconnected node(s)
Two nodes at the same location
Lack of necessary parameters
Negative plate area

The negative plate area warning occurs when the I, J, K node order represents a clockwise progression around the plate. In this case the program will prompt:

Do you wish to reverse the numbering of the plates with negative area?

■ Enter [Y] or Yes if nodes are known to be correct.

■ Enter [E] or Exit to return to the Create/Edit Geometry menu if there is an error to be corrected.

Reversing the order of the node numbers will not affect the node location or numbers, it will only change the order of the plate definition.

While the program is checking the geometry, the screen will display the messages:

Checking Materials
Checking Nodes
Checking Beam Properties
Checking Beams
Checking Plates
Checking Springs
Checking Restraints

If not all types of elements were defined, only the types used will appear in the check list.

In the event of any errors, the appropriate warning or error message will be printed. In general, error messages indicate a change is necessary. Warnings indicate a potential error which may in fact be an intentional modeling technique. In all cases, the cause of errors or warnings should be determined to ensure model integrity.

Check the geometry by:

■ Pressing the menu item number that corresponds to Check Geometry.

Renumber Nodes

Minimize the nodal bandwidth by having the program renumber nodes.

Caution: Node renumbering must be performed even if the bandwidth is initially minimized by the user.

The nodal bandwidth (actually half-bandwidth) is defined as the maximum difference between node numbers on the same element, plus one. For example, consider the structure shown below:

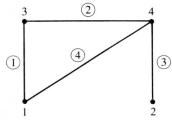

The maximum node difference occurs on element 4, and the bandwidth is $(4 - 1) + 1 = 4$.

Rearranging the node numbers to minimize the band would appear as:

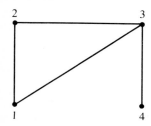

The band in this case is $(3 - 1) + 1 = 3$.

Although there is no significant difference in bandwidth in this example, renumbering saves significant computer storage and time in larger problems.

The screen shows the progress as node renumbering takes place. A typical display appears as:

Original Nodal Band   20
Pass 1   122
Pass 2   14
Current Nodal Band   14

Passes 1 and 2 indicate the number of iterations being performed. When Pass 2 begins, the renumbering is close to completion.

The printed output shows the correspondence between the original node numbering scheme (WAS) and the generated numbering scheme (IS)

All results are given in the original (WAS) numbering system. The IS system is used only to locate errors if they are encountered during the solution process. A description of error tracing and use of the IS numbering system is provided in Appendix B.

To automatically renumber the nodes:

■ Press the menu item number that corresponds to the Renumber Nodes option.

Plot Geometry

The program will display the following menu:

Menu:
1. Plot Nodes and Elements
2. Plot Nodes Only
3. Plot Elements Only
4. Plot Elements With Node Numbers
5. Return to Geometry Menu

Only the node points and elements are required to plot the geometry.

To plot the geometry:

■ Enter the appropriate menu number.

Plate elements will be plotted before beam elements, except in the case of the Plot Nodes Only option. The following message will appear with the plate element plot:

Press Return to Continue

- Press the [Enter] key to plot the remaining nodes and beam elements.

The Plot Elements With Node Numbers option includes the option of enlarging portions of a geometry plot.

To use the enlargement feature once the plot has been displayed:

- Press [B]. A cursor in the form of a plus sign will appear in the center of the screen.

- Press the appropriate cursor key on the numeric keypad to move the cursor to the corner of an imaginary box enclosing the area to be enlarged.

- Press the [Enter] key when the cursor is at the desired location.

- Move the cursor to the opposite diagonal corner.

- Press the [Enter] key. The enlarged section of the plot will be displayed.

Note: If the cursor does not respond to the cursor keys, press [NUM LOCK] to unlock the cursor function.

STATIC ANALYSIS

Static loading consisting of any combination of concentrated loads, gravity loads, axial thermal gradients, and enforced displacements can be applied to the model.

IMAGES-2D calculates deflections and, optionally, loads and stresses, at all points in the structure.

Static Menu

When you choose this option, the program displays the message

　　Remove Disk No. 1 and insert Disk No. 2.

If you are using a 2 floppy system, replace the IMAGES-2D Disk 1 in drive A with the IMAGES-2D Disk 2 and press [ENTER].

If you are using a hard drive system, simply press [ENTER].

The Static Menu, consists of the following options:

1. Create/Edit Loads
2. Assemble Stiffness Matrix
3. Solve Displacements
4. Solve Loads, Stresses, and Reactions
5. Sum Static Load Cases
6. Plot Deflected Shapes
7. Return to Program Menu

CREATE/EDIT LOADS

Specify load vectors for static analysis.

Menu:

1. Define Concentrated Loads
2. Generate Gravity Loads
3. Define Temperatures
4. Define Enforced Displacements
5. Define Distributed Loads
6. Change Active Load Case
7. Save Loads on Disk
8. Get Old Loads from Disk
9. Additional Options
10. Return to Static Menu

The seven additional options listed below expand the Create/Edit Loads Menu:

1. Zero Loads in Active Load Case
2. Delete Any Load Case
3. Print Load Cases
4. Define Load Case Titles
5. Add Two Load Cases (Loads Only)
6. Define Concentrated Loads between Nodes
7. Define Distributed Load in Global Direction
8. Return

IMAGES-2D allows up to five static load cases to be defined. At all times while in the Create/Edit Loads section, one of these load cases is termed the *active load case*. The active load case number is always displayed at the bottom of the screen. The active load case is the one which is being acted upon, either by your defining loads for it, changing loads, or having the program automatically generate gravity loads.

You can change the active load case at any time without using the Save Current Loads on Disk option. The only time the Save option is needed is before you return to the Static Menu because the working memory contains the information for all defined load cases.

Static loadings are described using the menu options listed above. A detailed description of each option is given on the following pages.

Define Concentrated Loads

Define concentrated loads at selected node points.

Concentrated loads are defined by entering the node number, direction, and load magnitude for each desired load location.

The direction (referred to in the program prompt) refers to the global coordinate system and is one of the following:

1 = Specified load is a force in the global X direction.

2 = Specified load is a force in the global Y direction.

3 = Specified load is a moment about Z.

Positive or negative loads are defined by the sign of the magnitude.

The following program prompt appears:

Enter Node, Direction, Load
?

■ Enter the node number, direction, and load magnitude for each desired load location. Separate each entry with commas.

■ Enter [E] or EXIT to terminate the load entry.

Generate Gravity Loads

Generate concentrated nodal loads from the element material and cross-sectional information. The direction (X or Y) and the multiplier, which is requested by the program, form the load case. The multiplier may be positive or negative.

Consider for example, a coordinate system with the positive Y axis pointing in the upward vertical direction. Deadweight loading would be specified by entering the Y direction with a multiplier of -1.

Gravity loads are generated by summing the contributions from each element attached to a given node. For plates, one third of the element weight is applied to a given node, for beams, the contribution is one half the element weight. Element weights are determined from geometric and material properties as illustrated below:

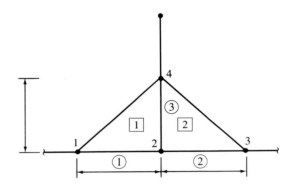

1, 2, 3, 4 are node numbers
①, ②, ③ are beam elements
[1], [2] are plate elements
l is the element length
A_p is the plate area
A_c is cross-sectional area
ρ is the material density
t is plate thickness

The weight applied to node 2 is given by

$$W_2 = [\tfrac{1}{2}[(\rho A_c l)_① + (\rho A_c l)_② + (\rho A_c l)_③]$$
$$+ \tfrac{1}{3}[(\rho A_p t)_{[1]} + (\rho A_p t)_{[2]}]]*\text{Multiplier}$$

Where the multiplier is entered by the user.

The following program prompt appears:

Existing loads in this case, if
there are any, will be zeroed.
Do you wish to continue (Y/N)
?

■ Enter [Y] to continue.

The following prompt appears:

Direction X or Y
?

Multiplier
?

■ Enter the proper direction, either [X] or [Y], and the multiplier.

Define Temperatures

Specify axial thermal gradients. Thermal gradients are frequently used to define beam preload and prestress conditions.

The following program prompt appears:

All loads in the active load case will be zeroed. Do you wish to continue? (Y/N)

■ Enter [Yes] or [No] to continue.

Note: If the desired load case includes gravity and thermal loads, define these loads in separate load cases and combine them using the Sum Load Cases option from the Static Analysis menu or the Add Two Load Cases option within the Create/Edit Loads additional options section.

The following program prompt appears:

Enter Reference Temperatures
(Default is 70)
?

■ Enter the reference to which all temperatures are compared to calculate thermal gradients.

■ A blank entry or carriage return causes 70 degrees to be the reference temperature.

For beams, the program prompts:

Enter Beam Temperatures
?

■ Beam temperatures are entered in the following format:

From Element, To Element, Temperature

where *From Element* and *To Element* are entered as a pair of beam numbers defining the beginning and ending beam in the range, or as the word ALL. For example, if there were a total of 10 beams in a model, and you wanted to specify a temperature of 200 degrees for all of them, the following two entries would be equivalent:

1, 10, 200
ALL, 200

A single beam temperature can be defined by specifying the same beam number twice as the beam range, or by setting the 'To Element' to zero.

For plates, the program prompts

Enter Plate Temperatures
?

■ Plate temperatures are entered in the form

From Element, To Element, Temperature
See beam temperatures for details.

■ Enter [E] or EXIT to return to the Create/Edit Loads menu.

Define Temperatures Example 3

Consider a model with ten beams for all elements. Beams 1 through 7 are at 200 degrees, beams 8 and 9 are at 250 degrees and beam 10 is at 300 degrees. The reference temperature is 68 degrees.

Define Temperatures Screen Display

Enter Reference Temperature
? 68
Enter Beam Temperatures
? A, 200
1 200
2 200
3 200
4 200
5 200
6 200
7 200
8 200
9 200
10 200

Enter Beam Temperatures
? 8, 9, 250
8 250
9 250
Enter Beam Temperatures
? 10, 0, 300
10 300
Enter Beam Temperatures
? L
1 200
2 200
3 200
4 200
5 200
6 200
7 200
8 250
9 250
10 300
Enter Beam Temperatures
? E

In the above example, all user entries are preceded by the program prompt, ?. The user first set all beams to 200 degrees, then went back to change beams 8 and 9 to 250 degrees and beam 10 to 300 degrees.

Define Enforced Displacements

Define enforced displacements at selected node points.

Enforced displacements are defined by entering the node number, direction, and magnitude for each desired displacement location.

The direction (referenced in the program prompt) refers to the global coordinate system and is one of the following:

1. Specified displacement (in unit length) is in the global X direction.

2. Specified displacement (in unit length) is in the global Y direction.

3. Specified rotation (in radians) is about the global Z axis.

The following program prompt appears:

Enter Node, Direction, Displacement
?

■ Enter the node number, direction, and magnitude of the displacement for each desired location. Separate each entry with commas. Positive or negative displacements are designated by the sign of the magnitude.

■ Enter [E] or Exit to terminate the displacement entry.

Note: To obtain reaction loads at points where enforced displacements are applied, the node and corresponding degree-of-freedom must be defined as a restraint. (See Define Restraints Section).

Define Distributed Loads

Define distributed loads acting on beam elements only.

Distributed loads are defined by entering the beam element number and load magnitude for each desired load location.

The following program prompt appears:

Enter Element, Load
?

■ Enter the beam element number and load magnitude for each desired location. Separate each entry with commas.

The program will echo resulting forces and end moments in LOCAL coordinates. Optionally, a distributed load may be applied to a series of elements by entering the starting element number, load magnitude, ending element number, and increment as illustrated in the following example

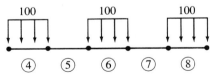

Enter Element, Load
? 4, 100, 8, 2

This entry defines a distributed load of magnitude 100 on elements 4 through 8 in increments of 2.

■ The ending element number, 8, in the above example, must be greater than or equal to the starting element number.

- The increment defaults to one if zero is entered.

The sign convention is illustrated below.

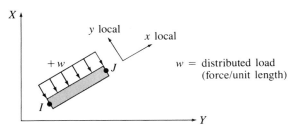

Note that a positive value of W results in a load in the negative local Y direction.

Tapered Distributed Loads Can Be Applied to Beams

Distributed loads are entered in the form:

Beam, Load at I end, Load at J end

or

From Beam, Load at I end, Load at J end, To beam, Increment, C

If To beam is zero then From beam is assumed.
If Increment is zero then 1 is assumed.
If a 'C' is entered in the Continuous field, a continuous load varying from the load of I end at From beam to the load at J end of To beam is generated. Load has units of force per unit length.

The sign convention is illustrated below.

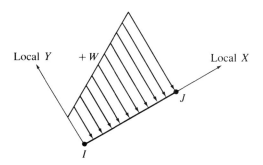

Change Active Load Case

Change the active load case in which data are being entered.

Menu:

1. Change to previously defined case
2. Define a new case

- Select number 1 to change to previously defined load case.

- Select number 2 for a new load case. The current load case number is incremented by one up to the maximum allowable number of cases.

Additional Options

Menu:

1. Zero Loads in Active Load Case
2. Delete Any Load Case
3. Write Loads to Printer
4. Define Load Case Titles
5. Add Two Load Cases (Loads Only)
6. Define Concentrated Loads between Nodes
7. Define Distributed Load in Global Direction
8. Return

Each additional option is described below.

Zero Loads in Active Load Case—All loads and displacements in the currently active load case are set to zero.

Delete Any Load Case—Any load case may be deleted by load case number. Since the numbers are sequential, all subsequent load case numbers are automatically decreased by one.

Write Loads to Printer—All load case data for all cases are written to the printer.

Define Load Case Titles—A 60-character descriptive load case title can be entered.

Add Two Load Cases—Any two load cases, selected by load case number, are algebraically added together. The final result may be stored under either of the original case numbers. If desired, the unused load case can be deleted with the Delete Any Load Case option.

Define Concentrated Loads between Nodes

Concentrated loads between nodes are entered in the form:

Beam, Local DOF, Offset, Load

where

- Beam = Element number of beam with load between nodes
- Local DOF = 1: Force acts in local X direction
  = 2: Force acts in local Y direction
  = 3: Moment acts about local Z direction
- Offset = Distance along the local X axis between node I and the load
- Load = Value of applied load

Up to 50 different concentrated loads between nodes may be applied in any load case. The sign convention is illustrated below.

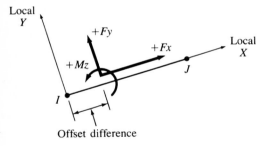

Define Distributed Loads Parallel to Global Axes

Global distributed loads are entered in the form:

Beam, Direction, Loading

or

From beam, Direction, Loading, To beam, Increment

where

Direction = 1: Loading applied parallel to global X direction
= 2: Loading applied parallel to global Y direction

Loading has units of force per unit length. Only one global distributed load can be applied to a given element in each load case.

The sign convention is illustrated below.

Return—Returns to the Create/Edit Loads menu.

To use any of these additional options:

■ Enter the menu number that corresponds to your option choice.

Assemble Stiffness Matrix

The program assembles the stiffness matrix in the following two separate steps.

The first step generates the structure stiffness matrix by assembling the element stiffness matrices into the global matrix.

The second step, termed PACK matrix, removes any restrained DOF or singularities from the matrix. (A singularity is a DOF with zero stiffness on the diagonal. This occurs when all of the moment-carrying ability is released at a joint as in the case of a truss.)

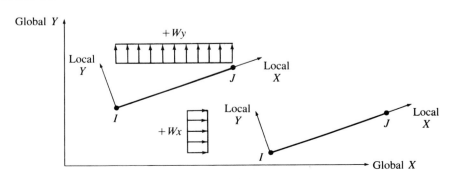

In addition, any DOF for nodes not connected to an element are removed at this time.

During the formation of the global stiffness matrix, the screen displays the element number currently being processed.

The PACK matrix phase lists any restraints or singular points removed from the stiffness matrix followed by a summary of the number of degrees-of-freedom and the bandwidth. This bandwidth is calculated in terms of degrees-of-freedom in contrast with the nodal band given in the RENUMBER portion of the geometry section.

The printed output shows the diagonal terms of the stiffness matrix in the IS numbering system. Finally, a list of degrees-of-freedom versus node number is provided for help in error tracing. Refer to pages 517 and 518 for more detail on error tracing using this list.

Solve Displacements

Displacements are solved in two steps:

1. Decompose Stiffness Matrix
2. Solve Displacements

If the applied loading consists of enforced displacements, the stiffness matrix is decomposed for each load case, otherwise it is only decomposed once.

Decompose Stiffness Matrix

The program automatically factors the stiffness matrix into a form suitable for solving a large number of simultaneous equations.

Decomposition is performed using the Choleski triangular decomposition method.

During decomposition the current DOF being processed is displayed.

Solve Displacements

The program automatically solves for static displacements.

Caution: Displacements can be determined only after the applied loads have been defined.

The nodal displacements are solved in a two-part solution termed forward and backward substitution. All nodal displacements are given in the structure's global coordinate system. Multiple load cases are solved in the order in which they were input. Remember, all displacements are given in the original node numbering system.

Two messages are displayed during the displacement displacement solution:

Forward Substitution Completed!
Backward Substitution Completed!

Once finished, following these messages, the final nodal displacements are displayed. Translational displacements are in units which are consistent with input units, and rotations are in radians.

The printed output lists the node numbers and their corresponding displacements as described above.

Solve Loads, Stresses, and Reactions

The program automatically determines the element loads and corresponding stresses and restraint reactions.

Prior to execution of this routine, the user selects print options.

Menu:

Set Print Options

	Print (Y/N)
Beam Global Loads	Y
Beam Local Loads	Y
Beam Stress	Y
Plate Corner Forces	Y
Plate Stress	Y
Spring Loads	Y
Reactions	Y

Use the cursor to change the default. Press ESC when ready.

■ Use the cursor keys or the [Y] and [N] keys to select the print options.

■ Press [ESC] when the print options have been selected.

The program now solves the loads, stresses, and reactions and displays them on the screen.

Screen Format:

Solve Beam Loads and Stresses
  Load Case 1
  Element 1

  Global Loads
  I node F_x F_y M_z
  J node F_x F_y M_z

  Local Loads
  I F_a F_v M_z
  J F_a F_v M_z

  Stresses
  Node I
      $\sigma_a =$
      $\sigma_b =$
      $\tau =$

  Node J
      $\sigma_a =$
      $\sigma_b =$
      $\tau =$

where

  F_x = Global X Force
  F_y = Global Y Force
  M_z = Bending Moment
  F_a = Axial Shear
  F_v = Shear Force
  σ_a = Axial Stress
  σ_b = Bending Stress
  τ = Shear Stress

Solve Plate Loads and Stresses
  Plate No. 1

  Corner Forces
  Node I F_x F_y
  Node J F_x F_y
  Node K F_x F_y

  Stresses
  $\sigma_x =$
  $\sigma_y =$
  $\tau_{xy} =$

$\sigma_1 =$ (principle stresses)
$\sigma_2 =$
Angle = Angle from X axis

Solve Spring Loads
Spring No. 1

Global Loads
Node 1
$F_x =$
$F_y =$
$M_z =$

Solve Restraint Reactions
Node 1
$F_x =$
$F_y =$
$M_z =$

Note: While the screen is presented in this unlabeled element-by-element format for a quick glance, the printed output is summarized and clearly labeled.

Printing does not begin until all elements have been displayed on the screen. **Reactions do include the applied loads at the restrained degree-of-freedom.**

Sum Static Load Cases

Superimpose the results from various static load cases.

When you choose this option, the program displays the following message:

    Remove Disk No. 2 and insert Disk No. 1

If you are using a 2 floppy system, replace the IMAGES-2D Disk 2 in drive A with the IMAGES-2D Disk 1 and press [ENTER].

If you are using a hard disk system, simply press [ENTER].

  Menu:
  1. Sum Displacements
  2. Sum Beam Loads
  3. Sum Plate Loads
  4. Sum Spring Loads
  5. Sum Reactions
  6. Return to Static Menu

After a Summation menu option is chosen the following menu appears:

1. Form Summation
2. Return to Summation Menu

The results of several static load cases can be summed using algebraic (ALG), root-sum-square (RSS) or absolute value (ABS) methods. IMAGES-2D is used much like a calculator for summing load case results, and includes STORE and RECALL commands for storing and recalling results of intermediate steps. Factors can be applied to any load case.

The procedure for summing load cases is best illustrated by an example.

Sum Static Load Cases Example 4

Denoting load case by F1, F2 and so on, the following combination is desired.

$$\left| (1.0*F1) \right| + \left| \left[(2.0*F2 + 0.5*F4)^2 + (1.0*F3 + 0.5*F4)^2 \right]^{1/2} \right|$$

where the inner bracketed terms are Algebraic summation, combined under Root-sum-square summation, all under Absolute-value summation.

Select option 1 from the menu and proceed as shown below:

Sum Static Load Cases Screen Display

Starting load case no.
? 2
Factor for load case 2
? 2
Next load case (or STORE, RECALL, END)
? 4
Factor for load case 4
? 5
Summation (ALG, ABS, or RSS)
? ALG
Next load case (or STORE, RECALL, END)
? S

Starting load case no.
? 3
Factor for load case 3
? 1
Next load case (or STORE, RECALL, END)
? 4
Factor for load case 4
? .5
Summation (ALG, ABS, or RSS)
? ALG
Next load case (or STORE, RECALL, END)
? R
Summation (ALG, ABS, or RSS)
? RSS
Next load case (or STORE, RECALL, END)
? 1
Factor for load case 1
? 1
Summation (ALG, ABS, or RSS)
? ABS
Next load case (or STORE, RECALL, END)
? E

In Example 4, load cases 2 and 4 are selected as a starting point. The load cases and their factors are defined first, then the method of combination, algebraic. This intermediate result is stored so that it can be used later. Next, load cases 3 and 4 are combined algebraically. This result is then combined with the stored data using a root-sum-square method. Finally, an absolute summation is made of this result with load case 1.

When E is entered, the program exits, echoes the load case combination, and then prints the full deflection, stress, and reaction summary.

No data are stored on disk and no plots of combined results can be made.

Plot Deflected Shapes

Plot the deflected geometry due to static loading.

Menu:
1. Plot Deflected Geometry
2. Return to Static Menu

The scale factor is the number of dots on the screen to which the maximum translation is equated. If a scale factor of zero is entered, the scale defaults to 10. The factor may be either plus or minus. A minus will reverse the displacement directions.

After choosing item 1, Plot Deflected Geometry, the following prompt appears:

Which load case
?

■ Enter the load case to be plotted.

Prompt:

Scale Factor
?

■ Enter the scale factor. The plot will then appear on the screen.

■ Press any key to return to the plot menu. No carriage return is required.

Program Limits

The following program limits are imposed due to fixed dimensions in IMAGES-2D:

Number of nodes	100
Number of beam elements	150
Number of plate elements	50
Number of spring elements	20
Number of material properties	50
Number of cross-sectional properties	50
Number of restrained degrees-of-freedom	200
Number of static load cases	5
Number of enforced displacements per load case	20

IMAGES-2D allows you to use any printer, with the IBM printer assumed as a default.

Theoretical Background

The beam element used in IMAGES-2D is derived using the following assumptions:

■ Linear elastic
■ Plane sections remain plane
■ Small deformations
■ Constant properties over the element length
■ Axial and bending are uncoupled
■ Cubic displacement functions

The element coordinate system and displacement degrees-of-freedom are shown below:

The triangular plate element in IMAGES-2D is derived using the assumptions listed below:

■ Linear elastic
■ Isotropic
■ Small deformations
■ Constant strain
■ Plane stress
■ Linear displacement functions

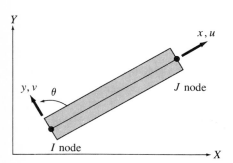

X, Y = global system
x, y = element system
u = axial displacement
v = normal displacement
θ = angular displacement

Displacement degrees-of-freedom and the positive directions of the resulting loads are indicated below.

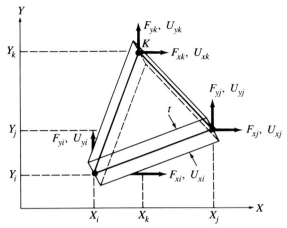

X, Y = global system
I, J, K = node numbers
U_x, U_y = displacement components
F_x, F_y = force components
t = thickness
x, y = coordinates

The material matrix for plane stress is:

$$D = \frac{E}{(1-v^2)} \begin{bmatrix} 1 & v & 0 \\ v & 1 & 0 \\ 0 & 0 & \frac{(1-v)}{2} \end{bmatrix} \quad (1)$$

The strain-displacement matrix is:

$$B = \begin{bmatrix} y_j - y_k & y_k - y_i & y_i - y_j & 0 & 0 & 0 \\ 0 & 0 & 0 & x_k - x_j & x_i - x_k & x_j - x_i \\ x_k - x_j & x_i - x_k & x_j - x_i & y_j - y_k & y_k - y_i & y_i - y_j \end{bmatrix} \quad (2)$$

The element stiffness is calculated using the equation

$$K = \iiint_V B^T C B dV \quad (3)$$

or $K = At B^T C B$ for constant strain (4)

where
A = area
t = thickness

Strains are determined from the strain-displacement relationship

$$\varepsilon = B\delta \quad (5)$$

and the stress-strain relationship is

$$\sigma = D\varepsilon \quad (6)$$

Principle stresses are determined by the Mohr's circle equation as

$$\sigma_{p1,2} = \frac{\sigma_x + \sigma_y}{2} \pm \sqrt{\left(\frac{\sigma_x - \sigma_y}{2}\right)^2 + \tau_{xy2}} \quad (7)$$

Springs (Stiffness Matrix Modification)

Springs in IMAGES-2D are stiffness terms added directly to the stiffness matrix.

The stiffness matrix is expressed as

$$K_e = \begin{bmatrix} K_1 & 0 & 0 \\ 0 & K_2 & 0 \\ 0 & 0 & K_3 \end{bmatrix} \quad (8)$$

where the subscript 1, 2, or 3 represents the degree-of-freedom which is input.

The transformation matrix is defined as

$$T = \begin{bmatrix} \cos(\text{angle}) & \sin(\text{angle}) & 0 \\ -\sin(\text{angle}) & \cos(\text{angle}) & 0 \\ 0 & 0 & 1 \end{bmatrix} \quad (9)$$

The global stiffness matrix is then determined from

$$K_g = T^T K_e T \quad (10)$$

The element stress resultants, or loads, and their positive directions are indicated below:

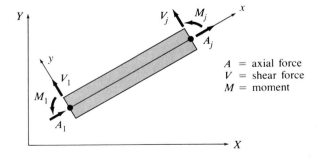

A = axial force
V = shear force
M = moment

The standard six degree-of-freedom element stiffness matrix is shown below:

$$K_e = \begin{bmatrix} \frac{AE}{L} & 0 & 0 & -\frac{AE}{L} & 0 & 0 \\ 0 & \frac{12EI}{L^3(1+\phi)} & \frac{6EI}{L^2(1+\phi)} & 0 & \frac{-12EI}{L^3(1+\phi)} & \frac{6EI}{L^2(1+\phi)} \\ 0 & \frac{6EI}{L^2(1+\phi)} & \frac{EI(4+\phi)}{L(1+\phi)} & 0 & \frac{-6EI}{L^2(1+\phi)} & \frac{EI(2-\phi)}{L(1+\phi)} \\ -\frac{AE}{L} & 0 & 0 & \frac{AE}{L} & 0 & 0 \\ 0 & \frac{-12EI}{L^3(1+\phi)} & \frac{-6EI}{L^2(1+\phi)} & 0 & \frac{12EI}{L^3(1+\phi)} & \frac{6EI}{L^2(1+\phi)} \\ 0 & \frac{6EI}{L^2} & \frac{EI(2-\phi)}{L(1+\phi)} & 0 & -\frac{6EI}{L^2} & \frac{EI(4+\phi)}{L(1+\phi)} \end{bmatrix} \quad (11)$$

where: $\phi = \dfrac{12EI}{G\dfrac{A}{(SSF)}L^2}$

Static Analysis

The equation of static equilibrium is defined as:

$$P = K\delta \quad (12)$$

where

P = applied load vector
K = global stiffness matrix
δ = displacement vector

The solution of this equation for δ, the displacement vector, is obtained as follows:

■ **Create/Edit Loads** One or more applied load vectors, P, are defined.

■ **Assemble Stiffness Matrix** The global stiffness matrix, K, is assembled from the individual element stiffness matrices, K_e, given in Equation (11).

■ **Decompose Stiffness Matrix** Choleski decomposition is used to factor the global stiffness matrix K as follows:

$$K = LL^T \quad (13)$$

where L is a lower triangular matrix.

■ **Solve Displacements** The displacement solution is obtained in a two-stage process termed forward substitution and backward substitution.

Before decomposition, enforced displacements are introduced by partitioning the equation as follows:

$$\left|\begin{array}{c} P_a \\ P_b \end{array}\right| = \left|\begin{array}{cc} K_{aa} & K_{ab} \\ K_{ba} & K_{bb} \end{array}\right| \left|\begin{array}{c} \delta_a \\ \delta_b \end{array}\right| \quad (14)$$

where

P_a = specified nodal loads
P_b = unknown nodal loads
δ_a = unknown nodal displacements
δ_b = specified nodal displacements

P_a in Equation (14) can be written as

$$P_a = K_{aa}\delta + K_{ab}\delta_b \quad (15)$$

or

$$P^* = K_{aa}\delta_a$$

where the modified load vector is given by

$$P^* = P_a - K_{ab} \delta_b \quad (16)$$

The equation of equilibrium can be rewritten substituting Equation (13) in Equation (12):

$$P = LR \quad (17)$$

where

$$R = L^T \delta \quad (18)$$

The vector R is obtained using forward substitution in Equation (17); then, the displacement solution is obtained by performing backward substitution on Equation (18) for δ.

■ **Solve Loads, Stresses and Reactions** Displacements in the global coordinate system are transformed to the element coordinate system, and are multiplied times the element stiffness matrix, K_e, to recover element loads.

Stresses are determined from the element loads at each node with the following basic engineering equations:

Axial Stress, σ_a

$$\sigma_a = \frac{(\text{Axial Force})}{(\text{Area})} \quad (19)$$

Shear Stress, τ

$$\tau = \frac{(\text{Shear Force})}{(\text{Area/SSF})} = \frac{\text{Shear Force (SSF)}}{(\text{Area})} \quad (20)$$

Where SSF is the shear shape factor

Bending Stress, σ_b

$$\sigma_b = \frac{(\text{Moment})(\text{Fiber Distance})}{(\text{Moment of Inertia})} \quad (21)$$

Reactions are determined from the difference in the element loads at each node.

ERROR TRACING AND MESSAGES

Error Messages

This section of this appendix presents a list of messages common to all portions of IMAGES-2D. This is not a listing of all messages within the program. Also, consult Appendix A, "Error Codes and Messages", of Microsoft Corp. BASIC User's Guide.

Message	Description
Data Disk Is Full	All available storage space on the disk has been used. If you have several problems stored on one disk, erase or copy to a spare disk all filenames except your current problem.
	If the disk contains data on a single large problem, erase or copy to another disk, the files not used in the current analysis. (See the table provided in Appendix A of the BASIC User's Guide.)
Data Disk Is Write Protected	Attempt to write data on a write-protected disk. A disk is write-protected when the notch in the upper right corner is covered. Remove the sticker over the notch.
Disk or Drive Not Ready	Door to the disk drive is open or there is no disk in the drive. Check the drives.
Numeric Overflow	Attempt to operate on a number larger than the computer's capabilities. Carefully review all input to assure all numbers are reasonable.
Out of Paper	Printer has run out of paper or the paper has jammed the printer.
	Check the printer before continuing.
Printer Not Ready	The printer is off line.
	Push the on-line (or select) button.
Too Many Files on Data Disk	The limit on the number of files a disk can hold has been exceeded.
	To correct, refer to "Data Disk Is Full"
Data Disk Is Bad	Attempt to read information from a bad disk. Check if the disk has been formatted or try copying the files to another disk.

Error Tracing

To illustrate the use of various cross-reference lists for error tracing consider the following problem:

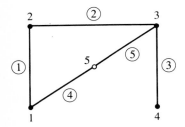

in which elements ④ and ⑤, the diagonals, are axial-only load carrying members (Pincode = 3).

We know by inspection, that releasing the moment-carrying ability at node 5 creates a mechanism and therefore, an unstable structure. During decomposition the following error message is printed:

   ***ERROR—Singularity at DOF 2

Referring to the Node versus DOF cross-reference list printed at the end of the stiffness matrix output:

Node Versus DOF Cross-Reference List

Refer to WAS/IS list to locate actual node number.

IS Node	DOF-Direction		
2	1-X	2-Y	
3	3-X	4-Y	5-Z
4	6-X	7-Y	8-Z

DOF 2 corresponds to IS node 2. The IS number refers to the internal numbering system and not the actual node number. Going to the Node Renumbering Cross Reference List:

Node Renumbering Cross Reference List

WAS	IS	WAS	IS	WAS	IS
1	1	2	3	3	4
4	5	5	2		

Original Nodal Band 5
Final Node Band 3

IS node 2 is equivalent to node number 5 in the original (WAS) numbering system.

When the problem area is located the error is corrected, in this case, by replacing elements ④ and ⑤ with a single element connecting node 1 to node 3. Alternatively, pincodes for elements ④ and ⑤ could be changed so that the moment at node 5 was not released.

APPENDIX TWO

Geometric Properties of Areas

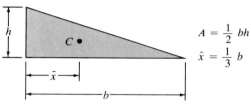

Triangle

$A = \frac{1}{2} bh$

$\bar{x} = \frac{1}{3} b$

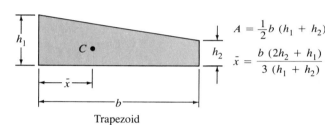

Trapezoid

$A = \frac{1}{2} b (h_1 + h_2)$

$\bar{x} = \frac{b (2h_2 + h_1)}{3 (h_1 + h_2)}$

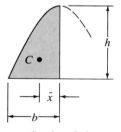

Semiparabola

$A = \frac{2}{3} bh$

$\bar{x} = \frac{3}{8} b$

Semisegment of nth degree curve

$A = bh \left(\frac{n}{n + 1} \right)$

$\bar{x} = \frac{b (n + 1)}{2 (n + 2)}$

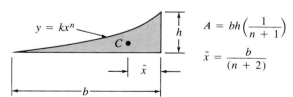

Spandrel of nth degree curve

$A = bh \left(\frac{1}{n + 1} \right)$

$\bar{x} = \frac{b}{(n + 2)}$

APPENDIX THREE

Answers to Selected Problems

CHAPTER 2

2.1 $H_L = 3$ kips; $V_L = 9.5$ kips; $V_R = 5.5$ kips
2.2 $H_L = 0$; $V_L = 6$ kips; $V_R = 8$ kips
2.4 $H_L = 0$; $V_L = 5.9$ kN; $V_R = 4.1$ kN
2.6 $H_L = 0$; $V_L = 34$ kips; $M_L = 394$ kip·ft
2.8 $H_L = 27$ kips; $V_L = 4$ kips; $T = 28.5$ kips
2.12 $H_L = 10$ kips; $V_L = 4$ kips; $V_R = 16$ kips
2.19 $H_L = 360$ lb; $V_L = 264$ lb; $V_R = 336$ lb
2.23 $H_R = 16$ kips; $V_L = 13$ kips; $V_R = 5$ kips
2.26 $H_a = 0$; $V_a = 5$ kips; $V_c = 16.8$ kips; $V_d = 4.2$ kips
2.29 $H_a = 3.9$ kips; $V_a = 8.4$ kips; $H_c = 2.1$ kips; $V_c = 0.4$ kip
2.32 $H_a = 10$ kips; $V_a = 13.3$ kips; $H_c = 10$ kips; $V_c = 16.7$ kips
2.35 $H_a = H_d = 63.8$ kips; $V_a = 15$ kips; $V_d = 17$ kips; $T_{max} = 66$ kips; $s = 7$ ft

CHAPTER 3

3.27 $F_a = +22.5$ kips; $H_b = -6.25$ kips; $V_b = -12.5$ kips; $F_c = -17.5$ kips; $H_d = +1.25$ kips; $V_d = +2.5$ kips; $V_e = +2.5$ kips
3.29 $V_a = H_a = -3.3$ kips; $F_b = H_c = V_c = 0$; $H_d = -3.3$ kips; $V_d = -3.3$ kips
3.33 $V_a = H_a = +4$ kips; $H_b = V_b = -8$ kips; $H_c = V_c = +12$ kips
3.35 $V_a = H_a = -93.3$ kN; $V_b = -38.3$ kN; $H_c = +80$ kN; $V_c = +40$ kN
3.37 $V_a = +2.1$ kips; $H_a = +8.3$ kips; $H_b = V_b = -15.4$ kips; $F_c = +2.5$ kips

521 Answers to Selected Problems

CHAPTER 4

4.1 $Y_a = 5$ kips; $X_b = 12.5$ kips; $X_c = 32.5$ kips; $Z_a = 12.5$ kips; $Z_b = 8.33$ kips; $Z_c = 20.83$ kips; $F_{ab} = -7.07$ kips; $F_{ac} = +5$ kips; $F_{bc} = -18.2$ kips; $F_{ad} = +13.6$ kips; $F_{bd} = -10.9$ kips; $F_{cd} = -28.7$ kips
4.3 $X_a = X_c = 20$ kN; $Z_d = 0$; $Y_a = Y_d = 10$ kN
4.5 Zero force members: $F_{58}, F_{56}, F_{15}, F_{78}, F_{37}, F_{67}$

CHAPTER 5

5.1 $M_{max} = +120$ kip·ft; $V_{max} = -14$ kips
5.3 $M_{max} = +196$ kip·ft; $V_{max} = -32$ kips
5.7 $M_{max} = -91$ kip·ft; $V_{max} = +17.5$ kips
5.12 $M_{max} = -380$ kN·m; $V_{max} = +60$ kN
5.20 $M_{max} = -327$ kip·ft; $V_{max} = +35$ kips
5.21 $M_{max} = -730$ kip·ft; $V_{max} = -64.6$ kips
5.25 $M_{max} = +570$ kN·m; $V_{max} = +97.6$ kN
5.26 $M_{max} = +223$ kip·ft; $V_{max} = -36.7$ kips
5.32 $M_{max} = -958$ kip·ft; $V_{max} = -72$ kips

CHAPTER 6

6.4 Load at A: $R_B = +1.5$; $V_C = +0.5$; $M_C = -7.5$; $M_D = -2.5$
6.7 Load at D: $V_A = V_B = V_C = +1$; $M_A = -20$; $M_B = -14$; $M_C = -10$
6.10 Load at D: $R_C = +1.3$; $R_A = -0.3$; $V_C = 0$; $M_C = -3$
6.13 Load at B: $R_A = 0$; $R_D = +2/3$; V_B maximum $= +2/3$; $M_B = +8$
6.21 Maximum $+V_B = 27$ kips; maximum $M_B = +380$ kip·ft
6.23 Maximum $R_B = 152$ kips; maximum $V_C = +56.3$ kips; maximum $M_C = +443$ kip·ft
6.27 Maximum $V_{AB} = +36.3$ kips; maximum $M_C = 594$ kip·ft
6.28 Maximum $+V_{B-C} = +13.2$ kips
6.30 Maximum $M_C = 404$ kip·ft
6.31 Maximum $M_C = 43$ kip·ft
6.32 Absolute maximum $M = 976$ kip·ft
6.34 Absolute maximum $M = 1605$ kip·ft

CHAPTER 7

7.3 Load at L_2: $U_1 U_3 = -8/9$; $L_2 L_4 = +4/3$
7.5 Load at L_2: $L_2 L_4 = +5/4$; $U_3 U_5 = -1.0$
7.18 Maximum force $= +180$ kips (tension)
7.20 Maximum tensile force $= 341$ kips
7.22 Maximum compressive force $= 353$ kips

CHAPTER 8

8.1 At supports: $M_L = 153$ kip·ft; $M_R = 126$ kip·ft
8.3 (a) At supports: $M_L = 450$ kip·ft; $M_R = 350$ kip·ft
(b) At supports: $M_L = 225$ kip·ft; $M_R = 200$ kip·ft
8.6 $V_{max} = +102.5$ kips; $M_{max} = +1053$ kip·ft
8.7 $V_{max} = +103.3$ kips; $M_{max} = -1024$ kip·ft
8.12 Bottom floor: $V = 8.75$ kips, $M = 87.5$ kip·ft for exterior columns
8.15 Bottom floor: $V = 8$ kips, $M = 48$ kip·ft for interior column
8.19 Bottom floor: $V = 5.85$ kips, $M = 58.5$ kip·ft for left column

CHAPTER 9

9.1 $\delta = 2PL^3/81EI - 11wL^4/972EI$
9.3 $\theta_a = wL^3/6EI, \delta_a = y_a = wL^4/8EI$
9.5 At midspan: $y = -6.51 \times 10^{-3}(wL^4/EI)$
9.7 $\theta_a = \theta_b = 0.0045$ radians; $\delta_a = 0.75$ in; $\delta_b = 0.43$ in
9.10 $\theta_b = 0.0094$ radians; $\delta_b = 1.73$ in
9.11 $\theta_a = 0.00246$ radians; $\delta_a = 25.9$ mm
9.15 $\delta_c = 1.08$ in
9.17 $\delta_c = 2250$ Kn·m³/EI
9.20 $M_a = 8.7$ kip·ft; $M_b = 31$ kip·ft
9.22 $\theta_a = 0.0083$ radians; $\theta_b = 0.0073$ radians; $\delta_a = 1.3$ in; $\delta_b = 0.54$ in
9.26 $\theta_a = 0.00215$ radians; $\theta_b = 0.0015$ radians; $\delta_a = 0.338$ in; $\delta_b = 0.106$ in
9.31 $\delta_b = 4560$ kip·ft³/EI; $\delta_d = 44{,}692$ kip·ft³/EI
9.32 $\theta_c = 281$ kN·m²/EI; $\delta_c = 2250$ kN·m³/EI
9.34 $M_a = 71$ kip·ft; $M_b = 35.7$ kip·ft
9.36 $\theta_b = 0.0058$ radians; $\theta_d = 0.0044$ radians; $\delta_d = 0.662$ in
9.38 $\theta_a = 0.0034$ radians; $\theta_c = 9 \times 10^{-5}$ radians; $\delta_c = 0.018$ in

CHAPTER 10

10.1 0.71 in down and 0.98 in to the right
10.4 0.11 in down and 1.06 in to the left
10.5 (a) 0.368 in down; (b) 0.201 in down
10.11 $\theta_a = 0.0063$ radians; $\delta_a = 86$ mm
10.14 $\theta_c = 0.003$ radians; $\delta_c = 0.245$ in
10.15 $\delta_b = 0.343$ in; $\theta_d = 0.0078$ radians
10.19 $\delta_b = 0.136$ in; $\theta_b = 0.00206$ radians
10.22 $\theta_b = 0.001$ radians; $\delta_d = 0.083$ in
10.23 $\delta_d = 27.6$ mm
10.24 $\delta_C = 0.156$ in; $\delta_E = 1.98$ in
10.26 $\delta_B = 6.6$ in downward
10.32 $\theta_c = 6.4 \times 10^{-4}$ radians; $\delta_c = 0.041$ in
10.35 $\theta_b = 1.7 \times 10^{-3}$ radians; $\delta_b = 100$ mm
10.38 δ at free end $= 3.42$ in down
10.40 Horizontal displacement at $C = 1.94$ in to the right

Answers to Selected Problems

CHAPTER 11

- **11.2** $R_B = 22.5$ kips
- **11.4** $R_B = 65$ kips
- **11.6** $R_B = 35$ kips
- **11.7** $R_B = 25.7$ kN
- **11.9** $R_B = 47.4$ kips
- **11.15** $H_D = 19.4$ kN; $H_A = 0.6$ kN; $V_A = 69.4$ kN; $V_D = 50.6$ kN
- **11.16** $H_A = 0$; $H_C = 20$ kips; $V_C = 0.8$ kip, $M_C = 16$ kip·ft
- **11.18** $H_D = 23$ kips; $V_D = 12.1$ kips; $H_A = 9$ kips; $V_A = 27.9$ kips
- **11.19** $H_E = 2.75$ kips; $V_E = 6.8$ kips; $V_A = 3.2$ kips; $H_A = 0.75$ kips
- **11.21**   Vertical reaction: $L_2 = 22.8$ kN; $U_0 = 5.7$ kN; $U_3 = 1.5$ kN
- **11.25**   Force in member $L_1U_2 = -30$ kips (compression)
- **11.27**   Force in member $U_1L_2 = +16.2$ kips (tension)
- **11.30** $T = 17.5$ kips
- **11.32** $T = 30.6$ kips

CHAPTER 12

- **12.1** $M_A = 44.8$ kip·ft; $M_B = 64$ kip·ft; $M_C = 83.2$ kip·ft
- **12.3** $M_A = 22.5$ kip·ft; $M_B = 45$ kip·ft; $M_C = 112.5$ kip·ft
- **12.5** $M_A = 51.5$ kN·m; $M_B = 47$ kN·m; $M_C = 10.6$ kN·m; $M_D = 5.3$ kN·m
- **12.8** $M_B = 90.9$ kip·ft; $M_C = 136.4$ kip·ft
- **12.12** $M_A = 222$ kN·m; $M_B = 101$ kN·m; $M_C = 50$ kN·m; $M_D = 1$ kN·m
- **12.13** $M_A = 111$ kip·ft; $M_B = 78$ kip·ft; $M_C = 51$ kip·ft
- **12.16** $M_A = 56$ kip·ft; $M_B = 27$ kip·ft; $M_C = 34$ kip·ft; $M_D = 0$
- **12.20** $M_A = 417$ kip·ft; $M_B = 160$ kip·ft; $M_C = 316$ kip·ft; $M_D = 0$

CHAPTER 13

- **13.1**   Support moments: Left 80 kip·ft; center 80 kip·ft; right 110 kip·ft
- **13.6** $M_A = 208$ kip·ft; $M_B = 95$ kip·ft; $M_C = 6$ kip·ft; $M_D = 3$ kip·ft
- **13.8**   Support moments: Left 150 kip·ft; center 150 kip·ft; right 0 kip·ft
- **13.10**   Interior support moments: Left 23 kip·ft; right 65 kip·ft
- **13.14** $M_A = 0$ kip·ft; $M_B = 219$ kip·ft; $M_C = 180$ kip·ft
- **13.18** $M_A = 0$ kip·ft; $M_B = 307$ kip·ft; $M_C = 180$ kip·ft
- **13.19**   Part (a) $M_A = 188$ kip·ft; $M_B = 67$ kip·ft; $M_C = 15$ kip·ft; $M_D = 2$ kip·ft
  Part (b) $M_A = 186$ kip·ft; $M_B = 74$ kip·ft; $M_C = 7$ kip·ft; $M_D = 56$ kip·ft
- **13.20**   Support moments: Left 14.7 kip·ft; center 3.6 kip·ft; right 48.5 kip·ft
- **13.21**   Support moments: Left 7.3 kip·ft; center 11.3 kip·ft; right 0 kip·ft
- **13.25**   Joint moments: upper left = upper right = 35.6 kip·ft

- **13.27** Upper joint moments: Left 39 kip·ft; right 55 kip·ft
  Support moments: Left 23.5 kip·ft; right 40.5 kip·ft
- **13.28** Upper joint moments: Left 32.8 kN·m; right 31.3 kN·m
  Support moments: Left 15.2 kN·m; right 16.8 kN·m
- **13.30** Upper joint moments: Left 50.7 kip·ft; right 56.5 kip·ft
  Support moments: Left 29.7 kip·ft; right 23.9 kip·ft
- **13.35** $M_A = 84.6$ kip·ft; $M_B = 31$ kip·ft; $M_C = 9.8$ kip·ft; $M_D = 4.6$ kip·ft
- **13.37** Midsupport moment: 891 kip·ft

CHAPTER 14

- **14.7** Reactions: $X_1 = 0$; $Y_1 = Y_3 = -5$ kips
- **14.8** Reactions: $X_1 = 5$ kips; $X_2 = 0$; $Y_1 = -Y_3 = -5$ kips
- **14.9** Reactions: $Y_3 = 4.12$ kips; $M_3 = 22.5$ kip·ft; $Y_1 = 1.88$ kips
- **14.10** Reactions: $Y_3 = 3.41$ kips; $M_3 = 21.4$ kip·ft; $Y_1 = 4.59$ kips
- **14.12** $X_1 = -0.497$ kips; $Y_1 = 8.42$ kips; $M_1 = 27.4$ kip·ft; $X_3 = -4.5$ kips; $Y_3 = 0.58$ kips; $M_3 = -37.2$ kip·ft
- **14.16** $X_1 = 5.8$ kips; $Y_1 = 9.1$ kips; $X_2 = -10.8$ kips; $Y_2 = 0.9$ kips; $M_1 = 129$ kip·ft; $M_3 = -129$ kip·ft

Index

Absolute maximum moment, 179, 181–183
Absolute stiffness, 385, 392, 417
American Association of State Highway and
    Transportation Officials (AASHTO), 16, 24, 27, 170
American Concrete Institute (ACI), 15, 27, 213, 236
American Institute of Steel Construction (AISC), 15, 27, 28, 213, 235
American Institute of Timber Construction (AITC), 236
American National Standards Institute (ANSI), 15, 19, 22
American Railway Engineers Association (AREA), 25, 27
Antisymmetry, 369, 395, 396, 488
Approximate methods of analysis, 204
Arch, defined, 10
Average-load method, 177–180

Bandwidth minimization, 490
Basic Building Code, 15, 22
Beams:
    cantilever, 119;
    continuous, 119, 213, 450
    simple, 119
Beer, Ferdinand P., 267
Bending moment:
    defined, 120, 122
    diagrams, 125–138, 394
    envelope curves, 341–343
    functions, 122
Bendixen, Alex, 351
Bergin, Thomas J., 412, 434
Bernoulli, John, 268
Betti, E., 291
Betti's law, 291
Building codes, 15

Cable, 12
Cable construction, 59–61
Carryover factor, 385, 413

Carryover moment, 384, 385
Cantilever construction, 55
Cantilever method, 222
Castigliano, Alberto Carlo, 293
Castigliano's first theorem, 293, 299
Castigliano's second theorem, 294–299
Castigliano's theorem of least work, 328
Cholesky method, 490
Compatibility conditions, 242
Compatibility methods, 306, 450
Complex structures, 14
Conjugate beam:
    method of, 253
    supports, 257, 258
Consistent deformations (method), 306, 309–327, 450
Cooper, Theodore, 25
Cooper loads 25, 168–170
Critical form, 93
Cross, Hardy, 351, 381, 382

Dead loads, minimum units for design (table), 16, 17
Degrees of freedom, 453, 458, 460, 461, 475, 490
Deflection compatibility, 453
Deflection methods:
    conjugate beam, 253–263
    double-integration, 236–242
    elastic weights, 254
    moment-area, 242–252
    virtual work:
        beams, 278
        truss, 273
        frame, 286
Design codes, 15
Determinacy, 36, 106–108
Determinant, 441–444
Direct stiffness method, 460
Displacement method, 307, 352, 381, 449, 452, 453
Distribution factor, 385, 386, 388, 392, 413
Distributed moment, 386
Double integration (deflection) method, 236–242
Dynamic pressure (graph), 22

Epstein, Howard I., 213
Equations of condition, 90
Equivalent nodal forces, 480
External work, 269, 270

Fabrication errors, 471
Factors of safety, 4, 27
Fixed end moments, 261, 353, (table of) 355, 382, 412
Flexibility coefficient, 309, 327, 449
Flexibility method, 306, 449–452
Flexural rigidity, 237
Force method, 307–309, 449
Form factor, 292
Frame:
   defined, 6
   deflections, 286–288
   diagrams (V & M), 132–137
   reactions, 41
Frames without sidesway, 365–369, 399–401
Frames with sidesway, 369–373, 402–408
Frames with sloping legs, 409–411
Frames with support settlement, 373, 374
Free-body diagram, 37

Gallagher, Richard H., 437, 492
Gauss elimination method, 447–449, 451
Gauss-Jordan method, 449
Geometric stability, 36, 50
Girder, defined, 160, 161
Global coordinates, 452
Greene, Charles E., 242

IMAGES, IMAGES-2D: 2, 30, 45–50, 59, 94–96, 139–143, 159, 198, 225, 288–290, 343, 344, 427, 490
Impact factor, 27
Influence coefficient, 309, 326, 453, 475
Influence line, defined, 148
Influence lines:
   for beams, 148–160
   for bridge decks and building floor systems, 160–164
   for building girders, 164–166
   for indeterminate structures, 333–343
   for plane trusses, 189–199
Influence line applications:
   concentrated loads, 164–166, 172

   highway bridge loads, 170, 171
   railway bridge loads, 169–170
   series of concentrated loads, 172–181
   uniform loads, 166, 167, 172
Increase-decrease method, 172–177
Internal strain energy, 271
Internal work, 270–272
Inverse of a matrix, 444, 445

Johnston, Jr., E. Russell, 267
Joints (method of), 77–82

Kinney, J. S., 293

Laplace expansion, 442, 443
Laursen, Harold I., 115, 412, 434
Lefter, James, 412, 434
Least-work theorem, 293, 328
Limit state design, 27
Live loads, minimum values for design (table), 18
Load factors, 27, 28
Loads:
   dead, 16, 17
   earthquake (seismic), 24
   highway, 24, 25
   live, 17, 18, 26, 148
   railway, 25, 26
   rain, 22
   snow, 22
   uniform lane load, 25
   wind, 18, 216, 234
Local coordinates, 452

Maguire, William, 437, 492
Manderla, Heinrich, 351
Maney, George, 351
Martin, Harold C., 457, 492
Matrices:
   column, 438
   diagonal, 438
   identity, 438
   null, 439
   row, 438
   square, 438
   symmetric, 438
Matrix algebra, 437

Matrix operations:
   addition and subtraction, 439
   adjoint, 444
   cofactor of an element, 442
   minor of an element, 442
   multiplication, 439–441
   inverse, 444, 445
   transpose, 441
Matrix structural analysis, 435–490
Maxwell, James Clerk, 290, 307
Membrane, defined, 11
McCormac, Jack, 343
Mènabrèa, L. F., 328
Method of joints (trusses), 77
Method of moments (trusses), 84
Method of shears (trusses), 83
Mill buildings, 206–211
Mohr, Otto, 253, 308, 351
Moment-area method, 242–252
Moment distribution method, 381–434
Moment resisting connection, 6
Morris, C. T., 411, 434
Moving loads (placement for maximum effect), 172
Müeller-Breslau, 157, 333, 339
Multistory frame analysis, 411, 412

Newton, Sir Issac, 32
Nonprismatic beams, 412

Panel shear (influence lines), 163, 164
Partitioning of matrices, 446, 447
Plates, defined, 11
Pin connection, 8
Pin codes (IMAGES-2D), 59
Point of inflection, defined, 208
Portal method, 217–221
Portland Cement Association (PCA), 417
Primary structure, 306, 308, 449
Principle of superposition, 137–139
Prismatic beam, 236, 242, 261, 353, 383, 385, 392, 417, 475, 480, 483
Purlins, 75

Reciprocal deflection theorem (Maxwell or Maxwell-Betti law), 290, 291, 327, 335, 338, 450, 457, 477
Redundants, 93, 308

Relative stiffness, 385
Resisting moment, 383

Seismic zones (map), 23
Serviceability, 3
Shear, defined, 120, 121
Shear diagrams, 125–139, 394
Shear functions, 122–124
Shells, defined, 11
SI units, 28, 29
Simple end support modification, 363–365, 392
Simultaneous equations (solutions of), 94, 115, 447
Slope deflection equations, 353–357
Slope deflection method, 351–376
Sloping leg frame, 374–376, 409–411
Smith, Albert, 233
Snow loads (map), 23
Space trusses, 103
Spandrel beam, 161
Static equilibrium equations, 32, 33, 106
Standard Building Code (SBCC), 15
Stiffness matrix:
   beams, 474–477
   frames, 483–487
   trusses, 454–466
Stiffness method, 307, 452, 453, 474, 483
Stiffness modifications, 392
Stringers, 75, 76, 161
Superposition (principle of), 61, 137, 405, 449, 455
Support settlement, 314–317, 373, 397, 398, 471
Supports:
   cables, 35
   links, 35
   roller, 35
   rocker, 35
   hinge (pin), 36
   fixed, 36
Symmetry, 369, 395, 488

Temperature change, 471
Three-hinged arch, 50
Transformation matrix, 458, 459
Truss:
   defined, 6
   analysis of, 74
   idealizations, 6
   types
      bridge, 72–74
      complex, 72

Truss: *cont.*
    compound, 71;
    plane, 68;
    roof, 72;
    simple, 70
Truss geometric stability and determinacy:
    three-dimensional, 106–108
    two-dimensional, 90–93

Unbalanced moment, 383
US customary units, 28

Uniform Building Code (UBC), 15, 24

Vierendeel truss, 225–228
Virtual work, method of, 268–300

Wind speed (map), 20
Wind velocity pressure, 19
Wilson, A.C., 233

Zero force members (truss analysis), 81, 82

Fixed-End Moments

$\text{FEM}_{AB} = \dfrac{PL}{8}$ — point load P at midspan — $\text{FEM}_{BA} = \dfrac{PL}{8}$

$\text{FEM}_{AB} = \dfrac{Pb^2 a}{L^2}$ — point load P at distance a from A, b from B — $\text{FEM}_{BA} = \dfrac{Pa^2 b}{L^2}$

$\text{FEM}_{AB} = \dfrac{wL^2}{12}$ — uniform load w over full span — $\text{FEM}_{BA} = \dfrac{wL^2}{12}$

$\text{FEM}_{AB} = \dfrac{11\,wL^2}{192}$ — uniform load w over left half — $\text{FEM}_{BA} = \dfrac{5wL^2}{192}$

$\text{FEM}_{AB} = \dfrac{wL^2}{20}$ — triangular load, max w at A — $\text{FEM}_{BA} = \dfrac{wL^2}{30}$

$\text{FEM}_{AB} = \dfrac{5wL^2}{96}$ — triangular load peaked w at midspan — $\text{FEM}_{BA} = \dfrac{5wL^2}{96}$

$\text{FEM}_{AB} = \dfrac{6EI\Delta}{L^2}$ — support settlement Δ — $\text{FEM}_{BA} = \dfrac{6EI\Delta}{L^2}$

$\text{FEM}_{AB} = M(1-k)(1-3k)$ — applied moment M at kL from A — $\text{FEM}_{BA} = Mk(2-3k)$

$\text{FEM}_{AB} = \dfrac{wL^2}{60}\cdot k^2(10-10k+3k^2)$ — triangular partial load over length kL from A, max w at A — $\text{FEM}_{BA} = \dfrac{wL^2}{60}\cdot k^3(5-3k)$